Phenomenological Inquiry in Psychology

Existential and Transpersonal Dimensions

Phenomenological Inquiry in Psychology

Existential and Transpersonal Dimensions

Edited by
Ron Valle
Awakening: A Center for Exploring Living and Dying
Walnut Creek, California

PLENUM PRESS • NEW YORK AND LONDON

Library of Congress Cataloging-in-Publication Data

Phenomenological inquiry in psychology : existential and transpersonal dimensions / edited by Ron Valle.
p. cm.
Includes bibliographical references and index.
ISBN 0-306-45542-0 (hardcover). -- ISBN 0-306-45543-9 (pbk.)
1. Phenomenological psychology. 2. Existential psychology.
3. Transpersonal psychology. I. Valle, Ronald S.
BF204.5.P47 1997
150.19'2--dc21 97-38473
CIP

ISBN 0-306-45542-0 (Hardbound)
ISBN 0-306-45543-9 (Paperback)

A Division of Plenum Publishing Corporation
233 Spring Street, New York, N.Y. 10013

http://www.plenum.com

10 9 8 7 6 5 4 3 2 1

Printed in the United States of America

To **Rolf von Eckartsberg**
friend, mentor, and spiritual guide

Contributors

Scott D. Churchill, Department of Psychology, University of Dallas, Irving, Texas 75062

Constance T. Fischer, Department of Psychology, Duquesne University, Pittsburgh, Pennsylvania 15282

Paul Gowack, 1339 Milvia Street, Berkeley, California 94709

Steen Halling, Department of Psychology, Seattle University, Seattle, Washington 98122

D. Hanson, Rosebridge Graduate School of Integrative Psychology, Concord, California 94518

Bernd Jager, Department of Psychology, University of Quebec at Montreal, Montreal, Quebec H3C 3P8, Canada

Jon Klimo, Rosebridge Graduate School of Integrative Psychology, Concord, California 94518

Julie E. Lowery, Department of Psychology, University of Dallas, Irving, Texas 75062

Ourania Marcandonatou (Elite), School of Liberal Arts, John F. Kennedy University, Orinda, California 94563

Craig Matsu-Pissot, 15303 Northeast 166th Lane, Woodinville, Washington 98072

Owen McNally, Department of Psychology, University of Dallas, Irving, Texas 75062

Kathleen Mulrenin, Butler Hospital, Providence, Rhode Island 02906

Patricia A. Qualls, 26405 Valley View, Carmel, California 93923

Aruna Rao, Department of Psychology, University of Dallas, Irving, Texas 75062

Faith A. Robinson, 26485 Carmel Rancho Boulevard, Suite 6, Carmel, California 93923

Jan O. Rowe, Department of Psychology, Seattle University, Seattle, Washington 98122

Tania Shertock, Institute of Transpersonal Psychology, San Francisco, California 94131

Ron Valle, Awakening: A Center for Exploring Living and Dying, Walnut Creek, California 94598

Valerie A. Valle, St. Alban's Episcopal Church, Brentwood, California 94513

Damian S. Vallelonga, 201 Sherbourne Road, Syracuse, New York 13224

Rolf von Eckartsberg, Late of Department of Psychology, Duquesne University, Pittsburgh, Pennsylvania 15282

Thomas B. West, Franciscan School of Theology, Berkeley, California 94709

Timothy West, 512 Malobar Drive, Novato, California 94945

Foreword

This fine new book, the third in a series, brings psychologists up to date on the advances of phenomenological research methods in illuminating the nature of human awareness and experiences. In the more congenial and welcoming intellectual climate of the 1990s, phenomenological methods have moved to the forefront of discourse on research methods that support and advocate an expanding view of science. In Valle and King (1978), phenomenological methods were presented as *alternatives* to behavioral methods. In Valle and Halling (1989), phenomenological methods were advanced to *perspectives* in psychology. This new volume is even less cautious, indeed bolder, in relation to conventional methods and epistemologies. By now, people knowledgeable about psychology, and most psychologists, have digested the criticisms directed against methods that operationalize, quantify, and often minimize human behavior. In bringing us up to date on the growing power of phenomenological methods, this volume brings welcome coherence and integrity to an increasingly harried science attempting to reenchant itself with meaning and depth, an endeavor artfully exemplified by phenomenological inquiries of the last several decades.

Since the late 1950s, phenomenological methods have provided descriptions and insights regarding the dynamic meaning of human experience. The well-honed and rigorous method of constituent analysis (not to mention the subtle philosophic vocabulary) of phenomenology has exhausted not a few researchers. It is hard work to carefully examine the descriptions and articulate the prereflective structures of meanings that humans give to unique experiences, let alone identify and bracket the researcher's preconceptions and biases. The fruitfulness of this approach to understanding the meaning and meaning-seeking nature of human experience is apparent in the numerous phenomenological inquiries described in this volume.

In my 25 years as a research psychologist, it has always concerned me that most eloquent speakers on the nature of human experience are poets, novelists, playwrights, storytellers, and theologians—and *rarely* psychologists. In claiming radical positivism and psychological behaviorism as our epistemological imprimatur, psychologists have ignored, and even trivialized, vast realms of fascinating human experiences. As pointed out by James F. T. Bugental in his foreword to the second volume in this series (Valle & Halling, 1989), "The objectivist view of psychology . . . regards all that is not familiar as dangerous, mythical, or nonexistent" (p. ix). Psychologists often seem content with meaning-diminishing methodologies. Marvelously rich topics such as the study of passion, making love, giving birth, dying, grieving, ecstasy, quietude, and mystical experiences have been largely neglected. There is a uniqueness to being human. So often psychological research methods fall flat before the fullness of being human, the extraordinary experience of being human day by day.

To honor this fullness, human awareness and experience must be studied comprehensively and imaginatively. In exploring these vast domains, our methods need to be as dynamic as the experiences studied. Unimaginative methods yield ineffectual portrayals of human experience. The field of transpersonal psychology, which studies psychospiritual development and transformation, especially warrants research methods that honor the vast and multifaceted nature of our experience. In exploring *both* the existential and transpersonal dimensions of human experience, such methods of inquiry often include narrative and storytelling, the social and relational context, unexpected insights, and a focused and comprehensive examination of the constituents of meaning, as they inform and reveal the subtle and complex nature of human experience.

The paradigms of science are shifting. The stage is set for change. Rollo May, in his foreword to the first volume in this series (Valle & King, 1978), commends the authors for presenting methods that he then called the "necessary bridge" to help move us beyond the elegance of psychological theories to experience ". . . the human being as he [she] exists, a living, acting, feeling, thinking phenomenon, at this moment in an organic relationship to us" (p. vii). Along with the efficacy of phenomenological methods and the existential–phenomenological critique, other methods and critiques have loosened the exclusive hold that the experimental method has enjoyed on psychological research. Some of these alternative views include the counterculture of the 1960s and 1970s, deconstructionism and the postmodern critique of culture, feminist research approaches, the epistemological insights of quantum and high-energy physics, narrative methods and discourse analysis, in-depth case studies, heuristic methods, and the concerns about external and effectual validity taking place within experimental psychology. To quote Adrienne Rich (1979, p. 35), we must get beyond the "assumptions in which we are drenched." Once thought of as an impregnable epistemology, behaviorism has been besieged by still more complete and far-reaching ideas and methods.

Some years ago, while I was still a young assistant professor and reluctant to speak up at faculty meetings, I was listening to a lengthy faculty discussion. We were talking about our undergraduate psychology majors and wanting to know what they wanted in the way of curriculum and career guidance. The conversation was lively and imaginative, as we went on about ways to trick or fool the students into giving us the information. Together, we thought up all sorts of clever ways to deceive them, by embedding pertinent questions in regular exams, creating subliminal cues for classroom films, setting up confederates to record conversations before and after classes, and so on. Finding myself puzzled by the conversation, I gathered my courage and gently inquired, "Why don't we just ask them?" A seemingly long and thoughtful silence ensued, followed by generous laughter at the obviousness, even innocence, of my suggestion. Years later, I came to realize that simply asking research participants to describe the fullness of their experience, and then rigorously analyzing this data, is at the heart of the phenomenological method of inquiry.

This book is unique in another way. Having explored the existential dimensions of human experiences such as anger, shame, and forgiveness, *phenomenological methods are extended to the study of transpersonal or spiritual experience* as it presents itself to human awareness. As a transpersonal psychologist, I study topics such as sacred weeping and the spirituality of landscape, and supervise doctoral students investigating topics that include the spiritual dimensions of pregnancy, ecstatic movement, identity and acculturation, healing through sleep and rest, and the story- and myth-making nature of children. I am acutely aware of the need for creative methods, such as those provided by phenomenological inquiry, to investigate experiences with clear transpersonal dimensions. The final section of this book provides the reader with a tempting smorgasbord of investigations into experiences with obvious transpersonal or spiritual elements. What transpersonal psychologist would not delight? Being with dying, the practice of loving-kindness, the experience of grace, the near-death experience, and the nature of suffering are fascinating explorations into the fullness, and the often mysterious and challenging nature, of human experience.

What is the nature of being *fully* human? We cannot know who we can be unless we push the edges of our potential and possibilities. In that way, we create our future in every aspect of our lives as we live them, including the dimensions of psychological inquiry. It is the nature of human experience, and indeed of evolution, to become more complex, finely tuned, and interrelated. The scientific endeavor and its attendant methodologies are not excepted from this progression. Our research methods and paradigms require more risk, adventure, subtlety, and imagination. If we don't do it now, when will we do it? Who will do it? And from a postmodern perspective, for the children of the 20th and 21st centuries, it is now or never. Modernity is running out of time to save us from our own successes.

Personally, I (Braud & Anderson, 1998) envision a science of psychology that extols the creativity and possibilities of human experience rather than one that consigns my rapturous, transcending, passionate, and painful human experiences to the status of epiphenomena. The concepts of psyche and soul, of course, have long been discarded by a disenchanted science. It is not surprising that many psychologists are now suggesting that it is time for the reenchantment of science.

It is ultimately a matter of integrity and, in conventional scientific terminology, validity. Are we studying something of value and studying it comprehensively? My commonsense definition of the validity of a study is that it must tell the whole truth about an experience. To study with integrity the full measure and depth of human experience, even our methodologies must face straight on the enormity of being *fully* human. I congratulate the editor and each contributing author for taking a significant step in this direction by providing us with this wonderful volume.

Rosemarie Anderson

Institute of Transpersonal Psychology
Palo Alto, California

REFERENCES

Braud, W., & Anderson, R. (1998). *Honoring human experience: Transpersonal research methods in the social sciences.* Thousand Oaks, CA: Sage Publications.

Rich, A. (1979). When we dead awaken: Writing as re-vision. In *On lies, secrets, and silence.* New York: Norton.

Valle, R. S., & Halling, S. (Eds.). (1989). *Existential–phenomenological perspectives in psychology: Exploring the breadth of human experience.* New York: Plenum Press.

Valle, R. S., & King, M. (Eds.). (1978). *Existential–phenomenological alternatives for psychology.* New York: Oxford University Press.

Preface

This volume is both a new book and the third in a series devoted to the application of phenomenological approaches to theory and research in psychology, its predecessors being *Existential–Phenomenological Alternatives for Psychology* (Valle & King, 1978) and *Existential–Phenomenological Perspectives in Psychology: Exploring the Breadth of Human Experience* (Valle & Halling, 1989). Since 1989, the phenomenological approach has continued to grow and has taken its place in the philosophical, theoretical, research, and applied clinical circles in contemporary psychology. Programs and individual psychologists at Duquesne University, Seattle University, the University of Dallas, and West Georgia College remain committed to the existential–phenomenological perspective. In addition to this volume, books published in the intervening years include titles such as *Entering the Circle: Hermeneutic Investigation in Psychology* (Packer & Addison, 1989), *Insight into Value: An Exploration of the Premises of a Phenomenological Psychology* (Fuller, 1990), *Experiential Method: Qualitative Research in the Humanities Using Metaphysics and Phenomenology* (Kidd & Kidd, 1990), *Being-in-the-World* (Dreyfus, 1991), *Emotion, Depth, and Flesh: A Study of Sensitive Space* (Cataldi, 1993), and *Hermeneutics and Truth* (Wachterhauser, 1994).

In this volume, 17 of 19 chapters are completely new to this series, compared to Valle and Halling (1989), in which nearly half the chapters were revised versions of those that appeared in the original 1978 book. Two other differences are also worth noting. Whereas the first two books were fairly balanced in terms of theoretical reflections and research, 14 of the 17 new chapters in this volume present new research studies, while the other three discuss phenomenological research and related issues. It should be noted that the research designs used in the chapters that report new findings are all of the empirical–phenomenological rather than the hermeneutic-process variety (Chapter 2 addresses this distinction).

Second, transpersonal–transcendent qualities of human experience now hold the same status as existential dimensions. In contrast, transpersonal psychology was discussed in a "special section" in the 1989 book, and not at all, at least not in any formal or explicit way, in Valle and King (1978). This more balanced distinction is reflected in the division of this volume into three parts entitled "Foundational Issues" (Chapters 1–4), "Existential Dimensions" (Chapters 5–11), and "Transpersonal Dimensions" (Chapters 12–19), each part with its own introductory commentary. Although a few chapters could each have been placed in one or more categories other than the one chosen, they are grouped according to their basic intent and the nature of their findings.

Even with these substantial changes, my basic intent remains unchanged: to provide an organized and accessible presentation of the phenomenological approach to a variety of relevant topics in psychology that will be helpful to students and colleagues alike. In the "Foundational Issues" section, for example, introductions to existential–phenomenological psychology and research are offered, along with new ways to explore the question of reliability in such research, and human subjectivity itself. The "Existential Dimensions" section offers new research addressing basic issues central to human life, including anger, shame, and forgiveness. Research studies relating to women's experience in three very different settings, as well as areas defined more specifically as clinical in nature, are also presented in this section.

The distinction I am making between these more purely existential issues and the transpersonal dimensions of experience has arisen directly from my own life experience. It is interesting to me that this very distinction grew out of a process or tenet that lies at the heart of the existential–phenomenological approach—that is, that experience is the ground of all knowing—and that this tenet is also held true by the world's great spiritual traditions. I invite the reader to reflect on those experiences and events in your life that were so powerful that you were left transformed, experiences that were impossible to explain to others, if not to yourself, experiences that altered the very fabric of your being-in-the-world.

For me, these experiences included a nearly fatal motorcycle accident (which involved a near-death experience, a three-month confinement in a body cast, a severe concussion, an extended loss of short-term memory, six operations in six months, and learning to read, speak, and walk again), the death of my father-in-law from cancer, my mother's suicide, the birth of three children, my wife's brain surgery, a number of shattered relationships, the loss of a business, and a dramatic change in my professional career as I had known it—all within a seven-year period. While I would not wish the suffering that occurred during this time on anyone, I can honestly say I am now thankful for the changes that went on within me as a result of what I experienced. I came to realize that self-transformation is quite real and, at least for myself, that events of this kind were quite necessary for me to become aware on a moment-to-moment basis of the spiritual/sacred dimensions of my life.

In attempting to understand and integrate these changes that seemingly touched every aspect of my life (death, sex, love, money, faith, pain, hope, God, grief . . .), I found it necessary to acknowledge and embrace perspectives that addressed realms of being that were somehow "beyond" the more day-to-day, habitual ways my mind had previously always responded to the people and things of my world. These are the transpersonal and transcendent dimensions I speak of in Chapter 12, and the kind of qualities addressed throughout the "Transpersonal Dimensions" section, more specifically in the findings of the phenomenological research projects described in Chapters 13–19, which include studies of synchronicity, being silent, being unconditionally loved, being with suffering, being with the dying, feeling grace, and encountering a divine presence during a near-death experience.

The relationship between the existential and sacred realms of experience has often been regarded as both mysterious and paradoxical in that, on one hand, they appear to be quite distinguishable (e.g., everyday versus mystical experience), while on the other hand, at the very same time, they appear inextricably intertwined (e.g., the intensely human and "other-worldly" qualities of any self-transforming experience). The results of the studies presented in these chapters, each situated in its own particular context, appear to collectively reflect this deep interrelationship between the existential and transpersonal dimensions of consciousness. It is to be hoped that careful and disciplined reflection on these findings will shed some light on the nature of this profound connection.

In professional psychology, the dialogue between existential–phenomenological and transpersonal approaches has taken on new dimensions, most notably at five different graduate psychology institutes, all located in the San Francisco Bay Area: the California Institute of Integral Studies, the Institute of Transpersonal Psychology, the Rosebridge Graduate School of Integrative Psychology, the Graduate School of Holistic Studies at JFK University, and the Saybrook Institute. Each of nine different research projects reported as chapters in this volume was completed at one of these schools. Recent books contributing to this dialogue include *Heidegger and Asian Thought* (Parkes, 1990), *The Religious within Experience and Existence* (Bourgeois, 1990), and *Phenomenology and Indian Philosophy* (Chattopadhyaya, Embree, & Mohanty, 1992).

In our preface to the 1989 volume, Steen Halling and I invited the reader to adopt a new attitude, to "look upon the world anew," this invitation echoing the phenomenological philosopher Edmund Husserl's renowned call for us to set aside our preconceptions and come "back to the things themselves," to life as we actually experience it. If we now open to the premise of a transintentional consciousness without an object, and to using the phenomenological approach to investigate transpersonal–transcendent experience, the call becomes "back to our true Self," to quieting our intentional minds and thereby becoming fully aware of who we really are. Here is the ground of a transpersonal–phenomenological psychology, and a perspective that expands the range of phenomenological inquiry to include all of the recognized dimensions of human experience and being.

> *I have been in that heaven, the most illumined, by light from Him and seen things which, to utter, he who returns hath neither skill nor knowledge. For as it nears the object of its yearning, our intellect is overwhelmed so deeply, it can never retrace the path it followed. But whatsoever of the holy kingdom was in the power of memory to treasure will be my theme until the song is ended.*
>
> —Dante, *The Divine Comedy*

Ron Valle

Walnut Creek, California

REFERENCES

Bourgeois, P. L. (1990). *The religious within experience and existence.* Pittsburgh, PA: Duquesne University Press.

Cataldi, S. L. (1993). *Emotion, depth, and flesh: A study of sensitive space.* Albany: State University of New York Press.

Chattopadhyaya, D. P., Embree, L., & Mohanty, J. (Eds.). (1992). *Phenomenology and Indian philosophy.* Albany: State University of New York Press.

Dreyfus, H. L. (1991). *Being-in-the-world.* Cambridge, MA: MIT Press.

Fuller, A. R. (1990). *Insight into value: An exploration of the premises of a phenomenological psychology.* Albany: State University of New York Press.

Kidd, S. D., & Kidd, J. W. (1990). *Experiential method: Qualitative research in the humanities using metaphysics and phenomenology.* New York: Peter Lang.

Packer M. J., & Addison, R. B. (Eds.). (1989). *Entering the circle: Hermeneutic investigation in psychology.* Albany: State University of New York Press.

Parkes, G. (Ed.). (1990). *Heidegger and Asian thought.* Honolulu: University of Hawaii Press.

Valle, R., & Halling, S. (Eds.). (1989). *Existential–phenomenological perspectives in psychology: Exploring the breadth of human experience.* New York: Plenum Press.

Valle, R., & King, M. (Eds.). (1978). *Existential–phenomenological alternatives for psychology.* New York: Oxford University Press.

Wachterhauser, B. R. (Ed.). (1994). *Hermeneutics and truth.* Evanston, IL: Northwestern University Press.

Acknowledgments

I give my heartfelt thanks to my parents, Elso and Mabel, for giving so freely to their son for so many years. I am deeply grateful to my dear wife Valerie and our now grown children, Demian, Alexa, and Chris, for not only accepting me as I am, but for encouraging me time and again to be all I can be. This book is a manifestation in the world of their love and support. It would not have happened without them. I wish to acknowledge my spiritual friends, Mary and Christine, for their personal commitment to integrating the sacred into every aspect of their lives. Their love and support for the spiritual vision from which this book emerged are truly unconditional.

I wish to thank Lester Embree, Ph.D., Department of Philosophy, Florida Atlantic University, for granting permission to reprint sections of *Life-World Experience: Existential–Phenomenological Research Approaches in Psychology* (von Eckartsberg, 1986), which appear as Chapters 1 and 2 in this volume.

Last, I wish to thank Rolf, whose insights and interpretations never failed to deepen my own understanding. His love and enthusiasm for phenomenology and psychology were so real and so infectious that 27 years later I still feel them each day. I miss you.

REFERENCE

von Eckartsberg, R. (1986). *Life-world experience: Existential–phenomenological research approaches in psychology.* Washington, DC: Center for Advanced Research in Phenomenology; University Press of America.

Contents

I

Foundational Issues

The first four chapters present foundational issues in existential–phenomenological psychology. They each offer material and raise questions considered basic to understanding the phenomenological approach in psychology.

Chapter 1, by Rolf von Eckartsberg, discusses the important concepts and individuals involved in the process that brought existentialism and phenomenology together as an existential–phenomenological psychology, with a particular regard for the nature of research in this context. In this way, Chapter 1 provides a philosophical and theoretical introduction to all of the chapters in this book and lays the groundwork for the description of approaches to research in existential–phenomenological psychology presented in the next chapter.

Chapter 2, also by von Eckartsberg, presents the two general approaches to research in phenomenological psychology: empirical–structural and hermeneutical–phenomenological. The works of van Kaam, Colaizzi, Giorgi, and W. Fischer (all from the Duquesne community during that time) are offered as examples of empirical phenomenological research, thereby illustrating a variety of ways in which to arrive at the structural characterization of the experience being investigated. Hermeneutical–phenomenological research is divided into two categories: actual life-text studies and studies of recollection and literary texts. The nature of life-text studies is discussed with an example from the work of von Eckartsberg.

Chapter 3, by Scott D. Churchill, Julie E. Lowery, Owen McNally, and Aruna Rao, continues the presentation of research issues by examining the question of reliability in phenomenological research. By presenting three different researchers' analyses of the same protocol, the authors bring to light several key factors involved in understanding the meaning of reliability in interpretive psychological research.

Chapter 4, by Bernd Jager, concludes this section with a unique perspective on the nature of human subjectivity. From a disciplined and creative process of self-and-world inquiry, he offers reflections and insights that are grounded in the understanding that consciousness always refers us to an implicit relationship between persons and things, and that this relationship forms the basis of our awareness of our world. In this way, one may regard Jager's insights—in and of themselves—as the "results" or "findings" of a hermeneutical process investigation.

1

Introducing Existential–Phenomenological Psychology

Rolf von Eckartsberg

Reflective analysis of life-world experience has been the focus of interest of philosophical phenomenology (Spiegelberg, 1960). Inspired by this work, psychologists and psychiatrists have begun to build a phenomenologically oriented approach to their disciplines (May, Angel, & Ellenberger, 1958; Spiegelberg, 1963). While much of this effort has been oriented toward clinical psychology and psychiatry, some of it is concerned with developing empirical and hermeneutical methods for doing phenomenological research on psychological phenomena. These research approaches are intended to study the meanings of human experiences in situations, as they spontaneously occur in the course of daily life. The emphasis is on the study of lived experience, on how we read, enact, and understand our life-involvements.

In this chapter and the next, I will review the evolution and creative proliferation of ways of doing such research on human experience and action as they have emerged from the collaborative efforts of phenomenologically oriented psychologists. This collective project, in which I have been involved as an active participant for 20 years, has been to rigorously develop and articulate a new paradigm: existential–phenomenological psychology. To this end we have built up a body of human science methods, researches, and theoretical statements that, we feel, do better justice to human experience than those psychologies that still restrict themselves to the paradigm of the natural sciences. According to Kuhn (1962), paradigm shifts involve a radical questioning and reworking of the basic presuppositions

Rolf von Eckartsberg • Late of Department of Psychology, Duquesne University, Pittsburgh, Pennsylvania 15282.

Phenomenological Inquiry in Psychology: Existential and Transpersonal Dimensions, edited by Ron Valle. Plenum Press, New York, 1998.

of a discipline. In our case, the advance of philosophical understanding during the last century has been a motivating force by virtue of which a new vision of human nature has emerged.

Whether a particular psychological paradigm recognizes it or not, its basic understanding of human nature, its philosophical anthropology, influences the whole conglomerate of problems, facts, and rules that govern its research. Giorgi (1970) has used the term "approach" to denote the way a science's basic presuppositions are intimately interrelated with the content it takes up and the methods it evolves. Thus, it follows that the new philosophical anthropology worked out by the existential and phenomenological philosophers of our century calls for new methods for the psychologists who hope to revise their field in the light of these new insights.

Foremost among these insights has been the effort to reject the notion that humans are merely biological objects whose every thought, feeling, and action can be said to be determined by a complex network of causes. This conception of human nature, borrowed from the natural sciences and ultimately from those philosophers who first extended the notion of causality to human being, is the implicit assumption of much of traditional psychology. These natural science psychologies have been unable to account for human freedom and the meaningfulness of human experience. Instead, they resort to quantitative, mechanistic, and computer models of human nature that, at best, record various regularities of behavior and make predictions and, at worst, do violence to our forms of self-understanding.

Existential–phenomenological psychology attempts to account for the fullness of human life by reconceiving psychology on properly human grounds. The model of the natural sciences, appropriate as it is for such fields as physics and chemistry, is nevertheless of limited usefulness when it comes to the study of the meaningful character of lived experience. Thus, it has been suggested (Giorgi, 1970; Strasser, 1963) that for psychology to fulfill its promise, this natural science model should be set aside in favor of a truly *human science* one. The human science approach recognizes that our privileged access to meanings is not by way of numbers but rather through perception, cognition, and language. Insofar as everyday human activity can be shown to be continuously informed and shaped by how we understand others and ourselves and by the meanings of the situations we find ourselves in, this is a most significant point. It indicates that the way for psychology to comprehend human behavior and experience as it is actually lived in everyday social settings is to begin by soliciting descriptive accounts of our actual experiences in such settings.

Thus, rather than hastily trying to quantify or abstract from everyday experience in the style of the natural sciences, we begin by more carefully attending to our actual living of that experience. This is the starting point for existential–phenomenological psychology—the arena of everyday life experience and action (experiaction)—for which Husserl (1962) has given us the metaphor of *life-world*. The life-world is the locus of interaction between ourselves and our perceptual environments and the world of experienced horizons within which we meaningfully dwell together. It is the world as we find it, prior to any explicit theoretical conceptions.

Of course, the world of experiaction does not stand still waiting for us to study it. Like time itself, life is forever streaming on and changing. Fortunately, by way of articulation and reflection, we can preserve our experiaction as narrative, as "life-text," and even submit it to rigorous and systematic investigation. While we live more than we can say, we can express more than we usually do if we make the effort, and nothing prevents us from describing our experiaction more carefully. With our ability to observe, remember, report, and reflect on both our own and others' experience and action, we have a rich source of material from which to build a truly human science psychology.

In the realization of this project, we acknowledge a heavy debt to the existential–phenomenological philosophers who have formulated the foundational insights for our psychology. These seminal thinkers have tackled many of the persistent and thorny problems that have always haunted psychology, such as that of free will versus determinism, the mind–body problem, the nature of perception, language, and action, and so on. Since in many instances we have taken up their thought and adapted some of their methods to the field of psychological research, a brief outline of the most vital contributions of existential–phenomenological philosophers follows.

EXISTENTIAL–PHENOMENOLOGICAL PHILOSOPHY

Existential phenomenology, as it has developed over the decades, has been primarily a philosophical endeavor. Numerous works, notably Spiegelberg's *The Phenomenological Movement* (1960), and Luijpen's *Existential Phenomenology* (1969), have traced its rich historical development, focusing upon such key figures as Husserl, Scheler, Jaspers, Marcel, Heidegger, Sartre, Merleau-Ponty, and Ricoeur, among others. Curiously, the exact origin of the hyphenated term existential-phenomenology is uncertain, as most of the philosophers mentioned identified themselves as either existentialists or phenomenologists. Although bringing both groups together under a single heading could possibly obscure some of their differences, we feel that the complementary movement of their thought, as discussed here by Luijpen (1963, p. 36), is a good justification for speaking of existential-phenomenology:

> The primitive fact itself of the new movement, of the new style of thinking, was reflected upon and expressed after Kierkegaard's existentialism and Husserl's phenomenology had, as it were, fused together in the work of Heidegger. At present it is realized that the new style of thinking uses as its primitive fact, its fundamental intuition, its all-embracing moment of intelligibility, the idea of existence or, what may be considered synonymous with it, the idea of intentionality.

As the originator of philosophical phenomenology, Husserl articulated the central insight that *consciousness is intentional,* that is, that human consciousness is always and essentially oriented toward a world of emergent meaning. Consciousness is always "of something." He argued that experiences are constituted by consciousness and thus could be rigorously and systematically studied on the basis of their appearances to consciousness—that is, their phenomenal nature—when an appropriate method of reflection—that is, phenomenology—had been worked out. Besides the explication of experiences, which Husserl considered to be a psychological project, it would also be possible to reflect upon and articulate the most essential structures of consciousness—that is, phenomena—such as intentionality, temporality, spatiality, corporeality, perception, cognition, and intersubjectivity, as he in fact did in *Ideas* (Husserl, 1962), and other later works. As philosophy, phenomenology had thus become the reflective study and explication of the operative and thematic structures of consciousness, that is, primarily a philosophical method of explicating the meaning of the phenomena of consciousness.

Husserl's methodology was to begin with the "phenomenological reduction," or "epoche," which involved the attempt to put all of one's assumptions about the matter being studied into abeyance, to "bracket" them. As Giorgi (1981) pointed out, to proceed without this step when reflecting upon personal experience leaves one open to the "psychologist fallacy," namely, the likelihood that one's judgments about such experiences will be biased by various preconceptions, wishes, desires, motives, values, and other influences. It was just this bias of one's uncritical "natural attitude" that Husserl wished to free himself from, in order to view a given topic from a position as free of presuppositions as possible. Only when the bracketing or suspending of such preconceptions had been achieved was the natural attitude said to give way to a more disciplined "phenomenological attitude" from which one could grasp essential structures as they themselves appear. As Giorgi (1981, p. 82) describes this process:

> Bracketing means that one puts out of mind all that one knows about a phenomenon or event in order to describe precisely how one experiences it. . . . Husserl introduced the idea of the phenomenological reduction, which after bracketing of knowledge about things means that one is present to all that one experiences in terms of the meanings that they hold out for consciousness rather than as simple existents.

The assumption of the phenomenological attitude thus implies that we do not describe something in terms of what we already know or presume to know about it, but rather that we describe that which presents itself to our awareness exactly as it presents itself. This movement is crisply formulated in the phenomenological imperative: "Back to the things themselves!"

By this dictum, the "things" toward which the phenomenological gaze struggles are no longer "objects" as such (in the sense of naive realism), but rather their meanings, as given perceptually through a multiplicity of perspectival views and contexts. Along with other presuppositions, the phenomenologist puts his or her existential belief "out of action," that is, dispenses with the belief that objects exist in and of themselves, apart from a consciousness that perceives them. When this belief is suspended, what remains is the phenomenon, the "pure appearance" that presents itself to consciousness. For example, when I eat an apple, I effectively destroy it as a physical object, and yet it remains as a phenomenon. Its various perspectival views—that is, its redness, its juiciness, its roundedness, and its other properties—can remain as a matter of contemplation for me, as what Husserl identified technically as *noema*.

Whether or not it is possible to put into abeyance all of one's presuppositons about an apple (or any other item of reflection) is questioned by some existential-phenomenologists, among them Merleau-Ponty. According to him, a totally presuppositionless vantage point cannot be secured, because as we put one presupposition out of action, we uncover more hidden ones beneath it. He believed that our vital interests and existential involvement with people and things in the world are of a fundamental character and will not allow themselves to be entirely undercut. Nevertheless, he considered the aim of the movement of the phenomenological reduction to be an extremely fruitful one, for by uncovering our presuppositions and interrogating them, we can clearly advance our understanding of the phenomenon under consideration.

The questions that guide research of philosophical phenomenology would be: What is the essence of this phenomenon? What are the conditions of possibility for the constitution of meaning by human consciousness? Because phenomenology had to do with the intuiting of essences, Husserl sometimes called it an "eidetic science." Like any science, it aimed to provide lasting and objective–universal knowledge, to separate the arbitrary and accidental from the necessary and the permanent, that is, the essential. To accomplish this aim, Husserl (1962) augmented the process of phenomenological bracketing with a procedure he called *free imaginative variation*. With this method, the noematic object was to be varied in imagination by altering its constituents in order to test the limits within which it retained its identify, so as to discover its variants. Applying this to our earlier example of the apple, one would begin by modifying its various aspects in our imagination, so as to engender a manifold of imaginary apples. Although some will be red, like the one that was eaten, others would be green, and even a purple apple could be imagined. That we don't find purple apples in actual experience is irrelevant at this stage, for what we want to discover is the essential structure and the essential constituents of an apple, and to do so we need to consider the possible alongside the actual. Already we can see that redness does not belong essentially to apples, although a skin that may be any of various colors does.

In principle, with a sufficiently thoroughgoing and deep-reaching imaginary variation, it should be possible to delimit the essence of a phenomenon such as our apple, or of a phenomenon of any other sort. Eventually, those aspects of the phenomenon that could be and could not be eliminated without altering its basic structure would become evident. As Gurwitsch (1964, p. 192) wrote:

> By means of the process of free variation, these structures prove invariant by determining limits within which free variation must operate in order to yield possible examples of the class under discussion. These invariants define the essence or eidos of this class, either a regional or a subordinate eidos. They specify the necessary conditions to which every specimen of the class must conform to be a possible specimen of this class.

In the main effort of his work, Husserl employed the phenomenological reduction and free imaginary variation for strictly philosophical pursuits. Nevertheless, he believed these procedures could be applied to other tasks, and he delimited some relevant domains of inquiry. Of the greatest concern to us are the distinctions he provided between phenomenology and psychology, which were conceived in terms of three separate and necessary domains of investigation. As summarized by Spiegelberg (1960, p. 152):

> *Pure phenomenology* is the study of the essential structures of consciousness comprising its ego-subject, its acts, and its contents—hence not limited to psychological phenomena—carried out with complete suspension of existential beliefs.
>
> *Phenomenological psychology* is the study of the fundamental types of psychological phenomena in their subjective aspects only, regardless of their embeddedness in the objective context of a psychological organism.
>
> *Empirical psychology* is the descriptive and genetic study of the psychical entities in all their aspects as part and parcel of the psychophysical organism; as such it forms a mere part of the study of man, that is, of anthropology.

We will see later in our presentation of the empirical phenomenological methodologies developed at Duquesne University and elsewhere that these distinctions have continued to play a contributing role, although unlike Husserl's reliance upon various forms of disciplined reflection on consciousness in general, the empirical phenomenological researchers have turned to the analysis of concrete descriptions of lived experience, gathered from the psychologist's most valuable source of data: other people.

Before we turn to the ways in which existential–phenomenological psychologists have created empirical and hermeneutical research methods, we want to review some of the results of the work of existential–phenomenological philosophical reflection that are relevant. From a methodological point of view, the work of philosophers is nonempirical. They do not design experiments, nor do they engage in systematic data-gathering methods or data analysis in a scientific sense. What philosophers do engage in is careful and systematic reflection upon and interpretation of human experience. What they reflect upon is life-experience in general as it is mediated by the accumulated tradition of philosophy itself. The accumulated biographical and historical stock of knowledge of the philosopher is the context of reference and source of examples and illustrations for philosophical work.

Phenomenology has been considered to be primarily a contribution of method applied to the phenomena of human consciousness. Existentialism, on the other hand, has been characterized as an effort to specify the essential and perennial themes of human existence in its broadest sense, as finite, embodied, mooded, in time, situated, threatened by death, capable of language, symbolism, and reflection, striving for meanings, values, and choices, self-fulfilling and self-transcending, as involving and committing itself to relationships, accountable and capable of responsibility. Existential-phenomenology thus means the application of the phenomenological method to the perennial problems of human existence.

Others have said that existential-phenomenology broadens the base of understanding of our discipline beyond acts of thematic consciousness by recognizing the importance of prereflective bodily components in the constitution of meaning (Merleau-Ponty); by emphasizing the existential choices a person makes about his or her life-situation, the "existential project" (Sartre); or by focusing on the totality of personal existence as being-in-the-world (*Dasein* [Heidegger]), including our dwelling in social relations and historical circumstances. There is still another opinion (Frings, 1965) that sees the unfolding of the phenomenological movement and the foundation of contemporary European philosophy as resting on the threefold foundation of phenomenology (Husserl), philosophical anthropology (Scheler), and the ontology of *Dasein* (Heidegger). Thus, there is much internal development within the phenomenological movement, as we shall see.

PHENOMENOLOGY: HUSSERL, SCHUTZ, AND SCHELER

Edmund Husserl

The phenomenological approach centers on the experienced fact that the world appears to us through our stream of consciousness as a configuration of meaning. Acts of consciousness, such as perceiving, willing, thinking, remembering, and anticipating, are our modalities of self–world relationship. They give us access to our world and that of others by reflecting on the *content* (i.e., its meaning or "the what") that we thus encounter and also by reflecting on the *process* (i.e., "the how"). Husserl, the founder of phenomenology, hoped to clarify in a descriptive–reflective manner the foundation and constitution of knowledge in human consciousness. Phenomenology became the study of human meanings as constituted by the stream of consciousness. Consciousness itself is understood as being *intentional*. It is as always *directed toward* something. As phenomenologists are fond of saying: Consciousness is always consciousness of something. Consciousness recognizes and creates meanings that subsequently inhere in the world as experienced.

With Husserl from 1900 on, we enter an era in philosophy and psychology that recognizes the participation of the subject in the creation of meaning. The subject's role is acknowledged even in physics through Heisenberg's uncertainty principle. In both physics and psychology, the assumption of "objectivity of reality" collapses under the realization that the observer as well as the actor are existentially and epistemologically implicated in the creation of meaning. There is no "really real" world of independent, objective facts. Rather, the world comes into being for us as meanings that we constitute and as political realities for which we fight. Husserl's fundamental contribution was to call our attention to the study of the meaning constituting power of the acts of consciousness. He developed systematic reflection as a research method. In working out some of the implications and strategies of this new reflective philosophical methodology, Husserl (1962) discovered the complexities of the *horizonal nature* of consciousness; that is, our field of awareness always extends beyond the factually given to that which is implied, remembered, anticipated, generalized, or otherwise processed. Husserl focused mainly on the *temporal horizons* of "inner time consciousness," that is, on the way people experience their embeddedness in the stream of time: past, present, and future. He explored how these horizons cooperate in creating the temporal meaning of the here and now. He also developed the notion of "inner and outer horizons"—what we might call *cognitive horizons*—which refer to contexts of knowledge playing on the here and now. Such horizons contextualize experience in terms of consensually available cognitive frameworks of perceived meaning. Husserl is the master of the articulation of the "mind space" within the larger sphere of the unified field of a person's consciousness and existence, "the psychocosm" (von Eckartsberg, 1981).

In his late work, Husserl (1970) developed his idea of the "life-world," the world of taken-for-granted everyday activities and commonsense meanings. Because we are embedded within the socially constituted meanings of our commonsense world, we are explicitly aware neither of the taken-for-granted nature of this reality nor of how we constitute it. Yet this unacknowledged realm of the life-world is the basis of all scientific activity. Scientific constructs are built on indubitable but taken-for-granted common activities and associated constructs of common sense. The life-world is the unexamined foundation and matrix of scientific activity, and phenomenology makes these commonsense constructs and phenomena its object of investigation.

Alfred Schutz

Whereas Husserl was concerned with how we construct our reality in general, Schutz (1962, 1964, 1966) focused more specifically on our construction of *social reality*. He took up the challenge of Husserl's phenomenology and related it to sociology and social psychology. He was primarily concerned with articulating the commonsense structures of consciousness, which he called *typifications* of consciousness, by means of which individuals comprehend the nature of social reality and are enabled to act in everyday life.

Temporal typifications articulate our experienced life-world in terms of the "world within restorable reach" (the past), the "world within reach" (here and now), and the "world within attainable reach" (the future). A related set of temporal constructs concerns our social partners in life as *predecessors, contemporaries,* and *successors,* in terms of which our biographical stock of knowledge is organized. For Schutz, the experienced scheme of temporality itself is formed from the interplay of lived-time, social calendar time, and cosmic time, which regulates the natural rhythms of days and seasons.

Schutz devoted much effort to an articulation of the biographical stock of knowledge organized in terms of what he called *hierarchical orders of typifications* (schemes of interpretations, recipes for action, or role conceptions) and in terms of schemes or orders of relevance that both express the interests and motivations of individuals and groups and thematize the world in a relativistic manner. His work opened up within phenomenology the phenomena of encounter, social interaction, and the reflective articulation of intersubjectivity. His important and fertile studies of the social structures of the life-world have been extended and deepened by the work of Peter Berger on the social construction of reality (Berger & Luckmann, 1966) and on problems of modernity (Berger, Berger, & Kellner, 1973) and by the work of Maurice Natanson (1970) on the structure of the "journeying self" engaged in self–other–world typifications. The work of phenomenology has become increasingly recognized and important for mainstream sociology.

Max Scheler

While the main concern of Husserl and Schutz was to articulate the purely rational structures of human being, Scheler (1961) was preoccupied with the phenomenological description and analysis of the nonrational essences in experience with invariant structures in emotional life. Scheler was the phenomenologist of values, feelings, social sentiments, and love. He forged a philosophical anthropology guided by the basic notion of personhood as a spiritual reality, by the belief in the essentially social nature of human existence, by the absoluteness of values and the eternal in human nature. His concern was to determine the place of human being in the cosmos. His starting point was the irreducibility of the person as *ens amans,* as a loving being and as the ethical being, and his method was a phenomenological one that he developed in an originary way.

Scheler explored the phenomena involved in the immediate apperception and emotional cognition of values—value-ception: value-awareness and value-perception—that he considered to be prior to and hence foundational for all other acts of cognition. Scheler was a passionate proponent of the primacy of the emotional and vital sphere. He worked out an influential phenomenology of ethics that articulated an objective hierarchy of values ranging from sensible to vital values—both values of life—and then to spiritual values and the value of holiness—both values of the person. Scheler made important contributions to the phenomenology of religion. He brilliantly described the key interhuman phenomena of love and hate, and the variety and forms of sympathy (Scheler, 1954a). He provided us with the exemplary study of the phenomenon of resentment (Scheler, 1954b) (resentiment), and he made important contributions to the sociology of knowledge, distinguishing three types of knowledge: knowledge of *control,* as in the aspirations of science and technology; knowledge of *essences,* as in the aspirations of philosophy, metaphysics, and phenomenology; and knowledge of *salvation,* as in the religious quest for spiritual fulfillment. Scheler's philosophical anthropology has been called *ethical personalism* within a Christocentric spiritual tradition emphasizing the multidimensional nature of human existence as bodily–vital, egoic–mental, and personal–spiritual. The highest good must be personal. Scheler emphasized love and the study of the *ordo amoris*—the configuration of love—as the core of the person and as the foundation for social relationships and societal forms. Seheler's work has great originality and masterful phenomenological subtlety. It is fertile and offers many challenges and invitations for corroborative psychological work (Frings, 1965).

EMERGENCE OF EXISTENTIAL-PHENOMENOLOGY: HEIDEGGER, SARTRE, AND MERLEAU-PONTY

The pure phenomenology of Husserl was later enriched by the "existentialist movement" in the tradition of Kierkegaard and Nietzsche. Expanded into *existential*-phenomenology, associated primarily with Heidegger, Sartre, and Merleau-Ponty, it recognized the importance of preconscious lived-experience, that is, the phenomenon of the "lived-body." It emphasized that being-in-the-world involves more than human consciousness and encompasses the total embodied human response to a perceived situation. Such insights led *existential*-phenomenologists to focus their research on *human situated experience*. Intentionality became redefined as a dialogal, relational dynamic of self–other interaction. Existence refers to the concrete, biographical, and embodied life of named persons who are characterized by uniqueness and irreplaceability. Existential-phenomenology studies existence in terms of the person's involvement in a situation within a world. It aims in its ultimate objective at "the awakening to a special way of life, usually called authentic existence" (Spiegelberg, 1963, p. 255).

Martin Heidegger

The main contribution of Heidegger (1962, 1971) lies in his radical questioning of the traditional Cartesian subject–object distinction, which leads to a dualistic universe and the dichotomy of subjective consciousness versus objective matter. With the subject–object split as an operative life-world assumption, there is always a gap and separation to be bridged between the two ontological realms of matter and consciousness, leading to unresolvable epistemological difficulties. But if we conceive of our existence completely in relational and field theoretical terms as a field of openness into which things and the world appear and reveal themselves in a dynamic way (*Dasein* as being-in-the-world), then we can avoid this problem. Persons are not selves separated from a world that is presumed to exist completely independently of them. Rather, they are personal involvements in a complex totality network of interdependent ongoing relationships that demand response and participation.

Heidegger advanced the thesis that the world comes into existence for us in and through our participation. He worked out the essential structures of being-in-the-world as grounded in *care,* that is, in concernful presence and openness to the world and others. He developed the general approach of phenomenology into an interpretative understanding of Dasein's total being. He called this approach the *hermeneutics of existence,* that is, the interpretative characterization of existence in the world.

Heidegger's work issues a call for action, personal movement, authentic participation, and a change in one's way of thinking from the calculative to the meditative mode. The movement depends on one's resoluteness to face basic existential contingencies, primarily the anxiety over one's own death. It requires one to acknowledge one's self as an illuminator and creator of one's world. Heidegger also talked about ultimate horizons and concerns. He postulated qualitative transformations in authentic moments and movements of personal existence. By doing so, he brought in a transpersonal context and went beyond a strictly rational world view. His attitude and concrete examples of existential–hermeneutic work (Heidegger, 1971) place him in kinship with the tradition of Zen (von Eckartsberg, 1981).

Jean-Paul Sartre

Sartre (1953, 1968) contributed greatly to the "existentialization" of phenomenology through his challenge that "existence precedes essence." According to Sartre, the person is the totality of his or her life choices, for which he or she is fully responsible. His idea of the *fundamental project* of a person's life refers to the way each person chooses himself or herself. This project can be disclosed by *existential psychoanalysis*, Sartre's method of personalistic reflection that was applied in his famous book-length case studies.

The concept of fundamental project is of great potential value to clinical and personality psychology. It refers to the unique configuration of meaningful existential choices, that is, the total web of existential moves a person makes. The project is the inner principle of coherence that we can perceive and articulate in our own and others' lives.

The study of individuals entails gaining an understanding of how they go about the actualization of possibilities, their "not yet." Sartre emphasized that one is always moving beyond oneself toward something else. He worked out what he called a "progressive–regressive" dialectical method that is said to be able to betray the "secret of the self," namely, the implicit purpose for which one strives. Sartre's method aims at *comprehension*, a mode of understanding wherein one lives the existence of the other in intuitive and empathic behaviors. To comprehend the action of another, we enter the original situatedness of that person biographically in terms of the operative historical and cultural conditions (regressive move); we seek to understand the purpose or goal choice that governs the direction of the action taken (progressive move) by means of which the person surpasses the givens in the direction of his or her possibles.

Maurice Merleau-Ponty

Merleau-Ponty (1962, 1968) widened the meaning of intentionality to include preverbal thought (thinking that exists in action) or the prepersonal dimension of bodily intentions and meaning. He maintained that the acting body always already understands its situation as well as its own possibilities quite before we pay any explicit attention to it. For Merleau-Ponty, intentionality, no longer merely a matter of cognitive consciousness, includes the life of embodied existence and interactive communication that precedes and is the foundation for explicit and thematic consciousness.

In the most global terms, Merleau-Ponty speaks of the mystery that I am part of the world and that the world is an extension of my body. Body and world mutually imply each other and are of the same nature. They stand in a relationship that he characterizes as *j'en suis* ("I belong to it"). This is the primordial ground of all of our awareness, a kind of prolongation of our body that Merleau-Ponty in his later writing expresses metaphorically as *the flesh of the world.* The subject–object dichotomy of traditional thinking is overcome by Merleau-Ponty. We are always in the midst of the world and have no vantage point outside of it. We can never achieve total clarity even in our reflective and critical orientation because we cannot fully penetrate the darkness of our primordial awareness in which meaning is always already constituted. We cannot attend the birth of meaning in our life. Bodily existence itself is a giver of meanings. Our body has the power of expression; it gives rise to meaning.

Our body gives us our power of motility, the "I can" or "I am able to." This happens on a prereflective level, on the level of *operative intentionality*, of *practognosie*, to use Merleau-Ponty's term. By virtue of our embodiment, we find ourselves always already situated and capable of meaningful interaction.

Merleau-Ponty has contributed greatly to our understanding of the person as a participant in and creator of meaning—even as a creature *condemned to meaning*. He also makes us aware of the limits of our power of reflection and of the fact that we find ourselves in a situation of essential ambiguity, of *chiaroscuro*, not being fully able to penetrate the sources and origins of our meaning making. This ambiguity is grounded in our bodily participation in being and on the paradox that we are ourselves constituted by the very being we become aware of. Merleau-Ponty rejects both materialism and idealism, that is, the reduction of the person's world to an idea. He establishes his own position of existential-phenomenology, which is the middle ground centering on one's *embodied* subjectivity and focusing on the *primacy of perception*.

FROM INTENTIONALITY TO RELATIONSHIP-BUILDING AND DWELLING

We have surveyed how Husserl's original inspiration of phenomenology has undergone significant development and change through the work of his successors. If we focus on the key phenomenological notion of *intentionaltty*, we can gain a measure of the development that has taken place in our thinking.

Originally, intentionality was metaphorized as an "intentional arrow" symbolizing the one-directional act of *ego–cogito–cogitatum*, an ego directing its attention toward an object and thereby revealing the sense or meaning of the object. Reflecting on intentionality revealed to Husserl the existence of horizons or halos extending from the perceptual or cognitive objects and linking those objects to their relevant contexts of interpretation and familiarity *cognitively*, that is, in terms of outer horizons; *perceptually*, that is, in terms of the horizons engendered by bodily movement; and *temporally*, that is, in terms of inner time-consciousness. Husserl's early phenomenological work focused primarily on *cognitive meanings*. In his later work, Husserl expanded the notion of intentionality to include *operative intentionality*, that is, the intentionality of spontaneous and competent autotelic body-movement. We could metaphorize this as "auto-pilot intentionality." Merleau-Ponty takes up and elaborates this idea in the context of his notion of the lived-body, the "body subject" (*le corps propre*) and motility. By means of our bodily insertion into reality, we are always already vitally responsive to the demands of our situation on our body. Our body moves in terms of prereflective intelligence and lived involvement that exceed our conscious awareness and control. Operative intentionality establishes and utilizes secret bonds of correspondence and interdependency: reciprocal involvement. In this way of thinking, the "intentional arrow" has become a two-way street of interaction, interexperience, and co-constitution.

For Scheler, the primordial human act is one of *value-ception*. He emphasizes the emotional and transrational nature of our relating to the world and to one another, and he concentrates on loving and hating, that is, value-laden acts by means of which we construct our lives. For Sartre, the notion of intentionality is linked with existential choices and radical freedom to make commitments and to choose our future. Sartre's key notion in this context is the *existential project*, characterizing the way a person chooses his or her long-range life commitments and life direction in and through all particular acts of involvement.

In Heidegger's hyphenated notion of *being-in-the-world,* which radicalizes the subject–object notion and bridges the subject–object split, Dasein's basic ontological structure is characterized as care, concernful presence, world-openness. The priority of the subject, or person, or ego, yields to the unitary and coequal relationship of mutual implication or "relational totality": caring–being-in-the-world. In his later work, Heidegger develops this notion into *dwelling,* by which he means our caring, sparing, spatializing, and temporalizing presencing and eventing of being, what we might call *culture-building.* Revealed by a new epistemological attitude, by meditative, responsive, "thanking thinking" (von Eckartsberg, 1981), dwelling is concerned with our authentic presence to our situations, our things, our people. Heidegger ushers in a normative dimension: our concern with the authentic and good life in deep relationship to the ground of Being.

In my reading, the concept of intentionality in the phenomenological tradition seems to continue to evolve. We can discern a spectrum that ranges from intentionality in the original Husserlian sense to relationship-cultivation and culture-building or dwelling in the sense of the later Heidegger. As we widen the context of understanding of the contributing dimensions in the constitution of meaning to include the role of our self-moving body, our essential intersubjectivity, and our embeddedness in language and culture, the meaning of intentionality changes from an emphasis on mostly cognitive understanding to one of existential engagement in the creation of a way of life (dwelling). Heidegger's understanding of dwelling is kin to the understanding and praxis of Zen and Taoism.

Thus, over time, our understanding of intentionality undergoes a shift in emphasis or focus, from *consciousness* to *culture-building acts,* from value-free phenomenological reflective analysis operating under the self-imposed disciplines of several steps of bracketing (epoches) to passionate value-engagement and existential commitment. We move from the primacy of knowing to the primacy of life praxis, to *enactment.*

THE EXISTENTIAL–PHENOMENOLOGICAL APPROACH IN PSYCHOLOGY

The existential–phenomenological approach to psychology proceeds on the assumption, as we all do in everyday life, that identically named experience refers basically to the same reality in various subjects. In the words of van Kaam (1966, p. 312):

> This basic identity of experience is an axiom in psychology. . . . This axiom seems to be confirmed by daily experience. . . . If we collect descriptions (of a named, i.e., linguistically specified experience) then we have to assume that others are focusing their attention on basically the same kind of experience that we are when we describe our experience. This statement is founded on the supposition that experience, with all its phenomena, is basically the same in various subjects.

In other words, we have to rely on the supposition that people in a shared cultural and linguistic community name and identify their experience in a consistent and shared manner. Our shared everyday vocabulary, including both ordinary language and those psychological terms that have filtered down from professional psychology, constitutes our access to experience, which is always to some extent already linguistically organized. Van Kaam refers to this basic identity of experience as an "axiom" underlying all of psychology. While we do not think it to be an axiom, at least it is a fruitful working hypothesis to be checked by empirical analysis. It gives us a starting point in working toward linguistic consensus and the establishment of communication communities.

The interrelationship of language and experience is a difficult psychological conundrum. How is it that we can say what we experience and yet always live more than we can say, so that we could always say more than we in fact do? How can we evaluate the adequacy or inadequacy of our expression in terms of its doing justice to the full lived quality of the experience described?

How are thought and life interrelated so that they can be characterized as interdependent, as in need of each other, as complementing each other, as interpenetrant? Living informs expression (language and thinking) and, in turn, thinking–language-expression reciprocally informs and gives a recognizable shaped awareness to living. Meaning, experience as meaningful, seems to be the fruit of this dialogue between inchoate living and articulate expression. Whereas living is unique and particular, that is, *existential*, thinking tends toward generalization, toward the universal, the essential, the *phenomenological*.

Our life takes place within this tension, "in between" these levels of participation. This tension and interdependence between life and thought is expressed in the very name of our approach as existential–phenomenological and constitutes what we might call the *existential–phenomenological paradox*. In *The Journeying Self*, Natanson (1970, p. 66) writes about this:

> If Phenomenology depicts the typical, Existentialism shows the unique; if Phenomenology deals with essence, Existentialism handles concreteness; if Phenomenology is interested in the structure of consciousness, Existentialism is concerned with the reality of the individual. It would seem these philosophies represent divergent standpoints rather than complementary approaches to human reality. That they are congenial, indeed intimately inter-sustaining and supporting views of man is the methodological thesis on which we have proceeded. What appears to be a paradox is a metaphysical duality which affects all philosophizing as it does the reality which all philosophizing is about. In classical and traditional terms the duality has been called the Universal and the Particular, or Thought and Life. The question is how the truth of the world in general is related to the truth of individual experience and how a theoretical system can understand the immediacy of the person.

We can express this formulation in terms of a diagram that places life and thought in dialectical opposition. It bespeaks the same tension as that between *existentialism*, which focuses on the problems and themes of life itself, of existence, and *phenomenology*, which focuses on the explication of the intentional structures of consciousness in general.

This tension is inherent in the very organization of language, which can move on either or both levels simultaneously, with descriptive specificity and uniqueness as well as in the mode of universalizing conceptualization and judgment. Language can encompass and interrelate both levels through mixed, multileveled discourse, and by means of such expressive tools as *metaphors*, *symbols* (Murray, 1975), and *Proper Names* (Rosenstock-Huessy, 1970). Each of these latter three linguistic structures constitutes a special class of "concrete universals" (Natanson, 1970), useful in aesthetic, religious, moral, and legal discourse.

Let us indicate the mystery of the interdependence of these two extreme levels of existence, this mystery of the intertwining and copenetration, by the ancient Chinese symbol of Yin–Yang, which visually dramatizes and represents the living reality of this mutuality, of "one within the other" (see Figure 1).

Figure 1. The yin–yang symbol representing the implicit interdependence/tension between the phenomenological/ as understood and existential/as lived realms.

With some justification, we can say that the mystery of existential-phenomenology is concealed in the *hyphen* itself. It indicates the difficult expressive problem of languaging the simultaneity and interpenetration of both living and thinking, of spontaneous enactment and reflective explication.

The systematization and "proceduralizing" of the movement from implicit to explicit by existential–phenomenological psychology is essentially a formalization of an everyday activity and general human ability: to reflect on experience. First we report and describe—narratization—and then we think further about something in order to conceptualize it. We keep asking: What does it mean? What does it say? What is concealed in it? What becomes revealed through dwelling with it patiently? What secret lies hidden therein? This is the *reflective attitude*, that is, openness and listening to Being in all its particular manifestations. This is the basic descriptive–reflective approach utilized by existential-phenomenologists. It is a skill anyone can acquire for himself or herself and integrate into his or her living as a meaning-enhancing depth dimension of self-development.

How to invite and initiate newcomers into this approach can be a serious pedagogical problem! It can be entered into if we begin gently with the first linguistic step of expression and description of the experienced flow of living, expressing a particular experience that we specify and identify by giving an account of it, by trying to bring it to as full and accurate an articulation as possible. We create a life-text. From there, we move toward the reflective study of this description, focusing on the essential meaning-constituents contained therein, the "experience moments" (van Kaam), the "meaning-units" (Giorgi), the "themes and scenes" (W. Fischer), or the "psychological plot" (von Eckartsberg), which can be inductively synthesized into an essential, structurally integrated description of the universally valid meaning of the phenomenon under consideration. This process involves a dynamic dialogue between two levels of description: the everyday level of *narrative language* and the more general and condensing achievement of reflective analysis, that is, *structural conceptual language*. Existential-phenomenologists refer to this general process of meaning-articulation as *explication,* as noted by van Kaam (1966, p. 305):

> By explication, implicit awareness of a complex phenomenon becomes explicit, formulated knowledge of its components.

By means of explication, we are to discover what the necessary and sufficient conditions and constituents of the type of event under study are, that is, what the structure of the phenomenon under investigation is. Explication, like any form of interpretive reading or hermeneutics, is a *qualitative research procedure* in that it wishes to arrive at an understanding and circumscription of *what* the phenomenon essentially is as a lived human meaning (structure) and *how* it is lived by individuals in their everyday existential lived contexts (style).

The existential–phenomenological approach in psychology tries to be *empirical* in the sense of basing itself on factual data that are collected for the purpose of examination and explication. As explained by Giorgi, Fischer, and Murray (1975, p. xi):

> Our research is empirical in that shareable, replicable observed events or personal reports are its data. Moreover, we remain true to each of the individual subjects' ways of embodying the general structure that we discover through examination of specific, situated instances.

In another formulation (Fischer & Wertz, 1979, p. 136):

> By empirical we refer to a) our reflection upon actual events and to b) our making available to colleagues the data and steps of analysis that led to our findings so that they might see for themselves whether and how they could come to similar findings.

In contrast to existential–phenomenological philosophers, who tend not to specify the "database" upon which their reflections are founded, the existential–phenomenological psychology researchers presented here are very explicit about the research design of their investigation and the methodological steps of data analysis and explication. In the tradition of experimental psychology, they are guided by strict considerations of verifiability and replicability. But in contrast to the traditional *quantitative* approach in psychological research, which uses statistical analysis of sets of data collected in terms of a strict hypothesis-testing research design, the existential–phenomenological approach in psychology does *qualitative* research, that is, meaning analysis and explication of descriptions of life-world experiences. Qualitative research is a form of content analysis covering a spectrum of approaches ranging from what has been called *empirical–phenomenological* psychology to *hermeneutical–phenomenological* psychology—depending on the "data source" that underlies the reflective–hermeneutic work. Giorgi, Knowles, and Smith (1979, p. 179) state:

> Empirical–phenomenological psychology has long been aware of this hermeneutical activity as intrinsic to research. The question remains, then, how does a hermeneutical–phenomenological psychology in the strict sense differ from the standard empirical–phenomenological research conducted at Duquesne University for the past 20 years? The essential differences would seem to relate to the grounds (database) upon which the interpretations are based and the procedures for arriving at these interpretations. Up to now, empirical–phenomenological psychology proceeded by collecting protocols descriptive of the subjects' experience (e.g., learning, envy, anxiety, etc.) and then systematically and rigorously interrogating these descriptions step-by-step to arrive at the structure of the experience. Hermeneutical psychology suggests another data source and a different method of analysis.

Hermeneutical–phenomenological psychology focuses on the works of great literature, on myths, autobiography, and works of art, including film, theater, and television, and thus widens the database and the understanding of the meaning of life-texts to include fictional and creative accounts as acceptable data for reflective–interpretative activity (Gadamer, 1975; Ihde, 1971; Palmer, 1969; Polkinghorne, 1983). In terms of the work of explication, the interpretative activities of hermeneutical–phenomenological psychology are as yet less well defined and more idiosyncratic than those of empirical–phenomenological psychology, although the attempt is always made to be as explicit as possible about the steps of the movement from concrete experience and description to universalizing conceptualization so that the reader can verify and replicate the process for himself or herself (Wertz, 1984).

The data source for existential–phenomenological psychological studies can vary from the collection of many protocols of described experience to the utilization of a single account of one subject. The data may also come from audio and videotapes and from speak-aloud protocols (Aanstoos, 1983) or from thinking out loud (Klinger, 1978). The focus of the investigation can be directed at arriving at the *general structure* of a phenomenon in terms of a synchronic formulation of essential constituents or at a *process structure* of a phenomenon that delineates the diachronic unfolding of the phenomenon in terms of essential stages aligned sequentially (Ricoeur, 1979).

Several empirical and hermeneutical models for the explication of phenomena using the existential–phenomenological approach have been described in the literature. Van Kaam (1966) led the way with the publication of *Existential Foundations of Psychology,* a wide-ranging work that emerged out of his doctoral studies and the research methodology developed for his 1958 dissertation. In 1970, another seminal work appeared, Giorgi's (1970) *Psychology as a Human Science.* In the same year, Duquesne also began publication of its *Journal of Phenomenological Psychology,* which presents the ongoing struggle for rigorous methodology. Notable among the publications that followed are the four volumes of *Duquesne Studies in Phenomenological Psychology* (Giorgi, Fischer, & von Eckartsberg, 1971; Giorgi et al., 1975, 1979, 1983), the most complete and detailed record of the work of the Duquesne group; Colaizzi' s (1973) book on methods, *Reflection and Research in Psychology;* and Keen's (1975) *A Primer in Phenomenological Psychology.* More recently are the collection of essays edited by Valle and King (1978), *Existential–Phenomenological Alternatives for Psychology; Conceptual Encounters* (de Rivera, 1981); and *Psychological Life* (Romanyshyn, 1982). The most recent publications on empirical existential–phenomenological research are *Exploring the Lived World: Readings in Phenomenological Psychology* (Aanstoos, 1984) and *Phenomenology and Psychological Research* (Giorgi, 1985).

This review of research on the analysis of life-world experience presents some prototypical examples to illustrate the range and diversity of the spectrum of human science methods as it has evolved in the work of the Duquesne group and other phenomenologically oriented psychologists. For a full and exhaustive representation of the accumulated research, we refer the reader to the original publications and the works cited above.

Our aim in the next chapter is to present the concrete ways in which this research and these kinds of qualitative analyses are made so that the reader will have a working familiarity with the design and methods of such work and with the types of language used to present the realities of such investigations. We hope that this presentation and illustration of the ways of doing analysis of life-world experience from a phenomenological perspective will encourage and enable the reader to engage in his or her own research into the meaning configurations of human experiences.

REFERENCES

Aanstoos, C. (Ed.). (1983). A phenomenological study of thinking. In A. Giorgi, A. Barton, & C. Maes (Eds.), *Duquesne studies in phenomenological psychology: Volume IV* (pp. 244–256). Pittsburgh, PA: Duquesne University Press.

Aanstoos, C. (1984). *Exploring the lived world: Readings in phenomenological psychology.* Carrolton, GA: West Georgia College.

Berger, P., Berger, B., & Kellner, H. (1973). *The homeless mind.* New York: Vintage Books.

Berger, P., & Luckmann, T. (1966). *The social construction of reality.* Garden City, NY: Doubleday.

Colaizzi, P. F. (1973). *Reflection and research in psychology.* Dubuque, IA: Kendall Hunt Publishing.

de Rivera, J. (Ed.). (1981). *Conceptual encounters.* Washington, DC: University Press of America.

Fischer, C. T., & Wertz, F. J. (1979). Empirical phenomenological analyses of being criminally victimized. In A. Giorgi, R. Knowles, & D. Smith (Eds.), *Duquesne studies in phenomenological psychology: Volume III* (pp. 135–158). Pittsburgh, PA: Duquesne University Press.

Frings, M. (1965). *Max Scheler.* Pittsburgh, PA: Duquesne University Press.

Gadamer, H. (1975). *Truth and method.* New York: Crossroads Publishing.

Giorgi, A. (1970). *Psychology as a human science: A phenomenologically based approach.* New York: Harper & Row.

Giorgi, A. (1981). On the relationship among the psychologist's fallacy, psychologism, and the phenomenological reduction. *Journal of Phenomenological Psychology, 12*(1), 75–86.

Giorgi, A. (Ed.). (1985). *Phenomenology and psychological research.* Pittsburgh, PA: Duquesne University Press.

Giorgi, A., Barton, A., & Maes, C. (Eds.). (1983). *Duquesne studies in phenomenological psychology: Volume IV.* Pittsburgh, PA: Duquesne University Press.

Giorgi, A., Fischer, C. T., & Murray, E. (Eds.). (1975). *Duquesne studies in phenomenological psychology: Volume II.* Pittsburgh, PA: Duquesne University Press.

Giorgi, A., Fischer, W. F., & von Eckartsberg, R. (Eds.). (1971). *Duquesne studies in phenomenological psychology: Volume I.* Pittsburgh, PA: Duquesne University Press.

Giorgi, A., Knowles, R., & Smith, D. (Eds.). (1979). *Duquesne studies in phenomenological psychology: Volume III.* Pittsburgh, PA: Duquesne University Press.

Gurwitsch, A. (1964). *The field of consciousness.* Pittsburgh, PA: Duquesne University Press.

Heidegger, M. (1962). *Being and time.* New York: Harper & Row.

Heidegger, M. (1971). *Poetry, language, and thought.* New York: Harper & Row.

Husserl, E. (1962). *Ideas.* New York: Collier.

Husserl, E. (1970). *The crisis of European sciences and transcendental phenomenology.* Evanston, IL: Northwestern University Press.

Ihde, D. (1971). *Hermeneutic phenomenology: The philosophy of Ricoeur.* Evanston, IL: Northwestern University Press.

Keen, E. (1975). *A primer in phenomenological psychology.* New York: Holt, Rinehart, & Winston.

Klinger, E. (1978). Modes of normal consciousness flow. In K. Pope & J. Singer (Eds.), *The stream of consciousness* (pp. 225–258). New York: Plenum Press.

Kuhn, T. (1962). *The structure of scientific revolutions.* Chicago: University of Chicago Press.

Luijpen, W. (1969). *Existential phenomenology.* Pittsburgh, PA: Duquesne University Press.

May, R., Angel, E., & Ellenberger, H. (1958). *Existence.* New York: Basic Books.

Merleau-Ponty, M. (1962). *The phenomenology of perception.* London: Routledge & Kegan Paul.

Merleau-Ponty, M. (1968). *The visible and the invisible.* Evanston, IL: Northwestern University Press.

Murray, E. (1975). The phenomenon of metaphor: Some theoretical considerations. In A. Giorgi, C. Fischer, & E. Murray (Eds.), *Duquesne studies in phenomenological psychology: Volume II* (pp. 281–300). Pittsburgh, PA: Duquesne University Press.

Natanson, M. (1970). *The journeying self.* Reading, MA: Addison-Wesley.

Palmer, R. (1969). *Hermeneutics.* Evanston, IL: Northwestern University Press.

Polkinghorne, D. E. (1983). *Methodology for the human sciences.* Albany: State University of New York.

Ricoeur, P. (1979). The human experience of time and narrative. *Research in phenomenology, 9,* 17–34.

Romanyshyn, R. (1982). *Psychological life: From science to metaphor.* Austin: University of Texas Press.

Rosenstock-Huessy, E. (1970). *Speech and reality.* Norwich, VT: Argo Books.

Sartre, J.-P. (1953). *Being and nothingness.* New York: Philosophical Library.

Sartre, J.-P. (1968). *Search for a method.* New York: Vintage Books.

Scheler, M. (1954a). *The nature of sympathy.* London: Routledge & Kegan Paul.

Scheler, M. (1954b). *Resentiment.* New York: Free Press of Glencoe.

Scheler, M. (1961). *Man's place in nature.* Boston: Beacon Press.

Schutz, A. (1962). *Collected papers: Volume 1.* The Hague: Martinus Nijhoff.

Schutz, A. (1964). *Collected papers: Volume 2.* The Hague: Martinus Nijhoff.

Schutz, A. (1966). *Collected papers: Volume 3.* The Hague: Martinus Nijhoff.
Spiegelberg, H. (1960). *The phenomenological movement: Volumes I & II.* The Hague: Martinus Nijhoff.
Spiegelberg, H. (1963). *Phenomenology in psychology and psychiatry.* Evanston, IL: Northwestern University Press.
Strasser, S. (1963). *Phenomenology and the human sciences.* Pittsburgh, PA: Duquesne University Press.
Valle, R., & King, M. (Eds.) (1978). *Existential–phenomenological alternatives for psychology.* New York: Oxford University Press.
van Kaam, A. (1966). *Existential foundations of psychology.* Pittsburgh, PA: Duquesne University Press.
von Eckartsberg, R. (1981). Maps of the mind: The cartography of consciousness. In R. S. Valle & R. von Eckartsberg (Eds.), *The metaphors of consciousness* (pp. 21–93). New York: Plenum Press.
Wertz, F. J. (1984). Procedures in phenomenological research and the question of validity. In C. Aanstoos (Ed.), *Exploring the lived world: Readings in phenomenological psychology* (pp. 29–48). Carrolton, GA: West Georgia College.

2

Existential–Phenomenological Research

Rolf von Eckartsberg

There are two general approaches or orientations to research in existential–phenomenological psychology: empirical and hermeneutical–phenomenological. This chapter will examine each of these in turn.

EMPIRICAL EXISTENTIAL–PHENOMENOLOGICAL STUDIES

The empirical existential–phenomenological branch of research has a *structural* orientation that aims to reveal the essential general meaning structure of a given phenomenon in answer to the implicit research-guiding question: What is it, essentially? Empirical existential–phenomenological studies focus on the analysis of protocol data provided by research subjects in response to a question posed by the researcher that pinpoints and guides their recall and reflection. There is a clear-cut general progression in the various genres of this type of research. We go first from unarticulated living (experiaction) to a protocol or account. We create a "life-text" that renders the experiaction in narrative language, as story. This process generates our data. Second, we move from the protocol to explication and interpretation. Finally, we engage in the process of the communication of findings.

Rolf von Eckartsberg • Late of Department of Psychology, Duquesne University, Pittsburgh, Pennsylvania 15282.

Phenomenological Inquiry in Psychology: Existential and Transpersonal Dimensions, edited by Ron Valle. Plenum Press, New York, 1998.

Steps in Conducting Empirical Existential–Phenomenological Research

Step 1. Problem and Question Formulation: The Phenomenon

The researcher delineates a focus of investigation. One formulates a question, a "hypothesis." One has to name the phenomenon, that is, the process in which one is interested, in such a way that it is understandable to others. Doing so is easy if the researcher names conventional and universally recognizable phenomena. It is difficult if the researcher studies phenomena that he or she has discovered and that have as yet no consensual meaning.

Step 2. Data-Generating Situation: Protocol Life-Text

In the empirical existential–phenomenological approach, researchers start with descriptive narratives provided by subjects who are viewed as co-researchers. The co-researcher reports on his or her own experience in writing. We query the person in his or her experiaction and engage in dialogue, or we combine the two, asking for a written description first and then engaging in an "elaborative dialogue" (von Eckartsberg, 1971).

Step 3. Data Analysis: Explication and Interpretation

Once collected, the data are read and scrutinized so as to reveal their "psycho-logic," that is, their structure, meaning configuration, principle of coherence, and the circumstances of their occurrence and clustering. In the traditional quantitative measurement-oriented approach, this step involves categorization and statistical treatment of group data organized in terms of sums, averages, percentages, and measures of dispersion and convergence. In the qualitative existential–phenomenological approach, emphasis is on the study of configurations of meaning in the life-text involving both the structure of meaning and how it is created. This explication brings out implicit meanings by means of systematic reflection. Reflection is the return in consciousness to scrutinize a particular event via its record in memory or as a life-text. As we reflect, we are guided by questions such as: "In what way is this description revelatory of the phenomenon I am interested in?"

In order for the life-text to reveal itself, we must approach it with an explicit concern expressed as a specific question. We call this the *explication-guiding question* because it gives focus as we question the text about its meaning.

The explication-guiding questions for the disclosure of situated and general structures are these: How is what I am reading revelatory of the meaning of the phenomenon in this situation? This question yields the *situated structure* of the phenomenon. What does the text and/or the situated structure tell me about the phenomenon in its generality and universality? What is its meaning essence? These more universalizing questions yield the *essential general structure* of the phenomenon.

Wertz (1984) discusses the problems of phenomenological sense-making or explication in conducting empirical existential–phenomenological research under the headings of (1) the general stance of the researcher, (2) the active operations of empirical reflection on the individual example of the phenomenon, and (3) the operations used to achieve generality of understanding.

As far as the stance is concerned, the important points are:

1. Empathic presence to the described situation.
2. Slowing down and patiently dwelling.
3. Magnification, amplification of details.
4. Turning from objects to immanent meanings.
5. Suspending belief and employing intense interest.

The use of active cognitive operations involves:

1. Recognition and utilization of an "existential baseline."
2. Distinguishing constituents.
3. Reflection on judgment of relevance.
4. Grasping implicit meanings.
5. Relating constituents.
6. Imaginative variation.
7. Conceptually guided interrogation.
8. Psychological languaging.

As to the explication and articulation of the phenomenological findings as universal insights into and constituents of the phenomenon, Wertz (1984, p. 24) considers the following features to be essential:

> *1. Finding general insights in individual reflections.* Since general features of the phenomenon are included in individual cases, the researcher may interrogate a single individual structure and find immanent generality.
>
> *2. Comparing previously analyzed individuals.* In comparing individual psychological structures of her phenomenon, the researcher discards divergences and retains convergences in her considerations of generality.
>
> *3. Generation and imaginative variation of new instances.* In order to extend general insights beyond the previously analyzed individuals, the researcher personally recollects and fancies new examples and counterexamples of the phenomenon to more incisively delineate the essential constituents and boundaries of her specific level of generality.
>
> *4. Explicit formulation of generality.* The researcher formulates her general knowledge of the phenomenon in a language which describes the specified diversity of instances included in the general class under consideration, whether it be a specific group, general type, or universality.

Step 4. Presentation of Results: Formulation

One's research findings must be presented in public form for sharing and criticism. These formulations present what we have called the "essential constituents" or the structure of a phenomenon, articulating what "it really is" as a human meaning.

Two different kinds of communication networks are involved in the presentation of results: the participants in the research themselves and fellow researchers. The "subjects" get a "debriefing" about the experiment in everyday language, while the "fellow experts" in the researcher's community, who share the professional relevancy structure and interest in the phenomenon, are communicated to in their shared expert language or professional idioms.

Examples of Empirical Existential–Phenomenological Research

Following are four examples of empirical existential–phenomenological research: the work of Adrian van Kaam, Paul Colaizzi, Amedeo Giorgi, and William Fischer. These examples are presented to show the variety that exists within this approach and, more specifically, to illustrate how we arrive at universal structural characterizations of the phenomenon. In order to present this overview of qualitative methods for the analysis of life-world experience, I have summarized their research work. For a full appreciation of the differentiatedness and subtlety of this kind of research, the reader must turn to the original publications.[1]

On Feeling Really Understood: Adrian van Kaam

Van Kaam (1966) is the founder and originator of the empirical existential–phenomenological psychological approach at Duquesne University. He characterized "empirical phenomenology" as a "return to the existential data." About phenomenological results in general, van Kaam (1966, p. 295) says: "Research performed in this way is pre-empirical, pre-experimental, and pre-statistical; it is experiential and qualitative." In Chapter 10 of *Existential Foundations of Psychology* (1966), entitled "Application of the Phenomenological Method," he reports on his Ph.D. research at Western Reserve University in 1958. Doing his dissertation at that time within a traditional quantitative research-oriented psychology department and within a Rogerian client-centered clinical orientation, van Kaam chose to develop his methodology as a variant of content analysis, a traditional quantifiable research methodology. Being also interested, however, in the qualitative and meaning-oriented philosophical approach of existential phenomenology, he chose to do his research both quantitatively and qualitatively using a large sample of subjects in his procedure. Let us look at the steps in his research procedure.

1. Problem and Question Formulation: The Phenomenon. Van Kaam (1966) took his topic and focus of investigation from within a clinical psychology context—research in psychotherapy—and the specific frame of reference of a Rogerian client-centered approach. Van Kaam (1966, p. 298) reports:

> I have selected one relatively simple concrete phenomenon that appears in psychotherapy. The patient at times "feels really understood" by the therapist. What is the fundamental structure and meaning of this experience of feeling understood?

In order to shed light on this problem, van Kaam needed to collect "data." Studying experiences, he needed descriptions of experiences—narrative descriptions. His first task, therefore, was to devise a research situation by means of which he could obtain such "data." He chose to formulate a specific question to be answered in writing by a large number of subjects—a question that he hoped would elicit responses (descriptions as "data") that he could then study as to their meaning-organization. Van Kaam (1966, pp. 320–321) thus devised the following instructions:

> Describe how you feel when you feel that you are really being understood by somebody.
> a. Recall some situation or situations when you felt you were being understood by somebody; for instance, by mother, father, clergyman, wife, husband, girl friend, boy friend, teacher, etc.
> b. Try to describe how you felt in that situation (not the situation itself).

[1]Von Eckartsberg's (1986) presentation of Giorgi's and W. Fischer's approaches in *Life-World Experience* contains examples of their particular styles of analysis using actual text from the protocol descriptions they each collected in their respective studies. These rather lengthy tables are omitted in the abridged version of his presentation offered in this chapter.—*Ed.*

c. Try to describe your feelings just as they were.

d. Please do not stop until you feel that you have described your feelings as completely as possible.

2. Data-Generating Situation: Collecting Descriptions. Several criteria had to be met: Van Kaam knew that he wanted verbal descriptions of experiences in response to a fairly delineated and directed question. In other words, he wanted descriptive narratives. To obtain them, he gave instructions so as to sharpen the focus of the descriptions—the phenomenon of being really understood by somebody. In terms of the writing of a response, van Kaam had to rely on the "natural story-telling ability" of his subjects (von Eckartsberg, 1979). This is a taken-for-granted expressive power that existential–phenomenological psychologists have to rely on. We have to rely on the everyday meaning-making skill of everyone—the ability to express one's experience in words, in the form of a story. As Crites (1971) claims: The very organization of human experience seems to be given to us in awareness as narrative.

In terms of the justification for using the procedure of written descriptions, van Kaam (1966, p. 310) states the following:

1. Feeling understood is a relatively common human experience.
2. Common human experience is basically identical.
3. This basically identical human experience is expressed under the same label.

3. Data Study Procedure: Explication. After the "data" were obtained as narrative descriptions of experiences, the next task was to "work" with these data, to "analyze" them—that is, to "study" and "process" them—so as to lead to results, findings, and conclusions. Van Kaam (1966) calls the procedure he uses *explication,* a term taken from existential phenomenology. This "scientific phase of explication" includes the following steps: "listing and preliminary grouping, reduction, elimination, hypothetical identification, application, and final identification" (p. 314).

Following standard praxis in content-analysis procedures, van Kaam divided his total sample into subgroups of approximately 80 each in order to arrive at results that could then be checked by fellow researchers and tested for interjudge reliability. The preliminary results were then tried out on the remaining samples in order to achieve further validation of the analysis. This multi-sample procedure and concomitant interjudge-reliability-check necessitate the complication of the steps called "hypothetical identification" and "application" in van Kaam's procedure. These intermediary steps have not been used by later existential–phenomenological researchers, who typically work with very small samples or even individual case studies and who have given up the process of obtaining interjudge reliability measurements, relying instead on full presentation of all steps of the explication and on an appeal to immediate intuition on the part of the reader's immediate personal experience for a sort of "existential validation," testing any statement or proposition against the personal stock of experience that serves as the ultimate criterion of evaluation.

Van Kaam (1966) distinguishes six *steps of explication:*

3a. Listing and preliminary grouping.

The first operation of the scientist is to classify his data into categories. These categories must be the result of what the subjects themselves are explicating. Therefore, the scientist makes his initial categories from empirical data, in this case, a sufficiently large random sample of cases taken from the total pool of descriptions (pp. 314–315).

3b. Reduction. The second step in the explication process is now undertaken and again tested for interjudge reliability.

Now that the elements are laid out for him in a quantitative and qualitative fashion, the researcher can proceed with the second operation of the scientific explication. He reduces the concrete, vague, intricate, and overlapping expressions of the subjects to more precisely descriptive terms.

3c. Elimination. The third step concerns a process of checking and elimination of non-relevant elements.

By means of the same operation, he now attempts to eliminate those elements that probably are not inherent in the feeling of being understood as such, but rather are complexes which include being understood in a particular situation, or which represent a blending of the feeling of being understood with other phenomena that most often accompany it (p. 316).

3d. Hypothetical identification.

The operations of classification, reduction, and element-elimination result in the first hypothetical identification and description of the feeling of being understood. The identification is called hypothetical because it was hardly possible to take into account at once all the details of all the descriptions during the element-elimination.

This is referred to by van Kaam as a hypothetical "formula."

3e. Application. We now enter a new phase, Application, which itself may have several steps within it. The work in this step consists in:

The application of the hypothetical description to randomly selected cases of the sample. This tentative application may possibly result in a number of cases of feeling understood that do not correspond to the hypothetical formula. It may be that the formula contains something more than the necessary and sufficient constituents of feeling understood. In this case, the formula must be revised in order to correspond with the evidence of the cases used in the application.

In other words, from the larger "data pool of description" ($N = 365$!) new samples are progressively drawn so as to check the "hypothetical identification formulas" against new evidence. This is continued until a satisfactory level of redundancy is achieved. This then gives rise to the last step, Final Identification.

3f. Final identification.

When the operations described have been carried out successfully, the formerly hypothetical identification of the phenomenon of feeling understood may be considered to be a valid identification and description (p. 316).

Van Kaam (1966) clearly stated the aim of the process of explication as being to answer the question: "What are the necessary and sufficient constituents of this feeling?" (p. 301). This question is to be answered by the Final Identification, which turns out to be a "synthetic description of the experience" that brings together, in a cluster, all constituents synthesized by the researcher into one description. On the way to this formulation, van Kaam reports on his findings.

4. Presentation of Results: Formulation. From step 3a, van Kaam (1966) reports that he arrived at a total of 157 different expressions listed under 16 different headings (p. 323).

From step 3b, Reduction, which he now refers to as the "Phenomenological Explication of the Data" (p. 323), he (p. 324) reports that:

In the operation of further "Reduction," each one of the 157 expressions had to be tested on two dimensions:

1) Does this concrete, colorful formulation by the subject contain a moment of experience that might be a necessary and sufficient constituent of the experience of really feeling understood?

2) If so, is it possible to abstract this moment of experience and to label the abstraction briefly and precisely without violating the formulation presented by the subject?

Next, all expressions discovered in this way, as either direct or indirect representatives of a common relevant moment of experience, were brought together in a cluster. This was labeled with the more abstract formula expressing the moment common to all.

The reduction resulted in nine probably necessary and sufficient constituents, each of them heading a certain number of expressions in which they were originally contained, and each of these expressions accompanied by the percentage of descriptions in which it was present.

The constituents which were identified in this way as being together necessary and sufficient for the experience under study had to be synthesized into one description which then identified the total experience of really feeling understood.

The results of this work of explication are then presented by van Kaam as a "Synthetic Description" with an amplification of each of the terms used in the description quoted here in full, so as to allow us to inspect the procedure close up. In this way, we arrived at the necessary constituents of the experience under study, with the following general operational definition: A necessary constituent of a certain experience is a moment of the experience that, while explicitly or implicitly expressed in the significant majority of explications by a random sample of subjects, is also compatible with those descriptions that do not express it. Nine constituents were finally identified as being together necessary and sufficient for the experience of "really feeling understood." These constituents are condensed in Table 1.

The synthetic description of the experience of really feeling understood, containing these constituents, is given below, followed by a justification and explanation of each phrase of the description (van Kaam, 1966, pp. 324–327).

The experience of / "really / feeling understood" / is a perceptual–emotional Gestalt: / A subject, perceiving / that a person / co-experiences / what things mean to the subject / and accepts him, / feels, initially, relief from experiential loneliness, / and, gradually, safe experiential communion / with that person / and with that which the subject perceives this person to represent.

The experience of: The term *experience* is preferred to *feeling* because the data show that this phenomenon, commonly called feeling, contains perceptual moments too.

really: The adverb *really* added to *feeling understood* emphasizes the distinction between objective and subjective understanding. The latter includes the "what it means to me" element and the emotional involvement of the subject.

feeling understood: This popular expression is maintained because it is used by most people when they express this experience spontaneously.

is a perceptual–emotional Gestalt: The data compel us to distinguish between perceptions and feelings (emotions), the former being predominantly object-directed, the latter subject-directed.

Table 1. Constituents of the Experience of "Really Feeling Understood" Finally Identified, with Percentages of 365 Subjects Expressing Each Constituent, Explicitly or Implicitly

Constituents of the experience of "really feeling understood"	Percentages expressing the constituents
1. Perceiving signs of understanding from a person	87%
2. Perceiving that a person co-experiences what things mean to subject	91%
3. Perceiving that the person accepts the subject	86%
4. Feeling satisfaction	99%
5. Feeling initially relief	93%
6. Feeling initially relief from experiental loneliness	89%
7. Feeling safe in the relationship with the person understanding	91%
8. Feeling safe experiential communion with the person understanding	86%
9. Feeling safe experiential communion with that which the person understanding is perceived to represent	64%

> But the perceptions and emotions are interwoven in experience; the term *Gestalt* implies that the distinction we make between perceptual and emotional moments does not correspond to a separation in reality.
>
> *A subject, perceiving:* The perceptual moment is mentioned first because of its priority in the explications obtained. The feeling of really being understood presupposes the perception of understanding as it is evidenced by various behavioral signs of understanding.
>
> *that a person:* The subject perceives that a "person," a fellow human being, understands him in a personal way. The understanding person is not experienced only as an official, a teacher, an adult, or so on, but as being-a-person.
>
> *co-experiences:* The understanding person shares at an emotional level the experiences of the subject understood. The prefix *co-* represents the awareness of the subject that the person understanding still remains another.
>
> *what things mean to the subject:* The subject perceives that the person understanding experiences the events, situations, and behavior affecting the subject in the way in which they affect him, and not as they might affect others.
>
> *and accepts him:* Even while sharing experiences of the subject which the person understanding does not accept personally, he manifests exclusively and consistently genuine interest, care, and basic trust toward the subject, whether or not the subject intends to change his views, feelings, or behavior.
>
> *feels, initially, relief from experiential loneliness:* The initial feeling of relief is the joyous feeling that experiential loneliness, a disagreeable perceptual–emotional Gestalt, is receding to the degree that real understanding is experienced. The adjective *experiential* specifies that it is not primarily a physical loneliness, but a being-alone in certain psychological experiences.
>
> *and, gradually, safe experiential communion:* This expresses that the subject gradually experiences that the self is in the relieved, joyful condition of sharing its experience with the person understanding. *Safe* emphasizes that the subject does not feel threatened by the experience of sharing himself.
>
> *with that person:* The deep personal relationship between the subject and the person understanding is prevalent not only in the perceptual, but also, and still more fundamentally, in the emotional area. Therefore our synthetic description not only opens, but also closes with a reference to this person-to-person relationship.
>
> *and with that which the subject perceives this person to represent:* When the person understanding typifies for the subject a certain segment of mankind, or perhaps all humans, or all beings, i.e. humanity and nature, or the all-pervading source of being, God, then the subject will experience communion with all those beings which are exemplified for him by the person understanding, and do this to the degree that this person is perceived as their representative.

If we compare the existential–phenomenological work of van Kaam with a more traditional approach in psychological content-analysis work, for instance, McClelland's work on the need for achievement (Brown, 1965; McClelland, 1953), we can note several similarities and several crucial differences. What is identical in the two ways of proceeding is the shared starting point: the collection of data as descriptive stories in response to a task given by the researcher.

McClelland uses adapted pictures of the Thematic Apperception Test (TAT) and asks several leading questions that are to be answered. Van Kaam uses many subjects, a "large sample," similar to the traditional measurement-oriented approach. The more subjects, the more confident the traditional researcher feels about the validity of his or her data and the generalizability of his or her findings.

Another identical commitment is the use of independent judges to establish interjudge reliability and thus achieve a degree of objectivity and measure of consistency in the tradition of the empirical scientific discipline. What is crucially different, however, and a consequence of the existential–phenomenological orientation—which is suspicious of attempts to bring in and in a sense "impose" a preconceived theoretical scheme on the data, although in any linguistic approach this can never be fully prevented—is van Kaam's insistence on letting the *experience moments* of the phenomenon reveal themselves from the description, as much as possible. Husserl gave us the injunction *"Back to the things themselves!"* Existential-phenomenologists generally are convinced and work on the assumption that the data, that is, narrative descriptions, reveal their own thematic meaning-organization if we, as researchers, remain open to their guidance and speaking, their disclosure, when we attend to them.

Empirical existential-phenomenologists therefore reject the praxis of using coding manuals, which allow a categorization and subsequent quantification of the data, the development of a measure that can then be applied, in the typical traditional psychological research campaign of correlational proliferation, to all other established psychological measuring tools.

Content analysis in the existential–phenomenological approach becomes a procedure of *qualitative analysis,* of "hermeneutics," that is, the study of implicit meanings that are to be explicated by verbal means and illuminated from within by virtue of their inherent experienced "psycho-logic" (meaning). Existential-phenomenologists shy away from the obvious impositions of explicit theoretical schemes in order to "interpret" the data because this procedure often covers up more than it reveals and, in its most extreme development, leads to the shortcoming of being a self-fulfilling prophecy, incapable of discovering anything new and not already contained within its conceptual definitional framework.

Traditionalists might object and say that they all go through an "inductive phase" of data study beginning with description or pilot data in order to construct their coding manual. In this sense, they can indeed be said to all go through the inaugural existential–phenomenological step: *to let human experience speak!* But that very initial inductive step, at the origin of all research, is all that traditional and existential–phenomenological researchers have in common. Very early, then, in the research work, in the study process, there is a decisive parting of the ways toward numbers and quantification on the part of *traditionalists,* who conceive of psychology in the model of a "natural science," and toward *meanings* on the part of existential–phenomenological researchers, who understand psychology to be a "human science."

We might add that psychological measures, scores, are also human meanings. They are fixed meanings, tied to operational definitions and specific acts using measuring instruments—test performance. In that context and as predictive and correlational possibilities, measurements have value. But determining the meaning of the psychological dimensions studied and how this meaning is lived in actual personal experience is a task that precedes any form of measurement, to which traditional psychology does not address itself and which has become the focus and subject matter of existential-phenomenology.

In this sense, van Kaam's study sheds light on the general experienced meaning of the lived process of "being really understood" as perceived moments in a person-to-person relationship. As van Kaam himself suggests, upon this clarified basis, one could develop all kinds of measurements.

On the Phenomena of Learning: Paul Colaizzi

Working under the direction of and in collaboration with Giorgi, Paul Colaizzi produced a master's thesis (1966) and Ph.D. dissertation (1968) devoted to an empirical phenomenological investigation of learning as experienced from the perspective of the learner.

Integrating this work in terms of a number of distinctions he made regarding the methodological and epistemological issues involved in doing existential–phenomenological research, Colaizzi (1973) published an interesting book entitled *Reflection and Research in Psychology: A Phenomenological Study of Learning.* Before its publication, many of the concrete issues involved in doing empirical existential–phenomenological research were only implicitly recognized and often overlooked. For instance, in reviewing van Kaam's (1958) work, Colaizzi (1973) observed that, properly speaking, it was indeed *empirical* and even *phenomenal,* but not yet fully *phenomenological,* because it did not penetrate into the implicit horizons of the descriptions gathered and stopped short of developing a complete *structural* explication of its phenomenon.

In Colaizzi's terms, what van Kaam's studies produced were *fundamental descriptions* (FDs), rather than *fundamental structures* (FSs). Each is a kind of research finding belonging to its own epistemological level, such that investigations at the phenomenal level yield FDs, whereas those at the phenomenological level yield an FS. This distinction of "levels of subject matter" is an important but difficult one. The phenomenal level is that at which subjects live through and describe their experience; it involves what happens explicitly for them. Keeping in accordance with Husserl's dictum "back to the things themselves," a reflection on the implicit and structural dimensions of subjects' descriptions shifts the researcher from the phenomenal level (that of the experience of the phenomenon) to the phenomenological level, where the focus is on the phenomenon that was experienced, the "what" that is being researched. In his book, Colaizzi (1973, p. 32) further sharpens this distinction:

> The structure of an experiential phenomenon need not coincide with a description of that same phenomenon as it is experienced by a subject, because the former is largely implicit and the latter is of a more explicit nature. After all, just as the description of a particular essence is not identical to the essence of a particular description, the FS and the FD of a single experiential phenomenon are not identical.

When a researcher focuses on the phenomenon *as it was experienced* and culls out the essence of that description, he has a fundamental description (FD); in order to arrive at a fundamental structure (FS), he or she must interrogate the implicit dimensions of his or her data, focusing on the essential elements that *constitute the phenomenon* through their internal relations with one another.

According to Colaizzi, the fundamental structure of a phenomenon may be accessed via two different types of reflections: (1) *individual phenomenological reflection (IPR),* in which the researcher uses only his or her personal experiences as "data," amplified through "free imaginative variations," and (2) *empirical phenomenological reflection (EPR),* in which the researcher performs his or her systematic reflections upon a body of descriptive protocols that have been collected from subjects. As will be recalled, the phenomenological philosophers (who did not "trouble" themselves with the descriptions of subjects) preferred individual reflection, believing this would be sufficient to uncover the essence of phenomena. Whereas the more empirically minded phenomenological psychologists had been skeptical of the completeness of such an approach and thus eliminated self-reflection, Colaizzi reintroduced it into empirical research, considering it a necessary first step that would reveal the researcher's own precomprehension of the phenomenon. This would be of help to other researchers who might wish to follow up on one's work, as well as provide a starting point for the reflections upon subjects' protocols.

In his book on learning, Colaizzi organized all his various distinctions and arrived at an order in which he felt empirical phenomenological research logically ought to proceed, namely:

1. Discovering an FS by IPR.
2. Obtaining an FD by the method of phenomenal study (PS).
3. Obtaining an FD via EPR.
4. Discovering an FS via EPR.

The following is a summary version of how Colaizzi applied these steps in his work on learning. Since the details of his findings in many instances are too extensive to reproduce, the interested reader is referred to Colaizzi (1973). What is included below is for illustrative purposes only and is not meant to do justice to the phenomenon of learning itself.

1. *Discovering an FS of learning by IPR.* Colaizzi began his research by setting himself a learning task—to learn the content of Spengler's book *The Decline of the West.* Reflecting on his experience of this task (which is the method of IPR), and performing the appropriate imaginary variations, Colaizzi (1973, p. 59) was able to articulate the following FS as a preface to his empirical investigations: "Learning is . . . that activity whereby the learner extracts from material his learned content, which is a meaning-idea of which he had no previous knowledge and which he posits as true."

2. *Obtaining an FD of learning by PS.* In this phase, Colaizzi set about to collect descriptive data on learning from subjects who had been given an actual learning task. Working in a laboratory setting, Colaizzi used tasks that were both traditional to learning experiments (e.g., memorizing nonsense syllables, solving problems) and nontraditional (e.g., disassembling and reassembling a gun, learning to walk on crutches). The data gathering was organized such that each of 50 subjects would fill out a questionnaire after having satisfactorily completed one task. Criteria (such as memorizing all ten nonsense syllables or correctly assembling the gun within two minutes) were employed for each task. Since there were ten tasks in all, Colaizzi distributed his subjects so as to receive five descriptions for each task.

The questionnaires employed consisted of various open-ended items inquiring about the "changes that occurred" during the course of doing the activity. In order not to prejudice the subjects' responses, the word "learning" was never mentioned. Instead, the questions centered around requests for detailed descriptions of any changes that did occur, asking the subject to distinguish between changes that occurred in himself or herself and those that occurred in the activity, wherever this distinction was possible. Moreover, to account for the constituting role played by the experimental situation (i.e., "demand characteristics"), subjects were also asked to describe fully what they believed the purpose of the investigation and the questionnaire itself to be.

With his data gathered, Colaizzi began the reflective work of interrogating the subjects' descriptions, with the aim of deriving an FD. The general process by which Colaizzi accomplished this was organized in terms of the following explication-guiding steps:

1. Each statement and expression contained in the original protocols was considered with respect to its significance for the fundamental description of the phenomenon in question (i.e., learning). Those that were relevant were retained; those that were clearly irrelevant were discarded.

2. All relevant statements were then classified into naturally forming categories, and all repetitive statements were eliminated.

3. The remaining relevant statements were then translated from the raw form in which they were presented in the original protocols to clear and succinct expressions or components.

4. Finally, the components were arranged into a series of statements that were accepted as the FD of the phenomenon of learning obtained by the method of PS.

The FD yielded in this segment of Colaizzi's research is given on page 71 of his 1973 book. Excerpting from that articulation (Colaizzi, 1973, p. 71), we find that learning involves

> . . . an increase of intellectual knowledge, or of bodily or intellectual skill in a task activity, or the discovery of an efficient though possibly slow method for executing the task activity, which is manifested in . . . cognitive changes and/or performance changes, these changes being affected by the methods employed which methods themselves change, and are accompanied by affective components that are interrelated with the cognitive and performance changes, by personal components, by bodily components, by temporal components, and by situation components, and there is an awareness of all the changes and of all their accompanying components.

Colaizzi then went on to elaborate these changes, detailing cognitive, performance, affective, personal, bodily, temporal, and situational components, as well as the awareness the subjects experienced of all the changes taken as a whole.

3. *Obtaining an FD of learning by EPR.* It soon became apparent to Colaizzi that different emphases could be given to the FD of a phenomenon, or, put differently, he realized that a phenomenon never has just one FD that captures it completely. For him, FDs could always be brought toward either of two ideal limits, one aiming at an *extensiveness of detail,* the other at an *intensiveness of substance.* Applying the method of EPR to the questionnaire data gathered from the learning tasks described above, Colaizzi (1973, pp. 75–77) sought to generate these two different types of FDs via the following phases:

> 1. The first step was simply the realization that the search for an *extensive FD* involved sacrificing substance for the sake of including as much detail as possible from the relevant source statements, whereas opting for a *substantial FD* necessitated sacrificing detail in order to achieve as substantially intensive a description as possible. Colaizzi's recognition was of the importance of "both poles of the continuum" and of the need to present not one but two FDs of learning.
>
> 2. Having made this procedural decision, Colaizzi next reflectively interpreted each relevant statement from the questionnaires to determine the meaning that it expressed and then listed each unique, that is, nonrepetitive meaning-expression.
>
> 3. Next, each meaning-expression was interrogated with respect to its significance for every other meaning-expression, thus producing interrelated clusters of meaning-expressions.
>
> 4. Next, the recognition that the clusters of meaning-expressions exhibited an interdependency demanded that they be synthesized into a single theme, constituted equally by each of them, which was thus accepted as the *extensive FD* of learning obtained by the method of EPR.
>
> 5. Abstracting all specifics from the interrelated clusters of meaning-expressions obtained in phase 3 above resulted in the *substantial FD* of learning obtained by the method of EPR.

The interested reader may locate Colaizzi's FDs of learning in his book (Colaizzi, 1973, pp. 81–82). The extensive FD seems to focus most upon the general progression from inept to efficient participation in the learning activity (i.e., the task begins as awkward, error-ridden, and forced, and progresses to automatic, easy, and errorless, whereas the subject's body loses its initial focus as an uncomfortable, aching burden, and eventually blends effortlessly into the performance), whereas the substantial FD uses a more psychological terminology to discuss participation in a learning activity as no longer an isolated event, but instead as "the nexus of a temporally extended situation that is constituted by emotional, mental and physical relations to previous and subsequent activities."

4. *Discovering the FS of learning by EPR.* This last step in the procedure of examining the phenomenon of learning entailed reflecting upon the FD of learning obtained by the preceding operation in order to reflectively explicate its implicit structure. This structure, as a *fundamental structure (FS),* was obtained by the method of EPR, focusing on the data provided by the *extensive FD* because of the richness and amount of empirical data it contained.

Regarding the explication-guiding question for the FS via EPR (Colaizzi, 1973):

> This involved a dialogue constituted on the one hand by each of the themes, and on the other, the reflective interpretation of them in terms of what they express as fundamental for learning (p. 83).
>
> *Execution and results of the method of Empirical Phenomenological Reflection directed to the discovery of the Fundamental Structure of learning.* In general, the themes extracted from the extensive FD are related to one another as follows. That to which learning participation is directed is first the total situation and then the content of the learning situation. Inasmuch as this content, initially unknown, becomes known and this participation is manifested in improved performance, the methods employed by the learner to execute this participation are determined by whatever motivational factors are present, and these motives in turn are determined by the learner's acceptance of and attitudes towards the content as it is integrated in what is viewed by the learner as the total learning situation, inclusive of its temporal references and emotional qualities.
>
> However, even if it is acknowledged that the above themes as they are related to one another are indispensable conditions for participation in a learning activity, it must be realized that they are indispensable only because and to the extent that they directly refer to learning participation which itself is prereflective. For example, the progression of performance from aching and awkward, etc., is not itself learning, but is the manifestation of engaged participation in learning activity. Thus the manifestation by performance of this participation does not in itself define learning since learning is constituted by participation as well as by performance. Moreover, since this participation is disrupted by the learner's awareness of it, and yet since it does nevertheless proceed, then learning participation occurs at a prereflective level. Likewise, with regard to his presence to the content of the learning situation, the learner's movement from involvement in the total learning situation expresses that the involvement is but a necessary prerequisite for the learning participation proper.
>
> Accordingly, although learning participation is impossible without these conditions, they are not intrinsic aspects of the participation itself: they are the necessary conditions for participation but are not themselves participation. In order, then, to capture that which sufficiently distinguishes and identifies learning, these conditions must be set aside in order to focus on actual participation in learning.
>
> Regarding this participation, it is essentially the learner's relation to the content of the learning situation. The relation of the learner to the content defines the very existence of the learning participation. That is, learning participation is directed to the content, specifically, the content as it is constantly given to the learner in new perspectives, these perspectives being co-constituted by the learner as opposed to being restricted to the physical characteristics of the content and of the learning situation. This co-constitution is determined by what the content means to the learner, whether it is personally, functionally, or emotionally relevant or insignificant. Furthermore, since the perspectives thusly determined in turn determine the motivation of and methods employed by the learner to co-constitute further perspectives of the content, learning participation thus expresses the circular process between the development of perspectives of the content and the learner's co-constituting of these perspectives.
>
> If, then, organization of the content is defined as "the learner's co-constituting of new perspectives of the content, regardless of whether the nature of the content is physical, cognitive, or bodily, by a process of disclosing its references to and contexts within which it can be harmoniously integrated," then the FS of learning discovered by the method of EPR may be formulated as follows: the prereflective organizing of a to-be-learned content whereby that content is apprehended by the learner from a new perspective (pp. 84–85).

Colaizzi presents the variety of research options and operations in terms of a schematic representation. Each operation is valid in its own right and yields a particular set of results so that the fruit of the labor is a pluralistic characterization of essential dimensions and facets of the phenomena that are interdependent and mutually enriching. All of the results together shed light on the meaning of the phenomenon of learning as experienced by the subject from the actor point of view.

> *Schematic presentation of operations.* At this point, the function of the four operations may be rendered more clearly by depicting them schematically; thus a schematization of the relationships among (a) the various methods; (b) their corresponding procedures; (c) types of subject matter with a typical finding concerning the nature of the learned-content that were established in this investigation is provided in Table 2 (p. 85).

In this complex study of the issues, levels, and strategies of existential-phenomenological research procedures regarding the phenomena of learning, Colaizzi clarifies two important issues: the *plurality–individual difference* and the *empirical–reflective difference.* We can approach the study of any experience by drawing on our individual stock of knowledge and personal experience, which includes all that we know including our socially mediated theoretical knowledge and our power of imagining human possibilities on the basis of this knowledge. We can also utilize other people's reports and descriptions or answers to a questionnaire to gather experiential protocols or data from a *plurality of subjects* to be reflectively analyzed. In some sense, we are always dealing with a combination of these two in that personal experience is always social and also in that event the study and interpretation of others' experience occurs within the at least implicit context of the researcher's own precomprehension and understanding. Similarly, although the empirical–reflective differ-

Table 2. Methods, Procedures, and Types of Subject Matter Involved in Research in Phenomenological Psychology[a]

Method	Procedure	Subject matter
IPR	Individual reflection without empirical data	Perspectival FS; learned-content as, e.g., an affirmed meaning-idea
PS	Analysis of empirical data without recourse to reflection	Organization of empirical data, but no definitive results; learned content as, e.g., network of facts
EPR	Reflection on organized, explicitly empirical data	Multiple FDs; e.g., personal involvement in learned content
EPR	Reflection on implicit dimensions of organized empirical data	Perspectival FS; learned content as, e.g., co-constantly emerging prereflective meanings
Combined IPR and EPR	Individual reflection without empirical data and reflection on implicit dimensions of organized empirical data	Comprehensive FS; learned-content as, e.g., an affirmed meaning-idea and co-constituted, constantly emerging preflective meanings

[a]From Colaizzi (1973, pp. 85–86).

ence can be established as an ideal postulate, it can never be sustained in a completely successful manner because empirical–factual reporting already implies some reflective-organizing, and even the most universal and abstract of formulations appears within horizons of concrete–factual–empirical references in the mutually interpenetrating dialectic of living and thinking.

It is instructive to keep these conceptual distinctions in mind when formulating a phenomenologically based qualitative research strategy. What is the source of our data? Is it one's own experience or the reports of a plurality of subjects? What is the level of analysis? Are we talking about empirical summary descriptions or the results of content analysis (the phenomenal level of fundamental descriptions) of the phenomenological level of structural analysis, which articulates the minimally necessary constituents and defining characteristics of a phenomenon: the fundamental structure, as articulated by the reflective work of the phenomenological researcher.

In the provocative comparative discussion of the results in the last part of his book, Colaizzi presents a detailed and sophisticated argument both for the distinctiveness of each of the operational steps of the method and for the interdependency and complementarity of the results. The difference between the findings regarding the *fundamental description* (FD) and the *fundamental structure* (FS) is seen to arise from a difference in focus on two layers of existence: *reflective* and *prereflective* life. Whereas the FD is a description of the subject matter in its reflective dimension, which is thematically present to the subject's experience and can thus be directly reported upon request, as an appearance, the FS represents a structural elucidation of the prereflective dimensions, which cannot be directly accessed by the subjects but can be revealed to the investigator and elucidated by him or her by reflective explication through an act of interpretive reading.

Colaizzi's statements regarding the issue of the relationship between reflective and prereflective dimensions of human experience raise some important methodological issues. Can one never be present and privy to the prereflective dimensions of one's experience? Is it necessary to articulate the prereflective dimensions after the fact, in the manner of the theory of the retrospective constitution of meaning (Schutz, 1962)? Does the prereflective have to be articulated in phenomenological–conceptual terms as fundamental structures? Can there be an experiential alternative to reflective phenomenological work in the direction of an "increasing effort of attention" or of an "intensification of consciousness"? Even if a new structural meaning dimension is discovered reflectively, and retrospectively, does not this insight become a permanent prospective horizon of meaning that irrevocably enriches the experience of the subject's thematic field (Gurwitsch, 1964)? Note the differences between the fundamental structure (FS) via the method of individual phenomenological reflection (IPR) articulated as ". . . that activity whereby the learner extracts from material his learned content, which is a meaning-idea of which he had no previous knowledge and which he posits as true" (Colaizzi, 1973, p. 100) and the fundamental structure (FS) via the method of empirical phenomenological reflection (EPR) articulated as ". . . the pre-reflective organization of a to-be-learned-content [which] is apprehended by the learner from a new perspective which discloses different aspects of the total phenomenon of experienced learning" (p. 100).

We will see in the following sections how the issues formulated by Colaizzi were selectively emphasized and elaborated by others who subscribed to a basically *empirical–*

reflective strategy. They all collect "data" and they all "analyze" the data reflectively, although there is much variation with regard to the particular steps taken in the actual research procedures. In this development, some of the rigorous distinctions worked out by Colaizzi between FD and FS and between IPR and EPR, between a phenomenal and a phenomenological study, were given up or transformed. Giorgi (1975a), for instance, worked out a series of steps of analysis that progress from "raw data," that is, descriptions of experiences, to an articulation of constitutive *meaning units* characteristic of a particular protocol description, to a formulation of the *situated structure* of the phenomenon, that is, the reflective articulation of the meaning-configuration for each subject, and, finally, the *general structure,* that is, the reflective–universal characterization of the phenomenon across all subjects. In Colaizzi's terminology, this involves working out the FS by means of the method of EPR and utilizing the distinction between individual experience expressed by the situated structure and that of the "plurality of subjects" expressed by the general structure.

Compared to Colaizzi's,Giorgi's research progression aims at a more clear-cut focus of results in that all the operations are designed to ultimately cull out the essential, general structure, that is, a characterization of what "it" (e.g., learning) essentially is, an aim that Colaizzi (1973, pp. 98–99) considers unattainable when he says:

> According to Heidegger, when man discovers any phenomenon he simultaneously co-discovers the world. What is meant by this is that there is nothing that is wholly isolated unto itself, but rather it is always related to an infinite horizon or an unlimited system of references. Each insight into something is accompanied by new areas of opacity concerning it. Expressed otherwise, man's knowledge is essentially finite. Yet the fact that man cannot possess infinite knowledge of something, or that he cannot even possess all of his potential knowledge of something immediately does not imply that his knowledge is "absolutely relative." It means only that what he knows is always and necessarily contingent, constantly in tension, awaiting further though never completed fulfillment. Accordingly, he must accept the idea of "absolutely absolute" knowledge as a chimera and satisfy himself with "relatively absolute" knowledge. In terms of a criterion for having articulated fully the fundamental structure of a phenomenon, this means that the fundamental structure is an absurdity. All that an investigator can hope to accomplish is to articulate how a phenomenon fundamentally is revealed to him from his finite perspective and submit this articulation to other concerned investigators who then reject it, modify it, complement it or temporarily accept it, and so on. Thus the endeavors of an individual investigator stimulate a dialogue between his results, a community of scholars and reality. There is no higher court of appeals for establishing a criterion for a fundamental structure than this dialogue. The reason for this is that a fundamental structure is an expression of an aspect of man as a transcendent being; and as a transcendent aspect, that which is expressed by the fundamental structure can never be fully captured by the laws which define, i.e., which definitely establish criteria for a natural event. In the meantime, during the unending dialectical development and evaluation of the fundamental structure, it can be measured against the criterion of its fruitfulness. For example, does it accurately and intelligibly reveal the phenomenon under investigation from within the perspective of the current comprehension of it? Does it eliminate confusions generated by the prevalent established system of knowledge concerning it?

In this way of speaking Colaizzi expresses and anticipates the attitude of a hermeneutical–phenomenological and of a dialogal approach—to be discussed later in this chapter—that is forever cognizant of the relativity of its own perspective and of its role as a contributing voice to an ongoing and ever-changing dialogue. The fundamental structure of the meaning of a phenomenon is the expression of a moment in the researcher's vision and understanding based on tacitly accepted presuppositions and situational circumstances. It is part of the hermeneutical work to critically examine, acknowledge, and discuss these operative assumptions. Colaizzi's work gives us a sophisticated and differentiated vocabulary and conceptual distinctions that can help us in defining these terms, tasks, and difficulties of a qualitative human science research approach.

In his later work on phenomenological research and the phenomenology of reading and existentially significant learning, Colaizzi (1978) moves into a dialogal–existential position. The emphasis is placed on the dialogue between the co-researchers in which the relevant contexts of operative presuppositions and the disclosure of existential contexts are explicated, revealing how they shape the person's understanding For Colaizzi, phenomenological human research becomes *existential therapy* (Colaizzi, 1978, pp. 69–70):

> Genuinely human research, *into any phenomenon whatsoever*, by seriously including the trusting dialogal approach, passes beyond research in its limited sense and occasions existential insight. This is nothing other than therapy. All human research, particularly psychological research, is a mode of *existential therapy*.
>
> Viewed as a mode of existential therapy, it can be understood why the phenomenologist has for so long maintained that human research into any particular phenomenon should shed light on the totality of the human situation, since it is clear that existential therapy should draw in the totality of the human person, e.g., his perceptions and cognitions, emotions and attitudes, history and predispositions, aspirations and experiences, and patterns, styles, and contents of behavior.

Colaizzi (1978, p. 70) argues that all research, including natural science research, should be human research:

> All research should be initiated by, engaged in with, and directed toward the clearing of existential significance.

The co-researchers, by uncovering their presuppositions, have to discover and articulate the humanly significant context in which they conduct their investigation. "They must be able to make explicit the existential significance of their research" and take full responsibility for its self- and world-transformative consequences.

On the Learning Experience: Amedeo Giorgi

Giorgi (1975a) has developed another form of empirical existential–phenomenological research that has been very influential at Duquesne. Giorgi cautions us, however, that there are many ways in which one can develop an empirical existential–phenomenological psychology. Giorgi (1975a, p. 83) states:

> In light of our introductory comments, it should not have to be emphasized that the method we are using as a demonstration is neither exclusive nor exhaustive and it should not be considered paradigmatic for all phenomenologically based research. It should be taken for what it is, one example of the application of phenomenology to psychology with some limited but valid generalizations of the value of phenomenology for methodological issues.

Being a theoretical and experimental psychologist with a traditional experimental psychology background, Giorgi has focused primarily on the topics of experimental psychology, perception and learning, and attempts to reformulate these problems from an existential–phenomenological perspective leading to the articulation of "psychology as a human science" (Giorgi, 1970) and to develop an alternative human scientific and qualitative approach to psychological research. Giorgi has done much pathfinding and programmatic work to bring this about.

In a series of theoretical and critical papers on the relationship of phenomenology and experimental psychology, Giorgi (1965, 1966, 1967) pointed out the need for a new qualitative, human scientific research methodology in psychology. In close but critical dialogue with the tradition, Giorgi argued that a rigorous phenomenological science of human experienced meanings is possible, but that it has to develop its own methodology, which involves the creation and collection of "unprejudiced verbal descriptions" as data, and the explication of meanings from the protocol in order to arrive at the formulation of the essence of the phenomenon under study. Giorgi explicitly accepts and affirms the goals of philosophical phenomenology to be an eidetic science of essences as valid for psychology also, and he proposes to apply this approach and method to the content area of psychological phenomena. Giorgi's own effort centered on the phenomenon of learning. Giorgi (1967) published an early example of the phenomenological approach to the problem of meaning and serial learning in which he demonstrated "the usefulness of applying a phenomenological interpretation to an empirical study" (p. 98), adding a qualitative dimension—experiential reports and their analysis—to the traditional focus on behavioral acts. Thus, in this early paper, Giorgi argued for the complementarity and mutual enrichment of the two approaches, using traditional methods and content areas, but complemented them with experiential data and the perspective and meaning of the subject, that is, he added a "phenomenological corrective." Comparing the two approaches, Giorgi summarized the two realms of the natural and human sciences (Table 3).

The "new vision for research in the human sciences" is based on a dialectic of the approach, that is, the fundamental viewpoint toward the person and the world that the scientist brings with him or her and that provides the foundation—the *method* and the *content* of the phenomenon that is being studied. Giorgi (1970) argued that the method is always developed in dialogue with the content of the phenomenon under study.

As compared to van Kaam and Colaizzi, with whom he has dialogued, Giorgi, by 1975, as compared with his earlier work on learning (Giorgi, 1967), has moved the focus of the content area of existential–phenomenological research psychology into life-worldly situations. The phenomenon is studied in the world and as part of the unfolding existence of the individual, and the research is moved away from the contrivances of the experimental psychology laboratory design, which is an abstraction and an artificial construction of

Table 3. Factors Important for Research in the Natural Sciences and Suggested Correctives for a Parallel Paradigm for Research in the Human Sciences

Realm of natural sciences	Realm of human sciences
Experimentation	Experimentation and other forms of research activity
Quantity	Quality
Measurement	Meaning
Analysis	Explication
Determined reactions	Intentional responses
Identical repetition	Essential phenomenon known only through manifestations
Independent observer	Participant observer

real-world experience. We can study the phenomena when and where they occur in the person's experience, and we can access this experience legitimately and scientifically in and through a retrospective, narrative account, an experiential protocol, a story, that constitutes our data.

We can say that empirical existential–phenomenological psychologists study life in and through texts, as "life-texts." What are we to do with these texts—how do we read them, study, analyze, and amplify them, and disclose other, latent meanings within them? These hermeneutical questions and issues come into play when one does this kind of empirical phenomenological psychology. How does one get *results* from stories? It is the contribution of the work of the Duquesne group to have developed and published detailed and documented step-by-step procedures for doing explication of meaning-essences called *structures* from reports, thus "operationalizing" hermeneutic activity, the aim of which is to enrich and deepen the meaning that is given by disciplined dialogue with the text. Whereas hermeneutic studies usually address artistic and cultural products, we have ventured to create the life-texts to be studied in the framework of psychological research.

In two publications, "An Application of Phenomenological Method in Psychology" (Giorgi, 1975a) and "Convergence and Divergence of Qualitative Methods in Psychology" (Giorgi, 1975b), Giorgi reports on his empirical methodology. In the second paper, he outlines his data-study and explication procedure (Giorgi, 1975b, pp. 74–75):

> The procedure for qualitative analysis is as follows:
>
> 1) The researcher reads the entire description straight through to get a sense of the whole. A phenomenological interpretation of this process would be that the researcher is present to the situation being described by the subject by means of imaginative variation, or by means of the meanings he apprehends through written language, and not that he is merely present to words on a page.
>
> 2) The researcher reads the same description more slowly and delineates each time that a transition in meaning is perceived with respect to the intention of discovering the meaning of learning (the research example discussed). After this procedure one obtains a series of meaning units or constituents. A constituent is not an element; the former means differentiating a part in such a way that one is mindful of the whole, whereas the latter implies a contextless discrimination.
>
> 3) The researcher then eliminates redundancies, but otherwise keeps all units. He then clarifies or elaborates the meaning of the constituents by relating them to each other and to the sense of the whole. This process also clarifies why the specific meaning units constituted in Step 2 were perceived.
>
> 4) The researcher reflects on the given constituents, still expressed essentially in the concrete language of the subject, and transforms the meaning of each unit from the everyday naive language of the subject into the language of psychological science insofar as it is revelatory of the phenomenon of learning. In other words, each unit is systematically interrogated for what it reveals about the learning process in that situation for that subject. It is at this point that the presence of the researcher is most evidently present, but he is needed to interpret psychological relevancy.
>
> 5) The researcher then synthesizes and integrates the insights achieved into a consistent description of the structure of learning. The structure is then communicated to other researchers for purposes of confirmation or criticism.

Giorgi's description parallels closely the steps of explication as outlined by van Kaam, except that Giorgi no longer considers it fruitful to attempt any form of quantification in terms of frequencies of statements, nor does he think it necessary to collect data descriptions from a large number of subjects. As a matter of fact, the actual description of the qualitative research method provided (Giorgi, 1975b) tells us how to read and work with the individual protocol in its biographical integrity, in its situatedness. Unlike van Kaam, who pools his data after he has identified the "experience moments," Giorgi keeps the "meaning units" he has identified in each description together for each subject and then explicates them in terms of a "situated structure" (see below).

It is therefore not surprising that Giorgi (1975b) presents a single case study of learning in his paper "An Application of Phenomenological Method in Psychology." He differentiates several levels of reflective analysis (p. 75):

> The end result of the application of this procedure is one of a number of descriptive statements that capture the naive description in a more clarified and more psychologically relevant way. The descriptive statements can differ in terms of level, e.g., a Situated Structure of learning, which presents the situation as learned by the subject in concrete terms, or a General Structure, which describes the learned situation irrespective of concrete situations in which the learning took place. Each level of description has both strengths and weaknesses. The descriptive statements may also differ with respect to type. That is, most concrete descriptions fall between the universal and the individual and cluster at a level of generality that can be differentiated with respect to type, e.g. bodily language vs. cognitive learning, and each type would necessitate a different descriptive statement. In sum, each descriptive statement, ideally, should both comprehend a large set of facts as well as deepen the understanding of the phenomenon under investigation.

Thus, we obtain results on *two structural levels of meaning and generality,* as *situated structure* and as *general structure,* both answering "What is it?" questions, and several *types* of structures answering "How did it take place?" questions.

1. Problem and Question Formulation: The Phenomenon. The research example presented comes from a research project on learning (Giorgi, 1975a, p. 84):

> The basic idea is to try to discover exactly what constitutes learning for ordinary people going about their everyday activities and how the learning is accomplished.
>
> The eliciting question to obtain experiential data was:
>
> Could you describe in as much detail as possible a situation in which learning occurred for you?

2. Data-Generating Situation: Descriptions. In the actual research presented by Giorgi, the narrative description in response to the cited question was not elicited in writing (as a written description), but through an interview procedure with a lead-off question, that is, by means of *questioning dialogue.* This dialogue of the researcher with the subject was tape-recorded and then transcribed. This transcription constitutes the "data."

3. and 4. Data Study Procedure (Explication) and Presentation of Results (Formulation). Giorgi (1975a, p. 87) writes:

> First one reads through the protocol to get the sense of the whole. The first step of the analysis itself is to try to determine the natural "meaning units" as expressed by the subject. The attitude with which this is done is one of maximum openness and the specific aim of the study is not yet taken into account. After the natural units have been delineated, one tries to state as simply as possible the theme that dominates the natural unit within the same attitude that defined the units.

The issue of natural "meaning units" deserves some reflection and discussion. It is not clearly specified by Giorgi, but it involves the articulation of the *theme*—a verbal statement formulated by the researcher that states the essence of meaning of the unit.

Cloonan (1979, p. 117), in his thesis work on decision making under Giorgi, discussed this concept under the title "Intentional Units":

> By "intentional unit" is meant a statement made by the S which is self-definable and self-delimiting in the expression of a single, recognized aspect of S experience. For example, a S might reply to the question of how he arrived at choice of preferred task (e.g. math), "Since I have always enjoyed working at numbers, I derive pleasure from math logic problems, reasoning problems that involve numbers, and math brain teasers." The intentional unit here is: "S has always enjoyed math problems." This statement contains the essence of the statements thus far; it is not a reduction of S's experience. Were S to have added, "I also find math challenging," this would constitute a second intentional unit. An "intentional unit," therefore, is a statement of subject's experience in which there are collapsed redundancies of an aspect of S's reported experience. Every aspect of the experience is an "intentional unit."

Cloonan refers to class lecture notes (taken in 1966) to establish a reference. Over the years, however, natural *meaning units,* Giorgi's (1975a) coinage, has become the preferred and unquestioned term used. It remains a problematical issue: What do you consider to be an aspect of an experience? Is it a unit of the behavioral act-intention? Is it a psychological act? What kind of event is a "meaning unit"? The methodological step involved represents a transformation of the narrative first person experiential report into a declarative third person summary statement in the researcher's language formulation, which becomes the basis for further reflections aiming at higher levels of universalization and more abstract meaning-comprehension.

Giorgi (1975a, pp. 87–88) continues with the delineation of the steps of the analysis and explication procedure:

> The next step is to understand the specific purpose of the study. If a study has a number of questions, these questions should be put to the data consecutively and should not be confused. To demonstrate this, let us first ask the question "What is learning?" and then follow up with "How was learning accomplished?" Therefore, the second step of the analysis is to look at the themes and the raw data from which they were taken with the specific attitude that asks "What does this statement tell me about learning?" or "How does this statement reveal significance about the nature of learning?" If there is nothing explicit about learning within a given natural meaning unit, which is possible, then one simply leaves a blank. For some purposes it is important to know the meaning of the statement anyway or what function it serves in the total narrative, but this step is not followed here. The results of these interrogations are the Expressions of Central Themes in Terms Revelatory of the Structure (What) and of the Style (How) of Learning.
>
> Once the themes have been thusly enumerated, an attempt is made to tie together into a descriptive statement the essential, nonredundant themes. This can be done a number of ways, but it seems that at least two ways are valuable for general communication. One is a description of what we can call the situated level which means one that includes the concreteness and specifics of the actual research situation employed. The second one is a description at the general level. The general statement leaves out the particulars of the specific situation and centers on those aspects of learning that have emerged which, while not necessarily universal, are at least transsituational or more than specific. "Structure" is the term used to describe the answer to the "what is learning?" question and "Style" is used to describe the "how did learning take place?" answer.

For Giorgi, the systematic elaboration and answering of these two reflective questions—"What is the essential structure of?" (Structure) and "How did this experience take place?" (Style)—constitutes the work of explication by means of which we arrive at the "final identification" (van Kaam) or the "synthesis and integration into a consistent description of the structure of learning."

Comparing Giorgi's approach to van Kaam's, we can note the following: Giorgi does not use any quantification of his data or any interjudge reliability checks. Like van Kaam, he engages in a series of steps of explication from data to themes or "meaning units" ("experience moments" in van Kaam's language) to structures. These steps constitute linguistic transformations of the materials from the subject's own words into professional psychological and explicitly phenomenological language. The description and rationale for this reads (Giorgi, 1975a, p. 95):

> I tried to read the description provided by the S without prejudice and tried to thematize the protocol from her viewpoint as understood by me. The interrogation that provided the above tables proceeded within the same phenomenological perspective. For me, again, this means that there were certain kinds of meaning that I allowed to emerge, and that I expressed these in the nascent language of phenomenological psychology (e.g., structure, style, meanings, situation, etc.), with all of the nuances implied by that particular context. Thus, these factors do not vitiate the findings but rather set the limits of the context in which they are valid.

While van Kaam proceeded immediately from many descriptions to one general characterization to which all the original descriptions contributed, Giorgi introduces several intermediary steps in order to distinguish the levels of situated structure, which respects the integrity of each individual, and general structure, which characterizes the implicit universal structure of meaning true of all the individual descriptions. While this differentiation is plausibly demonstrated by Giorgi using a single case that is an account of a personal situated experience, it remains unclarified and undemonstrated how this two-level analysis is to be done beginning with several descriptions, a procedure often recommended and insisted upon in order to achieve representativeness and redundancy and to provide the opportunity to differentiate subtypes or varieties of types and styles of experiences reported by different contributors.

Comparing Giorgi's way of studying human phenomena with Colaizzi's, we find that there are several differences: Giorgi is concerned with fundamental structures (FSs) arrived at by means of empirical phenomenological reflection (EPR), but he does not explicitly use individual phenomenological reflection (IPR). In recent discussions, Giorgi proposes the use of imaginative variation (part of EPR) in addition to structural analysis, although he has not explicitly operationalized this procedure. Giorgi does not sustain the level-of-discourse distinction between fundamental description (FD) and that of FS. Giorgi's fundamental themes are analogous to description on the FD level, but themes focus on part-meanings within the implicit context of the whole, and not on the integrated meaning description itself as in Colaizzi's FD.

Giorgi brings in another important distinction and order into the methodology by identifying the *situated structure* and the *general structure.* He works with individual experiences and protocols until he reaches the level of articulation of situated structure. Only then does he "universalize" or "essentialize," that is, transcend the existentially situated specificity in favor of an essential transsituational understanding. Another far-reaching change and feature of Giorgi's methodology is its emphasis on *life-world experience*. He studies learning as it happens to persons in their unique life-contexts, as a natural biographical task and experience. With the use of this approach, the importance of the intersubjective dimension or the interpersonal context of learning has emerged as an important facet and structural feature of the phenomenon. As Pollio (1981, p. 163) states regarding Giorgi's work:

> Learning "happened" for E.W. when she was directly told something quite specific by an important other person about a problem situation that had been bothering her for a long time. The

> learning took place when she could apply that information to a new situation after having seen a few prior examples.
>
> Even though E.W. is only one person, a careful reading of her approach to learning yields some interesting results. For one, the interpersonal nature of human learning comes through loud and clear: Not only does E.W. learn about vertical and horizontal lines from her friend Myrtis, she finds confirmation for her new way of looking at things by comparing it with her husband's less precise statement that the "room looks different." A second important point is that learning is sensed by the learner as a "new way of looking at things" that leads to new, previously impossible changes. Learning, from a first-person point of view, always seems to include a person's experiential history, for "a new way of looking at things" implies that there was an "old way." In addition, learning outside the laboratory always seems to involve an interpersonal context. For learning to have occurred, the learner must perceive or behave in a new way with respect to his or her own personal history, a way that was not previously possible.

Giorgi's multi-level analyses are quite complex. In his view (Giorgi, 1975a), they closely parallel the complexities of the multiple quantitative analyses that can be conducted in terms of numerical operations: as statistics, which yield "measures." The various qualitative analyses occur in terms of linguistic operations: as explications, which yield "end products." As Giorgi (1975a, pp. 78–79) says of this parallelism:

> I would describe what I have done as the development of a phenomenologically based procedure for the analysis of linguistic descriptions—as opposed to numerical descriptions. If there is a parallel with the achievements of numerical procedures it would be because a phenomenology of mathematics would lead to similar findings. In other words, the root or ground of both linguistic descriptive analysis and numerical descriptive analysis is ultimately in the perceptions and thoughts of man. It is this reflexive or self-referential movement that phenomenology tries to comprehend.

Developing the implications of this parallelism, it would seem that the type of statistic used—the level of qualitative analysis pursued—would depend on the intention of the researcher and the "appropriateness" to the problem studied. In the quantitative paradigm, this depends on the "design of the experiment," set up as the procedural guideline designed in order to answer questions and test hypotheses. In the qualitative paradigm, it depends on the guiding questions that the researcher already has about the phenomenon he or she investigates, that is, what he or she considers most important about the process under investigation: the phenomenon. We have to distinguish two levels of questioning:

1. The data-generating questions: This is the question asked of the subject, what the description is to be about, what phenomenon is to be selected and described.

2. The explication-guiding questions: This refers to the set of questions that the existential–phenomenological researcher addresses to the data, that guide his or her reflective questioning of the data, that is, explication.

The different researchers presented in this section ask different questions on both levels. They are interested in different phenomena and they do explications differently.

Van Kaam (1966, p. 323) asks:

> 1) Does this concrete, colorful, formulation by the subject contain a moment of experience that might be a necessary and sufficient constituent of the experience of really feeling understood?
>
> 2) If so, is it possible to abstract this moment of experience and to label the abstraction briefly and precisely without violating the formulation presented by the subject?

He calls the answers to these questions "necessary constituents." Van Kaam then composes a "synthetic description of the experience," his "end product."

Giorgi's (1975a, p. 88) explication-guiding questions are:

> 1) What is learning? What does this statement tell me about learning? How does this statement reveal significance about the nature of learning?

This leads to two levels in the description of structure: "situated structure" and "general structure."

> 2) How did learning take place?

The answer to this question is given the term "style." It is also written for two levels, situated style and general style, depending upon whether the specifics of the subject's existential situation are taken into account or not.

On the Experience of Being Anxious: William Fischer

In his book *Theories of Anxiety,* Fischer (1970a) attempts an integration of anxiety theories. In his phenomenological characterization of the "faces of anxiety," Fischer (1970b) presents a free-flowing informal but carefully argued report on his phenomenological reflective analysis of an actual report of "healthy anxious experiencing" and an imaginal example—created by Fischer himself—of "neurotic anxious experiencing." He analyzes both examples through a palpating reflective questioning in which he highlights the significant phenomenological themes and structures in dialogue with the descriptions. There is not yet at this stage in Fischer's work—around 1970—a full formalization of the method in specified steps, which, however, emerges fully articulated in 1978.

In his paper entitled "On the Phenomenological Mode of Researching 'Being Anxious,'" Fischer (1974) provides us with another explicit example and chronicle of doing research in a phenomenological mode in a more clinical psychological context. He studies the phenomenon of "being anxious."

Fischer (1974, p. 410) explicitly acknowledges a close collaborative relationship with the method worked out by Giorgi:

> Casting about for a place to begin, I discovered that Giorgi had already described, at least in its basic outlines, a three-stage method for analyzing qualitative, experiential descriptions or protocols. While the method that I came to use and am still evolving is not an exact, carbon copy of his, it is certainly based upon his suggested procedure to a significant degree.

1. Problem and Question Formulation: The Phenomenon. Regarding his general approach to research as a phenomenological psychologist, Fischer (1974, p. 408) states:

> One of the more important implications of all this is that every research enterprise, every effort to systematically understand and/or explain some phenomenon, becomes graspable as a project that actually seeks to further some perspectival access to that phenomenon. The researcher utilizes his own as well as others' experiences of the researched, in order to bring to increasingly articulate intelligibility his own already operative, personally evolving, semi-articulate preconception of the researched. Thus, the history of the researcher's relations with the researched is an unfolding, evolving dialectic, one pole of which is constituted by the researcher's presuppositions and preconceptions pertinent to the researched—what Giorgi has called his approach—while the other pole consists of the researched content, that is, the phenomenon that is under study as it has been solicited, illuminated, and articulated by the researcher's questions and methods.

In his research, Fischer reports on two related studies: one on the experience of one's being anxious and the other focusing on "the experience of another person being anxious." In this report, we will limit ourselves to a discussion of the first: "one's own being anxious." Fischer (1974, p. 408) formulated the following question to gain access to the experience of being anxious:

> Please describe in detail a situation in which you were anxious. To the extent that you can recall it, please include in your description some characterization of how your anxiousness showed itself to you as well as some statement of how you were, that is, what you experienced and did, when you were anxious.

2. Data-Generating Situation: Descriptions. Regarding the situation in which the data descriptions for the research were generated, Fischer (1974, p. 408) reports:

> I asked the students in my undergraduate personality theory class—about seventy per semester—to submit, as their term papers, descriptions of situations in which they themselves had been anxious. Needless to say, all the students were told in advance that I was interested in the phenomena of being anxious and that I wanted to read descriptions of other peoples' experiences with this human possibility. Further, all the students were asked if their descriptions could be read to others and/or used for publication.

Fischer thus obtained a large collection of descriptions of the experience of being anxious from college students in a psychology class. The demand-character of the situation was that the descriptions served as a term paper, in the context of an evaluative classroom situation. The students presumably made their best effort under those conditions, and, indeed, as we look at the single case example of a descriptive narrative reported by Fischer, we find an unusually articulate, perceptive, and dramatic rendering of an experience of being anxious, an excellent protocol.

It is clear that the richer, the more detailed, the more dramatic the experiential descriptions are, "the better the data," the better will be our opportunity for explications on the basis of these data. Empirical existential–phenomenological researchers are very much aware of this issue and the associated methodological problems. Giorgi (1975b, p. 74) comments:

> Lastly, I would like to say a few words about description itself. All approaches in sympathy with phenomenology agree that one must begin with naive description. The discipline enters in when one has to analyze what has been described. Secondly, from a phenomenological perspective, description or language is access to the world of the describer. Descriptions, of course, can be better or worse or even enigmatic, but they always reveal something of the world of the describer, even if it is only the fact of an enigmatic world. The task of the researcher is to let the world of the describer, or more concretely, the situation as it exists for the subject, reveal itself through the description in an unbiased way. Thus it is the meaning of the situation as it exists for the subject that descriptions yield. While detailed knowledge concerning the criteria for better or poorer descriptions would be helpful, it is equally clear that good descriptions communicating the intentions of the describer do exist.

In our own work, we have found it most congenial and productive to begin with written descriptions in the context of an explicit invitation to a collaborative research venture and then, in a second step called *collaborative dialogue*, to engage in a clarifying interview–dialogue with the contributors so as to amplify the personal meanings expressed in the initial descriptions (von Eckartsberg, 1971). We thus start from a twofold "database": description plus interview transcript.

3 and 4. Data-Study Procedure (Explication) and Presentation of Results (Formulation). Fischer (1974, p. 410) comments upon his explication procedure in a first characterization:

> Each student's description was a relatively circumscribed chronicle; each characterized the unfolding of a particular event—the person's own anxiousness or his experience of another being anxious—from a particular perspective.

He then gives the full protocol of one particular student (Fischer 1974, pp. 413–414) and comments:

> As I continued to read and reread this, as well as other protocols, I gradually realized that each consisted of a series of interrelated scenes, that is, shifts in the focal attention of the one providing the description. Further, the central themes of each of these scenes was derived from and yet uniquely contributory to the overarching sense of the whole event that was being described and unfolded. In other words, each protocol constituted a structure; the meanings of the whole were given in the meanings of the interrelated parts or scenes, but the meanings of these were dependent upon the sense of the whole to which they pointed. Finally, I could see that the meanings of the whole were given in the ways that the successive scenes blended one into the next.

Proceeding with the work of explication, Fischer (1974, p. 414) reports:

> I took as my first task an articulation of the central themes that characterized the respective, unfolding scenes of each protocol. Further, wherever possible, I tried to state these themes in the student-subject's own language. Thus, in this manner I began the process of trying to situate myself vis-à-vis the totality of each protocol's structure so that it would speak to me in its own terms, so that it would show me all of its constituents, and so that in the unfolding of its scenes, I might grasp its inner logic and sense.

He thus articulates the descriptive narrative, the "chronicle," the series of "interrelated and successive scenes" in terms of central themes.

Although Fischer reports that he "tried to state these themes in the student-subject's own language," a look at the wording of the central themes shows that he actually first transformed the characterization from a first person singular "I-report" (Actor report) into a third person singular narration, that is, into a psychological scientific observer report: "S did this and this. . . ." Fischer closely retains, however, the narrative sequence and the variety of described experiences well articulated in the rich protocol that served as the starting point for the explication.

Continuing the work of explication—now working with the central themes as representative of the original protocol—Fischer (1974, p. 416) reports:

> Having thus articulated the central themes that seemed to characterize each shift in focus, that is, each scene of the protocol, I turned my attention to an explicit interrogation of the meanings of being anxious that were lived, experienced and described therein. The questions that oriented this reflective enterprise were: what was the anxious situation that this person found herself in, and what did it mean for her to live that situation anxiously?

The aim of these explication-guiding questions is to arrive at a characterization and "understanding of the meanings of being anxious as she lived, experienced, and described them." Fischer comments upon the difficulties involved in moving from the subject's descriptions to a characterization of the situated meanings of what Giorgi calls the "situated structure." Writes Fischer (1974, p. 416):

> At this juncture, I must disappoint those readers who are looking for some procedure that would be analogous to traditional experimental design or statistics. I cannot offer a method of

analysis that allows one to proceed strictly by the numbers. Instead, I would suggest that the movement of my reflections was quite similar to that which is utilized in the following everyday life situation: Imagine that a person is seeking to secure from his friend a characterization of a particular other whom the friend has recently met. The friend is graciously willing to repetitively describe his experience of this other without limit. However, despite the person's persistent questions, the friend confines himself to the exact same words. Still, to the extent that this description is broad and yet detailed, there begins to emerge a more or less coherent sense of the other. In going back to the friend's words again and again, the person eventually came to realize a stable, though perspectival grasp of the other.

In going back to this student-subject's description again and again, I was able to achieve the following understanding of the meanings of being anxious as she lived, experienced and described them.

The next move in explication, in the method Fischer describes, is to go from situated specificity of a unique person's existential situation to the level of "essential or invariant meanings," to the *general structure.* Fischer (1974, pp. 417–418) reports in this context:

While the foregoing constitutes a characterization of the meanings of being anxious as they were lived, experienced and described by a particular person in a particular situation, the ultimate goal of my phenomenological analyses is to move towards a description of the essential or invariant meanings of being anxious as a fundamental human possibility. Thus, by analyzing instances of being anxious in their particularity, I hope to come to an understanding of being anxious in its suchness, in its universality. To this end, I took the characterization of the meanings of being anxious as they were lived by that particular girl in that particular situation, and I brought it together with other such characterizations based upon other protocols. Reading and rereading these, I sought to describe what seemed to be essential meanings of being anxious as they were revealed through the variant characterizations of these particular instances.

Again, I cannot say that I followed some straight-forward, mechanical, by-the-numbers procedure. However, I would ask the reader to think about what he sometimes does when an other asks him to describe, for example, a humanist, or a behaviorist, or a psychoanalyst, or a schizophrenic, or a representative of whatever group he might like. When he responds positively to this kind of request, he soon becomes involved in an inductive movement that cannot be adequately characterized as a simple abstracting of common elements. Rather, in gradually coming to a sense of a particular kind of person, in his or her typicality, he imaginatively and creatively reunderstands and surpasses his own characterizations of the various styles of being anxious that were revealed in the descriptions.

Similarly, my efforts at illuminating the essential or invariant meanings of being anxious are based upon my readings and analyses of how this fundamental human possibility has been incarnated in the particular instances that were described by the student-subjects. In each of their respective protocols, meanings of being anxious were discovered as more or less explicit themes. As an increasingly articulate sense of this phenomenon began to emerge, it was seen that these themes, incarnated in the variations of each protocol, constituted the skeletal cores or infrastructures of the respective protocols. Hence, in reading and rereading each student-subject's description, as well as my own characterizations of the various styles of being anxious that were revealed in the descriptions, I came to realize the following sense of being anxious in its suchness.

In his "portrayal" of the essential invariant meanings of the experience of being anxious, Fischer distinguishes "two varying modes" or "types." While he does not comment on the exact origins of this differentiation, it appears from the context that all the descriptions studied—"about 70 per semester"—follow either one or the other mode.

Fischer found it necessary to distinguish two modes or types of being anxious to account for the full existential spectrum of situated experiences of being anxious. Reading Fischer's characterization of "being anxious in its suchness," that is, in terms of its "general structure," its "skeletal core" or "infrastructure," it is very noticeable that the final version is written from the personal, actor-perspective, that is, in the first person singular: "I am . . . ," "I do . . . ," My . . . ," etc. Who is "I"?—presumably the universal human, *homo universalis.* It is as though the reader, when he or she reads this description, is talking to him- or herself about him- or herself. Indeed, if the description has achieved universality of meaning, then it must be possible to say it in the manner of a "universal I." This is a rather interesting and powerful narrative device, and it avoids the implied distance and observer-orientation that inheres in any third person singular structural characterization, that is, a structural description in the voice of the scientist, speaking about the subject matter over there, speaking "objectively" about it (which is another form of narrative manner, another level of discourse).

If we look a little more closely at the metaphors used by Fischer, both in his descriptions and in his characterization of the work of explication—for example, "chronicle," "scenes," "successive scenes," "portrayal," the "event organized in terms of scenes"—we find literary metaphors and the imagery of literary criticism. A type of analysis and formulation presents itself that speaks in terms of actors, of situations and scenes, themes of moves, of decisions, of responses, of attempts, and of failures. It is a characterization in terms of dramatic action, much as one would describe the unfolding of a plot in a novel or a play. We would like to suggest here that the underlying and implicit motif of the characterization of the experience of being anxious is to disclose the "psychological plot" of the experience, the "essential experience story-line," that makes being anxious what it is.

We want to make another observation concerning the internal organization of Fischer's general characterization of being anxious. In comparison with the summated situated description, the final characterization is much more elaborate and informed by a strong implicit theory of personality and motivation from within the existential–phenomenological tradition. More specifically, it is the utilization of the notion of one's personal existential project (Sartre, 1953)—"my project to be a competent human being," "to become a certain kind of person," "effortful commitment," "to perform some self-saving act"—that gives Fischer's characterization its dramatic power, its internal cohesion and psycho-logic, which blends nicely with the choice of the grammatical "universal-I-form." Every person can recognize himself or herself in this familiar psychological plot, provided he or she is familiar with the literature of phenomenology and phenomenological ways of speaking. The level and type of language in which the characterization of the "essential or invariant meanings" are couched is of a professional and phenomenological nature and goes beyond commonsense contents of everyday language.

We have already referred to Giorgi's statement that explication, particularly the step of "reduction" (à la van Kaam), involves a translation from the "everyday naive language of the subject" into the language of psychological science. There are many schools of thought in psychology, each with its own specialized conceptual framework and language. Giorgi also states that it seems inevitable that the existential–phenomenological researcher will express meanings he or she discovers in the "nascent language of phenomenological psychology (e.g., structure, style, meanings, situations, etc.)." Thus, we cannot really escape our approach, "where we are coming from," our theoretical preconceptions. This means that as part of our approach, we bring with it always an implicit conceptualization of "Who is the human person?"—that is, personality structure—and "What makes a person move?"—that is, motivational dynamics. "Humankind as existential project" is a core theoretical concept of existential-phenomenology, a real theoretical contribution, the discovery of a hitherto neglected phenomenon in the light of a new starting point, a new view of human being, which elucidates the meaning of our experience in general, and our experience of being anxious in particular.

In the research by Giorgi, reported before, it was the notions of "awareness" (of the new), "meaning," and "interpersonal context" that constituted the core insight and line of articulation of the experience of learning. Again, it was the utilization of core concepts of existential-phenomenology—awareness, meaning, level of functioning, consciousness, interpersonal context—that opened up the understanding for us. Giorgi, however, utilizes a more commonsense level of everyday language to express his existential structure, although his way of speaking is complex, informed by a phenomenological approach. For van Kaam, the guiding notions are "perceptual-emotional Gestalt," "co-experience," "personal meaning," "relief from experiential loneliness," and "experiential communion," which place him in communion with the existential–phenomenological framework of thinking and languaging. We cannot escape our theoretical presuppositions, our "approach." All we can do is to try to make our approach as explicit as possible.

Approach always implies a theoretical preconception, an implicit operative hierarchy of what we consider most important and illuminating. This preconception is expressed and revealed through the particular psychological concepts employed. In this sense, existential-phenomenology is not "a-theoretical" but self-consciously theoretical. As van Kaam (1966, p. 88) states:

> Observation is thus really perceiving what is "out there." But observation is also perceiving according to a theory of an observing subject. Therefore, existence as observation is necessarily theoretical and empirical at the same time.

HERMENEUTICAL–PHENOMENOLOGICAL STUDIES

The hermeneutical–phenomenological approach to research investigates human experience as it becomes expressed in spontaneous productions of speech, of writing, or of art. It emphasizes the understanding or model or paradigm that human actions and interactions can be conceived of as texts that are yet to be interpreted. The "database" on which hermeneutic work is based is conceived more broadly than in empirical protocol studies. The steps of explication and the questions that are addressed to the text also vary more widely in the hermeneutic approach and are less clear and operationalized.

In organizing the research of hermeneutical–phenomenological psychology, we set up two divisions:

1. Actual life-text studies: In this group are the works that create data, life-texts, by tape-recording actual verbal expressions in situations, ranging from a "speak-aloud" protocol on the situated stream of experiaction (von Eckartsberg, 1972), to the study of therapy transcripts (Fessler, 1983), to research on interview transcripts from a professional role-ensemble and clients (Kracklauer, 1980), to dialogal research in the framework of a "conceptual encounter" (de Rivera, 1981).

2. Studies of recollection and literary texts: A second group of hermeneutical–phenomenological studies comprises the work that draws on personal experience in an intuitive and illustrative way to engage in hermeneutical reflection (Buckley, 1971), work that utilizes established masterworks of dramatic literature as texts (Halling, 1979), hermeneutical work on the meaning and origins of the theoretical attitude that guides our activities as researchers and scholars (Jager, 1975), and hermeneutical reflective work on a theme in a historical context (Romanyshyn, 1982).

What follows is a discussion regarding the nature of actual life-text studies, with an example from my own work in hermeneutical–phenomenological research.[2]

Actual Life-Text Studies

The database for hermeneutical research is typically broader than that for empirical existential–phenomenological studies. Its contents range from personal documents, to literature, to works of art that present themselves for interpretation. There does not seem to be a set starting point or a clearly outlined procedure to follow in hermeneutical–phenomenological investigations. One embeds oneself in the process of getting involved in the text, one begins to discern configurations of meaning, of parts and wholes and their interrelatedness, one receives certain messages and glimpses of an unfolding development that beckons to be articulated and related to the total fabric of meaning. The hermeneutic approach seems to palpate its object and to make room for that object to reveal itself to our gaze and ears, to speak its own story into our understanding.

In hermeneutic work, we grope for the single expression that will do justice to the integrity, complexity, and essential being of the phenomenon. We become spokespersons and messengers for the meanings that demand to be articulated. We become intrigued and entangled in the webs and voices of language and its expressive demands. In hermeneutic work, we become engaged in an expanding network of meaning-enrichment that contributes new meanings to the ongoing dialogue. It is a process of contextualization and amplification rather than of structural essentialization. Hermeneutic work is open-ended and suggestive, concerned with relational fertility.

In a more formal characterization, we can say with Ricoeur (1971) that the hermeneutic approach to the social sciences utilizes the notion of the text as the basic paradigm. Human action and all other products of human activities as expression—for example, artworks, rituals, institutions—can be understood as "text analogues" in need of interpretation.

Titelman (1979) discusses the relevance of Ricoeur's work for a hermeneutic phenomenological psychology. As compared to actual situated experience, as in a face-to-face dialogue, action as text has the following four discernible characteristics:

1. Fixation of meaning: We need some record or "protocol" of experience or some transcription or trace with which to work, whether it is produced from personal experience by direct investigation or taken from works of creative imagination.

2. Dissociation of the mental intention of the author from the text: In contrast to the face-to-face situation of dialogue, the text is independent of the author's consciousness. It stands in a new communicative situation. The intentions of the author have to be inferred and reconstructed by the hermeneutic investigator; they cannot be ascertained by asking the author. The text stands on its own.

3. Display of nonostensive references: As Titelman (1979, p. 186) phrases this concern:

> The meaning of a subject's descriptive protocol(s) cannot be comprehended by analyzing and reducing them atomistically into separate and unconnected parts; rather, the meaning of the subject's described experience(s)/behavior(s) must be understood in terms of the relation between the constitutive dimensions. . . . The descriptive protocol is more than a linear succession of sen-

[2] The reader is referred to von Eckartsberg's (1986) *Life-World Experience* for additional examples of actual life-text studies, and a discussion and examples of studies of recollection and literary texts.—*Ed.*

> tences—it projects a *world,* a cumulative and holistic process, the structure of which, like the text, cannot be derived from or reduced to the linguistic structure of the sentence.

4. Its universal range of address: The text is open for everyone who can read and gain access to it. Unlike the directed speech of two partners in a situated speech event, the text is there to be taken up and read in any way whatsoever. Titelman (1979, p. 186) writes:

> Like a text, a descriptive protocol can be read by an indefinite range of possible psychologist-readers. The descriptive protocol, like the text, is an "open work" that awaits fresh interpretations from different perspectives and as history, in its unfolding, sheds new light on the experiences and events that have been described. The meaning of a text-analogue is in "suspense" in the sense that its meaning is never totalized. The description of human experience and behavior, like a text, is open to everybody who is capable of reading.

While Ricoeur presents a generalized theory of action as text, Titelman's paper relates this work directly to the problems of a phenomenologically based psychology concerned with the analysis of life-world experience.

Another central issue in hermeneutical work is the so-called *hermeneutic circle.* This thought-figure describes the open-ended and continuously spiraling nature of the hermeneutic inquiry and sense-making process. We have foreknowledge about most aspects of life. We come to any phenomenon with a precomprehension of its meaning, yet in search of deeper understanding and more precise differentiation. As Titelman (1979, p. 187) states this:

> The circularity of the hermeneutic endeavor is not vicious in that it involves a passage from a vague pre-conceptual understanding of the meaning of a phenomenon to the explicit seizure of its meaning. There is no entrance to the hermeneutic circle, no beginning point. The psychological investigator must "leap" into the circle in order to elucidate it.

The hermeneutic stance acknowledges the perspectival nature and biographico-historical involvement of the researcher and makes the investigation of the implicit precomprehensions of the researcher part of the interpretative process.

An important consideration in all hermeneutic work is the problem of the validity of the findings. How can we evaluate the claims of truth made by the various interpretative readings? Titelman (1979, p. 190) discusses this issue in the following way:

> The validation of the particular interpretation in the area of the social sciences is not like proving a conclusion in logic, but rather it is closer to a logic of probability than to a logic of empirical verification. Showing that a particular interpretation is more probable, given the available knowledge, is different than proving that a conclusion is true.

Ricoeur (1971, p. 553) seems to regard hermeneutics as an argumentative discipline, as a subspecies of rhetoric, when he says:

> Like legal utterances, all interpretations in the field of literary criticism and in the social sciences may be challenged, and the question "what can defeat a claim?" is common to all argumentative situations. Only in the tribunal is there a moment when the procedures of appeal are exhausted. But it is because the decision of the judge is implemented by the force of public power. Neither in literary criticism nor in the social sciences, is there such a last word. Or, if there is any, we call that violence.

The validation of interpretation is a difficult issue. Ricoeur thinks that a new type of logic is involved here, a "logic of uncertainty" and of "qualitative probability."

For Gadamer (1975), the "hermeneutical experience" is one of becoming involved in a dialogue or a dialectic of question and answer. Its primordial medium is language or "linguisticality," the saying power that creates the world within which understanding and disclosure take place. Worlds appear between persons in dialogue and rest on the common ground of our tradition as shared language. In the hermeneutical experience and situation, an encounter takes place between the heritage that is sedimented in the text, revealing a world made text from a perspective through a particular experience and the horizon of the interpreter. Author and reader belong to the language, but differently through the distance of time and perspective. A "fusion of horizons" takes place in which the reader becomes a servant of the text, open to hearing what is said and articulating the message.

As Palmer (1969, p. 209) states this issue in commentating on Gadamer's work:

> He is not so much a knower as an experiencer: the encounter is not a conceptual grasping of something but an event in which a world opens itself up to him. Insofar as each interpreter stands in a new horizon, the event that comes to language in the hermeneutical experience is something new that emerges, something that did not exist before.
>
> The method appropriate to the hermeneutical situation involving the interpreter and the text, then, is one that places him in an attitude of openness to be addressed by the tradition. The attitude is one of expectancy, of waiting for something to happen.

All human understanding is historical, linguistic, and dialectical; each disclosure contributes to the task of developing our effective historical consciousness. According to Gadamer (1960, p. 350), the task of hermeneutics is

> to bring the text out of the alienation in which it finds itself (as fixed, written form) back into the living present of dialogue, whose primordial fulfillment is question and answer.

On the Social Constitution of the Stream of Experiaction: Rolf von Eckartsberg

Our awareness of our ongoing stream of consciousness, of experience and action, of "experiaction," is a fascinating phenomenon. We are always already bodily involved in a here-and-now situation. Experiaction is situated. Though some awareness has to be used by a person to relate bodily to his or her physical environment, our experience or conscious awareness also transcends the situation in order to make present other domains or landscapes of consciousness through our modes of psychological access: to remember or to anticipate, to imagine possibilities, to reconstruct, to think and reflect. In experiaction, we are not limited by or exclusively bound to the physical situation at hand. We live in a psychological life-space mediated by experiactions of which we are partially aware. We live in a mappable "psychocosm" (von Eckartsberg, 1981) or a personal "lifescape."

We can research the constitution of the psychocosm by collecting samples from the experiactional flow of a person. We select and specify the situation and duration of the observational period, thereby setting limits on what and how we can record the experiaction. We can make concurrent recordings in the form of tape recordings of behavior and dialogue, and we can create "speak-aloud protocols" by asking our respondents to verbalize everything that occurs to them in conscious awareness. In situations in which the researcher is alone, this poses no particular problem, although at first it feels unusual to verbalize aloud in a monologue all of what is "going through one's mind." Speak-aloud protocols (Aanstoos, 1983; Klinger, 1978; Wertz, 1983) are difficult or impossible to obtain in interpersonal situations except under special circumstances or in psychoanalytic work. Interesting attempts have been made to sample the stream of experiaction by the creative use of kitchen timers (Leary, 1970) and by random electronic beeping (Csikszentmihalyi, 1978) to alert the person to what he or she is experiencing at the moment.

In my own hermeneutical research, the focus was on a solitary situation in which I let my experiactional stream unfold itself without censorship, such that it could reflect underlying personal concerns and social determinants. I simply recorded about 20 minutes of my own experience as reportage on a Sony tape recorder. I spoke the experience as it was happening to the extent that I was capable of doing so: voicing my experience.

In the situation, I sat alone on our sunporch. I felt that it was a very everyday event; nothing "unusual" happened. I think I was my "usual self," whatever that means. Whenever I sit down by myself, some such train of thought and experience unfolds, although naturally it is always different. That particular day, however, as I started speaking out loud, my experience quickly became oriented and organized around a particular focus: I had to give a lecture about Experiential Psychology at New College, Florida, two weeks hence. This anticipated task and event became the major organizing principle—the theme—that shaped my experience during these 20 minutes. This is what happened, in verbatim excerpts selected in terms of the relevance to the topic: how an anticipated event enters and co-constitutes the flow of experience and how a social determinant, that is, the demand character of the social situation of giving a public lecture, exerts its shaping influence on the personal stream of experience. Although I was alone, social reality nevertheless permeates and shapes my "private" experience.

Transcript from Recording

It's Easter Sunday . . . just around noon . . . and I'm sitting here in my sunporch . . . gorgeous day . . . quiet, very quiet . . . looking out over the golf course there . . . sun is bright . . . can see the green, just starting. . . . The lawns are just beginning to be green here in Pittsburgh . . . Pittsburgh spring is late. . . . Now looking down there . . . see the Steel Building . . . U.S. Steel Building . . . just a massive . . . gray and brown . . . slab there, iron . . . steel slab. . . . Now I'm thinking about, uh . . . just sitting down . . . thinking about experience . . . just tuning in to experience . . . my own experience . . . what's happening right now . . . just this next few minutes . . . just settling down now . . . sort of have to relax, just sort of quiet . . . and uh. . . . What comes to mind is uh . . . that I have to go to Florida . . . and give a talk there at the New School . . . at the uh New College. . . . I got this letter . . . Dave, Dave Smillie . . . he called up and asked me to come down and speak . . . and uh speak about experience . . . what I'm doing, studying experience, experiential psychology . . . and uh. . . . New College . . . I get drawn into that. . . . What is that? I only have an image of that; I have some knowledge of what this means . . . but looking forward or just thinking about it. It's just a concept more or less. An image, maybe faintly . . . New College. But this is sort of an image, visually. I know about it through Dave . . . Smillie who came down there, to teach there, two years ago. He visited, he told me a little bit, he called me up for this talk . . . and uh . . . from the conversation I got a feeling for . . . New College, a small school, bright students, very bright students—I heard that also, from a colleague of mine, Jack Rains. We've had a few conversations about New College, nothing specific. I don't have a specific image, but this concept now fills up—New College, small school, very bright students, interesting faculty probably. People who want to get close with students, work intensively with students. I know what that's like. That's what we're trying to do also . . . at Duquesne. I know what that is. But when I turn and think of what is my knowledge, what is my experience . . . of New College, looking forward . . . empty . . . anticipations . . . just pointers, just . . . inklings. Nothing substantive, no concrete experience. And uh . . . so I can just flow with the idea of Florida, knowledge. . . . New College, how a concept . . . fills in. Here I got lost. . . .

But somehow I come back to this talk, because uh . . . I'm tuning in to this talk. I've been thinking about what I should be doing. I haven't written out anything yet. I made a few sketches, in my diary a couple of weeks ago . . . having to do with knowledge and uh . . . objective knowledge, subjective knowledge. I was reading Roszak then, the counter-culture and . . . about the myth of the objective consciousness . . . and uh . . . tuning in and going around consciousness, more as personal consciousness, private consciousness, the flow just as I've been describing. How unique this is, how uh . . . you know, interwoven, intermeshed, flowing, jumpy, a little bit. Just going along . . . I was thinking a little bit about that . . . yeah. . . . So I'm drawn back to: I have to give this talk. So uh I've been thinking, here and there it comes up. I sit down, I think about it. Anytime it comes upon me. What am I gonna do there? And uh as it has shaped up, Experiential Psychology, this, this is the title. Dave asked me for a title, what shall I talk about . . . and I said Experiential Psychology. That's, that's the term I use now for doing what I'm doing. . . . I'm recording it as it happens right now and . . . you know, my eyes sometimes drift off, swerve off the thrust of attention that I had focused on this talk and . . . and . . . then I shift back into, let my eyes connect me with the world, because that's what I'm talking about, the . . . you know, what I'm conscious of, the intentional aspect of consciousness, as phenomenologists would say . . . that consciousness is of something. It's always what you pay attention to as it presents itself, and it presents itself in different . . . levels, guises. Close up, I'm connected. I'm in touch . . . with the eyes, hands, sitting in this blue chair, heavy, I'm just relaxed, sitting, and . . . I can feel my body, my arms on the chair here and . . . looking out, that view, I like that view; that view called me so I was getting away from . . . describing what I was doing, recording the experience, just giving an account just like a radio announcer who would describe an event, so I was describing the event of sitting here in this room right now . . . quiet, as I said, you know, it's this atmosphere and . . . I let my mind go, you have to just, just let it glide, let it flow, let it go around and in doing so it's an exploration in experience, it's sort of like the movement, a journey, an experiencing. It always comes back to: I'm thinking about this talk and . . . how I'm recording it now, because that's one of the things I came up with . . . just thinking about what I should do, why not record experience, and then comment upon it, the second stage, and reflect upon it, take it into a more thoughtful, reflective generalizing stance, and then those three steps would be what I would present—not talking about Experiential Psychology. I don't like the "about," talking about something out there anymore. I'd really like to just get into it. So, so this is, I feel, getting into it. Just a short recording, a few minutes. It's only been ten minutes or something. Just a few minutes and then uh I'll take this as the data and begin to talk about it, think about it, show how consciousness is manifested. It's just an everyday event. . . . So much for the structure of this. . . .

But what intrigues me is the knowledge, the way in which I am connected . . . to . . . New College—people here, the ideas. I have sort of two personal connections—Jack Rains

. . . and Dave Smillie—conversations that continue with these two, which put me in touch with this place there and . . . it's at a distance, and how we know at a distance, and . . . how we really know at a distance . . . through the experience of other people . . . it's through these conversations that . . . the world becomes richer, it grows, it's challenged. There's a dialectic there so that . . . experience . . . it is not just personal but it involves others. It's interpersonal experience basically. And so even sitting down here by myself with this Sony machine. Yet, nevertheless, I'm also already anticipating this event in the presence of . . . this event that I have to perform, or that I have to be at, or that I have to conduct. So it's an interpersonal situation.

Reflection

Now, I would like to go over this experience to amplify and to reflect upon it in a contemplative attitude of openness, of letting the phenomenon—the meaning constitution of my stream of experience—speak to me. The material is quite rich and lends itself to reflection in many different ways, answering many explication-guiding questions. I will develop one particular theme that has to do with the way in which a socially determined event—the task to give a lecture—in the future, as anticipated, enters and helps constitute the ongoing stream of experiaction in a meditative moment and setting. The future event seems to cast its "shadow" on the here and now. The task ahead exerts a demand character on me; it has a claim on me to heed it, and to fill my time in anticipation and preparing and structuring it in my consciousness prior to its enactment in tangible reality.

As I go along, reflecting on this experience, I try to discover essential or universal aspects of the process of experience. These are exemplified in this unique-concrete reported event, but they can be said to be valid for experiencing in general. This constitutes the reflective work, looking back and thinking about this experience, discovering meaningful patterns and structures, universal features that are lived out concretely in a unique fashion.

This is perhaps a first universal we find concerning experiaction: that it is situated and hence in a close interface with the material-and-meaning-organization of the situation. We cannot study experience apart from the situation in which it originates. As individuals, we participate in a large though never unlimited repertoire of situations, and hence the domains of experience to be studied are quite expansive. In interpersonal terms, experiaction varies, of course, depending on who the people in dialogue are in relationship to each other. With an intimate friend, you obviously tend to experiact spontaneously and in a spirited manner, whereas with strangers a more formal style prevails.

Reflecting on the first part of my experience, it became situated within the context of an implied presence of an audience—a sophisticated college audience in my anticipation, as my movement into a thoughtful and psychological–experiential language indicates. Also, the choice of topics within the "monologue"—knowledge, concepts, images, conversational community, perspectivity—makes sense within the context of my envisioning the type of my anticipated audience.

As we can readily see from this, my experience was embedded in multiple contexts of meaning, such as to "record and experience," "make it relevant to the talk," "something for the New College," "representing the approach of experiential psychology," which are all part of my definition of the situation. To be concrete and faithful to what happened, I would have to differentiate the physical situation (sunporch) from the experiaction definition of the situation (recording an experience for my talk), which in a sense constitutes a situation within a situation. The experiaction is held together and derives its coherence and meaning from the future event (giving this talk) to which it refers and for which it is preparatory. In motivational terms, we can speak of my "awareness and commitment of giving this talk" as being the project that then ties together many strands of my experience that contribute to setting up the event. With these experiences, I prepare for the event.

Within the recorded experience, I noted that I had thought of this occasion several times and that slowly a strategy had evolved. Recording my experience was part of the procedure. It makes sense only as part of this overall event that got set off in Dave Smillie's awareness before he called me on the phone, then entered my life through the call and has occupied my attention intermittently throughout this period . . . involving sometimes sustained thought-work as I am writing this reflection here and now. It is entering a new phase right this moment—one week after the description—and it will continue to call on my attention for some time to come, depending on what happens with it in relationship to other aspects of my living. Any event has its own unfolding life within the rest of a life.

Reflecting on this experience so far, I have highlighted how there is a situated context and focus and guideline that pulls the strands of experiencing together. This continued focus or project allows me to "stay on track" and to become aware of distraction or detours that pull me in another direction. The implication for our understanding of the nature of experiaction is that experience is always something (the content of consciousness) that typically occurs in the service of something else, be it the guiding intention of the conscious person or the habitual and hence anonymous biosocial programming. We might call this the *double intentionality* of consciousness. Naturally, we may not always be conscious of our intention in this double sense.

Questioning the meaning of an experience leads to the widening of the horizon of understanding, to a broadening of the context. We see this particular situation in the broader context of an array of situations that are held together in their intermittent and yet continuous and sequential unfolding by an overarching intent or project. The ultimate context for an individual is his or her life-project (the way Sartre works this out in his existential psychoanalysis) and the projects that society and culture provide. Each moment of experience is therefore also embedded in a particular sociocultural matrix of meaning. Let me reflect on how this works.

In my recorded experience, I accept the "social," that is, "professional collegiate," definition of the situation to a large degree. I choose a certain level of language. I accept the etiquette of formal presentation. I had been screening out imaginative variations. I tended toward the serious atmosphere prevailing in academic circles in my monologue; there was little humor, which tends to sparkle easily in the presence of others and not so readily "in solitary." It seems that projects—not necessarily rational projects exclusively, to be sure—hold situations together and that situations are the matrix of all particular experiactions that may occur within them.

But "situation" is itself a term that needs clarification. In the context of this recorded experiaction—this situated event—it becomes clear that we are involved in a complex structure. Using the thought-dimension of time, we can see that my situation—sitting bodily on the sunporch in Pittsburgh—was transcended in awareness both forward in time, to my anticipated presence at New College in Florida two weeks hence, as well as backward in time, to my remembered-presence to conversations with Dave Smillie back in Pittsburgh a few years ago. In other words, any situated experiaction spills over its physical space-time dimensions, unless one is able to limit one's awareness by concentration. In experiaction, we can move freely and usually extend ourselves beyond our bodily here and now. This is why an exclusive focus on the observer attitude is misleading to a large extent.

We can describe this by saying that experiaction, even though it presents itself to us as orchestrated and unified, can be recognized as occurring at different layers. Psychological language has names for these different experiential processes, such as perception, thought, memory, anticipation, and imagination. Although it makes some sense to distinguish these levels or modalities of experience—particularly since we as persons have some measure of control over where to direct our attention and what modality to use—there is usually a mixture of modalities occurring in any concrete experienced event. They run together.

Going back to my recorded experience, it is apparent that in my experiaction I moved back and forth in time, and that I also moved from immediate perception to recalling to conceptualizing to questioning. In a sense, everything is connected within the unity of my own consciousness, and we are witnessing an organic process of interpenetration and growth with continuous evolving transformation—informed and shaped by efforts at making sense—giving rise to complex understanding and awareness of the functioning of consciousness. By comparison, computer memories are terribly static and dead.

Human awareness as it is accessible to description and reflection is intriguingly complex, subtle, and challenging to understanding. If there are any root motives to human life, certainly one of them is to understand and to make sense of experiencing one's world, although not everybody might feel this urge, which is the beginning of scientific and philosophical interest.

Studying experience is different from studying physical or even biological reality out there. The moment people are involved—human experiaction—we are dealing with interpreted reality, with human meanings. For me, it's the U.S. Steel Building the way I see it, in terms of what it means to me, and this is dependent upon what my extended relationship with the U.S. Steel building is, my knowledge, my attitudes, my values, etc. In one act of awareness, looking at this building—as I sustain my focus of attention, as I tune in and just hold steady in this determinate direction—all my past and future involvement somehow assembles relevant experiences to fill in and amplify the meaning of what I am looking at. An organic structuring of the total stock of my experience occurs around the focus of my attention, the U.S. Steel Building.

In phenomenology, we speak of experienced horizons that open out from the spectacle. The focus of attention acts as a sort of vortex or vacuum field that draws into itself relevant items from our stock of experiaction established through previous contact. Thus, associated meanings are awakened and begin to dwell in the invisible and hovering presence around and through the topical focus of interest, modulating its meaning. For example, my wife and I, though we may be looking at the same identifiable structure (the U.S. Steel Building), nevertheless experience two different realities: what this spectacle means to her and what it means to me. We agree on the same object of reference—on this level we have common understanding—but we differ in what it means—in its idiosyncratic meaning. Using Gurwitsch's phenomenological language, what we are aware of is the total meaning co-constituted by the thematic object (the U.S. Steel Building) and the thematic field (the U.S. Steel Building meaning-horizons for me). Experience is thus a constantly changing and evolving cumulative meaning manifold, shifting from aspect to aspect and from domain to domain. Yet, despite the personal subtleties and nuances in the unique flavor of experiacting, there are the common foci about which we can establish consensus—what we are talking about—and there are universal shared features of experiacting—the way in which something appears to a person—always as contextualized, from a perspective, in time, and other factors.

The experiactional stream of a person can be seen to be organized. It takes place in a situation as a "situated event." Events have a duration through time; they could be called Time-Gestalten Events (such as giving a public lecture) that tend to appear first in the stream of experience in imaginary anticipation, usually as a result of a social invitation or as a self-initiated project. In anticipatory experience, projected events are developed, worked on, thought about on repeated occasions. The event takes shape in experience as an anticipation (what I thought the lecture to be like) before it takes place (what it turns out to be in actuality). This process has consequences and is important because it results in strategy-decisions that actually create the future event. During the recorded experience, I actually decided what I was going to do for the lecture. (Needless to say, one can make mistakes and proceed from an inadequate anticipatory image. I imagined myself to be speaking to interested psychology students, and the audience turned out to be senior citizens in retirement in Florida who like to hear a lecture at the college on any topic. They were very kind to me.)

The meaning of this anticipated event, which has personal (what shall I do) as well as social–professional horizons (what is expected of an existential–phenomenological psychologist and of a lecturer in a college setting, generally), sets limits as to what is allowed into the imaginary anticipation or the projecting. These limits could be said to act as "charges" upon my experiaction here and now. It seems that the event in anticipatory elaboration is richer and allows more play in experience. As we approach the actualization proper (actually giving the lecture), the range of possibilities narrows. Anticipation has to become embodied: It springs into action and enters the domain of public and shared social reality. We enact and we witness the event.

Any component of the experiactional stream—particularly in the anticipating phases of event-structures—is built up in complex ways. This may be described as "interwovenness." Whatever we focus on (e.g., the U.S. Steel Building in Pittsburgh, Pennsylvania) can be seen to arise on a string of previous encounters reaching back into one's at-bottom indeterminate biographical stock of experiaction. This stock hovers horizonally around the theme—what I pay attention to at the moment—and codetermines its meaning. This makes for idiosyncratic meanings. But our past experience is also in a good measure shared (social) because it is based on experiences of communication to which I had access through language or images. To some extent, therefore, there is also a public and shared meaning to a perceived theme.

Components of the experiactional stream are also gathered up and connected through a higher and unifying intention, the project. We take many steps to actualize a project, and we also seem to have more than one project going on at any one time. This tends to "mix up" our stream of experiaction in surprising and complex ways.

In this study, I took a small strip (20 minutes of continuous experiaction: sitting down and recording what occurs to me) that turned out to be connected to the larger sequence of being involved in giving a lecture. This event as an overarching project casts its shadow ahead of itself and colored my anticipatory experience in clearly definable ways.

The general structural features uncovered here through reflection were present and at work during my experiaction although I was not explicitly aware of them at the time, while living them out. Nevertheless, my prereflective experiaction turned out to have been structured and organized. In what I thought was a unique-private experience, general–universal structures were revealed and operative. The personal–existential and the general–phenomenological components are inextricably intertwined and collaborative in the creation of human experience in a particular situation. They constitute a tension inherent in the flow of experiaction. In living, we constantly move between these levels, and as researchers in this field, we always try to bridge this "gap" that arises in understanding and dichotomizes "lived immediacy" and "theoretical understanding." Either extreme—"just living" or "just thinking"—if emphasized unduly is an aberration. The human way is to experiact and to make sense of it so that mere living-out is humanized into living-with-awareness. This can never be fully accomplished, but is always in process, ongoing. Everyone has to unfold his or her own awareness and understanding, recognizing that the depth or ground of living eludes objectification and final conceptual grasp. We can only tune in and become involved.

CONCLUDING THOUGHTS

In offering this overview and illustration of existential–phenomenological research in psychology, I have been aware that there may be other ways of organizing these diverse analyses of life-world experience and that there are types and areas of research not included in this presentation. We are discussing "work in progress," and the examples offered should be understood as discoveries and formulations along the way, rather than as ironclad, Procrustean schemes of data analysis. The personality, style, and world view of the researcher–interpreter enter into both empirical and hermeneutical phenomenological work, as in the continuous questioning and examination of implicit assumptions that is itself part of the hermeneutical enterprise and that makes it sometimes difficult to specify the methodological "steps" involved in doing hermeneutical research. There is an unanalyzable, idiosyncratic component operative in all interpretive–phenomenological work that is associated with an original vision of human reality held by the researcher. It is this realization of the implicit interconnectedness between the researcher and that which he or she explores that characterizes any research study, however categorized, as phenomenological.

REFERENCES

Aanstoos, C. (Ed.). (1983). A phenomenological study of thinking. In A. Giorgi, A. Barton, & C. Maes (Eds.), *Duquesne studies in phenomenological psychology: Volume IV* (pp. 244–256). Pittsburgh, PA: Duquesne University Press.

Brown, R. (1965). *Social psychology.* New York: Free Press.

Buckley, F. M. (1971). Toward a phenomenology of at-homeness. In A. Giorgi, W. Fischer, & R. von Eckartsberg (Eds.), *Duquesne studies in phenomenological psychology: Volume I* (pp. 198–211). Pittsburgh, PA: Duquesne University.

Cloonan, T. (1979). Phenomenological psychological reflections on the mission of art. In A. Giorgi, R. Knowles, & D. Smith (Eds.), *Duquesne studies in phenomenological psychology: Volume III* (pp. 245–277). Pittsburgh, PA: Duquesne University Press.
Colaizzi, P. F. (1973). *Reflection and research in psychology: A phenomenological study of learning.* Dubuque, IA: Kendall Hunt Publishing.
Colaizzi, P. F. (1978). Psychological research as the phenomenologist views it. In R. Valle & M. King (Eds.), *Existential–phenomenological alternatives for psychology* (pp. 48–71). New York: Oxford University Press.
Crites, S. (1971). The narrative quality of experience. *Academy of Religion, 39*(3), 291–311.
Csikszentmihalyi, M. (1978). Attention and the holistic approach to behavior. In K. Pope & J. Singer (Eds.), *The stream of consciousness* (pp. 335–358). New York: Plenum Press.
de Rivera, J. (Ed.) (1981). *Conceptual encounters.* Washington, DC: University Press of America.
Fessler, R. (1983). Phenomenology and "the talking cure": Research on psychotherapy. In A. Giorgi, A. Barton, & C. Maes (Eds.), *Duquesne studies in phenomenological psychology: Volume IV* (pp. 33–46). Pittsburgh, PA: Duquesne University Press.
Fischer, W. F. (1970a). *Theories of anxiety.* New York: Harper & Row.
Fischer, W. F. (1970b). Faces of anxiety. *Journal of Phenomenological Psychology, 1*(1), 31–49.
Fischer, W. F. (1974). On the phenomenological mode of researching "being anxious." *Journal of Phenomenological Psychology, 4*(2), 405–423.
Gadamer, H. (1975). *Truth and method.* New York: Crossroads Publishing.
Giorgi, A. (1965). Phenomenology and experimental psychology I. *Review of Existential Psychology and Psychiatry, 5*(3), 228–238.
Giorgi, A. (1966). Phenomenology and experimental psychology II. *Review of Existential Psychology and Psychiatry, 6*(1), 37–50.
Giorgi, A. (1967). The experience of the subject as a source of data in a psychological experiment. *Review of Existential Psychology and Psychiatry, 7*(3), 169–176.
Giorgi, A. (1970). *Psychology as a human science: A phenomenologically based approach.* New York: Harper & Row.
Giorgi, A. (1975a). An application of phenomenological method in psychology. In A. Giorgi, C. Fischer, & E. Murray (Eds.), *Duquesne studies in phenomenological psychology: Volume II* (pp. 82–103). Pittsburgh, PA: Duquesne University Press.
Giorgi, A. (1975b). Convergence and divergence of qualitative methods in psychology. In A. Giorgi, C. Fischer, & E. Murray (Eds.), *Duquesne studies in phenomenological psychology: Volume II* (pp. 72–79). Pittsburgh, PA: Duquesne University Press.
Gurwitsch, A. (1964). *The field of consciousness.* Pittsburgh, PA: Duquesne University Press.
Halling, S. (1979). Eugene O'Neill's understanding of forgiveness. In A. Giorgi, R. Knowles, & D. Smith (Eds.), *Duquesne studies in phenomenological psychology: Volume III* (pp. 193–208). Pittsburgh, PA: Duquesne University Press.
Jager, B. (1975). Theorizing, journeying, dwelling. In A. Giorgi, C. Fischer, & E. Murray (Eds.), *Duquesne studies in phenomenological psychology: Volume II* (pp. 235–260). Pittsburgh, PA: Duquesne University Press.
Klinger, E. (1978). Modes of normal consciousness flow. In K. Pope & J. Singer (Eds.), *The stream of consciousness* (pp. 225–258). New York: Plenum Press.
Kracklauer, C. (1980). *The drug problem problem.* Unpublished doctoral dissertation, Duquesne University, Pittsburgh, PA.
Leary, T. (1970). The diagnosis of behavior and the diagnosis of experience. In A. Mahrer (Ed.), *New approaches to personality classification.* New York: Columbia University Press.
McClelland, D. (1953). *The achievement motive.* New York: Appleton-Century-Crofts.
Palmer, R. (1969). *Hermeneutics.* Evanston, IL: Northwestern University Press.
Pollio, H. (1981). *Behavior and existence.* Monterey, CA: Brooks-Cole.
Ricoeur, P. (1971). The model of the text. *Social Research, 38,* 529–562.
Romanyshyn, R. (1982). *Psychological life: From science to metaphor.* Austin: University of Texas Press.
Sartre, J.-P. (1953). *Being and nothingness.* New York: Philosophical Library.
Schutz, A. (1962). *Collected papers: Volume 1.* The Hague: Martinus Nijhoff.
Titelman, P. (1979). Some implications of Ricoeur's conception of hermeneutics for phenomenological psychology. In A. Giorgi, R. Knowles, & D. Smith (Eds.), *Duquesne studies in phenomenological psychology: Volume III* (pp. 182–192). Pittsburgh, PA: Duquesne University Press.
van Kaam, A. (1958). *The experience of really feeling understood by a person.* Unpublished doctoral dissertation, Case Western Reserve University, Cleveland, OH.
van Kaam, A. (1966). *Existential foundations of psychology.* Pittsburgh, PA: Duquesne University Press.
von Eckartsberg, R. (1971). On experiential methodology. In A. Giorgi, W. Fischer, & R. von Eckartsberg (Eds.), *Duquesne studies in phenomenological psychology: Volume I* (pp. 66–79). Pittsburgh, PA: Duquesne University Press.

von Eckartsberg, R. (1972). Experiential psychology: A descriptive protocol and reflection. *Journal of Phenomenological Psychology, 2,* 161–173.

von Eckartsberg, R. (1979). The eco-psychology of personal culture building: An existential–hermeneutic approach. In A. Giorgi, R. Knowles, & D. Smith (Eds.), *Duquesne studies in phenomenological psychology: Volume III* (pp. 227–244). Pittsburgh, PA: Duquesne University Press.

von Eckartsberg, R. (1981). Maps of the mind: The cartography of consciousness. In R. Valle & R. von Eckartsberg (Eds.), *The metaphors of consciousness* (pp. 21–93). New York: Plenum Press.

von Eckartsberg, R. (1986). *Life-world experience: Existential–phenomenological research approaches in psychology.* Washington, DC: Center for Advanced Research in Phenomenology; University Press of America.

Wertz, F. J. (1983). From everyday to psychological description. *Journal of Phenomenological Psychology, 14*(2), 197–241.

Wertz, F. J. (1984). Procedures in phenomenological research and the question of validity. In C. Aanstoos (Ed.), *Exploring the lived world: Readings in phenomenological psychology* (pp. 29–48). Carrolton: West Georgia College.

3

The Question of Reliability in Interpretive Psychological Research[1]

A Comparison of Three Phenomenologically Based Protocol Analyses

Scott D. Churchill, Julie E. Lowery, Owen McNally, and Aruna Rao

---

The "Achilles heel" of interpretive research is that two or more researchers, confronted with the same data, posing the same question(s), will invariably express their findings differently. Different words will be used to express the "same" (i.e., congruent) meanings; it is also likely, of course, that different meanings will be thematized as well. Thus, there arises the question of the *reliability* of interpretive research methods. In what sense can a method (when applied by multiple researchers to the same set of data) yield results that are not "consistent" yet can still be considered "reliable"? Should we alter our definition of reliability to accommodate qualitative methods, or should we adhere to definitions of reliability

[1]An earlier version of this chapter appeared in *Methods: A Journal for Human Science,* 1992 Annual Edition.

Scott D. Churchill, Julie E. Lowery, Owen McNally, and Aruna Rao • Department of Psychology, University of Dallas, Irving, Texas 75062.

Phenomenological Inquiry in Psychology: Existential and Transpersonal Dimensions, edited by Ron Valle. Plenum Press, New York, 1998.

that have been developed with respect to quantitative methods? Should we abandon interpretive research altogether if it cannot stand up to existing scientific conventions? These questions are merely rhetorical, for it is clear to us that definitions of methodological criteria are themselves a function of human understanding and interpretation; thus, they cannot be carved in stone and are always open to reconsideration.

Let us begin with a preliminary definition of the issue we are addressing: "Reliability refers to the consistency of a measuring procedure or instrument. A method of measurement is reliable if it always produces the same result under the same conditions" (Lewin, 1979). Reliability assumes that one can establish an *equivalence* of measurement. In the modern sciences, "measurement" has come to mean quantification according to an established standard or scale. When measurement is in numbers, then equivalence can readily be established in view of the principle of identity: In observations of the behavioral performances of white rats, for instance, 10 seconds equals 10 seconds, just as 10 copulations equals 10 copulations. Thus, in quantitative research, we have a criterion for "equivalence" that is both formal and objective and that makes the establishment of reliability a matter of "fact," something we can "count" on. However, when measurement takes the form of a narrative description that employs connotative language (imagine, for example, two college laboratory assistants separately describing the relative *vigor* of the copulatory behavior of two white rats), then the equivalence of description is less easy to establish. Indeed, not only is the criterion for agreement between two verbal descriptions not clearly defined, but also an agreement among judges regarding the equivalence of descriptions becomes equally difficult to establish.

In the interest of providing a concrete demonstration to serve as a basis for reflection on the issue of reliability in phenomenologically based interpretive research, we will present below one protocol and three analyses. The protocol was drawn from a thesis by the second author, a study of current sexual experiences described by subjects who previously had been victims of date rape (Lowery, 1991). The protocol was analyzed subsequently by two other researchers (who were not told about the subject's prior experience). All three researchers were trained by the senior author in the employment of an empirical existential–phenomenological method—that is, they had been taught to perform analyses of self-report data, grounded both methodologically and conceptually in the literature of existential-phenomenology. Each researcher was free to bring his or her own set of questions and interests to the data.

METHODOLOGY

The process of empirically based phenomenological research generally proceeds through three discernible phases. First, there is the formulation of a research problem; second, the collection and analysis of data; third, the communication of one's findings in the form of a written or oral presentation. Much of the literature on phenomenological method deals with the essential *preparatory* moments of interpretation: (1) *intuition*—the process by which one familiarizes oneself with the phenomenon: living it, reexperiencing it, coperforming it, taking it up—and (2) *analysis*—the process by which one actually discerns the constituents of the phenomenon. The *results* that emerge from the preparatory phases of intuition and analysis can be described as a "culminating description" (Spiegelberg, 1982, pp. 681ff). What is not very often discussed is the *form* and *content* of phenomenological findings, which will vary according to the interpretive framework and discursive style of the researcher.[2] (Indeed, it is the aim of this chapter to provide the reader with a basis for better appreciating how differing styles of analysis can coherently and consistently converge upon common reference points of data.)

The actual moments or "steps" of an existential–phenomenological psychologist's reflection upon protocol (or self-report) data have been described by numerous authors, most notably the initial formulators of this "family" of methods, including Colaizzi (1973, 1978), Fischer (1974, 1978, 1985), Giorgi (1975, 1985), Van Kaam (1966), and von Eckartsberg (1971, 1986).[3] In addition, one can find elaborations of the empirical–phenomenological method in Churchill and Wertz (1985), Moustakas (1994), Polkinghorne (1989), and Wertz (1983a, b, 1985). The culminating description, insofar as it is the *articulation* of the researcher's understanding, is an "interpretation" in the Heideggerian sense of the term: "... interpretation is grounded existentially in understanding" (Heidegger, 1962, p. 189); moreover, "that which can be Articulated in interpretation ... is what we have called 'meaning'" (p. 204). Interpretation is, then, the actual working out of *possible* meanings that have been projected *in advance* by one's understanding *in the face of* a manifest experience. In other words, the researcher is not simply a *tabula rasa* receiving impressions from the data; rather, he or she is present to the data guided by what Heidegger (1962) calls a "fore-sight" or what Husserl (1973) called "anticipation." Note here that the *articulated meaning* is considered only a *possibility* in relation to the subject matter of one's understanding. This means that the "equivalence" of two or more interpretations will not be based upon a necessary *identity*, but rather upon a *congruence of possibilities*. We shall return to this issue later in the chapter.

Procedure Employed

The method employed by the three researchers in this study was drawn from the literature cited above and consisted of reading the subject's protocol, then breaking it down into "meaning units" (each of which expresses a discriminate theme). These themes were then reflected upon and articulated in the form of a moment-by-moment analysis that was subsequently organized by each researcher into a coherent statement that expressed the psychological meaning of the experience as a whole. Following Churchill's (1990, 1991a) approach to teaching phenomenological methodology, each researcher's intuition and analysis of meaning was guided by these essential guidelines:

1. Empathic Dwelling

The phenomenological approach requires that the researcher enter into direct, personal contact with the psychological event being studied. One brings oneself to the encounter with the research phenomenon by patiently "listening to" or "staying with" the subject's description. In doing so, one becomes ever more *open* to what is being communicated.

---

[2]In Chapter 8 in this volume, Churchill presents different models for the presentation of empirical existential–phenomenological research results.

[3]See Chapters 1 and 2, by von Eckartsberg, in this volume.

Heidegger (1962) has written that "listening to . . . is [our] existential way of Being-open" in our concernful and understanding presence toward others (p. 206). In listening to someone's self-expression, "a co-state-of-mind gets 'shared'" (p. 205) in which "we proximally understand what is said, or—to put it more exactly—we are already with him, in advance, alongside the [experience] which the discourse is about" (p. 207). By means of this resonating attunement, one begins to understand the other's *position* within the situation described.

Far from being some kind of cerebral attitude, this empathy is fully lived through one's body. Husserl (1989, p. 176) writes: "In order to establish a mutual relationship between myself and an other, in order to communicate something to him, a Bodily relation . . . must be instituted." The ideal bodily relation here would be the face-to-face encounter, but in principle one can institute a bodily relation to the other even if this relation remains one-sided, as in the case of reading self-report data or listening to a recorded interview. What is essential is that the researcher be capable of "co-performing" the subject's intentional acts: "In empathy I participate in the other's positing" (Husserl, p. 177). To "posit" is to take a stand in relation to something, to "position" oneself in such a way as to illuminate certain meanings within one's situation. Thus, *in empathy I participate in the other's positioning himself or herself from a unique perspective within a situation.* Husserl (1989) described this as "trading places." In empathy, while maintaining one's own position as researcher, one gradually allows oneself to feel one's way into the other's experience. A heightened sense of critical awareness accompanies this act so as to avoid any jumping to conclusions regarding the other's experience. In this manner, the researcher can begin to acquaint himself or herself with the "essence" of the experience described by the subject.

2. Concentrated Focusing and Disciplined Fascination

When one allows oneself to "dwell" with an experienced situation, details become magnified, as there are fewer distractions on the periphery of one's consciousness. The concentrated focusing of the phenomenological attitude allows one to marvel at what lies closest. In this attitude, the researcher is free to "seek the leisure of tarrying observantly" (Heidegger, 1962, p. 216). In the employment of intense interest, the researcher listens to every nuance of a subject's self-presentation with the aim of sensing possible significations. Single words or phrases open up constellations of meaning. This is perhaps best described as a kind of *fascination.* As one becomes more and more absorbed in the world of the subject, there is a loosening of the hold on one's own world; the researcher thus begins to "inhabit" the existential field of the subject. In being spellbound by the other's self-presentation, one becomes attentive to various imaginable (as well as self-evident) meanings of that presentation. The imagined meanings are not groundless, but rather erupt out of the "bodily felt sense" (Shapiro, 1985) that emerges when one is fascinated. To quote Wertz (1985, p. 174): "When we stop and linger with something, it secretes its sense and its full significance becomes magnified or amplified. What to the subject was a little thing becomes a big deal to the researcher, who hereby transcends the mundanity of the subject's situation. The slightest details of the subject's world become large in importance for the researcher." *The most important realization of the phenomenological researcher is that what the subject presents are not "brute facts" but dynamic meanings that are haloed with intentions.*

3. *Thematizing Meanings*

The way that a person acts in a situation or expresses his or her understanding of an experience typically reveals (even if only implicitly) that the person believes his or her actions, perceptions, or feelings to be reflections of the way things "really are." This is what phenomenologists call "the natural attitude"—the belief that the world "really is" the way it appears. In the phenomenological attitude, there is a suspension of this belief. In its place, a different belief comes into play—a belief that experienced meanings are the correlative of particular attitudes that are assumed wittingly or unwittingly by the research subject. There is a turning, then, from "given facts" to "intended meanings"—from the simple "givenness" of the situation in the subject's experience to a reflective apprehension of that situation's meaning as constituted by the subject's consciousness. "What had hitherto been simply accepted as 'obvious'. . . is now recognized as a *performance of consciousness* and subjected to analysis" (Natanson, 1973, p. 59 [emphasis added]). Here we are interested in understanding the *intentionality* of human conduct—not its "causes" or "contingencies," but rather its concerns and contexts. This requires the cultivation of a *sensitivity to meaning.*

4. *Attending to the Motivational Context of Experience*

In particular, the researchers were trained to look for *project-horizons* within human experience (Churchill, 1986). Drawing from Sartre's (1956, pp. 557–575) paradigm for an existential psychoanalysis, one looks for the more general "life projects" that underlie a given "situational project." Sometimes the underlying motives or interests are so implicit that one must engage in what is clearly an act of interpretation in order to bring these horizons to light (e.g., see Churchill, 1997, in press). At other times, as in the protocol presented below, an essential life project is stated "up-front" by the research subject; in such cases, attending to the individual's project-horizons simply means paying attention to the relation of a particular moment of the subject's experience to self-revealing material expressed elsewhere in the data. Concretely, for each meaning unit, the researcher poses this question to the data: What is the subject's "project" here? What is the "in-order-to" motive (Schutz, 1967, pp. 86–91) that helps us to understand what the subject is up to? The researcher must also be on the lookout here for self-deception in the subject's account of his or her experience, which might be described as a project to disclaim one's agency within a situation (see Churchill, 1987, 1991b).

Data Analyzed by All Three Researchers

The following protocol was obtained from a college student who was a member of a campus support group for victims of sexual assault. She was asked to provide a written description of any recent sexual experience in as much detail as she could remember. The aim of the original study (Lowery, 1991) was to investigate the woman's current sexual life, with an interest in understanding how her past might still be "present." (Material obtained from the subject in a follow-up interview aimed at clarifying the original protocol is included in brackets.)

When I think of the term "sexual experience," I think of a sexual mingling among or between a male and a female. Such an experience ought to be shared and reciprocal, but this has not been the case in many of my sexual experiences. I feel like something is missing, forgotten, or lost in the whole affair and that I am the outsider searching for an entrance back in.

One time I met someone at a nearby function. We talked for a while, and then he asked me for my phone number so that we could "get together some time." I didn't think much of it at the time. He seemed like a nice person, so I consented and gave him my phone number. He began to call me several times a day for about a week. I was elated this entire week, walking on cloud nine. I felt that I had finally found someone who had taken an interest in me. [I hadn't dated anyone in a long time. It had been so long since someone was interested in me. He paid attention to me. But, he only talked about himself He kept calling. He was persistent.]

We talked for hours on the phone, and he seemed genuine and sincere. He was open and told me things about himself and his past that I considered to be personal—things that I myself wouldn't share with anyone until I knew them for a long time and felt comfortable sharing such intimate and personal experiences with. We seemed to get along great. I began to feel like there actually might be something in this relationship that may last for awhile.

Then one evening he asked me if he could see me. I had only seen him once and was excited to see him again. I thought that he wanted to take me out somewhere on a date, but he said that he just wanted to come over and spend some time with me. This was fine with me. I figured that he would take me out some other time. I hung up the phone and hurried to get ready. Although we were not going anywhere special, I wanted to look nice for him. I wanted to impress him so that he would like me. I got the impression over the phone that he liked me, but I wanted him to like not only my voice and how I talked over the phone, but also how I looked. [When I first met him, we saw each other. I was wondering if he remembered what I looked like. Looks are important. I wanted to keep his interest. I wanted to make him see that I was a person, too. I wanted to make a good impression.]

My heart raced the entire time I was getting ready while thoughts of the evening filled my mind: We would talk as we did over the telephone and get to know each other better and everything would just be really special. As I sat down to calm my nerves, there was a knock on the door. I opened it up and there he stood, looking better than I remembered. He stretched out his arms and I gave him a big hug. It was as if we were old friends who hadn't seen each other in years. I felt warm being in his arms and almost loved—although I knew it was way too early to classify our relationship as "love." [It was a feeling like I was secure. But I knew that I wasn't supposed to feel that way because I hardly knew him.]

He came in and sat down and I got a burst of nervousness again. Suddenly, I didn't know what to say or how to act. [It felt like I knew something was going to happen that I didn't want to happen. I didn't know how to act. I didn't know what to say. It was like the real me was gone. I couldn't open up. I couldn't be who I was.] I kept telling myself to be myself and act naturally, like I had been both over the phone and the first time we met. I figured that I had nothing at all to lose. [I wasn't so attached right now, so if it didn't work out I wouldn't be devastated because I hardly knew him. Actually, he hardly knew me.] Although I thought that a relationship may evolve from this, it was still early enough so that if I really screwed up I wouldn't get hurt. But I did.

The conversation ended before it started, and suddenly I found us kissing each other, which was fine with me. Soon he picked me up and carried me over to my bed, where we continued to kiss each other. I tried not to let him pick me up by trying to hold my ground on the floor, but I gave in. It was inevitable. I did not want to be on the bed, as I knew it would lead to more than just kissing. I felt uncomfortable on the bed. His hands began to wander all over my body and I kept pushing them away. I wanted for him to like me, but I did not want to be put in this position where I felt obligated to succumb to his desires. I was satisfied just kissing with him. Why wasn't he? What did he want from me? It became obvious that he wanted to do more than just kiss, but how much further? I knew that there was no way that I was going to have sex with him. I didn't want to. It would not mean anything to me. I didn't want to have a sexual relationship with him, I wanted a meaningful one. Sex at this time in our relationship would not mean anything to me because I did not care for him in a sexual way. I did not want to share this part of me with him until I was ready . . . until I said so, not him. But still, I wanted for him to like me because there was a feasible relationship in the future . . . a relationship where I could feel safe and secure. I liked him a lot except for the fact that he was trying to force me to do more than I wanted to do.

THREE STYLES OF ANALYSIS

Following are excerpts from each researcher's analysis of the data presented above. Statements pertaining to each researcher's *approach* to the analysis, as well as any particular questions or interests that were addressed to the data, are presented first so that the reader will have an idea of the *perspective* that was constitutive of each researcher's presence to the data.

Researcher 1: Julie Lowery

Research Question

The research question is: What is the lived meaning of a date rape victim's/survivor's sexuality? That is, the researcher wants to understand how date rape victims/survivors live their sexual bodies and how they personally understand their sexuality. Two subjects, each a victim/survivor of date rape, were asked to describe in a written protocol any recent sexual experience in as much detail as they could remember. Both subjects chose to describe negative sexual experiences. Although the facticities of their descriptions were different, there were common themes between the two protocols, comprising the essential characteristics of the phenomenon according to the researcher.

Researcher's Approach

The researcher adopts a phenomenological approach as her fundamental viewpoint toward human experience. A phenomenological approach to the research question here—namely, "What is the lived meaning of a date rape victim's/survivor's sexuality?"—opens a variety of thematic horizons to shed light upon the phenomenon. By asking what is the intentionality of the subject's sexual experience, the researcher can illuminate the meaning of the research subject's actions. Sexuality is ambiguous, for the touched is also touching. Because the phenomenological approach asserts that experience is a legitimate ground for knowledge, the researcher can allow for the ambiguity of the subject's sexual experience and arrive at an understanding of her as a human being in the life-world. In reading and reflecting on the protocol, the researcher asked herself: What is the subject's existential "project" with regard to her relationships with men, and how do her intentional acts manifest this project?

Results

There are six themes common to both the protocol presented above and a second protocol included in my original research (Lowery, 1991). Although the focus of this chapter is on the *reliability* of the empirical existential–phenomenological method, it is worth noting that the results of the analysis of the protocol presented above have been cross-validated with the results of another set of data. Although the themes are discussed separately below, they are not independent of each other. Just as meaning units can be viewed as "interrelated scenes," these themes are also interrelated. To illustrate each theme, the researcher tried to include the best example from the protocol; however, the themes necessarily stem from the entire protocol.

Detachment. Allison became detached from her body once her acquaintance with the man in her description moved from verbal contact to physical contact. When her relationship with the other was basically a telephone relationship, Allison was aware of her body: She felt her heart racing and felt a need to calm her nerves as she was getting ready to see him for the first time since they had met. When he came over, she gave him a hug and felt "warm being in his arms and almost loved," but she discounts her feelings by saying, "But I knew that I wasn't supposed to feel that way. . . ." Allison stated, "The conversation ended before it started, and suddenly I found us kissing each other. . . ." She "finds" her body engaged in sexual behavior. There is no "moment" before the physical contact: Allison does not "actively" experience that first moment of being touched, but simply discovers that she has "been" touched. In her experience, the touched is not touching; that is, there is no bodily reciprocity.

Ambivalence. Allison is ambivalent about her sexuality. In one sense, she presents herself as an object of desire to be seen, but in another sense, she does not want to be an object of desire to be touched. Allison stated, "Although we were not going anywhere special, I wanted to look nice for him. I wanted to impress him so that he would like me . . . . I wanted to keep his interest. I wanted to make him see that I was a person too." In order to make a "good impression," Allison feels that she must make herself into a desirable object to be looked at. She had stated about the telephone conversation that "he only talked about himself." She was missing from the conversation. In order for him to *see* her, she must make herself into an object of desire to be looked at. Only if she makes herself into such a desirable object will he be able to *see* her as a person, too. However, Allison states, "His hands began to wander all over my body and I kept pushing them away." She wanted to be *seen* but not *touched.* She states, "I did not want to share this part of me with him until I was ready . . . until I said so, not him. But still, I wanted for him to like me because there was a feasible relationship in the future." Allison wanted him to like her in order to have a relationship in the future. Her ambivalence in this situation is clear in the "but still." Even though he was forcing her to do more than she wanted to do, she wanted that feasible relationship with him. "I liked him a lot except for the fact that he was trying to force me to do more than I wanted to do."

Fatalism. Allison sees her unwanted sexual experiences as something "inevitable." Within Allison's horizon of having been date raped and having had many sexual experiences that were not shared and reciprocal, Allison understands forced sexual experiences as being inevitable. When the man began pushing for more sexual behavior than she wanted to engage in, she did not say no." Rather, Allison resigned herself to the other's sexual aims. She states, "Soon he picked me up and carried me over to my bed, where we continued to kiss each other. I tried not to let him pick me up by trying to hold my ground on the #oor, but I gave in. It was inevitable. I did not want to be on the bed, as I knew it would lead to more than just kissing." Allison did not want to engage in sexual behavior beyond kissing, but she never told him that; she never said "no." Allison did not want to be on the bed since she "knew it would lead to more than just kissing," but she gave in. "More than just kissing" was inevitable, so she never told him "no."

Passivity. In a metaphorical sense, Allison became a corpse, an objectified body without a will or voice. On the one hand, becoming lifeless and passive resolves the ambivalence about resigning to the sexual aim of the other or stopping unwanted sexual behavior. On the other hand, as a date rape victim/survivor, Allison *has* said "no" to unwanted sexual intercourse, and that "no" was discounted by the other. Allison has become an object of desire whose lack of choice or efficacy in the past serves as a context for her "choice" in the current situation not to say "no" to what appears to her as "inevitable" unwanted sex. It is as if she has lost her will and her voice, and in these one-sided sexual experiences, she intentionally becomes passive and lifeless in order to bear "the inevitable." Allison states that she has had many sexual experiences that were not "shared and reciprocal." She states, "I feel like something is missing, forgotten, or lost in the whole affair and that I am an outsider searching for an entrance back in." When something is not reciprocal, it is one-sided. What is missing in Allison's unwanted sexual experiences is her will. She is an outsider who is intentionally not present during the sexual encounter.

Intentionality. Allison understands her sexuality as something she must give to the other in order to get a relationship. She does not *seek* men with which to have sexual intercourse in order to "trap" them into relationships; however when an acquaintance "pushes" for more sexual behavior than she wants to engage in, she feels that she *must* engage in sex in order for a relationship to evolve. This theme further illustrates the previous theme in which Allison becomes passive and lifeless during sexual acts. By becoming corpse-like, she can resolve the ambiguity of not wanting to engage in sexual behavior, while conceding that she must do so in order to attain a desired relationship. Allison states, "I wanted for him to like me, but I did not want to be put in this position where I felt obligated to succumb to his desires." Further she states, "I didn't want to have a sexual relationship with him, I wanted a meaningful one." In order for him to like her, Allison makes herself into an object for him to look at and feels put in a position of indebtedness, a position in which she must die a metaphorical death of her will and resign herself to unwanted sexual behavior. In order to resolve the ambiguity of the situation, of not wanting a sexual relationship yet feeling obligated to have sex in order to have a relationship at all, Allison intentionally becomes an object. Allison describes this feasible relationship as being one in which she could "feel safe and secure," yet this relationship would be with someone who was forcing her to engage in sexual activity. Allison's desperate search for a relationship seems to have nothing to do with the present moment or the man himself. She will sacrifice that safety and security as well as herself in order to have a relationship.

Repetitive Behavior. The experience that Allison describes, in which she has engaged in unwanted sexual behavior in order to attain a desired relationship, is not a one-time occurrence. Interviews with the subject have revealed that her behavior is repetitive, and this leads the researcher to understand that the themes presented above transcend their reference to the specific situation described and address the more overarching question concerning the lived meaning of a date rape victim's/survivor's sexuality. Allison states that a sexual experience ought to be something "shared and reciprocal, but this has not been the case in many of my sexual experiences." Allison's expectation that having unwanted sex is inevitable is understandable in light of her previous experience of date rape, as well as her unshared and unreciprocal sexual experiences.

Researcher 2: Owen McNally

Researcher's Approach

In this study, I will examine a transcript of a woman who concretely experiences the laborious process of defining herself as a sexual person with a capacity for agency, in order to explicate the implicit meanings contained therein. Specifically, I seek to understand how her sense of agency and autonomy, versus a sense of passivity and objectification, corresponds to her defining what is for her a meaningful psychosexual relating to the "other."

The phenomenological way of doing empirical research embraces several distinct but interrelated component activities that concretize the theoretical insights of existential–phenomenological psychology and distinguish it as a "human" (as opposed to "natural") science. According to Churchill (1991a), there are three primary steps:

1. In diametrical opposition to the objectivistic rationalism of traditional Cartesian science, the phenomenologist instead enters into a sensitive and profound (but not wholly contiguous) relationship with the subject, seeking to understand him or her on the subject's own terms, a process known as empathy or empathic dwelling.

2. In doing so, the researcher is attentive to the subtleties of the lived description and of the nuances of meaning it contains. The details of the subject's experience are dwelt upon intensely as the researcher focuses attention upon the array of significances.

3. The researcher then steps back from the data and reflects upon it, employing that hallmark of phenomenology known as the *epoche.* What occurs here is the suspension of the "natural attitude" (or the basic, naive belief that the world *is* as it seems to us) and a shift of perspective whereby the researcher holds in check his or her own preconceptions, interpretive structures, and biases, and makes the crucial distinction between the objective events and the subject's experience, keeping in mind that, psychologically speaking, "perception defines one's reality." One is not really interested in the abstract notion of what really might have happened in a situation; rather, one thinks about the reality of the lived event and the meanings that arise from having experienced it.

Analysis

The protocol of concern here is the description of an experience of a date by a young woman who was asked by a researcher to describe a recent sexual experience. She describes how she meets a young man who begins to call her and comes over one evening (all with her consent). They begin to kiss, and soon he attempts to escalate the sexual intimacy, while she is unhappy about this. "Factually," this is probably all that occurred (that we know of, at least). What is worthy of analysis are the meanings involved.

She begins her description with a definition of what a sexual experience is for her, describing it as a heterosexual event that is ideally "shared and reciprocal." However, this ideal is not realized in many of her encounters. She explains this further by noting that "something is missing, forgotten, or lost in the whole affair," adding that "I am the outsider searching for an entrance back in."

Right at the outset, she describes the real in terms of the ideal. The research question inquired specifically about a recent sexual experience. She complies, but not before orientating her description in relation to a project, which consists of a mutual and co-created dynamic. Thus, we are immediately presented with a revelation about her emotional life, before the requested data are proffered. When she thinks of a recent sexual encounter (or at least is asked to reveal the contents of her introspection about the experience), she does so in relation to a sometimes unfulfilled need to have the physical intimacy occur in a context of reciprocity and agency expressed by both partners. For her, the question of sex immediately raises the issue of how she will be able to actively participate in the event. Her experience of sexual interplay is defined initially as a function of the extent to which she (as well as the "other") can share and reciprocate.

By contrasting the ideal sexual event against what for her is often a very different reality, we are at the very beginning aware of her dissatisfaction with many sexual events in her life, and of her desire to realize the desired mutuality and reciprocity. She informs us of her intentionality realized as project-consciousness by contextualizing the data soon to be provided in terms of that project. Before we can access the data she was asked to provide, we are to understand her as "searching for an entrance back in." The sexual experience is "spun" as a concrete instance of her alienated and rather peripatetic status.

At this point, one immediately homes in on the implication of her initial statements: What is missing, forgotten, or lost for her? What inside is she outside of? At this point, a clearer focusing of the research question is appropriate and necessary. What is of concern here is how a sense of her activity/passivity structures her experience of sex as a meaningful mode of experience. Approaching the data with this question in mind, I could begin to effectively illuminate the implicit meanings of the protocol.

She has defined her project as a movement toward a coequal, reciprocal sexual relationship. We know that many of her sexual encounters lack some crucial element. What apparently is lacking in these situations is some sense of agency on either her part or the part of the "other." If her encounters fall short of her project goal, then the sharedness, mutuality, and/or reciprocity is what is lacking. Thus, the question of her activity/passivity comes into play. Perhaps the problem for her is an inability to act in sexual situations. The other immediately obvious possibility would be that the "other" is unable or unwilling to reciprocate or share. Certainly, the capacity for agency is at least of penultimate concern. Regardless of whether or not passivity/agency is the definitive, core issue behind the disparity between what her project aims for (e.g., mutuality) and what she says she often experiences (the feeling of incompleteness, alienation, and so on), we cannot understand her project divorced from the aforementioned question regarding agency or a lack thereof. At this point, it would be tempting to posit one or the other of the above possibilities (she or the "other" being deficient in providing the reciprocal character), but at this point very little is actually known (though that is indeed important and interesting).

Returning to the protocol: She has met someone at a nearby function who asks for her phone number for a future date. She finds him to be a "nice person," and she agrees, though she notes that it seemed insignificant at this time. This prosaic description of an event "I didn't think much of . . . at the time" is nevertheless revealing to the phenomenologist. Her tone of (relative) disinterest reinforces the sense of his agency predominating in a nonetheless mutual event. They talk, he makes a move to which she responds. Her description downplays her agency. The overall impression is vaguely positive, and more so in the light of her just previous description of the profound inadequacy of many of her erotic encounters.

He begins to call her, "several times a day for about a week." Her response is euphoric; she feels that she has finally found someone who has taken an interest in her (which, she notes, has not happened for some time). The felicitude she expressed as a result of this attention sounds somewhat discontinuous when compared with the previous lack of enthusiasm for his attentions. Perhaps she is warming up now that his interest in her is at least so solid as to require him to follow through with his initial attention, but the contrast in tone certainly seems to conflict with her earlier description.

Upon returning to the protocol, this disparity is not resolved, given another apparent paradox: He pays attention to her and yet only talks about himself. She passively enjoys his activity toward her, but frets at its reflexive character. This can be understood as a threat to her project, the possibility of which otherwise was in the process of being nicely reinforced. Her tone has been very positive as she basks in the warmth of his attention, but a brief phrase of doubt occurs when she mentions that he speaks only of himself. This would lend weight to the interpretation that what is missing for her in her sexual relationships is the other's reciprocal response. She senses the possibility that he will not provide this, and the change of tone may reflect the correlative lessening of her project's attainment.

Interestingly, the striking positivity of her response to his phone calls is accompanied by a sense of passivity: "I was elated" (which would literally mean she was "picked up"). While she has been decidedly nonaggressive and passive, she explains her happiness due to her finding someone who would take an interest in her. Here we see signs that there is a reciprocal relationship possible, that her project may be accomplished. Thus, she maintains an active complementarity in regard to his activity, but only her feelings have been active.

She goes on to mention his persistent calls and the great length of time they spend on the phone (talking about him?). She tells us that he reveals things about himself she considers personal and too intimate to reveal to anyone except someone known for a long time with whom she was quite comfortable. This can be interpreted as reinforcing her project (perhaps he feels the same way about such things). Again, her passivity occurs in the context of a positive appraisal of his activity. She then expresses the situation in terms of their mutual reciprocal activity: "We seemed to get along great." It is at this point that she tells us she takes seriously the prospect of her project's realization. She already refers to it as a "relationship." Except for the very brief mention of the actual content of his talking (and with no explicit articulation of how that is significant), she has constructed in her protocol thus far the implication that she may be on the verge of having the reciprocity–mutuality component of what, according to her, would be necessary to have the ideal sexual experience.

He then requests an opportunity to see her. Having seen him only during their initial encounter, she says she "was excited to see him again." She thought that this meant a date, but he just wanted to come over to her home for a visit. She sounds slightly disappointed despite her glib "This was fine with me." She figures the date will happen another time. A date would have best dovetailed with the means by which the "relationship" had progressed (according to her criterion) to the level it had; that is, he would initiate their mutuality. His dropping by leaves her without the aforementioned goal of mutuality established as it might otherwise have been in a date situation.

As she prepares for the evening, she attempts to amplify her physical attractiveness because she wants to impress him so that he will like her. Their phone conversations had hinted at this, but she wanted him to like not only her voice but her looks. Her agency occurs in the context of his, and has his agency as its raison d'être. Her capacity for action seems to be demarcated by and derives significance according to his activity. Again she expresses activity and mutuality whenever possible, noting that at their first encounter they saw each other. She states the importance of appearances immediately after wondering if he remembers her appearance. Her looks are understood as vital to the continuation of his interest in her. She also wants him to notice her personhood as well. This implies a certain doubt as to whether this will take place. He is interested in her, perhaps, but she has yet to be noticed as a person. This implies that the mutuality is on unsteady grounds, but she does not linger and, instead, sums up her previous description by stating, "I wanted to make a good impression." She nervously awaits his presence while thinking of how they would talk and get to know each other better and "everything would just be really special." Even her cognitive acts are talked about in a detached manner—"thoughts of the evening filled my mind." Again, her activity is in reference to the mutuality that he prompts. Thus far, the function of her agency is a response to his.

As he knocks, she feels a sensation of dread, that something would happen against her desires. She momentarily loses her capacity for activity—nor does she know how to act. The whole authenticity of her being is here called into question: "It was like the real me was gone." She could neither open up (a fundamental expression of agency) nor even be herself (the very predicate of any agency). When she opens the door, he is there, appearing better than she recalled. He stretches out his arms and she responds to his activity by giving him a big hug. Before he confronts her with his agency, she seems incapable of any self-referential activity. The sensation of being held by him is reminiscent of greeting an old friend, and she feels almost loved. This is the high moment for her in the description. She can reconcile the activity with her need for mutuality, though she can do so only insofar as he catalyzes the mutuality (which, to be precise, begs the very question of whether it is indeed mutual). She concedes the prematurity of feeling loved in terms of it being "too early." She describes it as a feeling of security, though she admits she was aware of not being supposed to have such feelings because she didn't know him very well. This points to a situation where her feelings no longer match her ideals. If she doesn't know him very well, he will not see her as a person, nor will a genuinely shared sexual relationship be a possibility for the two of them.

After the hug ends, she again is unable to express her agency. She desires to regain the naturalness of the phone conversations and the initial meeting. She estimates that she has nothing to lose from this potential "relationship failing" and that at this stage, if she "screwed up," she would not suffer emotional damage because of her lack of attachment to him and her not really knowing him: "Actually, he hardly knew me." This suggests that her feeling of "almost being loved" is less than genuine. She seems to recognize that what is lacking here is the more authentic context for physical intimacy that she in principle would like to experience.

Their talking "ended before it started." This is a mutual experience, but not an active one. She suddenly "found us kissing each other, which was fine with me." This is an outright hiding from her activity, cloaked in passivity. This activity violates her ideal standards, and thus she shies away from it. This statement goes a long way toward explaining what she feels she lacks in sexual experience. She refuses to acknowledge her agency, which at most is tentative and responsive to that of the "other's" when she participates in the erotic intimacy of kissing.

Kissing is acceptable until he picks her up and carries her to the bed, which she is uncomfortable with, as she knows that it will lead to more. Instead of actively resisting his move toward the bed, she tries to "hold my ground on the floor." She could have presumably made him stop another way, but all she could do was to try to hold her ground. She gives in to his agency, and she describes it as inevitable. She is a passive recipient of his activity. The most she can do in this situation is push his hands away. The mutuality has disintegrated, and now they have warring projects. One of the most revealing phrases states how she wants him to like her, but "I did not want to be put in this position where I felt obligated to succumb to his desires." His agency is a given, but hers is frail. This statement exemplifies her dilemma. The most she can do is to feebly resist now because she otherwise is backed into a corner if events continue. She was never able to truly act beforehand, and now fights hard to not let his project overwhelm hers.

This leads to a bewildered, almost "battle-weary" inability to deal decisively with the situation: "What did he want from me?" "How much further?" She knows what she does not want, but cannot or will not make a definitive move to rectify the situation accordingly.

She ends the protocol as she began it, by defining what sex is for her. She wants a meaningful relationship with him, to which sex now would be anathema. Sex now is out of the question, it would be meaningless to her. She must assert her own desire for it and not merely go along with his. She still wants him to like her and can still see the relationship she was hoping for, "a relationship where I could feel safe and secure." Her project is defined again as passivity. She is uncomfortable with her agency. She must at times assert it, but she prefers to let the "other" provide it. She concludes by noting that she likes him, except for his trying to force her into greater sexual intimacy than she wanted.

Her inability to act does not present obvious problems at the onset. When presented with the strong agency of the "other," she remains passive but attempts to compensate by referencing what occurs as mutual. In the protocol, the most positive language is used when the situation putatively consists of mutual, reciprocal activity. But her languaging of these activities as mutual does not make them so. She has a project and attempts to interpret her rather passive responses to the initiations of the "other's" as shared activity. She does not have to confront her lack of agency until their projects become mutually exclusive. Here, sex for her lacks meaning because she cannot initiate it. She nowhere possesses a sense of actively apprehending the situation and altering it according to project. At most, she defends (responds to attack) her project.

Her alienation stems from her lack of agency. In order for such a project as hers to be concretized, a passive response to the projects of the male "other" is not sufficient. She does not realize her own lack of agency and thus is at a loss in dealing with the demands of psychosexual politics. Because she can interpret her passivity as mutual activity, she does not confront the quite different projects of the male "other," which in the short term tends to revolve around immediate, sometimes indiscriminate sexual gratification more than it does meaningful, mutual, reciprocal, and shared sexual experience.

Researcher 3: Aruna Rao

Researcher's Approach

Embedded within any sexual experience lies the awakening of the self to the other. It is precisely this awakening that marks the onset of the journey upon which each individual will embark. To embark upon a journey means to choose a destination and a means of reaching it. In making the analogy between a sexual experience and a journey, I am acknowledging that experience, sexual or other, is not suspended in time devoid of life and animation, but rather is situated within one's past as lived in the present with the future in view. Hence, in my analysis I will employ an encompassing existential perspective through which the full potential of the data can be realized with regard to the subject's intentionality and her existence in the world (see also Rao, 1992).

Researcher's Questions Posed to the Data

How is sexual intimacy used as a vehicle, whether passively, willingly, or unintentionally, to begin the search for finding the person who will yield a full blossoming relationship? How does this individual prepare for the journey and what are the means she will employ in reaching it? Does her body become the means, and union with another her destiny? Moreover, how can one understand a sexual experience, in light of an understanding of the body-subject as a sexual being with many projects, meanings, perceptions, and intentionalities?

To the extent that phenomenological inquiry takes one back to the lived world of experience, it is essential to understand the psychological presence of the subject to the phenomenon described. In other words, it is essential that my understanding of the description emanate from an empathic presence to the experience as described by the subject. In order to "hear" the data speak, one must lend an empathic ear. Pursuing the data by way of a noetic–noematic analysis allows one to bracket and consequently abandon all preconceived notions of the phenomenon researched, facilitating a fresh approach to the data. With this being the point of departure, I read the protocol as one coherent piece of narration. Then I posed this question: How does the protocol reveal the intentionality of the subject? To make explicit the implicit intentionalities of the subject, however, also means to realize the sometimes distorted lens through which the subject tells her tale. One must render oneself free from the natural attitude and come to embrace expressions and narratives as events that are full of personal meanings—keeping in mind, however, that the way the world appears is always colored and animated by a unique presence to it. Thus, the researcher should approach the description by placing herself in a position to understand the kind of presence required to yield the kinds of statements contained within the protocol. Hence, the researcher must ask, "What type of lens is the subject perceiving through, such that these particular perceptions are made?"

To pursue this method means to pay great attention to the details of each statement, which led me to divide the protocol into individual statements or groups of statements, known as *meaning units*. From the meaning units, I was able to ascertain a clearer expression of the subject's sexual experience, keeping in mind that the experience as lived and the experience as recalled were two different realities. In my analysis of the meaning units, I came across many anticipatory, biographical, and motivational "horizons" that I had been trained to look for in self-report data.

Thematic Results

Intertwined within the subject's description of a sexual experience is a collage or network of themes. Each theme depicts the subject's awareness of herself to the other, in the face of a sexual encounter. Behind the many faces that the subject portrays lie the distinctive characteristics and patterns that contribute to her perceptual understanding of a sexual experience. It is precisely these inherent characteristics that I will broadly thematize in order to gain a structured, comprehensive picture of the phenomenon as lived and described by the subject.

Social conventionality. In several areas of the protocol, the subject tends to relinquish her agency to the hands of the other. She notes very early on that a sexual experience is a "sexual mingling among or between a male and a female." She allows the male to precede the female in a hierarchical manner. Later on in the protocol, when the subject states that "he asked me for my phone number. . . . ," a shift of agency occurs, where he takes the initiative of asking for her phone number as well as the responsibility of calling. After the subject's many phone conversations over the course of one week with him, which were indicative of someone finally taking an interest in her, she states that "he asked me if he could see me. . . ." Although the subject was "elated this entire week, walking on cloud nine," she made no attempt at arranging to meet with him. Soon to follow, the subject admits, "I thought that he wanted to take me out." The language chosen is of particular interest, for the subject did not say, "I thought he wanted to go out," or even, "I thought he wanted to go out with me," but rather, "take me out." This implies that "he" does the active taking and "she" does the passive going-along. Throughout her protocol the subject expresses her feelings and actions in a curiously detached way. I realize that the phrases I have cited above may not be idiosyncrasies of the subject, but rather incidences of colloquial language, yet they serve as an indication that the subject does buy into the schema of conventionality. Her portrayal of herself as the yielding woman may originate from the learned constructs of society, which constitute an essential part of the subject's awareness of herself in relation to the other.

A "relationship project" as an anticipatory horizon. While the subject surrenders her agency to the other, there occurs a shift from "who she is" to "who she will be." The term "project" makes reference to the goals or aims that the subject intends to fulfill. Perhaps the most overarching project that must be thematized to gain a holistic understanding of the data is the subject's project to enter into a relationship with an other. This particular "horizon" of the subject's intentionality (namely, her desire to be in a relationship) outweighs any other project, taking precedence even over her own ability to live out her agency within the immediate situation. The subject forfeits her agency as a way of arriving at her chosen destiny, a relationship. Several times in the protocol, the subject prematurely alludes to "having a relationship." For instance, "I began to feel like there might actually be something in this relationship that may last for a while." At this point in the protocol, the "relationship" was based on one week of phone conversations, which the subject believed to be the foundation for a relationship. It is here that the first glimpse of her intentionality is seen. Soon to follow, the subject states, "I knew it was way too early to classify our relationship as 'love.'" But because the subject places great emphasis on being in a relationship, I came to understand that the journey she had embarked upon was prior to this sexual experience, and that her chosen destiny had always been to enter into a relationship. It was with this frame of mind that the subject entered into this sexual experience. Her intention to be in a relationship dictated the sexual experience she described. In other words, the experience as described by the subject is colored with the intentionality of ultimately being in a relationship. With this in mind, one can better appreciate the psychological presence that the subject assumed throughout this particular sexual experience.

Temporality. To speak of the subject's temporal awareness means to relate or parallel the sequence of time to the sexual experience that transpired. Because of the subject's enthrallment with her project, the dimensions of her temporality had become enmeshed with the dimensions of her experience. Contained within the protocol lies a totalization of temporality, where no such thing as "early" or "late" exists. The sequence of time and experience has become one entity, void of their innate boundaries. Thus, the anticipatory project of having a relationship was lived in the present moment, with the future simultaneously present in the form of her project. It is precisely here that one sees the past, present, and future informed by the project.

The fantastic. The subject's description of her sexual experience is laced with many illusionary portrayals. She states that she was "elated . . . walking on cloud nine," as though this experience had somehow transcended all others. Later on she remarks, "It was as if we were old friends who hadn't seen each other in years." The subject lives out her intentionality behind the mask of the fantastic. She continues to state, "I felt warm being in his arms and almost loved." Once again, the subject situates herself within an illusionary world that eclipses the reality of the experience, and perceives through the lens of the fantastic.

Motivated perceptions. Reading the subject's description through the lens of the phenomenological epoche, one begins to see "motivated perceptions" revealing a horizon or dimension of the subject's experience in which she implicitly seeks the fulfillment of an existential interest. In other words, this "motivated perception" allows the subject to see what she needs to see. The researcher's interest here is not whether or not she perceives the other "correctly," but rather what makes her perceptions possible. Her presence to the experience yields her perceptual understanding. For example, the subject states, "He paid attention to me." If being "paid attention to" were not of interest to her, it would be highly improbable that she would ever have made this statement. But because the subject seeks attention, she sees his attention being given to her. She also states, "We seemed to get along great." In the process of living out the project of having a relationship, the subject needs to see the other, in relation to herself, as "getting along." As the subject continues the description, she states that there "was a feeling like I was secure." In her desire to "feel secure," she perceives that he was freely giving security. Toward the ending of the protocol, the subject judges that "there was a feasible relationship in the future . . . a relationship where I could feel safe and secure." Such statements are motivated by a particular interest that needs fulfillment by the other. The subject's needs for attention, security, love, and, most important, a "meaningful relationship" all become motivated perceptions that she seeks out from the other. I do not mean to imply that her perceptions of the other are inauthentic, but, rather, that each perception is perceived in the interest of fulfilling a need. In order for her to even perceive security, love, and attention from the other, these qualities must exist in her motivational horizon, or she would have perceived other qualities that needed to be fulfilled. Thus, it is the need to see a particular quality in the other that motivates the subject's perceptions, which ultimately make her experience one that is dictated by motivated perceptions.

A Contextualized Account of the Subject's Sexual Experience

In this section of the results, I will rework the subject's description of a sexual experience in a phenomenological manner in order to gain an all-embracing account of her experience as lived. (For this purpose, I have given the subject the fictitious name Helen.)

Helen seeks a shared and reciprocated relationship with a man. She laments that many of her past relationships have not met her expectations. She, a being-in-the-world, has been left out of her own relationships and is in search of a way back in. She is the outsider in her own relationships and can only peer into them, restrained from touching them. She feels her agency has been somehow stripped away.

Helen meets a man at a place close to home. They engage in a conversation, which Helen regards to be a sign of his taking interest in her. He inquires for her phone number and she obliges, for she feels safe enough to give him her phone number.

Soon, he begins to call quite frequently. Helen, seeking a relationship, begins to appropriate her relationship-project with him. She is overwhelmed at the numerous phone calls she receives, and becomes euphoric at the thought of him actually taking an interest in her. Helen feels that because he calls often, he must have a genuine interest in her and who she is as a person. She feels that the phone calls and the extensiveness of the conversations that are taking place are indicative of him appreciating her for who she "is." In her quest for the attention of a man, Helen perceives this man to be fulfilling her aim.

Helen believes him to be sincere in his approach. Because she feels that they are getting along quite well on the phone, the possibilities of having a relationship have increased tremendously. She believes her project is being reinforced.

One evening he calls her and asks if he can see her. Helen, thrilled with the idea of not only seeing him but also "going out" with him, readily agrees. She is ecstatic, for she is eager to connect his voice to his face, to materialize the project. Soon she learns that he would prefer to simply come over, as opposed to going out. Helen thinks that this idea is also fine, although she secretly would have liked to be seen with him in the world.

Helen becomes slightly insecure of her appearance, as she believes that pleasing him with physical beauty is an essential part of the project. She wants to look desirable and be desired. As Helen proceeds with making herself up, the evening's project fills her mind. She imagines and intends it to be one of physical attraction and wonderful conversation. Helen waits anxiously in the face of encountering the man who will fulfill her project.

Helen is secretly preparing for a sexual encounter, but due to her past disappointing relationships and her firmly rooted views on the elements of a meaningful relationship, she finds herself to be faltering.

In the midst of this confusion, Helen loses her agency and her stance in the world. He comes to the door and she lets him in. Helen finds him to be physically attractive and proceeds to fill his outstretched arms with her body. She feels as though they are long-lost friends, perceiving warmth, security, and a secret feeling of love. Helen embraces the fulfillment of her project.

Helen is here confronted with both her projects and her passions, which appear to be incongruent. In the face of a situation that calls for a decision, she becomes anxious. The predicament in which she now finds herself is manifested bodily insofar as she experiences a loss of composure. In the face of the very situation for which Helen had meticulously prepared, she finds herself somewhat estranged from her "true" self: Helen is keeping her relationship-project in sight, but worries that something, perhaps she herself, will stifle the progression of it. Helen would like to believe that there is no attachment to him, but secretly knows that she is attached to the intention of his fulfilling her project.

Soon, Helen finds herself kissing him and is quite appeased. She is peering down from above, and receives a soothing sort of sensation. In saying that she "found herself" kissing, Helen implies that her agency in this situation is not one of volitional action but of spontaneous passion. Still, her action remains in service of her project-to-be-fulfilled.

Helen begins to realize that the kiss is going to lead to "more," and secretly wishes she could allow herself to continue, but is confronted both with the question of who she will be afterward and with the uncertainty as to whether her relationship-project will still be attainable. Helen experiences herself as being "picked up and carried" to the bed, and in this moment there is revealed a collapse of her agency, or at least of her responsibility within the situation. His hands explore her body, and although Helen pushes them away, she is "unable" to verbally object. At this point, Helen is face-to-face with both her immediate desires and her overarching project. She is confused. She is unsure of what the possible outcomes will be, if she does or does not have a sexual encounter. Helen wants a meaningful relationship and is afraid that sex might violate this project. Helen feels pressured by him and her own desires. She wants to protect her project. Helen finds her body acting without agency. She cannot decide whether a sexual relation is the way to fulfill the project. Helen is left feeling misconstrued and confused, but still hanging on to her project and the man who is supposed to fulfill it.

THEMATIC CONVERGENCES AND DIVERGENCES

Given the foregoing three psychological analyses of the lived-meaning of a sexual experience (as reported by a female college student), our task now is to consider in what ways the analyses can be seen to converge on a common ground of interpretation, as well as in what ways each analysis bears the distinctive mark of the analyst. To be sure, the three sets of results presented above display different emphases; nonetheless, they can be seen to cohere around a common focus on the ability of the subject to display agency, or the lack thereof.

Lowery's analysis thematizes how the subject experiences psychosomatic alienation, objectification, and ambivalence as a result of (1) a tension between wanting to be desired and not to be touched, (2) the inevitability and familiarity of unwanted sexual intimacy, and (3) a general lived-out detachment from her sexuality. McNally's investigation focuses on the psychosexual power relations of the subject and her date, and how she is able to refuse agency (relatively) comfortably, until the sexual-project of the date clashes with her own relationship-project. Rao's analysis examines how social constructs reinforce the subject's lack of agency, how the experience of receiving attention prompts a certain unreality, and especially how her project of wanting a relationship (the "anticipatory horizon") frames her perception of the event.

Biographical Presence of the Researcher

A striking difference between Lowery's and Rao's analyses stems from the different biographical contexts within which each approached the data. Lowery, personally acquainted with the subject (and herself a former victim of date rape), knew that the subject had been a previous victim of date rape and was particularly interested in seeing the subject's current experience of her sexuality in the light of her previous experience. In addition, the underlying political agenda of her original thesis (Lowery, 1991), from which the data were taken, shows itself in some of the unique ways she thematizes the subject's experience. For example, she writes:

> In a metaphorical sense, Allison became a corpse, an objectified body without a will or voice. . . . Allison has become an object of desire whose lack of choice or efficacy in the past serves as a context for her "choice" in the current situation not to say no to what appears to her as "inevitable" unwanted sex. It is as if she has lost her will and her voice, and in these one-sided sexual experiences, she intentionally becomes passive and lifeless in order to bear "the inevitable."

In contrast, Rao's subsequent analysis is colored by what Lowery later considered to be a "romantic tone": Rao thematizes the subject's "passion" (as opposed to inhibition) and describes the subject as

> . . . thrilled with the idea of not only seeing him but also "going out" with him. . . . She is ecstatic, for she is eager to connect his voice to his face, to materialize the project. Soon she learns that he would prefer to simply come over, as opposed to going out. Helen thinks that this idea is also fine, although she secretly would have liked to be seen with him in the world.

Later, Rao suggests that as "Helen embraces the fulfillment of her project, . . . [she] begins to realize that the kiss is going to lead to 'more,' and secretly wishes she could allow herself to continue. . . ."

To Lowery, Rao's account sounded as though the protocol were coming from a virgin trying to decide whether or not to lose her virginity, rather than from a woman who had been raped and had had many experiences such as the one described here. This only serves, however, to underscore the hermeneutical nature of this kind of research: The text is interpreted in terms of its context. The background information, to which Lowery was privy and the others were not, made a difference in the three researchers' interpretations, insofar as "the 'sedimented' structure of the individual's experience is the condition for the subsequent interpretation of all new events and activities" (Schutz, 1962, p. xxviii).

Lowery describes her own interest as "trying to uncover *the lived meaning* of a date rape victim's survivor's sexuality" [emphasis added]. Perhaps it is precisely such an "uncovering" that Rao has accomplished in her own analysis, not influenced by a knowledge of the subject's earlier experience or by a politicization of her own discourse in light of feminist concerns, when she sees beneath the veneer of the subject's ambivalence to a secret yearning. On the other hand, her feminist concerns enabled Lowery to explicate the power issues inherent in the subject's experience that might not otherwise have been uncovered. Each of these two researchers, in reading and reflecting upon the subject's description, *dramatizes* the action according to her own presence to the data. Yet each dramatization touches upon immanent possibilities that are "visible" within the subject's experience. Thus, Freud was right when he spoke about the "overdetermination"[4] of behavior. The three analyses appear together as different rays of insight intersecting upon a

---

[4]In his analysis of Dora, Freud (1963) wrote that "in the world of reality, which I am trying to depict here, a complication of motives, an accumulation and conjunction of mental activities—in a word, overdetermination—is the rule" (p. 77). The term occurs earlier in the same text in a footnote (p. 47). Other references can be found in Freud's 1896 paper on the etiology of hysteria (Freud, 1989, p. 108), as well as in his more extensive *Case of the Wolf-Man* (Freud, 1991, p. 200), written 20 years later. See Moore and Fine (1990, p. 123) for further discussion of Freud's use of this term.

common subject, each adumbrating a different sensibility or "immanent possibility of meaning" as figural against the background of other meanings. Together, they amount to variations on a theme. Like the reversals of ambiguous figures, each act of interpretation reveals even as it conceals: The power issue falls into the background, thereby enabling the romantic motive to come to light—just as the political issue becomes figural against the backdrop of sexual ambivalence.

Rao's analysis reflects the cultural definition of the female role to which she had been personally socialized. More specifically, Rao's Indian sensibility regarding the relations of the sexes leads her to see the profoundly co-constitutive nature of gender interactions and at the same time to be puzzled by the tendency of American women to disclaim agency in seeing the male as the sole perpetrator of sexual exploitation. The subject writes: "I felt warm being in his arms and almost loved—although I knew it was way too early to classify our relationship as 'love'. . . . It was a feeling like I was secure." How, asks Rao, can such statements be coerced? How can one say this, except from one's own volition?

Even if she had been previously date raped, it is clear (to both Lowery and Rao) from the subject's protocol that she is still seeking male desire and affirmation. Rao therefore believes that by thematizing this in her analysis, she is not simply being blind to the power issues inherent in male–female relationships, nor is she over-romanticizing that experience. Rather she is acknowledging the research subject as a human subject who is capable in principle of transcending the limits of her former experiences (see also Churchill, 1993). The subject's keeping herself open to horizons of interpersonal encounters is precisely what enables her to perceive "warmth . . . an almost love . . . and security." That which was lacking in her previous experiences shows itself in the form of desire. After all, does Lacan (1981) not teach us that desire is fundamentally related to lack? Unfulfilled desire lives on in project form until it is met, which, to Rao, only goes to underscore the whole idea that to perceive love is already to understand it, and thereby to "live" this understanding as an anticipatory horizon. In other words, we do not always "stumble" into situations or experiences; we co-constitute and to some degree direct ourselves toward them. And in directing ourselves, we get lost, confused, and even misunderstood, but that does not mean we have been exploited by the other. On the contrary, the "beautiful" thing here is that the subject surpasses her past date rape experience by allowing herself to be open to the other (see Rao, 1992). And by even attempting to write this protocol, the subject expresses herself in an articulate manner, demonstrating the ability to be open to maybe even herself.

Interestingly, McNally's analysis appears to be neutral with regard to both the political and the romantic interests of the other two researchers, even while he embraces the desire and power dynamics situated within the subject's experience. He *sees* the data differently (perhaps less sympathetically?) than his female colleagues do, in a way that suggests a *gender horizon* of interpretation coming into play in the three analyses. (Is it a coincidence that both female researchers gave the subject a *name,* possibly to "personify" the experience and thereby facilitate a "first person" identification with the female subject, while the male researcher maintains a third person distance, referring to the subject simply as "she"?) Merleau-Ponty (1968) writes in *The Visible and the Invisible:*

> Thus he who speaks (and that which he understands tacitly) always codetermines the meaning of what he says, the philosopher is always implicated in the problems he poses, and there is no truth if one does not take into account, in the appraising of every statement, the presence of the philosopher who makes the statement (p. 90).

> [Another] witness . . . is not a pure gaze upon pure being any more than I am, because his views and my own are in advance inserted into a system of partial perspectives, referred to one same world in which we coexist and where our views intersect (p. 82).
>
> What is given, then, is not the naked thing, the past itself such as it was in its own time, but rather the thing ready to be seen, pregnant—in principle as well as in fact—with all the visions one can have of it, the past such as it was one day *plus* an inexplicable alteration, a strange distance . . . (p. 124).

The last part of this quotation expresses a truth not only about the researchers' collective adumbrations or dramatizations of the subject's experience, but also about the subject's own changing relationship to her past experience. Even if "the 'many knowings' of the 'one phenomenon' never perfectly coincide with each other" (Wertz, 1986, p. 201), they still can be seen to cohere around a common nucleus of meaning.

Although all three analyses above are distinct and focus on somewhat different concerns, a unifying theme does emerge: The subject does not "own" her sexual experience actively. All three researchers thematize a vacillation within the subject's experience from active to passive agency, with passivity emerging precisely at those moments when a decision is called for on the subject's part. Likewise, all three see her as "disowning" her body—disconnecting her "self" from her actions when her integrity is at stake. Finally, all three see that her integrity within the situation is a function of her "relationship project"—her desire for a sexual experience that is "shared and reciprocal." Perhaps these consistencies among the three sets of results lend some credence to the claim that human science is indeed *science,* since it appears that an at least somewhat coherent set of themes can be gleaned from three different interpretive research efforts. Moreover, the findings presented here lend credibility to the claim that human science is indeed *human,* insofar as it appears that we must recognize and affirm that perceptions are always partial, yet never closed in solely upon themselves. Our picture of reality is never taken one frame at a time, but as a montage that effectively adumbrates the subject of perception from *always varying perspectives.* (Was this not the very point of cubism in painting?) As Wertz (1986) eloquently puts it:

> Perceptual consciousness is nothing other than a meaningful perspectival variation and as such it is the very movement of transcendence through which we genuinely contact something that exists (p. 191).
>
> In recognizing the intrinsic role of subjective perspectivity in the way reality is "reliably" established in life, we are required to give up the goal of absolute or exact coincidence between perception and the thing perceived, language and what is spoken about (p. 193).

Perceptual Adumbration and the Relativity of Interpretation

Existential-phenomenologists, following Heidegger, assert that all description is already interpretation. Interpretive or "hermeneutical" research involves the seeing (or "retrieval") of *possible* meanings from data. As Laing (1967, p. 62) has observed, it would be more appropriate to speak of *capta* here than of *data,* for the interpretive findings are not simply "given" but are "captured" or *"taken* out of a constantly elusive matrix of happenings." As such, possible meanings are revealed as a function of the researcher's questions and perspective on the data. Research results are no more, and no less, than *possibilities* of interpretation of the data. Giorgi (1975, p. 96) writes:

> Thus the chief point to be remembered with this type of research is not so much whether another position with respect to the data could be adopted (this point is granted beforehand) but whether a reader, adopting the same viewpoints as articulated by the researcher, can also see what the researcher saw, whether or not he agrees with it. That is the key criterion for qualitative research.

What typically happens when one reads a phenomenological research report is that one finds oneself going back and forth between the results and the data, trying to gain a sense of (1) the insights being offered by the author, (2) the data from which those insights were generated, and, most important, (3) a sense of the *relation* between the insights and the data. The critical moment comes when the reader asks whether the interpretive findings are indeed inherent within the data. It is also in this moment of contemplative presence to the research presentation that one finds oneself really learning something from the other's work. What one learns is not just *what* others have revealed but also something about *how* others see human events psychologically. As Spiegelberg (1982, p. 694) observes, "The main function of a phenomenological description is to serve as a reliable guide to the listener's own actual or potential experience of the phenomena."

In both doing and reading interpretive research, it is important to allow ourselves the time and space to experience the other's presentation. To experience a structure "is to live it, to take it up, assume it and discover its immanent significance" (Merleau-Ponty, 1962, p. 258). In empathy, one gradually allows oneself to feel one's way into the other's position. *Just as the researcher listens to every nuance of a subject's self-presentation with the aim of sensing possible significations, the reader listens to the nuances in a research analysis that reveal the researcher's own unique presence to the data.* (Empathy is thereby involved in both the "production" and the "consumption" of interpretive research.) Single words or phrases open up constellations of meaning. It is the researcher's job to facilitate the reader's understanding by presenting the results in a way that opens the reader to data and to the researcher's movement through those data. The style of such facilitation will always vary from one researcher to another, but it should at least be a fundamental consideration within any research endeavor. If the research results are always *relative* to the researcher's questions and interpretive frame of reference, then questions regarding the reliability of the interpretive procedure and verification of results must be posed while keeping in mind this characteristic of the interpretive process.

Interpretation and Truth

When we raise the question of verification in interpretive research, we are asking a question about the experience of truth. Sokolowsi (1974) writes:

> Verification, etymologically and in fact, is bringing about truth (p. 233).
>
> If I make a statement and you say you agree, you and I have used our ability to be truthful. What other people say, and what we ourselves later experience and state, may confirm our truth or force us to reappraise it; but some truth has been achieved and its claims cannot be disregarded (p. 3).

Concretely, what does it mean to experience a sense of "truth" in the findings communicated to us by interpretive research? Understood phenomenologically, the meanings expressed in the culminating description of a piece of interpretive research can be seen to exist "in a movement of constant relativity of validity" (Gadamer, 1975, p. 218), depending always on the meaning context or interpretive framework of the individual looking at the data.

Whether one is judging the coherence of a single narrative description or the congruency of two or more narrative descriptions, the question of coherence remains a matter of personal judgment; reliability thus becomes a second-order question of agreement or "fit" among observations that are presumed to have achieved a "fit" at the more primary level with regard to validity. Since no one narrative description can in practice embrace the "whole" phenomenon, the adequacy of a particular description is judged in view of its limited grasp of the phenomenon, and so an equivalence of described observations is either merely coincidental or is established by the perceived "hanging together" (*Zusammenhang*) of different descriptions from different viewpoints.[5]

One of the problems in establishing the reliability of interpretive methods in the human sciences is that the researcher is often privy to more data and background information than can be shared with the reader or judge. (As mentioned earlier, in the case of our three researchers here, the original researcher *knew* the subject and was able to interpret the data from the perspective of what she already understood about the subject's earlier sexual trauma.) Whether two descriptions or measurements agree is thus a matter of perceiving a *coherence* between the two descriptions rather than merely a direct *equivalence*. A "coherence" criterion for reliability acknowledges the perspectival nature of qualitative description, and therefore does not presume that two valid descriptions will be equivalent, much less the same. Rather, a coherence among descriptions (i.e., among different perceptions of the "same" data) asks only that these varying perspectives be able to "fit" together.

The distinction between validity and reliability thus becomes blurred in qualitative research precisely to the extent that, in the process of determining whether or not one set of findings is congruent with another, one must make one's own assessment of how well, if at all, the findings have illuminated their target. The two issues are intimately related insofar as the procedure for assessing the adequacy of the findings is at the same time an assessment of the adequacy of the "measure" used to obtain the findings, namely, the researcher's perspective on the data. In the end, the value of the findings depends on their ability to help others gain insight into the ever mysterious realities of human life.

REFERENCES

Churchill, S. D. (1986). *A phenomenological approach to motivation: The contribution of Jean-Paul Sartre.* Symposium paper published as: "An existential-phenomenological framework for consumer psychology." In W. D. Hoyer (Ed.), *Proceedings of the division of consumer psychology* (pp. 34–38). Washington, DC: American Psychological Association.

Churchill, S. D. (1987). *Seeing "through" self-deception in protocols: How to find methodological virtue in problematic data.* Paper presented at the Third International Symposium for Qualitative Research in Psychology, Perugia, Italy, August 17–21, 1987.

Churchill, S. D. (1990). Considerations for teaching a phenomenological approach to psychological research. *Journal of Phenomenological Psychology, 21*(1), 46–67.

Churchill, S. D. (1991a). *Empathy and intuition: Doing psychology phenomenologically.* Unpublished manuscript, University of Dallas, Irving, TX.

Churchill, S. D. (1991b). Reasons, causes, and motives: Psychology's illusive explanations of behavior. *Theoretical and Philosophical Psychology, 11*(1), 24–34.

Churchill, S. D. (1993). The lived meanings of date rape: Seeing through the eyes of the victim. *Family Violence and Sexual Assault Bulletin, 9*(1), 20–23.

Churchill, S. D. (1997). The alchemy of male desire: Femininity as totem and taboo. In S. Marlan (Ed.), *Fire in the stone.* Chicago: Chiron Press.

---

[5]"The validity of one's findings is therefore not contingent upon whether they are consistent with other viewpoints; for, according to the phenomenological approach, it is not possible to exhaustively know any phenomenon. In other words, other perspectives—perhaps rooted in different research interests and their corresponding intuitions—are always possible" (Churchill & Wertz, 1985, p. 554).

Churchill, S. D. (in press). Forbidden pleasure: Male desire for (and) identification with the female body. *Contemporary Psychodynamics.*
Churchill, S. D., & Wertz, F. J. (1985). An introduction to phenomenological psychology for consumer research: Historical, conceptual, and methodological foundations. In E. C. Hirschman & M. B. Holbrook (Eds.), *Advances in Consumer Research: Volume XII* (pp. 550–555). Provo, UT: Association for Consumer Research.
Colaizzi, P. F. (1973). *Reflection and research in psychology.* Dubuque, IA: Kendall/Hunt.
Colaizzi, P. F. (1978). Psychological research as the phenomenologist views it. In R. S. Valle & M. King (Eds.), *Existential–phenomenological alternatives for psychology* (pp. 48–71). New York: Oxford University Press.
Delius, H. (1953). Descriptive interpretation. *Philosophy and Phenomenological Research, 13,* 305–323.
Dilthey, W. (1977). *Descriptive psychology and historical understanding.* The Hague: Martinus Nijhoff.
Fischer, W. F. (1974). On the phenomenological mode of researching "being anxious." *Journal of Phenomenological Psychology, 4,* 405–23.
Fischer, W. F. (1978). An empirical–phenonomenological investigation of being-anxious: An example of the meanings of being-emotional. In R. S. Vale & M. King (Eds.), *Existential–phenomenological alternatives for psychology* (pp. 166–181). New York: Oxford University Press.
Fischer, W. F. (1985). Self-deception: An existential–phenomenological investigation into its essential meanings. In A. Giorgi (Ed.), *Phenomenology and psychological research* (pp. 118–154). Pittsburgh, PA: Duquesne University Press.
Freud, S. (1963). *Dora: An analysis of a case of hysteria.* New York: Collier Books.
Freud, S. (1989). The aetiology of hysteria. In P. Gay (Ed.), *The Freud reader* (pp. 96–111). New York: Norton.
Freud, S. (1991). The case of the wolf-man. In M. Gardiner (Ed.), *The wolf-man.* New York: Noonday Press. (pp. 153–262).
Gadamer, H. (1975). *Truth and method.* New York: Crossroad.
Giorgi, A. (1970). *Psychology as a human science: A phenomenologically based approach.* New York: Harper & Row.
Giorgi, A. (1975). An application of phenomenological method in psychology. In A. Giorgi, C. Fischer, & E. Murray (Eds.), *Duquesne studies in phenomenological psychology: Volume II* (pp. 82–103). Pittsburgh, PA: Duquesne University Press.
Giorgi, A. (1985). Sketch of a psychological phenomenological method. In A. Giorgi (ed.), *Phenomenology and psychological research* (pp. 8–22). Pittsburgh, PA: Duquesne University Press.
Heidegger, M. (1962). *Being and time.* New York: Harper & Row.
Husserl, E. (1973). *Experience and judgment.* Evanston, IL: Northwestern University Press.
Husserl, E. (1989). *Ideas pertaining to a pure phenomenology and to a phenomenological philosophy—Second book: Studies in the phenomenology of constitution.* Boston: Kluwer.
Lacan, J. (1981). *The four fundamental concepts of psycho-analysis.* New York: Norton.
Laing, R. D. (1967). *The politics of experience.* New York: Ballantine.
Lewin, M. (1979). *Understanding psychological research.* New York: Wiley.
Lowery, J. E. (1991). *The lived meaning of a date-rape victim's/survivor's sexuality.* Unpublished bachelor's thesis, University of Dallas, Irving, TX.
Merleau-Ponty, M. (1962). *Phenomenology of perception.* London: Routledge & Kegan Paul.
Merleau-Ponty, M. (1968). *The visible and the invisible.* Evanston, IL: Northwestern University Press.
Moore, B. E., & Fine, B. D. (Eds.) (1990). *Psychoanalytic terms and concepts.* New Haven, CT: Yale University Press.
Moustakas, C. (1994). *Phenomenological research methods.* London: Sage.
Natanson, M. (1973). *Edmund Husserl: Philosopher of infinite tasks.* Evanston, IL: Northwestern University Press.
Polkinghorne, D. E. (1989). Phenomenological research methods. In R. S. Valle & S. Halling (Eds.), *Existential–phenomenological perspectives in psychology: Exploring the breadth of human experience.* New York: Plenum Press. (pp. 41–60).
Rao, A. (1992). Experiencing oneself as being beautiful: A phenomenological study informed by Sartre's ontology. *Methods: A Journal for Human Science,* 73–96.
Sartre, J.-P. (1956). *Being and nothingness: An essay on phenomenological ontology.* New York: Philosophical Library.
Schutz, A. (1962). *Collected papers I: The problem of social reality.* The Hague: Martinus Nijhoff.
Schutz, A. (1967). *The phenomenology of the social world.* Evanston, IL: Northwestern University Press.
Shapiro, K. J. (1985). *Bodily reflective modes: A phenomenological method for psychology.* Durham, NC: Duke University Press.
Sokolowski, R. (1974). *Husserlian meditations.* Evanston: Northwestern University Press.
Spence, D. P. (1982). *Narrative truth and historical truth: Meaning and interpretation in psychoanalysis.* New York: Norton.
Spiegelberg, H. (1982). *The phenomenological movement.* Boston: Martinus Nijhoff.

van den Berg, J. H. (1972). *A different existence: Principles of phenomenological psychopathology.* Pittsburgh, PA: Duquesne University Press.

Van Kaam, A. (1966). *Existential foundations of psychology.* Pittsburgh, PA: Duquesne University Press.

von Eckartsberg, R. (1971). On experiential methodology. In A. Giorgi, W. F. Fischer, & R. von Eckartsberg (Eds.), *Duquesne studies in phenomenological psychology: Volume I* (pp. 66–79). Pittsburgh, PA: Duquesne University Press.

von Eckartsberg, R. (1986). *Life-world experience: Existential–phenomenological research approaches in psychology.* Washington, DC: Center for Advanced Research in Phenomenology; University Press of America.

Wertz, F. J. (1983a). From everyday to psychological description: Analyzing the moments of a qualitative data analysis. *Journal of Phenomenological Psychology, 14,* 197–241.

Wertz, F. J. (1983b). Some constituents of descriptive psychological reflection. *Human Studies, 6,* 35–51.

Wertz, F. J. (1985). Methods and findings in a phenomenolgical psychological study of a complex life-event: Being criminally victimized. In A. Giorgi (Ed.), *Phenomenology and psychological research* (pp. 155–216). Pittsburgh, PA: Duquesne University Press.

Wertz, F. J. (1986). The question of the reliability of psychological research. *Journal of Phenomenological Psychology, 17,* 181–205.

4

Human Subjectivity and the Law of the Threshold

Phenomenological and Humanistic Perspectives

Bernd Jager

THE WORKADAY WORLD AND THE FESTIVE WORLD

One of the key assumptions of phenomenological psychology is that human consciousness is structured by an *intentional* relationship. Human consciousness is openness to a world, in the same sense that subjectivity is necessarily intersubjectivity and that personhood necessarily implies interpersonality. Our problems in understanding these psychological dimensions are attributable in part to our use of nouns to denote what are essentially actions that bridge the distance between one person and another or between a person and a thing. Consciousness is not something that can be found among things, not even among the thinglike bodies described by biology and medicine. Consciousness refers to a fundamental relationship between persons and things, and it is this relationship that forms the basis of our awareness of our world.

This way of understanding consciousness in particular, and human reality as a whole, is not a recent philosophical or psychological discovery, but represents a way of thinking

---

Bernd Jager • Department of Psychology, University of Quebec at Montreal, Montreal, Quebec H3C 3P8, Canada.

Phenomenological Inquiry in Psychology: Existential and Transpersonal Dimensions, edited by Ron Valle. Plenum Press, New York, 1998.

that is probably as old as mankind[1] itself. There is an ancient Greek proverb that says very simply and directly *Aner oudeis aner,* meaning, literally, "None without another," or, more fully, "One single human being, considered in the absence of a relationship to another, is in fact no human being at all." Being human means standing in a relationship to others, to things, and to a world. In what follows, we will explore two very different types of relationships to the world that always occur together and that in their dynamic interplay open for us a truly human world.

The first attitude opens *a world of work.* It seeks to transform the natural world in such a way that it conforms more closely to our particular bodily and material needs. In our Western societies, both technology and the natural sciences developed out of this workaday need to transform our natural environment and to make it conform to our needs. The central theme of the workaday world, of technology and the natural sciences, is the transformation and ultimately the complete appropriation of a natural environment.

The second attitude opens *a festive world* and cultivates close alliances with human, divine, and natural beings in such a way that both the self and the other are thereby revealed to one another. In our Western societies, religious practices, together with the arts, the humanities, poetry, and literature, form together a festive world that is structured around the fundamental desire to stand in a mutually revealing relationship to a natural, a human, and a divine world.

The various cultural forms to which these basic two attitudes give rise vary from place to place. They also vary from time to time and from one situation to another. The world of work of New Guinean Papuans clearly differs from that of the Bantus in South Africa, and the festive world of medieval France is not the same as that of modern France. Yet these differences, no matter how great or small, do not contradict the fundamental given that all societies have to a greater or lesser extent recognized and practiced the difference between a time and a place for work, centered on acquiring life's necessities, and a time and place devoted to thanksgiving, to celebration and the festive revelation of the self and the other.

We will later return to this subject for a closer examination of these two fundamental attitudes. For the moment, however, we want to focus our attention on the mysterious *transition* that occurs when we move from a workaday relationship to the world to one that is essentially a festive one.

To study this transition in some detail, we will make use of a thought experiment in which we follow the thoughts and feelings and behavior of a fictional person as he moves from a *workaday* attitude, in which he seeks to understand and mentally appropriate the natural world, to a *festive* attitude, in which he seeks a personal manifestation of an other.

We should realize that each of these fundamental attitudes toward our world has its own inherent possibilities and limits so that neither should be thought of entirely in isola-

[1]I have a particular affection for words such as *man* and *mankind,* which express so well what we as human beings are, or at least ought to be. The word "mankind" is made up of two words, "man" and "kind." The word "kind" relates to "kin" and thus to the entire complex of "race," "species," "family," "relatives," and the like. Thus, the word "man," as used in this context, has nothing to do with a biological masculinity or femininity, but refers to all human beings as a whole insofar as they are differentiated from inanimate things by their ability to think. We use a different form of the same word in German as *der Mensch* or in Dutch as *de mens,* always to refer to human beings without regard to their particular sex. Latin gives us the related form *mens, mentis* to speak of the mind. Our concepts of the "mind" and our ideas of what is "mental" spring from the same source that gives us our generic "man" and "mankind." What these words exclude is not femininity but *mindlessness.* To exclude oneself from "mankind" can only mean that one excludes oneself from the realm of thought and of spirit.

tion from the other. No human culture can prosper, or indeed survive, without some practical and intellectual understanding of the laws that govern the natural world. But neither can a human society last for long without learning to cultivate the festive dimension and finding effective ways in which to frame a mutual revelation of what is self and other, of what is human, natural, and divine being.

All functional cultures possess an understanding of what it means to work and to appropriate natural resources. They all possess also, to one degree or another, an understanding and a practice of festive disclosure in which they witness in celebration the generous, uncoerced appearance of self and other.

A THOUGHT EXPERIMENT

In seeking to explore the mysterious shift from a workaday to a festive attitude, we will want to know not only what sets these two fundamental attitudes apart or in what respects they differ, but also how these attitudes and the worlds they create stand in relationship to one another, and how together they form a whole. The world of everyday, mundane tasks and the world of the festive each has its own integrity, its own essential conception of what is real and important, its own sense of what is fitting and unfitting, of what it is right to do and of what should be avoided. Moreover, each of these worlds is surrounded by a horizon that announces the imminent arrival of the other world. While we participate in a festive gathering, we remain aware that we will have to return to the mundane world of work, and while we work, we comfort ourselves with the prospect of festive revelations. Only severe pathology could limit us merely to one perspective and deprive us of a counterbalacing perspective of the other.

Let us now turn to a concrete example in which we can study in some detail the experience of a person at the very moment when he turns from the world of work in which he had been pursuing his natural scientific interests and begins to shift to a festive attitude in which he seeks a mutually revealing encounter with an other.

Let us imagine a geologist on a scientific expedition in a very remote and uninhabited region of the world. Let us imagine him just after he has climbed the last mountain range to arrive at his planned destination, which is a small plateau overlooking a vast expanse of barren and uninhabited wasteland. We assume that he has come to study a important geological feature of this particular landscape that he had discovered while studying a satellite photograph of the area. At the moment when we begin to take an interest in his work, he already has behind him a very long journey that began with a flight to the capital of the remote country in which the wasteland is situated and then continued from there, first by car, then by camel, and finally by foot.

At the point in time when we begin to follow his adventures, he is very near complete physical exhaustion, and we see him struggle to hoist himself atop a large boulder from where he can look out upon most of the surrounding landscape. What he sees before him is an enormous expanse of barren sand and rocks that shows nowhere a sign of animal or even vegetative life. Such variety as is offered by the landscape is that of countless boulders of all shapes and sizes that lie strewn over the desert landscape as far as the eye can see. He finds himself thus completely alone in a world that appears actively hostile to every form of life.

In the course of his already long and distinguished career, our geologist has become used to barren landscapes, and he feels buoyed by the thought of having reached his final destination after having triumphed over so much adversity. He takes a little food and drink from his diminishing supply and then pulls out his notebook to begin to sketch out the physical features of the terrain.

As he surveys his surroundings, the geologist notices ancient traces of what once, several millennia ago, must have been a forceful stream running down the mountains and crossing in a wide sweep the entire length of the valley below. The river not only carved a still visible path from the mountain down through the valley, but also left behind a trail of variously sized boulders with rounded and smoothed features that testify to the shaping power of sand and water.

The geologist observes other, rougher stones that do not bear the mark of flowing water and that must have reached their present location by other means. He traces the path of their descent back to the eroding sides of the mountain, and he can read from their weathered surfaces the corrosive impact of rapid temperature changes combined with that of water, wind, and sand. Here and there, the combined action of these natural forces has shaped some of the stones into truly fantastic forms. Sometimes these strangely shaped stones appear in clusters, some leaning against each other as if embracing or fighting, others heaped together in bizarre formations that defy description. If an ordinary citizen were suddenly confronted with this sight, he might guess that he was seeing an intergalactic sculpture garden.

Our geologist is not given to such reveries, and the thought of sculpture is furthest from his mind. He treats the appearance and the precise location of each stone as a kind of material archive, containing the record of all the natural forces that have left their imprint on the landscape since it first came into being. When he sees the rounded form of a boulder in the desert, he thinks of the forces that brought it there, about those that broke it loose from the mountain and brought it to its present location, all the while scraping away its edges and exposing it to a new array of natural forces. He sees the stone as broken loose from its base, as carried away from where it was, as smoothed and rounded by water, wind, and sand. He reads the presence of natural forces from what is *missing* from the stone, from the fact that it no longer forms part of the mountain, that it no longer appears as jagged and huge, as it was when it first broke loose, but as smaller and rounder. The presence of the natural forces is thus entirely identifiable with what the stone is no longer, *with what is missing* from it.

Sculpture proceeds on a very different principle. The sculptor also removes parts of a block of stone, and in this his labor superficially resembles the natural action of heat and cold, of water, sand, and wind in all their naturally occurring combinations. Yet there remains a profound and unbridgeable difference between the two kinds of actions. The sculptor removes material from the surface of a stone not in a completely accidental way, not as determined by a pure interplay of physical forces, but as a means of *revealing a personal presence*. What is missing stands thus in an active relationship to what is thereby revealed.

It was Michelangelo who pointed out that the sculptor working on a stone can be understood as gradually revealing a presence, meaningful to him personally, that in retrospect can be thought of as having been imprisoned in the stone. In this way, sculpting reveals itself as essentially different from the simple and mechanical process of erosion, which in retrospect reveals nothing except the simple fact of its own operation. Unlike such simple natural action, sculpting can be understood as knocking on the door of an already inhabited stone in order to bring the inhabitant out of his hiding and into the light of day.

The physical act of artful carving, understood as the revealing of a subjective presence, cannot take place within the Euclidean space of natural science, but requires an intersubjective space in which it becomes possible for one subjectivity to reveal itself to an other. Artful carving and sculpting thus resembles speaking and conversing closer than it does the process of erosion. It is as erroneous to look upon the essential purpose of sculpting as an action that removes parts of stones as it is to regard the purpose of speech as that of moving our lips or that of forcing air outward through our oral or nasal passages. Both speaking and sculpting serve the same ultimate purpose of cultivating and revealing natural, human, and divine subjectivity.

As a natural scientist, our geologist would certainly shrug his shoulders at the suggestion that the weathered boulders before him could be seen as art objects. This is not because the geologist is hard of heart, or lacks artistic sensibilities or cultural refinement, but because his scientific training and his current mission place him in a culturally constructed, naturalistic, and workaday frame of mind that for the moment screens out all aesthetic and subjective experiences as being irrelevant to his present field of vision.

It is not that the geologist is incapable of the latter, but rather that he has learned to set these experiences aside for the moment in order to pursue his scientific work. His very training as a scientist permits him to make carefully cultivated distinctions between a workaday natural scientific landscape and a festive interpersonal one. By dint of much effort, he has learned to perceive a physical world solely in terms of natural forces, and the scientific task he seeks to accomplish is one of reconstructing the natural history of a given geological formation. This *natural history* is principally and methodically distinguished from ordinary *human history* by the fact that it scrupulously avoids every and all reference to either a natural, a human, or a divine subjectivity. This methodic principle that calls for the exclusion of subjectivity forms the very basis of natural science, and to transgress it means to step outside its perimeter and to forsake its explanatory power.

It is important to understand, however, that the geologist practices this methodic exclusion of subjectivity only as long as he is engaged in practicing his discipline. When he returns from his labors in this distant part of the world, he will be able to embrace his family and friends and respond with gladness and affection to their heartfelt welcome. Moreover, even here in this deserted outpost of the world and in the midst of his labors, he would easily step outside the naturalistic frame of his discipline if he heard a child cry out or saw a person approaching him, or if he were distracted from his work by an awe-inspiring sunset. While absorbed in his work, he thinks about a world of material forces, but at the fringe of his consciousness he remains ever ready to welcome a stranger, or to say a prayer, or to suddenly experience a landscape as radiant or majestic, or to think of the surging mass of a mountain as expressing grandeur. Only madness could condemn a person to become imprisoned in the landscape of geology or in the body of biology. Our essential humanity resides in the fact that we can shift perspectives from the realities of work to those of the festive, and from those of impersonal forces to those of subjective manifestations.

For the moment, our protagonist is obliged to practice the methodic exclusion of subjectivity that is demanded by his scientific discipline. He thus seeks to reconstruct the natural history of the landscape without in any way making reference to a human or a divine person. This methodic exclusion constrains him to write this history without at any time making use of personal pronouns. The scientific account of his natural observations therefore cannot make use of such attributions as "He created" or "She managed" or "They did."

This methodic exclusion places him, grammatically speaking, in an awkward position, since the usual construction of ordinary sentences in European languages generally demands that a verb be linked to a subject or to a subject and an object. In order to overcome this grammatical difficulty, our geologist takes recourse to a subterfuge by using the *impersonal* pronoun *it*. His account will therefore necessarily take the form of *"It* rained," *"It* fell," *"It* formed," *"It* froze," *"It* melted."

We should note here as an aside that the particular perspective of the natural sciences, which excludes all direct reference to subjectivity and which gives us access to the strange world in which "it" is the sole source of action, was developed first in the millennia-old discipline of astronomy. This perspective slowly developed starting from the day when the first eager amateur looked up at the night sky and began to take note of the orderly changes he observed in this realm.

In astronomy, the observer faced a realm that was at once clearly visible and observable, but at the same time also entirely beyond the reach of human beings. The astronomical observer studied a world that could become as familiar to him as the back of his own hand, but that for all its familiarity would nevertheless remain *inaccessible* to him.

Natural science always presents us with a natural world that we learn to observe and get to know in great detail, but from which we nevertheless remain substantially excluded. We may know the world of physics and chemistry, but we shall never be able to fully make it our own. Nobody can bathe himself in H_2O, nobody can dine on chemical compounds or make love to a biological organism. The world in which my friend Mike becomes a biological entity and where a cup of tea becomes a chemical compound is like the starry sky of the night: We can see it and study it, it may induce us to make marvelous discoveries that in turn have extraordinary implications for our life on earth, but we can never *inhabit* it.

It was the first systematic knowledge of the starry sky that gave us the calendar and that made it possible for mariners to trace a path across the seas. The great scientific revolution of the modern world took place when we learned to view our familiar earth with eyes that had been trained to observe the sky and with a mind formed by the study of the unhabitable regions of astronomy. We learned to see the earth as though it were itself an uninhabited planet, and we learned to set the course of our daily lives guided by the rapidly increasing knowledge derived from the study of this new planet. But then we began to confuse the planet that we studied from the perspective of astronomy with the familiar world that we must inhabit. The era of modernity is profoundly marked by this confusion, and we will not be able to leave it behind us until we have reestablished the primacy of the intersubjective world of host and guest. It is only in such a world, founded in hospitality, that we can humanly live and die, love and pray. And it is only in such a world that we can ever be truly at home.

It is interesting to note how Freud introduced the natural scientific world into the innermost recesses of our being with his second theory of the psychic apparatus. He spoke here of *das Es* (the "it") as an unconscious psychic region governed entirely by impersonal material forces. His English translator, perhaps thinking that to speak of the unconscious in such a simple term as the "it" might detract from the dignity and the professional image of psychoanalysis, rendered it in latinized form as the *Id*. Freud wrote that his use of the term "it" to describe the unconscious had been inspired by his reading of Georg Groddeck's (1987) book bearing the German title *Das Buch vom Es,* which appeared in an English translation as *The Book of the It* (Groddeck, 1961). Freud (1975) continued to use the plain German term *das Es* till the end of his life (p. 292). He defined that term as "an impersonal psychic entity that unavoidably forms part of our psychic make-up" (Freud, 1964, pp. 72–75).

Here is not the place to discuss in detail Freud's understanding of what he referred to as the "psychic apparatus," or that part of it that he understood as the domain of the "it," except to point out that the fundamental attitude within which he analyzed the human mind and soul had been developed long before his time by Babylonian, Egyptian, and Greek astronomers scanning the starry dome of the night. By adopting their attitude and applying it to the study of the "psychic apparatus," he transformed the starry heavens of the early astronomers into the subterranean, dark, and impersonal realm of the human soul.

But let us return to our geologist in his lonely desert outpost, where he continues to contemplate the land of "it" of the early astronomers in the manner in which he has learned to project it upon an earthly landscape. Let us assume that there comes for him a moment when, wearied from his calculations, he allows his eyes to wander a bit aimlessly over the barren surroundings. Let us suppose that his glance is drawn toward some particular feature that does not seem to fit in with the rest of the geological landscape and that from that moment on keeps attracting his attention as a source of difference within the vast expanse of sameness. The place to which his attention is drawn and from which this difference emanates is a strangely regular, geometrically formed rectangular clearing that is bordered on all sides by small round boulders of about equal size. On one side of the rectangle, the geologist notices, is a five-foot-high, pyramid-shaped mound of smaller rocks that appears to form part of the composition.

It is difficult to say which of the many differences from the rest of the landscape was the one that made it stand out. Was it the regular shape of the small pyramid or the rectangular form of the clearing, or was it perhaps the rhythmic pattern of the border, made up of nearly equal-size round stones carefully laid out in such perfectly straight lines?

No matter how hard he tries, the geologist cannot find a satisfactory place for this particular configuration within the world of geology. At the end of all his attempts to find a natural scientific explanation for the phenomenon, he still remains faced by a stubborn and unresolvable difference that refuses to disappear in the geological landscape of "it."

To say that this particular formation does not fit in the landscape of natural science is not to say that this part of the landscape, on which he is now focused, is not subject to the identical natural forces that govern the rest. The sun shines on these symmetrically placed stones as it does on all the others. What sets it apart from the surrounding world of "It stands," "It falls," "It rains," "It shines" is the fact that it insistently evokes a purposeful and expressive world of "He built," "She stood," "We live," and "We die." What the geologist perceives at the outer limit of his neutral, natural scientific landscape is the upsurge of a very differently constructed *inhabited* world in which it is possible to express a *judgment* and to incorporate a *decision.* After spending several hours in the uninhabited regions of the world of natural sciences, the geologist, like his predecessor the astronomer, finds himself suddenly called back to earth by the presence of another human being. Where only a moment ago he was lost in the rarefied world of planetary systems, he is suddenly confronted by a human presence that calls him back to a very different world that demands of him a radically different manner of approach. His humanity does not reside in his ability to use one approach or the other, but rather in his capacity to make the shift from one fundamental perspective to the other. It is precisely madness and loss of our humanity that imprisons us in one attitude or the other and that prevents us from making the shift back and forth between the world of work and the world of celebration.

The first two letters of the word *judgment* refer us to the Latin *jus,* meaning "law" or "what is right"; the second two refer to the Latin *dicere,* meaning "saying," "telling," "informing" (Klein, 1971, [see under *Judgement*]). A judgment speaks to us of a subjective presence capable of differentiating between what is right and what is wrong and gifted with the means to make this difference known to us.

The word *decision* refers us to the Latin verb *caedere,* which means "to strike down," "to cut down," "to beat," or "to cut off." To encounter someone's decision means to enter the arena of fateful human action, to the place where the Gordian knot is cut, where choices are made, where one path is pursued instead of another. This arena of fateful choice, of judging, and of choosing to do one thing rather than another does not fit inside a geological landscape that is entirely constructed on the principle of impersonal forces. "It" can rain, but "it" cannot be made to decide, to judge, or to choose!

This *odd* appearance in the *even* landscape draws the geologist out of the neutral landscape of his academic discipline. He closes his notebook, climbs down from his perch, and begins to walk in the direction of the mysterious rock formation. His entire outlook on the world is now transformed. The same landscape that only a moment ago was, from the perspective of the geologist, a field of resisting natural forces holding the promise of some day becoming entirely transparent to natural scientific reason now has transformed to become *the domain of an other.* This metamorphosis of the landscape takes place at the exact moment when the geologist begins to suspect that the little mound of boulders and the small rectangular clearing before him is not a natural but an artificial formation and that in all likelihood he is facing a human grave.

From this fateful moment on, the geologist begins to assume a very different emotional and intellectual stance toward his surrounding world. His thinking of even just a moment ago had been formed by notions of physical causality and by an exclusive logic of natural forces and material interchanges. Just a moment ago, he still sojourned in the country of "it." But now that he has come face-to-face with a *monument,* his thinking and feeling enter a very different register and become restructured along the primordial fault line that divides the "self" from the "other."

He now enters an ethical realm of "right" and "wrong," a sexual domain of "he" and "she," and a generational domain of "older," "contemporary," and "younger," of the "unborn," the "living," and "the dead." He rediscovers motive and desire in relationship to an other, he enters an aesthetic world of beauty and ugliness, and a religious world of what is permissible and not permissible to do, of what is sacred and what is profane.

In more concrete terms, his thoughts now turn to the possible identity of the person buried beneath these carefully arranged stones. Was he, like the geologist himself, perhaps also an explorer, and did he meet here with an accident or perhaps fall ill? (The geologist's imagination follows here the path of *literature,* and specifically that of the heroic adventure novel.)

Or was he perhaps an aboriginal hunter confused by a sandstorm? (His imagination now turns to *anthropology* and to questions about possible early inhabitants of what is now a desert.)

Perhaps the buried person was once a ruler of an ancient city that prospered long ago on the banks of the now extinct river? (He now turns to *archeology* and the *history* of succeeding civilizations.)

What became of this ancient people and what is now left of their civilization? (He remembers a haunting painting by Caspar Friedrich of a lone man in a dark 19th-century overcoat standing atop a mountain overlooking an empty landscape of rocks and clouds. Here he summons his experiences of *art* and *poetry.*)

Could it really be true that at one time children played along the fertile banks of a river where today one sees nothing but stone and sand? Is this how all civilizations end? (His sensibilities now draw nearer to questions concerning the meaning of life and evoke his experiences with *philosophy* and *religion.* He hears a fleeting fragment of a melancholy *poem* by Goethe about a frightened maiden consoled by death, and he hears strains of the melody Schubert created while setting the poem to *music.*)

We see thus how the geologist, in facing the monument, has returned to the inhabited world of intersubjectivity. His mind has now turned away from the scientific calculations that kept him tied to the uninhabited universe of the astronomer's sky, and he has returned to the inhabited world in which he has learned to love and cultivate the humanities and the arts. The world of geology and the land of "it" has not disappeared from his consciousness; it still forms a horizon around his present consciousness of a festive world in which it is possible to encounter another human being. But he now sees the grave site and the landscape surrounding it, not with the eyes of astronomy, but with sight ready to discover a painting and with ears attuned to the human voice and to music.

This different way of seeing the world does not alienate him from concrete worldly reality, but it does makes him experience it in a different way. He now has entered the intersubjective and festive world of host and guest. He is no longer the lone astronomer studying the sky with all the world around him sunk in sleep, but is now the guest walking up to the house of his host in eagerness to make his acquaintance. He temporarily has left behind the natural scientific struggle with inert obstacles and is now eager to enter the world of the humanities in which it is possible to have a conversation. His reflections on the fate of the unknown person in the grave inevitably provoke in him thoughts about his own precarious situation, about his own dwindling supplies, about his own exhaustion, about the dangers he still must face on the way back home, perhaps even about his own mortality.

His melancholic thoughts about the fate of civilizations, aroused by the presence of the stranger, make him reexamine the meaning of his current mission and even make him briefly question his utter devotion to his discipline. His present awareness of this grave and of this death, even if later it would prove to have been based on an error in perception, constitutes at this time his awareness of an *other.* And it is this *other* who has opened his eyes and his heart to the intersubjective world of the humanities, of the arts and religion. It is this *other* who has opened to him a world of hospitality and conversation. It is the presence of the other that has opened to him the portals of reflection that led him to meditate on the meaning of life and the fate of civilizations.

It was thus the evocation of a *personal presence,* incarnate in a artifact, that at first disturbed the geologist's ongoing natural scientific preoccupation. It was this awareness of a personal presence amid the stones of geology that transported him from the neutral geological world of "It rains," "It falls," and "It happens" to another world in which it was possible to think of mortality, of truth, of beauty, of good and evil, of motivation, of life and death, and the fate and ultimate purpose of what we do or fail to do. His entire train of thought, which drew sustenance from philosophy, theology, art, history, architecture, literature, and music, was set in motion the very instant he felt himself called upon by someone.

If, instead of discovering a grave, the geologist had suddenly heard a child cry out, or had he been surprised by the song of angels, the result would not have been different. In all these instances he would have been called away from his geological preoccupations in order to confront an other.

Note here how the contemplation of physical nature introduces us to a generic, a-historic time of "it," prior to the emergence of a "he" and a "she," a "self" and an "other," while the contemplation of a monument, even though it ostensibly belongs to the past, introduces us to a temporal order in which we are invited to a relationship of mutual revelation of self and other. *A monument addresses us.* It demands of us a personal acknowledgment; it opens to us a space that is hospitable to conversation; it invites us to enter into a dialogue with the past. We are thus *present* to a grave in ways that we can never be present to the rocks of geology, to the forces of physics, or to the substances of chemistry.

THE WORLD OF THE BARRIER AND THE WORLD OF THE THRESHOLD

We have seen thus far how the awareness of the presence of another subject takes the geologist from the neutral and indifferent world of natural scientific pursuits and transports him to a very differently structured world of hospitality and dialogue in which it becomes possible for him to encounter his own and the other's subjectivity. His situation differs in no essential respect from that of someone who in the midst of his daily labors is surprised by a telephone call or by a knock on the door. Such a person interrupts his absorption in the world of work, where he is in the habit of removing one obstacle after the other, and enters a very differently structured world that leads him to a threshold and fills him with the hope of a hospitable and personal encounter.

As we observe the geologist walking toward the grave, we notice that he no longer moves and acts in the forthright and businesslike manner of the experienced field geologist. He no longer is surveying a geological terrain in the manner of the astronomer scanning the skies, but is now on his way to meet a stranger. This shift in perspective introduces a certain note of reticence in his manner and a certain hesitation in his footsteps. His entire body now moves in a way that testifies to his mental and physical understanding that he is approaching a *threshold* that leads to the mysterious domain of an *other.*

When he reaches the grave site, he kneels down beside the monument to inspect it at close range. His demeanor is now respectful, his movements and expressions more tentative and ready to respond to the presence of an other. As he begins to examine the conical headstone, he is very careful not to place his foot upon the rectangular clearing that forms part of the grave. And as he bends down to inspect the individual stones that form part of the monument, he no longer studies them in the manner of a geologist who is primarily interested in their physical and chemical composition; rather, he studies them as an anthropologist might study a mask or a historian an ancient manuscript. Within this new world of festive encounters with others, he now approaches these stones as markers or as letters that spell out and symbolize facets of a *personal* identity and a *social* history. He now finds himself in the position of someone calling upon a stranger. We expect such a person to be alert and to behave so as to open himself up to possible surprises.

We might think here of an insurance man, for example, as he makes his way along the path through flower beds to the front door of the house of a prospective client. We assume that this is his first call, and we imagine him thus as eagerly looking around for clues that may help him frame the right approach to his new client. He certainly will take note of the size, the stature, and the state of upkeep of the house and flower beds. He may be glad to discover a prosperous bed of roses, because he himself cultivates roses, and the topic of rose cultivation may serve as a bond between him and his client. If he notices a child's bike on the lawn and is greeted by a cocker spaniel wagging his tail, he will gauge his sales pitch differently than he would if he were to come upon a growling pit bull fiercely tugging on a chain. All these details inform the salesman about the identity of his client and help him find a proper manner to approach him.

Within the given frame of mind, the salesman is not likely to inspect the flower beds as would a biologist or a professional gardener. Nor would he be likely to scrutinize the structural details of the house as might an architect or a real estate salesman. He would more likely take notice of all those details of maintenance and construction that would afford him insight into the tastes, preferences, and habits of those he is about to meet.

What the visitor seeks in the flower beds, the shrubbery, the garage, and the steps leading up to the front door are the unknown faces of those he seeks to get to know. He seeks not the house or the car or the garden. He seeks not the instrument, but the user of that instrument, his state of mind, his character. This search for a face, for an identity, for a state of mind has little in common with the research of the technologist or the natural scientist, but rather resembles the activities of the humanist historian, the literary scholar, the religionist, and the humanist–psychologist.

As the geologist cautiously inspects the grave, all material objects that fall under his scrutiny awaken from their slumber within a neutral, material world of natural forces and transform themselves into a kind of material *adjectives* that add concreteness to what is at first only a blank portrait of a stranger, Within this new attitude in which he approaches the threshold, all objects that draw his attention reveal themselves as *belonging to an other*. These objects now point to an other subject and begin to reveal to him that other subject's world.

The geologist has stepped from one attitude into another, and the world around him stands now revealed in a very different light. It is as though the objects that he now encounters have undergone a sudden and miraculous transformation from indifferent natural things to symbols capable of describing a human and a cultural reality. This transformation or transubstantiation takes place at the very moment when the geologist steps outside the uninhabited world of the natural sciences and enters the inhabited world of *personal encounters*.

This moment of transformation thus marks the transition from the cultural workaday sphere of the natural sciences to the festive sphere of the humanities, the arts, and religious practices. It also marks the shift in perspective from a naturalistic psychology that dissolves the human presence into the world of "it" to a humanistic and descriptive psychology that belongs to the world of the humanities, the arts, and religious practices.

The progressive world of the natural sciences is one of relentless advances in which we move from discovery to discovery in an unending search for intellectual and material mastery of the natural world. Entering this world is like beginning a long march on a road where every step on the way demands the removal of a physical or mental *barrier* and where each new breakthrough, as soon as it is achieved, brings into view some new obstacles on the way to an ultimate mastery of nature. Each barrier incarnates a facet of the natural world's resistance to human dwelling, each manifests its essential indifference to human needs and desires, and the removal of each barrier represents a weakening of this resistance and a further step on the road of the natural scientific and technological conquest of nature.

Both the world of work and the world of the festive offer us an ultimate prospect in which all our needs and desires would at last be met and put to rest. The ultimate dream of the workaday world is that of an absolute appropriation of nature in which all of natural reality would have been made subordinate to our will. The ultimate dream of the realm of the festive is that of an absolute revelation of self and other in which all our love and desire for the other would be stilled and in which our passion to give recognizable form to our experience would have been not just calmed, but silenced. Our desires unrestrained would thus lead us in the direction either of a total appropriation of the natural world or of a total revelation of self and other. In the end, we would be forced to make the impossible choice between *being* and *having*. Clearly, our desire to gain mastery over nature must be tempered with the desire for festive manifestation, and our desire to gain material possession of a natural landscape must be tempered by our desire for a festive revelation of that landscape. Equally, our desire to control and dominate the other and to use him as an instrument in the service of our workaday projects must be counterbalanced by a festive desire to witness his free and spontaneous self-manifestation.

The world revealed to us within the attitude of mental or physical appropriation and the world revealed to us in the festive revelation of self and other resemble each other insofar as both are structured by a *difference* between what is self and what is other than the self. Neither world can maintain itself in the absence of the other, so that both are characterized by an internal and an external difference. Another way of saying this is that both the world of work and the world of the festive can be understood as two different ways of cultivating the difference between what is self and what is other.

Within the world of everyday work, of technology and natural science, each task and each problem presents itself as an obstacle to progress, as a *barrier* against which we pit the strength of our bodies and the agility of our minds in order to overcome and remove it. But the festive world of revealing encounter is structured, not by natural *barriers* that must be opposed and removed, but by *thresholds* that must be left in place and that demand to be respected. The world of the festive demands that we regard the threshold as an inviolable limit before which we bring to a halt all progress of a workaday, technological world and where we do no more than announce our presence and await the manifestation of the other.

The path of scientific inquiry into the natural world is that of Columbus sailing uncharted waters to the unknown corners of the earth. To progress along that path means to do battle with an endless succession of obstacles in which each obstacle represents a measure of our ignorance and inexperience with the world of natural forces. Within this workaday world, we progress by relying on resources of body and mind and by cultivating the virtues of courage and steadfastness that make us persist in the face of an indifferent natural world.

But the festive world of inquiry in which we seek a personal revelation of self and other sets us on a path that leads us to altars, portals, doorways, and monuments. Our inquiry here begins with accepting the first law of the threshold, which forbids us to make use of force or trickery to gain our way and places on us the demand to arrest our workaday progress, to make known our presence, and to await the appearance of the other. This path of inquiry explores the festive self-manifestation of the world of host and guest in which we live and die and feel at home.

THE ALTAR AS THRESHOLD IN HESIOD'S *THEOGONY*

Hesiod's (1959) *Theogony* tells the story of Prometheus and details for us the important role he played in the emancipation of mankind from an original state of dependency and confusion to that of an independent people inhabiting their own domain. The myth tells us that in a very distant past, human beings lived among the gods without any awareness of their own identity as mortal human beings and without perceiving the essential differences that set them apart from the gods. For Hesiod, the mystery of human origins is not that of the material creation of a particular type of creature and does not concern even a process of making or fabricating. The creative act whereby humankind comes into being takes here the form of an acknowledgment and subsequent celebration of what is initially a painful truth. Within this vision, mankind came into being the very day it began to realize its own distinct nature and began to properly orient itself in respect to the other forms of being. The birth of humankind came thus in the form of a discovery of the difference between self and other, and about the place that human beings should properly occupy within both the natural and the supranatural world.

Ontological distinctions are distinctions made on the level of being. Thus, we think of the difference between divine and human being, or of that between human beings and animals, as ontological distinctions. It appears that in the *Theogony,* the birth of mortal human beings takes the form of their discovery of their own ontological difference from the immortal gods. We find in this text nothing concerning the physical making of mortals, nothing that would make us believe that human beings had from the start been set apart from the gods. The story of the creation of mankind should logically recount two different stages, the first of which does not appear in this story. That first stage should tell about the coming into being of human beings and the creation of a difference in the universe of divine beings. The second stage, which forms part of the Prometheus story as it is told here, concerns mankind's discovery and final acceptance of that fatal and glorious difference. This discovery and acceptance comes to light under the tutelage of Prometheus, the great emancipator and benefactor of humankind. It is for this reason that the story of the birth of mortals is told here in the form of a story concerning Prometheus.

Hesiod (1959, verses 535 and 536, p. 155) tells the story of the discovery of the fateful difference in a very succinct way:

> It was when gods and mortal men
> took their separate positions at Mekone.

The story gives us no specific details about what it was that caused the separation or how the discovery of man's separate nature came about. All we know is that it was made and that the best way to understand man's emancipation is to follow the story of Prometheus, who incarnates the human spirit and its movement toward independence.

What sets the story of mankind and Prometheus in motion is thus not a quarrel, not an accidental misunderstanding, but the intellectual and spiritual discovery of an ontological difference. When Zeus was told about mankind's decision to set up separate households at some distance from the gods, his initial reaction seemed to have been favorable. He ordered that a great feast be prepared to celebrate the impending separation so that mortals and immortals would be able to enjoy a last supper together and embrace each other for a last time. He charged Prometheus with slaughtering a magnificent bull and with preparing two portions, one to be offered to the gods and the other to the mortals.

The choice of Prometheus for this task shows the central role he played in the drama of human emancipation. Throughout the complex weavings and windings of the story, he plays the role of the benefactor of mankind, which he serves with his ability to anticipate the future—his name derives from the Greek *prometheia,* meaning "forethought," "foresight," and "caution"—and with his extraordinary ability to make appropriate distinctions.

After killing and butchering the bull, Prometheus proceeded to divide the portions, and he did so in such a way that mankind appeared to gain the better part. But however we may interpret the division of the portions, it appears clear that it was made to reflect the different natures of mortals and immortals. After Mekone, it was still possible for gods and men to eat together, but it no longer was suitable for them to eat the very same food.

Prometheus took the massive bones of the slaughtered animal, craftily covered them with layers of fat, and placed this offering before Zeus as the portion destined for the gods. He then hid the meat inside the unsightly stomach, covered it with the entrails, and placed it next to the other portion. Zeus declared himself satisfied with the seemingly more desirable portion for the gods, and his doing so raises the question whether he was truly taken in by Prometheus' deception.

If we read the story with the understanding that a quarrel had led to the fateful separation, then we must assume that Zeus was unaware of the clever trick being played on him, and this hardly seems plausible. But if we approach the story as a myth concerning human emancipation, then we see Zeus' apparent gullibility in a very different light. The deception practiced by Prometheus would then represent a first step in the emancipation of mankind, and Zeus' apparent failure to notice it should be read as his quiet, unstated approval of the growing independence of the human spirit. The deception could then be seen as the end point of a period during which mortals had remained morally and intellectually transparent to the gods.

Something very similar can be observed in the developing relationship between young children and their parents. There comes a time when the child begins to outgrow a relationship of complete transparency to the parents and begins to move in the direction of a more autonomous life in which it becomes possible to hide something from the parents and to keep a secret. The child thereby accepts a first fruitful distance from the parents, and this distance, if it is properly respected, lays the foundation upon which all further stages of emancipation are built.

Yet, this first distance and this fateful first step in the direction of independence invariably give rise to conflicting feelings in parents and child alike, feelings that must be borne with patience, like the pain of birth itself. Few parents would fail to recognize themselves in this portrait of a half-irritated and half-pleased Zeus, as he sees mortal beings take their first fledgling steps in the direction of independence. In any case, if we place Zeus' silent suffering of Prometheus' tricks within the larger context of the emancipation of mankind, the whole story begins to make sense.

There is another good reason to read the myth in this manner. All the events of which the myth makes mention occurred in a place named Mekone. Some have tried to link this name to that of an ancient town in Corinth, but such realist interpretations do not advance our understanding of the myth (Lamberton, 1988, p. 98). A more promising approach would be to read the place name as a hidden revelation of a central aspect of the myth. The Greek noun *mekoon* translates as "poppy" or "head of the poppy," or, in botanical terms, *Papaver somniferum* (Liddell & Scott, 1966 [see under *Meekoon*]). As such, it refers to the realm of sleep and dreams and to a preconscious realm prior to conscious human existence. To escape from that realm thus constitutes a first step on the way to full humanization and self-possession.

It is perhaps useful here to recall that Aristotle (1967, p. 587) used the word *meconium* almost the same way we still do in modern English, namely, to refer to either the juice of the poppy or the fecal discharge of newborn infants. If we read these ancient meanings back into the place name, we begin to understand that the site of the last banquet with the gods was also, at the same time, the true birthplace of the human spirit.

It was thus at Mekone that the human spirit, guided and symbolized by Prometheus, came to recognize its essential difference from the gods. Characteristic of that recognition was mankind's adopting for itself the name *mortals* and learning to speak of the gods as the *immortals*. It thus seems that human confusion disappeared and that sensible action became possible when human beings began to recognize their nature as essentially fragile, mortal, and limited. This would imply, at the same time, that the condition of confusion and lack of sense would return whenever hubris overwhelmed common sense and decency, and blur human understanding of the lines of distinction between mortals and immortals. A corollary of this proposition would be that the soundness and the clarity of the human mind would in some important measure depend on the human ability and willingness to assume a proper stance in regard to the gods and, by extension, in respect to fellow human beings, other living creatures, and, finally, to natural and artificial objects. A fully human rationality developed beyond the stage symbolized by Mekone would thus be based on the lived recognition of the threshold that both separates mankind from and unites us to the gods.

But let us return to the scene of the farewell dinner and observe what next took place. Most likely, the two groups made complimentary speeches to each other following the dinner and, most likely, sang poems and made merry and then embraced each other in a last farewell. The mortals then gathered themselves under the leadership of Prometheus and set out for their new homeland.

The gods remained on Mount Olympus, and perhaps watched the departing mortals begin their long journey from Mekone to the distant and as yet unsettled human world. Let us remind ourselves that the road traveled by the mortals opened in only one direction, so that mortals could never retrace their steps. One is reminded in this context of the title of Thomas Wolfe's (1940) great novel, *You Can't Go Home Again.* Only Prometheus was able, now and then, to return to the abode of the gods. Yet this inability of ordinary mortals to physically return to their origins did not, for all that, make them forget or scorn their origins, and one of the first things they did after reaching their new homeland was to build an altar.

In building that altar, they created a threshold between the realm of the gods and that of mortal men and thereby established a definition of the human condition that both separated it from and linked it to the domain of the immortal gods. In this way, they situated themselves in a meaningful way within a larger cosmos.

Hesiod (1959, verses 556 and 557, p. 156) writes:

> Ever since that time the race of mortal men
> on earth have burned
> the white bones to the immortals
> on the smoking altars.

Greek ritual sacrifice required that a victim be killed and slaughtered in a carefully prescribed fashion. Much as Prometheus had done in preparation for the last supper with the gods, the priest officiating at a sacrifice would burn the bones and the fat of the victim on the altar as a gift to the gods while he offered communion to the celebrants in the form of roasted meat.

The sacrificial ritual would thus be, first of all, a commemorative event in which mortals would remember the time when they had lived without distinction among the immortal gods. At the same time, they would also remember the painful day of their separation at Mekone, together with the proud day when they built an altar to delineate their own domain. The sacrifice thus evoked not only nostalgic feelings about a lost closeness to the gods, but also pride and confidence in a newly established position within the whole of the cosmos. It was in final instance this establishment of a domain of their own that enabled mortals to host a meal and to make a sacrifice to the gods.

In Aeschylus' (1976) *Prometheus Bound,* verses 444–458, we find a passage in which Prometheus describes the condition of humanity prior to the exodus from Mekone, and this portrayal is anything but flattering. He describes that condition in the following words:

> In those days they had eyes, but sight was meaningless;
> Heard sounds, but could not listen; all their length of life
> They passed like shapes of dreams, confused and purposeless.
> Of brick-built, sun-warmed houses, or of carpentry,
> They had no notion; lived in holes, like swarms of ants,
> Or deep in sunless caverns; knew no certain way
> To mark off winter, or flowery spring, or fruitful summer;
> Their every act was without knowledge, till I came.
> I taught them to determine when stars rise or set—
> A difficult art. Number, the primary science, I
> Invented for them, and how to set down words in writing.

The Greek ritual of sacrifice thus offered an opportunity to think back upon a time of great intimacy with the gods, while at the same time rejoicing in the blessing of living in an intelligible world in which it was possible to make moral, aesthetic, and intellectual distinctions. It was above all a time to celebrate the gift of a circumscribed identity, of a place and a time of one's own that, in turn, would make it possible to extend hospitality to mortals and imortals. It was all these gifts that together provided a sense of direction to what otherwise would have been a merely scattered, episodic life.

As noted before, it was Prometheus who led mankind out of its state of primordial confusion and who guided humanity in the direction of a more autonomous and separate existence at some distance from the gods. It was again Prometheus who presided over the original division of the portions and who gave a lasting form to the institution of ritual sacrifice. He introduced distance and difference into the life of the human race, and the Greek altar remains a lasting monument to his spirit.

In Aeschylus (1976), we read how he taught humans to distinguish themselves from the gods and how, once that distinction was properly made, he succeeded in teaching them the difference between the seasons, between the constellations of the stars, and between one natural or artificial sign and another. His ultimate claim (verse 506) is that he laid the foundation for

> All human skills and science. . . .

We might ask ourselves what relationship we might be able to discern between this abundance of Promethean gifts and the establishment of the first altar. To begin to understand this relationship, we must first realize that the altar, understood as a first threshold between mortals and immortals, placed human beings for the first time *in a proper relationship* to the gods. The essential Promethean gift did not take the form of a particular thing or a particular circumscribed ability. It was not something added to the bill of particulars of a human life; rather, it consisted in a fundamental reorientation of human existence that placed it in a right perspective, first, in regard to the gods and, second, in regard to each other and the natural world. Prometheus thus did not endow human beings with intelligence—Hesiod and Aeschylus are both clear on this point—but he placed them in a relationship to their surrounding world in such a way that their intelligence could become properly engaged and bear fruit.

What we can say about human intelligence applies equally to the human senses, to the worlds of sight and sound. Just like human intelligence, the senses of sight, of taste, of touch, of smell,and of hearing were never absent from the human condition. But these senses could make their rich contributions to a human world only after mortals had found a right perspective and a proper relationship to the immortals and, by derivation, had found the proper stance from which to approach themselves, others, and their surrounding natural world. This manner of understanding the birth of mankind and the essential gift of Prometheus as that of an emancipating orientation brings us back to the central issue of the altar, understood as the archetype of all thresholds.

The Greek rite of sacrifice can be understood as a journey back in time and space to the outer limits of the human realm marked by an altar that points to the adjoining realms of mankind and the gods. The journey thus brought the celebrants into the presence of a marker that pointed to a difference that was understood to be the very source of all subsequently discovered differences. The ritual repeated the journey of the ancestors that had ended in the founding of a separate human realm and the establishment of the first altar. At the same time, the journey commemorated the last supper humans and mortals had enjoyed in each other's immediate company. The ritual thus revived the memory of an old relationship of confusion in which mortals still lived in ignorance of their own mortal nature. But it also offered grounds for pride in a new alliance, in which mortals proved capable of making and bearing distinctions, of holding and keeping apart, while at the same time being able to build bridges, to forge meaningful connections. In final instance, the ritual spoke of the dignity of the human condition as deriving from a dual, and only seemingly contradictory, movement. Mortals stood their own ground at some respectful distance from the gods, but with unceasing efforts to create and maintain meaningful links between the two realms. Between the two realms there stood as symbol, of both their separate status and their alliance, the altar founded by Prometheus. Such an altar may be understood as the very archetype of all thresholds.

The rite of sacrifice can thus be seen to *symbolize* the entire emancipatory myth that tells the story of the birth of mankind. We use the word "symbol" here in the literal and ancient meaning of *a bringing together a host and a guest within a hospitable realm*. The word *symbolon* originally formed part of the vocabulary of ancient Greek customs of hospitality (Liddell & Scott, 1966). If two strangers befriended each other while on a journey away from home, they would upon parting break a piece of pottery or a coin in two pieces and have each guard a half as a permanent token of their friendship. This token could then later serve as a sign by which to recognize either each other or each other's descendants. The *symbolon* thus always signified at the same time the actuality of a separation and the promise of a hospitable return. But note that the promise contained in the two broken pieces of pottery, like that contained in the altar, referred to a hospitable meeting and not the cancelling of the original breach.

The rite of sacrifice made visible the primordial separation of mortals and immortals through the dramatic act of killing and slaughtering the victim. This separation was then further emphasized by burning the bones and eating the meat. But the same rite also made visible a new union growing out of this separation, and that union was symbolized by the altar, understood here as a threshold that opened up a realm of hospitality.

We are thus thinking of the Greek ritual of sacrifice as an opportunity for the celebrants to revisit and reaffirm those fundamental and fateful choices that they experienced as having constituted their humanity. The sacrificial rite offered them a dramatic and reflective space in which it became possible to reexperience the birth of humanity as an entering into a new relationship to self, to others, to other forms of being, and to the cosmos.

The dramatic and reflective space in which this revisiting of the ontological past became possible was and remains typically associated with hospitable thresholds. Hospitality in its fundamental form is always a meeting at the threshold, where it opens a particular type of conversation that begins with this question: "Who are you? Please declare and identify yourself, please manifest your being, your truth, your nature!" There are obviously many levels on which this question can be asked and answered, but the meeting at the threshold cannot avoid this question. As guests, we walk up to the threshold of the host to announce who we are, and in response the host comes to the door to manifest his presence and to offer the hospitable space in which we can further speak and give form to our world, and visit the perspectives offered by our host.

By guiding mankind to take its proper distance from the gods and by inducing them to establish their own domain, symbolized by the establishment of an altar, Prometheus brought mankind to the fundamental discovery of hospitality. By establishing a hospitable threshold between their own domain and that of the gods, human beings at the same time established a hospitable enclosure within which all aspects of the experienced world could make their uncoerced appearance. The first ritual sacrifice was thus also the first true encounter, not only between mortals and immortals, who now stood sufficiently apart to come into the presence of each other, but also between neighbor and neighbor, between mankind and beast, and between mankind and nature. The altar thus made possible a first comprehensive outlook upon self and other and upon the world as a whole, We may think of it as a kind of theater upon the stage of which the diverse aspects of the human world could move to the fore and make themselves known to a waiting audience.

A human world perceived from the festive perspective of a hospitable threshold differs essentially from the same world revealed in the hand-to-hand combat with nature that characterizes the profane and workaday world. A merely vital contact between the hunter or the trapper and his quarry provided him with the means to still his hunger and to prolong his biological life. A purely utilitarian, profane, and workaday contact with animals would provide information relevant to the hunt, or to the raising of livestock, but only the festive and hospitable enclave of the cult site could bring the early hunters into the uncoerced presence of reindeer, boar, and bison.

The festive threshold provides a fullness of access, both to the self and to the other, in a way that a profane breakdown of barriers can never attain. At the end of our complete mastery of the other, we find a compliant or evasive slave. But the invocation of the threshold and performance of the rites of hospitality transforms both self and an other into inexhaustible and mysterious sources of vital interest. Profane and practical life provides us with tantalizing glimpses of the human world, but only the realm of the festive and the sacred can create for us the hospitable enclosures from which we can attain coherent insights into our human condition.

We saw how Prometheus described mankind's early state of confusion prior to the original separation from the gods. He portrayed them as effectively blind and deaf, as bereft of common sense and lacking in any understanding. In that early condition of confusion between self and other, there was as yet no altar, no distance, no difference, and therefore no hospitable means to bridge distance and difference.

We thus learn from the tale of Prometheus that the acceptance of a festive difference and distance within the spirit of hospitality provides us with the only means we have to a full disclosure of self, of other, and of our world. Without that vital and humanizing ingredient, all existing differences, all clarifying and productive distances between things and beings begin to disintegrate; all actions, all manner of objects, all distinctions begin to lose their contours, begin to invade each other, begin to fuse with one another, thereby draining the world of sense and purpose. Only a relationship of hospitality, based on the sacredness of the threshold, opens us to a meaningful perspective upon the world so that all manner of objects and beings can appear in their full and true dimensions. It is thus only within the embrace of hospitality that anything at all can emerge as fully visible, fully tangible, audible, and sensible. It is only within a pact of hospitality that the world can be properly queried and understood and that it can fully show itself for what it is.

The threshold is a place where differences are hospitably received and acknowledged in such a way that they can be placed into a relationship to one another. It is here that they find mutual reconciliation and emancipation within the wide embrace of hospitality. The threshold should be seen as essentially the entryway to a place of *festive disclosure* that bids all those who are gathered within its embrace to manifest themselves and to endeavor to come fully into each other's presence. Seen in this way, only the threshold holds out for us the prospect of a fully human world.

THE THRESHOLD AND THE STUDY OF RELIGIOUS LIFE

As we have seen thus far, thresholds constitute the ultimate borders of the human domain and indicate the very limits beyond which it is no longer possible to forge ahead with the otherwise powerful instruments of progress that the world of work puts at our disposal. These ultimate borders demand from us a response different from that elicited by barriers and obstacles. They demand that, for the moment, we lay down our weapons, abandon our instruments of progress, abstain from strategies, forsake all designs for mental or physical conquest, and await instead the manifestation of an other as an *other*. Where we encounter the threshold, we turn from our ordinary daily concerns, we suspend the struggle for life in order to assume a festal or religious attitude and await, whether in hope and joy or in fear and trembling, the one who is *other* and irreducible to ourselves.

We may think of the *other* whom we await at the threshold as the one who completes us as an individual person, as our "better half" who is our wife or husband, or who, as a child or grandchild, completes us as a parent or grandparent, or who, as a neighbor or close friend, completes our social being. But it is also possible to think of the other as the *other* of humanity as a whole, as a god or venerated ancestor whom we await in a festal attitude at our communal border. Such an *other* may also take the form of another biological species, of a bison, an ibex, or a boar, such as we see depicted on the cave walls of the paleolithic sanctuaries of Lascaux and Altamira. We may think of such a cave wall as a threshold or as the communal border of the tribe, and may imagine the painter as an inspired priest welcoming the mysterious appearance of the other-as-animal from beyond the threshold.

There is a clear difference between seeking to approach an animal in the context of a hunt and welcoming the appearance of that same animal by means of the inspired action of song or dance or mime, of shadow play or theater, or via a dramatic retelling of an ancient myth. The animal we conquer and appropriate in the hunt is the animal we meet across obstacles. The mysterious animal that appears to us in dreams and visions, in dance and in pictorial representations, is an appearance of which we ask nothing more than that it show itself fully and clearly. It is the abundance of that appearance that renders us thoughtful and perhaps more grateful than we could be for any quarry. The animal we catch in the hunt feeds the hunger of the belly, but the animal that shows itself in response to our dancing it, to our singing or painting it, feeds the hunger of the soul. We pursue the animal of the belly past obstacles, we await the animal of the soul near the thresholds where the domain of human beings ends and that of animals begins. We hunt animals that live "in the wild," in field and forest. But we paint, dance, and sing animals as creatures that dwell with us on earth like neighbors.

On the walls of Lascaux, the animals appear, not as prey, not as something to be stealthily approached and conquered, not as a source of physical danger, not as a mere visual or auditory trace apprehended amid the excitement of the hunt, but as *others* that step completely out of hiding and openly approach us to meet us at a sacred border. These paintings thus show the *other* side of the life of the hunter—not when he is in active pursuit of his quarry, but when he turns from the world of obstacles and contemplates his world from the site of the sacred border and threshold. He experiences the animals that make their appearance here as manifestations of an *other* form of life that complements his own, that forms a whole with it. He sees these now as a necessary complement to human life, a life that without that complement would be fatefully altered and impoverished.

But the realm of animals, or that of nature, is not the only *other* that completes our human life. Our life is inevitably touched and in part defined by the dead and by the gods who inhabit the realm at the other side of the threshold. To be human means to be inevitably bordered and defined by other creatures and other beings. We are inescapably neighbors to both natural and supernatural beings.

We have stressed the fact that the border between self and other must ultimately take the form of a threshold. Between neighbors and neighbors, we find the threshold that marks the end of the inhabited domain of the one and announces the beginning of the domain of the other. It is respect for and obedience to this threshold that makes possible the relationship of one neighbor to another. To remove that threshold from our world would at the same time remove from our lives the very possibility of being someone's neighbor.

Between mankind and the gods, there stands the threshold in the form of an altar, of a holy place, of a place of worship. To desecrate or to remove that threshold would mean not only to despoil the human world of its treasured religions, but also to rob humanity of an important means of self-definition and self-understanding.

Between the living and the dead, there stands the threshold in the form of the grave and the funeral monument. To rob mankind of that threshold would transform the dead into mere refuse and our memories of them into empty and idle chatter.

A monument is literally "something that reminds us," and as such it is the cornerstone of any civilization. A society bent on removing all thresholds and replacing them with barriers would by that act not only transform the gods into empty illusions, but also undermine the very foundations of the countless great works they inspired. Such a society would deprive itself not only of religion, of art and music, but also of all other forms of festive celebration. It would reduce the memories of our dead to ashes, deface all monuments, and transform home and hearth into mere shelters against rain and wind. In such a world of desecrated thresholds, love and friendship could no longer thrive, and in their place we would find only tasks to be completed, cravings to be assuaged, demands to be satisfied, procreation to be taken care of.

Such a self-destructive society, hostile to all thresholds, could endure for only a short time in the form of an unruly mob or a totalitarian state. In either instance, the marginal human life that it could offer would be joyless, and at the same time, nasty, brutish, and short. Our humanity depends on our ability both to work and to celebrate, to inhabit both quotidian and festive reality, both to be able to overcome obstacles and to obey thresholds, both to transform and master natural reality and to await the appearance of the *other.*

A natural science is inherently an instrument designed to remove barriers to our understanding and to overcome obstacles to our full use of natural reality. It is in the nature of natural science, as it is in the nature of our daily tasks, to overcome obstacles and to shape a natural environment in ways that fit our natural needs. But it is equally necessary to surround the task-oriented perspective of our daily life, together with that of the natural sciences, with a very different, festive perspective that enables us to recognize thresholds and to evoke the manifestation of an *other.*

Generally, our world is structured in such a way that the perspective within which we perform our ordinary, daily, technical or problem-solving tasks is always already surrounded by a festive perspective that awaits the end of our labors and that is eager to open us up to another world in which we celebrate the manifestation of the self as the self and the other as the other. All work days are surrounded by the prospect of coming feast days, just as all obstacles and barriers are surrounded by thresholds.

To translate this understanding to the field of psychology means for us to recognize the needs for both a natural scientifically oriented psychology, capable of removing obstacles from the world of work, and a humanities-inspired psychology of the festive, devoted to the study of the establishment, the care, and the maintenance of intersubjective thresholds. The task of such a humanistic psychology would be to contribute to our understanding of all practices, whether ancient or recent, whether indigenous or foreign, that invoke the sacred distance of the threshold while evoking the appearance of the *other.* Such practices include, besides prayer, meditation, and the remembrance of the dead, the craft of writing and the hermeneutical task of meditative reading. They include the arts of painting, sculpting, and drawing, together with those of pantomime and theater. They include singing and dancing and all forms of making music. Each of these practices places us before a door to which we have no key and that can be opened only from the other side and by an *other.*

REFERENCES

Aeschylus (1976). *Prometheus bound.* Baltimore: Penguin Books.
Aristotle (1967). *Historia animalia.* In *The Works of Aristotle: Volume IV* (p. 587). Oxford: Clarendon Press.
Freud, S. (1964). *New introductory lectures on psychoanalysis.* New York; W. W. Norton.
Freud, S. (1975). *Psychologie des Unbewusten: Studienausgabe: Band III.* Frankfurt am Main: Fischer Verlag.
Groddeck, G. (1961). *The book of the It: A revealing theory of Eros.* New York: New American Library,
Groddeck, G. (1987). *Das Buch vom Es: Psychoanalytische Briefe an eine Freundin.* Frankfurt am Main: Fischer Taschenbuch Verlag.
Hesiod (1959). *Theogony.* In *Hesiod* (pp. 155–159). Ann Arbor: University of Michigan Press.
Klein, E. (1971). *Klein's comprehensive etymological dictionary of the English language.* Amsterdam: Elsevier.
Lamberton, R. (1988). *Hesiod.* New Haven, CT: Yale University Press.
Liddell, G. L., & Scott, R. (1966). *Greek–English lexicon.* Oxford: Clarendon Press.
Wolfe, T. C. (1940). *You can't go home again.* New York: Random House.

II

Existential Dimensions

Each of the following seven chapters reports an existential–phenomenological investigation of an experience of particular interest to its respective author(s). The topics range from more everyday experiences (Chapters 5–7), to issues of a clinical and psychotherapeutic nature (Chapters 8–10), to a study that explores an experience that, although explicitly existential, has implicit transpersonal implications (Chapter 11).

Chapter 5, by Constance T. Fischer, presents a descriptive structure of being angry. Using four different examples as illustrations, Fischer focuses on being angry as "self-deceptive protest," a finding that challenges our cultural belief that anger is a natural force that we can only withhold or discharge. In this context, she then calls us to a new level of self-empowerment and personal responsibility.

Chapter 6, by Damian S. Vallelonga, offers a formulation of the lived structure of being-ashamed, a central result of his empirical–phenomenological study of "being-embarrassed and being-ashamed-of-oneself." Focusing on what he calls the "persisting ambiguity of shame," the author uses his findings to help us distinguish being-ashamed from related phenomena such as embarrassment, guilt, anxiety, shyness, and modesty.

Chapter 7, by Tania Shertock, takes us out of our own culture into the world of Latin American women. Shertock is especially interested in the experience of self-empowered women who succeed in a cultural setting in which men are far more likely to experience and manifest their power, namely, the experience of feeling able to move toward and accomplish a meaningful and challenging goal. Her discussion emphasizes the perspective that successful people perceive obstacles differently from unsuccessful people (regardless of gender), and that the women in this study evidenced an integration of the feminine and masculine principles, each within herself.

Chapter 8, by Scott D. Churchill, presents a phenomenological investigation of clinical impression formation, focusing on the psychologist's own experience within the diagnostic interview. Churchill offers a general structural description of psychodiagnostic seeing, providing a new and deeper understanding of how the psychologist is involved as a perceiving agent in the diagnostic task.

Chapter 9, by Faith A. Robinson, provides a rare look into the realm of dissociative experience. Designed to reveal the essence of nonsuicidal self-cutting behavior among persons with dissociative (multiple personality) disorder, her results indicate the complexity of the experience, including the unexpected dimension of ritual abuse. These findings not only deepen our personal understanding of this experience, but also serve to increase the sensitivity of professional psychotherapists who work with dissociative disorder clients.

Chapter 10, by Jan O. Rowe and Steen Halling, discusses the psychology of forgiveness in the context of the researchers' own phenomenological findings. Two dimensions of forgiveness, forgiving another and forgiving oneself, are presented. The authors then address the implications of their research for the practice of psychotherapy, especially in regard to being with clients with "deep-seated hurt."

Chapter 11, by Kathleen Mulrenin, presents research that examines what happens in women when they experience difficulty in praying to a masculine image of the divine. She concludes that this experience plays an important part in a woman's developing awareness of the sociocultural significance of gender, and discusses its impact on the psychospiritual development of women.

5

Being Angry Revealed as Self-Deceptive Protest

An Empirical Phenomenological Analysis

Constance T. Fischer

This chapter begins with four examples of being angry, and then characterizes an empirical phenomenological research method through which I developed a descriptive structure of being angry. A discussion of some personal and societal implications focuses on several aspects of being angry, namely, its being a self-deceptive, self-righteous protest against being demeaned and blocked in being who one is trying to be. I chose to emphasize these particular features of the overall structure in order to counter our culture's promotion of anger as being a natural force that we can only contain, dissipate, release, or discharge.

The chapter then returns to the beginning examples to discuss the power of the revised understanding of being angry, both for enhancing personal options and responsibility and for enhancing societal openness to people who are different from us. I remind us that language forms as well as expresses our thought and culture, and I suggest a relanguaging of anger.

INSTANCES OF BEING ANGRY

Frank

This account concerning Frank, who is about 40, is from my research notes.

As we enter the elevator, Frank, a fellow tenant in my office building, grumps back "Yeah" in response to my "Good morning!" To my quizzical look, he responds, "Up all

Constance T. Fischer • Department of Psychology, Duquesne University, Pittsburgh, Pennsylvania 15282.

Phenomenological Inquiry in Psychology: Existential and Transpersonal Dimensions, edited by Ron Valle. Plenum Press, New York, 1998.

night working on reports." We fall back in silence, looking at the elevator walls. He reads the taped announcement that the building will be closed for the Fourth of July and sighs, then scowls. As I exit at the next floor, he suddenly tears the announcement off the wall, crumples it, and tosses it to the floor, exclaiming, "Damn it, they can't keep me out—I need the computer to finish!"

Later in the day when I encounter him at the lobby snack bar, I say that it occurs to me that he can enter the building with his garage key even though the public doors will be locked. He says self-consciously, "Yeah, after I yelled at the [building management' s] secretary, she had Joan explain to me how it works. Actually, it will be great—air conditioning stays on, and no interruptions."

Susan

This account concerning Susan, who is in her late 20s, is from a handwritten interview.

It is a balmy day. Susan drives with the windows down, enjoying the wind flowing through her hair, smiling in pleased recollection of friends' recent comments about her agility and confidence in maneuvering through the traffic circles of Washington, DC. She finds herself obligingly leaning toward the passenger window as another driver gestures for her attention. He yells, "I'd like to suck your pussy!" He leers and shouts elaborations as Susan tries to maneuver into another lane.

She struggles with the passenger-side window handle, trying to roll the window up while steering the car. She opts to otherwise behaviorally ignore the "idiot." She finds that she is gripping the steering wheel tightly and that her heart is pounding. Her face feels flushed. Susan misses her exit off the circle as she concentrates on evading the goddamned bastard's car. She strikes the dashboard with her fist, swearing and muttering to herself about bashing him with a sledge hammer, slashing his genitals, etc. For more than an hour, she remains enraged, swearing and entertaining graphic counterviolence.

Several years later, while recounting the incident to the researcher, Susan remarks that she finds herself enraged again while recalling the event. She reports that demeaning sexual abuse by males during her girlhood was immediately evoked by the car incident and by recalling that incident.

Connie

This account is my own, from my late 50s.

I've just written a check on an emergency account to cover last month's office expenses. I was already behind in billings and in making requests to managed health care companies for additional psychotherapy sessions for my patients. Then back surgery forced a month's absence from my practice. I have just impatiently but laboriously filled out a four-page form asking for such unknowables as the date by which Ms X will no longer be depressed, along with an itemization of my short-term goals, methods for each, and anticipated dates of accomplishment.

Phrases often spoken by colleagues and myself collect silently like a fog or threatening downpour: "not what I earned my clinical degree for," "can't afford to maintain my private practice," "invasion of clients' privacy," "clerks judging my patients' needs."

Next in my foot-high stack of forms, I find a notice from another company that if I do not reply within 30 days, payment will be forfeited. The checkmark on the list of possible deficiencies indicates that the service code was missing from the receipts my client submitted. I sigh and fill in "90844" with ditto marks for each date. I wish I could argue with a person instead of helplessly complying with anonymous MHC demands. I find that I'm breathing deeply as though preparing to lift a barbell. Then I sit up straight, look up for a moment, and scrawl in capital letters across the bottom of the form: "PLEASE NOTE THAT THE RECEIPT STATED THAT THE SERVICE WAS 'INDIVIDUAL PSYCHOTHERAPY—50 MINS,' FOR WHICH THERE IS ONLY ONE CODE: 908441!" Now I feel rather foolish, knowing full well that I could have saved myself some of this sort of trouble if I had my receipts reprinted with service codes. Another patient's managed health company requires that I use its manual for coding level of interpersonal relationships, grooming, etc., all of this billable at only half my usual rate. I fight an image of forcefully shoving all this paperwork into the wastebasket. I'm clenching my teeth and breathing heavily. I smile abashedly when I recognize that I am experiencing the phenomenon that I've been researching.

John

This account concerning John, who is in his early 30s, is abstracted from descriptions presented over the course of several psychotherapy sessions.

John had felt reasonably comfortable during his interview with his Faculty Advisory Committee as they reviewed his progress through his Ph.D. requirements. After all, he was one of very few who passed all areas of the comprehensive exams at once. And he had obtained interviews at a couple of prestigious internship sites. And all of this was despite three years of being undermined by Professor Smith, who repeatedly had pointed out in classes that John was a "stolid, upwardly mobile fellow" who didn't think dynamically. John had endured Professor Smith's in-office "therapeutic" probings into his "rigidity," and had managed not to complain to the Division Chair about these intrusions.

Dr. Jones remarks that John's only grade below A– is a C from Professor Smith. Another professor observes that John's QPA is clearly high, and that the C is anomalous. John decides that the better part of valor is to say nothing; he nods. But when Dr. Jones says that John looks angry, even though John hadn't thought he was, he suddenly finds his throat to be tight and his eyes squinty. Feeling tricked by the committee, John blurts out that the C was an irregular grade but certainly was not anomalous, and that "that poor excuse for a professional, that asshole" had decided in advance to show him who had power, and he [John] darn well was not going to let him get away with it any more.

Weeks later, one of the committee members told John that they had been amazed by his outburst, that earlier they had thought that he might share a mature view of what had been going on between Professor Smith and several dissatisfied students.

Discussion of the Instances

These instances are intended to serve several purposes. First, they illustrate the sort of anger that this chapter addresses, namely, the angry outburst and the restrained thrust toward outburst. Second, along with situations that you may have recalled from your own life, the four examples will serve as concrete touch points for later discussion of the nature and consequences of anger. Finally, the instances will illustrate one form of results of empirical phenomenological analysis. They are what I call postresearch representations. That is, after having used participants' reports of these instances to work my way to an overall general phenomenological structure of being angry, I went back and selected material from each report to illustrate what turned out to be essential aspects of the general structure. In other words, knowing what I know now, but drawing only on each person's provided data, I crafted the instances as illustrations. These representations evoke and imply more than they say explicitly.

WHY AN EMPIRICAL PHENOMENOLOGICAL STUDY OF BEING ANGRY?

I undertook this study to explore what anger might look like in the absence of reductive theories. Now I can offer an alternative, a corrective, to the prevailing popular and professional conceptions of anger, a corrective that I hope will both encourage personal responsibility and discourage negative totalization of other people. I also intend this chapter to be an exemplar of the usefulness of an empirical phenomenological approach to research.

The prevailing (Western) popular notion of anger is that it is an internal thing with a life of its own, a force that inexorably grows and, if not given vent, either will erupt from its own pent-up power or will debilitate the body that houses it. Ulcers, heart attacks, strokes, depleted energy, and more are often attributed to blocked anger. Would-be helpers tell us to "let it out," "express your anger," "channel your anger," and so on. Both children and adults say, "I couldn't help it—I was angry."

Scientific psychology perpetuates this view despite alternative presentations such as Tavris's (1989) *Anger: The Misunderstood Emotion,* and despite the replacement of psychoanalytical drive and instinct theory by interpersonal dynamic orientations. The methods and philosophical approach of psychology conceived as a natural science primarily produce laboratory studies that map out the neurophysiological details of the flight or fight response to threat (e.g., Lang, 1995). Although important in its own right, that research has pretty much neglected the person who perceived the threat while going about his or her life. Both popular and scientific discussions of anger overlook possibilities of a person taking different meanings from being aroused, meanings that allow for choice.

Phenomenology addresses how human "consciousness" forms what we understand of the world. It is *the study of* ("ology") *what appears to us* ("phenomena," [as opposed to "noumena"—things in themselves]). In this context, phenomenology as a philosophical foundation for psychological inquiry is distinct from the North American use of the term, in which phenomenology refers simply to taking experience seriously, or in medical contexts to identifying similar patterns of symptoms with diverse etiology. Instead, phenomenology, as a philosophy of being and knowledge, grounds the empirical psychological study of meanings as they occur in and across situations (such as Frank's, Susan's, Connie's, and John's construction of their particular situations).

At a philosophical level, phenomenology addresses how all human endeavors such as science and theory building are constructions. That is, our body of knowledge is necessarily based in our living and in the approaches we develop to make sense of our world—as scientists, and as individuals. In empirical phenomenological investigation, we respect what appears to us (experience/perception), and undertake to describe that, without looking for external or reductive explanations. We stay at the level of the human order, even while acknowledging its being rooted in physical and physiological orders, as we ask *"What is* this phenomenon?" We attempt to form an answer holistically in terms of how persons take up and move through the particular situation. We can then reflect on how individuals might use these understandings to initiate shifts—for example, to bypass or move out of a state of being angry. Similarly, we can use a fairly pretheoretical descriptive structure of *what* anger is, to reflect on how different theoretical and research contributions address various aspects of the fuller phenomenon.

METHOD OF ANALYSIS

Many research procedures are possible within an empirical phenomenological approach. In this study, I varied procedures developed by the faculty and graduate students of the Psychology Department of Duquesne University over the past 30 years (e.g., Fischer & Wertz, 1979; Giorgi, 1985). Our work has been fundamentally empirical in that our data are first person descriptions of events as lived by that person; they are not second-order data such as measurements or categories. Nor are they only introspective on one hand or theoretically generated on the other.

I began my study of being angry years ago by asking five master's degree students to write responses to the following request: "Please describe a situation in which you became angry. Tell me what was going on before that, how you became angry, and what happened after that. Please write in enough detail so that I will know what it was like at the time." I typed the accounts myself and reread them many times. Then I marked with back-slashes each segment in an account that seemed to say something about the situation that the participants had not said before and that would be necessary to the coherence of the story. For example, I marked "I had been promised that we would go to Cheyenne," but dropped out "Tom—that's my cousin's middle child." I also inserted condensations above some segments; for example: "Now it must have been Wednesday the 23rd" became "3 days later."

Then, using the marked and condensed segments for each of the five accounts, I composed a temporally organized, compact description (something like the opening instances in this chapter). I was able to ask four of the five students to correct any distortions that I had unknowingly introduced. Mostly, they were surprised at how powerful the "summaries" were. With this more manageable material in hand, I asked each sentence of each account what it was telling me, both directly and implicitly, about the process of becoming and being angry, always in terms of how the person was relating to self, others, and world, and to past, present, and future. I sketched an outline of resulting themes, which over the years I have systematically corrected and expanded in light of further examples of interviews, observed and reported incidents, events described during psychotherapy, newspaper accounts, and other sources. The themes were both phenomenal (what it was like as told directly to me, e.g., feeling thwarted, blocked) and phenomenological (what else I could see about how the person was participating in and shaping the situation, e.g., the blocking implicitly being a blocking of one's way of being a particular kind of person). Gradually, through ongoing revisions, I composed the structure presented below, which is intended to evoke for readers what I came to be in touch with across all instances of becoming and being angry.

The initial disciplined work with written protocols has always served me well, keeping me from generalizing too quickly and from sweeping over variations. Considering nonprotocol incidents of anger helps me to see what my formal participants did not address explicitly. Over time, empirical instances have revealed and corrected many of my prior assumptions, refined my descriptions, and given me confidence that I am ready to share my representation of what I have learned so far.

As with all research, the process goes on from there. Readers will affirm much of what I have written from its fit with provided examples and from their own experience; readers also will suggest more apt wording, needed qualifications, and highlighting that did not occur to me. There might be productive disagreements, with empirical examples as the object of discussion.

FINDINGS: A PHENOMENOLOGICAL STRUCTURE OF BEING ANGRY

The following structure is reprinted, with modifications, from Fischer (1996). Note that the structure is indeed structural—no aspect is more essential than any other; all are necessary for a phenomenon to be anger; all the aspects belong to a relational whole. The title of this chapter highlights one constituent—self-deceptive protest—as timely for a particular purpose: changing the way we regard anger and our options when we are aroused. Still, all the aspects of the structure are necessary for an emotion to be that of being angry, and reflection on any of them can serve as a transformative pivot point out of anger. The current version of my evolving comprehension and expression of the structure of being angry is (Fischer, 1996, pp. 74–75):

> In the midst of going about activities, one finds him- or herself thrown back, progress blocked. Initially one tries again and/or quickly estimates availability of alternative means. A sense of being thwarted enlarges and becomes focal against a horizon of vague reminders of past obstructions and nemeses. Up to this point one has been perturbed *at* the disruption.
>
> As one becomes increasingly attuned to the personal importance of the activity—to its centrality for who one sees him- or herself to be, being thwarted becomes construed as being endangered. Whether fleetingly or remarkably, one's body pulls in (intake and holding of breath, tightened abdomen and throat, trunk and head drawn back). Attention likewise is pulled in, scanning less and less beyond the immediate scene.
>
> At a later moment, it may become peripherally or focally apparent to self or others that the personal meaning of being stopped was that of being made to be helpless to continue one's course, and of being demeaned (made to feel unable, shamed, unsure, awkward, unimportant, discounted, dumb, etc.). The blocked course is not just the one of getting out reports, maneuvering through traffic circles, processing forms, getting through an advisory committee meeting, and the like, but is also one of being some special aspect of oneself, as well as being on the way toward being a competent, appreciated, valued person in general. Implicitly one becomes angry *about* being lessened as a person through being derailed from his or her previously taken-for-granted project of being a particular kind of worthy person.
>
> However, the angry protest is directed *toward* the offender, who becomes the center of one's attention. Even when the obstacle is a material thing, one perceives it as intentionally villainous, or as witlessly getting in the way of one's obviously rightful course. The other ought to know better. Resentments at past wrongs become salient, even if subconsciously. One turns away from any glimpses of one's own role in these situations, instead sustaining or escalating the protest by self-righteously recounting the offender's arbitrary, unfair, unjustified intrusion.
>
> Promoting one's own rage, he or she now feels charged, pumped up, explosive, powerful. Nostrils flare, eyes widen; posture stiffens; jaw and hands clench. One glances about, to check on the reality and intransigence of the threat, and to identify any further risk. This pause may allow one to restrain aggressive desire, to de-escalate, to back away, and to regroup, perhaps to reflect.

Or one may give way to outward protests aimed at annihilating the obstacle in a defiant, seemingly sudden and powerful protest that conveys, "You can't do this to me!!" The protest may be verbal (expletives, commands, argument), behavioral (hitting, stomping out, breaking something), or it may be a visibly strained restraint of such impulses.

Following the strained restraint or the explosive protest, one may feel vindicated, and remain proudly and vigilantly ready to reassert the protest. Or, in the aftermath of tiredness he or she may catch clearer glimpses of what the protest was *about* (the meaning for one's sense of self and one's progress), beyond what it was *at* (the incident that blocked one's way), and *toward* whom the protest was felt or expressed. One may then feel abashed, embarrassed, defeated, enlightened, and so on, depending on one's sense of being able to revise the earlier course and still be one's self.

PERSONAL AND SOCIETAL IMPLICATIONS

Implications for Frank, Susan, Connie, and John

We can empathize readily with each research participant's initial dismay and irritation at being blocked. When we look more closely at their becoming angry, however, we can see that far from being liberating, empowering, or healthy, the urge toward angry protest serves as a self-lessening defense against feeling pain and against holding oneself accountable. The protest may fool both oneself and some witnesses, but ironically it leaves one locked into accepting that one has been disabled. The effort toward protective self-deception is self-defeating.

Frank wastes his time and remains helpless to solve his computer-access problem while he remains angry. His self-consciousness later implies that he feels he has lessened himself through his outburst. Susan, the driver, in effect sustains her victimization in that her prolonged protest buys into the continuing power of men to disable and demean her. Moreover, Susan perpetuates a violent personal environment through her own rageful retaliatory imaginings. As long as I (Connie) actively blamed the managed health companies for my being behind in my paperwork, I perpetuated my protested helplessness. Rather than rendering me powerful, my angry protests delayed coming to terms with my ineffectiveness and limitations and with the necessity of changing my office practices. Having finally hired a medical billing firm to help me get organized and to handle the details I no longer have time for, I am amazed at just how stultifying my prolonged self-deceptive protests were. And John's outburst with his clinical committee wound up putting him in the one-down position he resented. His "he can't do this to me" self-deceptive protest kept John down as he emoted to the committee.

Yes, of course, a moment of anger sometimes can "clear the air," allowing one to regain perspective. At a more profound level, experiencing one's own arousal to anger, or witnessing another person's arousal, can put us in touch with our existential condition: We must forever negotiate our life course while resisting, acknowledging, and taking up possibilities and limitations. Being angry is self-deceptive and self-defeating only to the extent that one turns away from what one does not want to acknowledge about one's options and limitations. Even then, sometimes in the drained aftermath of emotional restraint or outburst, one similarly can finally come to terms with one's own part in the situation and can reflect on options. It is unexamined, repetitive, or sustained anger that is problematic.

Contrary to current conventional advice, "letting *it* out" ("it" being a welled-up force) is likely to constrict one's sense of self and one's options. If instead we consider the structure of being angry outlined above, we can use any of its constituents as signs that we are heading into self-deceptive, possibly self-defeating protest. Body tensions, feeling thrown back, inclinations toward powerfully annihilating the obstacle, and revving up both memories of past offenses and feelings of self-righteousness are all particularly useful signs that we might want to find an alternative route, bypassing anger, toward where we were going. The signs of becoming angry are signs to step back and ask oneself several questions: Am I necessarily being rendered helpless? Are others necessarily doing this *to* me? Are there ways to get around the obstacle? Even though it is clear that I am becoming angry *at* the immediate blocking, and I know the target of my objection (the *toward which*), what is my growing protest *about?* How is my sense of myself and my overall journey through life being threatened, and is that necessarily so?

We can guess as to our research participants' disrupted, threatened journeys, "about which" they protested. Frank perhaps was on his way toward always finishing through persistence despite difficulties. Susan was continuing her course of having surpassed repeated childhood abuse and shame and of celebrating other achievements. I now realize that I (Connie) was into, on my own, doing it all (university commitments, clinical services, clerical tasks, daily life) despite surgery. John later described himself as having been up to doing remarkably well, while remaining imperturbable in the face of life's unfairness.

Yes, sometimes we can convert growing protest into positive determination to not be derailed. Sometimes we can decide to go ahead and display anger in order to intimidate potential blockers. To pursue these tactics successfully, one must take advantage of the pauses that occur as being angry evolves, such as being thrown back in the first place, and while scanning for further danger and for possible retaliation. A course toward outburst or even toward restrained outburst is not inexorable.

Frank might have noted the perceived threat to his progress (building to be closed) as announced by his incipient senses of being thwarted and of panic, and immediately inquired of fellow tenants on the elevator as to whether there were ways around the obstacle. In the future, when Susan recognizes that she is about to protect herself via rageful protest, she can instead honestly acknowledge to herself that she hates to be reminded of demeaning abuse, but can remain competent and autonomous by focusing on ways to continue her course. In other situations, she already has avoided denying the hurtfulness of men's sexual power plays, and has chosen relatively calmly either to bypass further interaction or to masterfully deliver a putdown. So, too, Connie and John each learned through reflection on their angry outbursts that they had been denying their situational limitations and had been keeping themselves stuck while self-righteously blaming others; escape from self-deceiving displays of power and protest got them back on reasonably modified courses.

Note that such shifts are personal transformations out of being angry; they are not a "rechanneling" of a force. Similarly, from the perspective of the structure of being angry, the high blood pressure that accompanies long-term angry stances is not due to "blocked anger," but to the person's sustaining an urge toward angry outburst.

Implications for Society

Being angry is a profoundly personal *and* interpersonal phenomenon, inevitably implicating basic moral issues and social consequences. Coming to terms with finding oneself becoming or being angry requires, to one extent or another, facing questions such as: What kind of person am I? Where do I want my life to go from here? Can I continue my own course if I make way for other people? These are morally foundational issues. The moral dimensions of sustained anger become clearer as we consider the impact on the persons toward whom it is aimed. The urge to override, if not to obliterate, those who get in our way obviously does not take their well-being into account. Even if restrained, that urge totalizes other persons into negative, detested "thems." Despite the angry person's feeling unjustly thwarted, it is the angry person who demonizes others and wishes them ill.

Unfortunately, much of American culture encourages and celebrates angry action. Children's "action heroes" and adults' action movies showcase self-righteous retaliation culminating in violent annihilation of the enemy. Just now in the United States, many radio call-in programs encourage angry pronouncements against various "thems"—gays, illegal immigrants, welfare recipients, non-Christians, members of the other political party, and on and on. Emoting, including against others, is provoked by many TV talk shows. These excesses are part of our me-first, feel-good era. We are concerned that others should always be fair to us (me), but we are not concerned with our responsibilities to others. If anger feels good at the moment, it's justified. Worrying about others' socioeconomic and personal well-being does not feel good and is eschewed.

These practices fuel racism, homophobia, and so on, when they aim at such identifiable groups and when those groups are moved to their own responsive anger. Being angry places one in a stance that disallows empathy, broader perspective, a sense of belonging to humanity at large, and being in touch with meanings beyond the immediate situation. Sustained anger leaves one perpetually trying not to recognize that autonomy based on castigation of others is false, empty. Being chronically or repeatedly angry dilutes one's humanity; at a cultural level, it undermines community and care. Repeated or sustained anger demoralizes both its source and its object. Demoralized people do not contribute to a moral society.

The standard literature on prejudice, and the contemporary literatures by feminists, African-Americans, and so on, strike me as validly pointing to violent words and actions as being grounded in perceived threats to how one has seen oneself as being a certain kind of valued person. I recall listening to a man rail against lesbians as not being real women. In the ensuing lengthy table discussion, which to his surprise included input from a bisexual woman, he became somber and asked what many people probably wonder less openly: "But if they can really be romantic with each other *and* satisfy each other so well sexually, doing just about everything a man does, how does a man define himself as a sexual man?" I believe that earnest exploration of such questions may allow this man to continue being a man, with deeper confidence and without anger at women who choose women as their partners. We will empower ourselves and our communities to the extent that we cut through self-deceptive protest and reflect on our own courses, on how they interweave with others' lives, and on how we might maintain our core goals while sometimes taking modified routes.

Languaging Anger

Our technological Western world is quick to *nounize*—my term for referring to processes as things or categories. Examples: learning slowly = being a "retard," or having a learning disability; solving cognitive tasks quickly = having a high IQ, being an intelligent person. Yes, use of nouns to stand for complex processes and contexts can be efficient. For example, identifying a patient as "having a major depression, overlaid on a dependent personality disorder" does immediately orient the involved professionals to probable history, symptoms and their severity, and treatment range. This efficiency, however, comes at the cost of attending to process, meanings, contexts, personal choice, and so on.

When we convert a person's being affected and taking an emotional stand into a construct, such as anger, we don't yet know what we mean in terms of that person's actual life. When we want to describe a particular person, we should change nouns into situated verbs and adverbs, thereby describing the person being on his or her way. For example, in a psychological assessment class, we explored possible meanings (from a student's report) of "he has a lot of anger." From the fuller report and consultation with the author, we generated individualized sentences, including these:

- William has been quick to become angry when he believed he was being held back by stereotypes.
- William is accustomed to blaming others' unfairness when he finds himself blocked. Most often, he has protested vehemently, instead of looking for ways around the obstacle.

From another assessment, we developed the following alternatives to "She has problems expressing her anger":

- She has been unsure of how to influence others to move out of her way.
- Being afraid that others would punish her for being angry with them, Jennifer instead has often blamed herself for being too passive to get what she wanted.
- However, Jennifer has discovered that when her daughter's welfare is at stake she has successfully squelched both her inclination to angrily demand her rights *and* her usual fearful backing off; instead she has politely but firmly stated Natalie's "needs."

Culturally available concepts and language shape experience. Psychologists can make an immediate contribution by encouraging use of language that acknowledges one's personal choices and options, even when angry. We can encourage reflection and responsibility, rather than validate self-deceptive embrace of self-righteous anger. Note that both languaging in terms of action and past tense, and including examples of bypassing angry protest, remind all parties that anger is not an "it" that runs its course through us, and that instead we can change how we cope with being thwarted.

CLARIFICATIONS AND ELABORATIONS

Many authors have already addressed many aspects of this study's representation of being angry. For example, Sartre (1975) has discussed emotion as the effort to magically transform situations; Stevick (1971) has described the perception of unfair blocking; de Rivera (1976) has pointed to the "the other ought to" dimension of emotion. The contribution of a phenomenological structural representation is that it portrays the temporal unfolding of a phenomenon as a whole. The portrayal only implies some of its aspects, but the integrity of the complex whole is respected. From there, we can better appreciate various authors' in-depth explorations of particular aspects of the phenomenon. We know not to pit one account against another, but rather to locate the pertinent moments within the structure.

For example, we can see several appearances of the "fight or flight" response in my portrayal of the structure of being angry. But the structural representation reminds us not to reduce being angry to a material, biological reaction; human meanings and bioneurological processes are co-present even though our language and conceptions separate them.

I do not denigrate the laboratory traditions of natural science. I *am* saying that such an approach, geared toward predicting the "behavior" of nonconscious material, is not suitable for studying uniquely human phenomena. On the other hand, phenomenological psychology cannot study the goings-on of our material bodies. A phenomenologically grounded human science psychology, however, can offer a philosophical framework within which we understand that (1) at this stage in our conceptualizing, we can focus alternately on lived meanings or on neurobiology with the other as background, and (2) even our natural science work is in part a construction and not just a revealing of nature in itself. And yes, of course, qualitative research too, despite its efforts to put aside assumptions, theories, personal background, and so on, also reflects the particular investigators and their interests, sensitivities, culture, and times.

Empirical phenomenological analysis of verbal protocols is not the only qualitative method for investigating how we humans live our worlds. Ethnographic, linguistic, and other methods are often equally, or more, appropriate (see Denzin & Lincoln, 1994). The type of analysis presented in this chapter is most appropriate when we are asking *what* a phenomenon is—how we experience and participate in that phenomenon such that it is *that* phenomenon rather than another.

Let me repeat that although I have emphasized the self-deceptive aspect of being angry, it is no more essential to the structure of being angry than are other discernible aspects. All are necessary for anger to be anger. I have chosen to highlight the self-deceptive constituent of being angry in order to suggest that the self-righteous urge to obliterate the blamed blocker does not confer a right to explode. Similarly, I hope that my emphasis on self-deception will counter the prevailing conception of anger as a force that one can only release, restrain, or rechannel.

I have not studied the many other ways of being angry (e.g., anger at assaults against society [Thomas Aquinas' (1964) "right anger"], chronic anger, temper tantrums, or angry displays intended to intimidate others and to build one's courage). Nor have I reflected sufficiently on the relations to and differences from similar states (such as being irritated, impatient, or frustrated). Nevertheless, I have developed some useful distinctions within the study of emotion, namely, that of finding oneself affected and taking an emotional stance toward that circumstance. It seems to me that, for example, finding oneself blocked, thrown back, is "being affected," which is always in terms of meanings for who one had been trying to be, usually only implicitly. For some people, being blocked is an interesting challenge; for others, it is an unquestioned sign of failure and a signal to quit. For others, who take up being blocked as an unfair, arbitrary threat to the kind of person they were on their way to being, the response is to protest vehemently against finding oneself being made helpless and demeaned. This effort to forcefully remove the offender is an "emotional" stance. Moreover, in my conception, the quiet responses as well as the angry protest are emotional stances toward how one was affected. Note that being irritated or impatient can be both how one finds oneself affected *and* an ensuing stance toward that circumstance. The angry protest is always a stance, never a state of being affected. How one is affected can also be celebrated by an affirmative stance, such as laughter, joyful hugs, or triumphant shouts of victory. In contrast, despair, for example, can be seen as a resigned stance. Mood can be seen as our background sense of how our general course is progressing. Our emotional life is the realm of being affected and taking stances toward that circumstance.

A stance of being angry is not bad in itself. Just as Freud (1936) said that initial anxiety is a signal of danger that should be attended to, so can we regard the urge to protest angrily as a signal to take stock of one's journey and options. Even acting out the protest can sometimes lead to positive transformation of self and situation. It is repeated or sustained, unexamined protests that can be "bad" for the person and for others.

SUMMARY

This chapter presented a temporal, descriptive, integrated representation of being angry. I chose to emphasize the defensively self-distracting, self-deceptive, and often self-defeating aspects of becoming angry, as well as the possibilities for pivoting into responsible optional courses. My purpose was to counter our culturally accepted notions of angry eruption being somehow both justified and biologically necessary. I pointed to the interpersonal character of becoming angry, and to the importance, for individuals and for society, of recognizing that the angry individual feels, whether focally or vaguely, that the kind of person he or she is trying to be is threatened. The display of powerful protest belies the person's feeling of being made helpless and demeaned. Acknowledging this "about which" of anger, which always accompanies precipitating "at which" and the "toward whom" of the protest, opens the way for understanding ourselves and others and for coming to terms with limitations and alternatives.

This chapter illustrated empirical phenomenological research findings, and I hope that thereby it also has illustrated that this research approach is uniquely appropriate for investigating *what a phenomenon is.* The chapter also highlighted some social and personal transformative possibilities of understanding how humans give and take meaning as they participate in "what happens to" them.

REFERENCES

Aquinas, T. (1964). *Summa theologica* (Questions 22–48). New York: McGraw-Hill.

de Rivera, J. (1976). *Field theory as human-science: Contributions of Lewin's Berlin Group.* New York: Gardner Press.

Denzin, N. K., & Lincoln, Y. S. (1994). *Handbook of qualitative research.* Thousand Oaks, CA: Sage.

Fischer, C. T. (1996). A humanistic and human science approach to emotion. In C. Magai & S. H. McFadden (Eds.), *Handbook of emotion, adult development, and aging* (pp. 67–82). Orlando, FL: Academic Press.

Fischer, C. T., & Wertz, F. J. (1979). Empirical phenomenological analyses of being criminally victimized. In A. Giorgi, R. Knowles, & D. L. Smith (Eds.), *Duquesne studies in phenomenological psychology: Volume III* (pp. 135–158). Pittsburgh, PA: Duquesne University Press.

Freud, S. (1936). *The problem of anxiety.* New York: Norton.

Giorgi, A. (Ed.). (1985). *Phenomenology and psychological research.* Pittsburgh, PA: Duquesne University Press.

Lang, P. (1995). The emotion probe: Studies of motivation and attention. *American Psychologist, 50*, 372–385.

Sartre, J.-P. (1975). *The emotions: Outline of a theory.* Secaucus, NJ: Citadel Press.

Stevick, E. L. (1971). An empirical investigation of the experience of anger. In A. Giorgi, W. F. Fischer, & R. von Eckartsberg (Eds.), *Duquesne studies in phenomenological psychology: Volume I* (pp. 132–148). Pittsburgh, PA: Duquesne University Press.

Tavris, C. (1989). *Anger: The misunderstood emotion.* New York: Simon & Schuster.

6

An Empirical–Phenomenological Investigation of Being-Ashamed

Damian S. Vallelonga

This chapter presents a formulation of the essential lived structure of being-ashamed derived from an empirical–phenomenological analysis of the descriptions of various subjects' situated experiences of the phenomenon. The research project and its results, from which this chapter is derived, were originally reported in my doctoral dissertation (Vallelonga, 1986). It has been ten years since that dissertation was completed and defended, and at least 12 books have been written on shame in the interim (Albers, 1995; Bradshaw, 1988; Broucek, 1991; Fossum & Mason, 1986; Harper & Hoopes, 1990; Kaufman, 1989; Lewis, 1992; Middleton-Moz, 1990; Nathanson, 1992; Nichols, 1991; Potter-Efron, 1989; Tangney & Fischer, 1995). Given the publication of so many books (not to mention articles) on the topic, the reader may legitimately ask if there is anything left to say about this phenomenon. There is much to say!

The fact is that what affirmed my original idea to study shame (in its undifferentiated sense) was the discovery in the shame literature of a persistent ambiguity regarding its essential meaning and how it is to be distinguished from cognate phenomena such as embarrassment, guilt, anxiety, shyness, and modesty. While progress has been made in the understanding of shame and the shame-family of phenomena, the situation remains essentially the same today, with widely differing definitions of the meaning of shame and widely divergent explanations of how it differs from its cognates. Furthermore, except for Miller's (1985) study of shame and Lindsay-Hartz's studies of shame and guilt (Lindsay-Hartz,

Damian S. Vallelonga • 201 Sherbourne Road, Syracuse, New York 13224.

Phenomenological Inquiry in Psychology: Existential and Transpersonal Dimensions, edited by Ron Valle. Plenum Press, New York, 1998.

1984; Lindsay-Hartz, DeRivera, & Mascolo, 1995), no author has attempted to articulate the essential structure of the phenomenon from the point of view of the experiencer. In addition, no one has addressed and unfolded the dialectical interplay between transformations of self and world or among the three dimensions of lived past, present, and future in the experience of being-ashamed.

This chapter begins with discussions of (1) the persisting ambiguity concerning the meaning of shame and (2) the relative absence of the agent's perspective on the experience of being-ashamed. It then proceeds, first, to an elucidation of the empirical–phenomenological method I utilized to comprehend the meaning of being-ashamed and, then, to a presentation of the results of that investigation, that is, the essential structure of the situated experience of being-ashamed. Since part of the purpose of my original study was to grasp not just the meaning of being-ashamed, but also how it is related to and differentiated from the other shame phenomena, the chapter then presents a discussion of how being-ashamed differs from being-embarrassed and being-guilty. Finally, at the end of the chapter, I explicate how it is that shame entails such persistent ambiguity, by explaining what I call the "equipotentiality of a shame-situation" and the role of the dialectical dimension in the shame experience.

PERSISTING AMBIGUITY OF SHAME

Multiple Definitions of Shame

As was stated above, a quick glance through the literature reveals a multiplicity of definitions of shame. Many of the differences in these definitions have to do with whether shame involves the presence of another person or not. Most acknowledge that shame, embarrassment, guilt, shyness, and the other shame phenomena entail some form of reference to exposure of the self. Authors differ, however, on whether that exposure is intrapersonal (i.e., to oneself) or interpersonal (i.e., to an other), as well as on several other points. The following discussion illustrates some of the varied definitions of shame.

Shame is variously characterized as: (1) a phenomenon of primarily or exclusively *interpersonal* exposure plus *public* disesteem similar to embarrassment, humiliation, or disgrace (e.g., Albers, 1995 [his "disgrace-shame"]; Broucek, 1991; Lewis, 1971; Schneider, 1977; Straus, 1966); (2) a phenomenon of primarily *intra*personal exposure and self-disesteem, but with some reference to interpersonal exposure (e.g., Fossum & Mason, 1986; Harper & Hoopes, 1990; Kaufman, 1974, 1989; Lindsay-Hartz, 1984; Lindsay-Hartz et al, 1995; Lynd, 1958; Middleton-Moz, 1990; Miller, 1985; Nathanson, 1992; Potter-Efron, 1989); (3) a phenomenon of *either* inter- or intrapersonal exposure or *both* equally (e.g., Edwards, 1976; K. W. Fischer & Tangney, 1995); (4) a form of anxiety (Freud, 1962; Mead, 1950); (5) a phenomenon prohibiting certain sexual acts (e.g., Albers, 1995; Freud, 1953a); (6) an attitude protective of privacy in general (similar to modesty) (e.g., Merleau-Ponty, 1962; Scheler, 1957; Shaver, 1976); (7) an attitude forestalling social disapproval (similar to shyness) (Darwin, 1965; Izard, 1977); and (8) the experience of the exposure of one's own privacy, particularly one's bodily privacy (Freud, 1953b; Merleau-Ponty, 1962; Scheler, 1957; Schneider, 1977).

It was my impression, upon reviewing the literature on "shame," that while authors use the *single* label "shame" to denominate the realm in question, what they are actually describing or discussing is *several related phenomena,* among which are the emotions of being-ashamed, embarrassed, humiliated, and disgraced, as well as the attitudes of modesty, shyness, and tact/discretion. The question that arises from the fact that shame appears to have so many faces is this: What about the phenomenon permits it to appear in so many different ways to different people? This is one of the questions that I posed when I did my investigation of being-ashamed.

Multiple Differentiations of Shame versus Embarrassment and Guilt

In addition to the lack of consensus on the essential meaning of shame, the literature also shows considerable disagreement regarding how shame is differentiated from embarrassment and guilt. The following are some of the discriminations made between embarrassment and shame by various authors. They are said to differ as: (1) a potential versus an actual negative judgment by the other (Freud, 1953b; Sattler, 1965), (2) a part of an affect versus the whole affect (Harper & Hoopes, 1990; Lewis, 1971; Lynd, 1958), (3) an interpersonal versus an intrapersonal exposure (Dann, 1977; Lewis, 1995), (4) a less versus a more intense/severe experience (Buss, 1980; Lewis, 1992, 1995), and (5) a lack of concurrence versus a concurrence with the other's negative judgment (Solomon, 1976). In addition, multiple authors (e.g., Izard, 1977; Schneider, 1977; Stierlin, 1974) declare that there is no difference between embarrassment and shame and thus explicitly equate the two; several other authors implicitly equate shame and embarrassment (e.g., Broucek, 1991; Isenberg, 1949; Riezler, 1943).

As stated above, a similar lack of consensus also exists in the literature regarding how shame is differentiated from guilt. Thus, shame and guilt are distinguished: (1) as interpersonal events versus intrapersonal ones (Emde & Oppenheim, 1995; Morano, 1973), (2) as involving a defect/failure versus a transgression (Friesen, 1979; Kaufman, 1989), (3) as a negative reference to one's self versus a negative reference to one's deed/behavior (e.g., Bradshaw, 1988; Fossum & Mason, 1986; Harper & Hoopes, 1990; Kaufman, 1974; Lindsay-Hartz, 1984; Lindsay-Hartz et al., 1995; Mascolo & Fischer, 1995; Miller, 1985; Nathanson, 1992; Potter-Efron, 1989), (4) as fear of the social group versus a fear of the introjected parents (Mead, 1950), and (5) as a part versus the whole (Becker, 1968).

Further, the lack of consensus regarding the meaning of shame is not restricted to the literature. It emerged among my pilot study subjects as well. Thus, of 54 subjects who contributed a description of being "ashamed," 15 described an experience of *intra*personal exposure plus objectification only, 13 described an experience of *inter*personal exposure plus objectification only, and 26 described an experience entailing both with a focus on one or the other.

Such a radical disagreement concerning the fundamental meaning of shame loudly proclaims the continuing need for a definitive exposition of its essential lived structure. I believe that the results of my earlier empirical–phenomenological study of being-ashamed (Vallelonga, 1986) provide such a definitive exposition.

Bilevel Ambiguity of Shame

Lewis (1992) aptly and accurately speaks of the "chameleon nature" of shame. However, the ambiguity of the term *shame* and the confusion over its usage in the literature are due not only to a multiplicity of specific referents (i.e., the "shame variants"), but also to a confusion of logical types or levels. In effect, the term "shame" is used to denote *both* a *class* of cognate phenomena *and* the specific *members* of that class. Thus, Lewis (1971), Lynd (1958), and Miller (1985) explicitly characterize shame as a "cluster" or "family" or "nonunitary" category but at the same time choose one of the specific members of the class as the "real shame" (without any empirical justification for doing so), or they blend the features of all the specific shame-phenomena (such as being-ashamed, embarrassment, guilt, shyness, and modesty) into a single omnibus-type concrete phenomenon. Lindsay-Hartz (1984, p. 704) recognized this blending of several shame-phenomena into one when she described Lewis (1971) as studying "a complex that can only be termed 'shame-embarrassment-humiliation-shyness.'"

Thus, on one logical level, shame is a *generic phenomenon.* This "generic shame" is either the *class* constituted by the possession of common features by multiple specific members of the class or the *common features* possessed by each member of the class. Tomkins (1963), Kaufman (1989), and Nathanson (1992) all identify their "shame affect" (i.e., the lowering of the head, the averting of the gaze, and blushing) as the feature that all the specific shame variants share in common (For a description of my candidates for the common features, see Vallelonga, 1986, pp. 769–771.)

On the other logical level, "shame" has been used to denominate each of the *specific,* concrete shame variants including being-guilty. On this level, "shame" has been used to denominate each of the phenomena that share the features of the common denominator and that, together, form the shame family. On this level, modesty, shyness, being-ashamed, being-embarrassed, being-humiliated, and the like are all *indiscriminately* referred to and denominated simply as "shame" by one author or another. In addition, "shame" is used on this level of specific variants to refer, not just to affects or emotions, but also to "attitudes."

As a result of the confusion created by the bilevel use of "shame," I chose to refer to the specific, concrete phenomenon as *being-ashamed* or *being-ashamed-of-oneself* and to use the term "shame" to refer to *either* the generic phenomenon, that is, the shame-family (i.e., all of the shame-phenomena taken as a group), *or* the common denominator of all the shame-phenomena *or* the topic as it is referred to in the literature. This was also done to convey that the phenomenon being described (i.e., an affect/emotion) entails the interaction of an embodied subject and his or her world over the course of time. In other words, it points to *lived temporality,* a feature that is absent from most of the descriptions of shame in the literature with the possible exception of K. W. Fischer and Tangney's (1995) "prototypical script" for shame and Lindsay-Hartz's (1984) "structures" of shame and guilt. But, even K. W. Fischer and Tangney and Lindsay-Hartz fail to grasp and articulate the dialectical relationships among the three dimensions of lived time and how they determine the unfolding of a typical "shame situation."

Besides the existence of multiple definitions of shame and multiple discriminations between shame and embarrassment and between shame and guilt, the ambiguity of shame is further revealed in the fact that certain authors use "shame" in several distinct even if related senses. Thus, Freud, at different times, uses "shame" to denote: an attitude that prohibits specific sexual behaviors (Freud, 1953a), a self-consciousness about bodily exposure (Freud, 1953b), and a form of anxiety in the face of public exposure and ridicule (Freud, 1962). Lewis (1971) and Mead (1950) also used "shame" in a similar multivalent fashion. In addition, many authors (e.g., Ausubel, 1955; Benedict, 1946; Lynd, 1958) use "shame" solely to denote affects, while multiple authors (e.g., Albers, 1995; Bradshaw, 1988; Lewis,

1971; Riezler, 1943; Schneider, 1977; Straus, 1966) characterize "shame" as having both an attitude- and an affect-like structure.

All of these varied usages of "shame" indicate that, as a label, "shame" is *inherently ambiguous or multivalent.* Yet no author succeeds in clarifying what makes it so. Any project to comprehend the meaning of shame must explain what is the ground for that ambiguity. Furthermore, no author—not even Lindsay-Hartz (1984), who performed a qualitative analysis similar to my own—describes the temporal unfolding of a shame-situation, that is, how a person moves from conditions, through experience, to resolution in the living of a situation of being-ashamed. Further, no author articulates the relations between a person's present experience and his or her lived past and lived future, especially not in terms of movement into related/cognate affects.

Failure to Empirically Access the Agent's Perspective

With the literature on shame, we are confronted with the anomaly of multiple authors seeking to explicate the essential meaning of an experiential phenomenon without attempting either systematically or rigorously (i.e., empirically) to gain explicit access to the agent's or experiencer's perspective. Only Davitz (1969), Miller (1985), and Lindsay-Hartz (1984) elicit from subjects descriptions of aspects of shame. However, only Miller and Lindsay-Hartz solicit descriptions from others of their actual situated experiences of shame and allow these descriptions to constitute the raw data from which they derive their essential definitions of shame.

Several authors (e.g., Becker, 1968; Buss, 1980; Izard, 1977) conduct experiments on shame but *not* to determine its essential meaning. They derive their definition of shame from a combination of literature review and generally covert "individual phenomenological reflection" or what some might call "armchair reflections." They use their experiments, not to define the phenomenon, but to correlate their elsewhere-derived definition with other phenomena. Davitz (1964) is the only author who uses the experimental method to derive his definition of shame, but his definition lacks any inherent structure and is therefore of little real value for our purposes. There is no sense of how the phenomenon is lived.

Further, no other author surveyed, except for Lindsay-Hartz (1984), comes close to articulating a detailed structure of how shame is lived. They do not identify the concrete lived structure of being-ashamed or how the experience of being-ashamed unfolds over time. Lindsay-Hartz's (1984) study does identify four "structural aspects" of shame: (1) the *instruction* or "desire to act in a particular way" involved in shame; (2) the *transformation* or "specific way in which [the] emotion transforms the experience of self, other, and environment"; (3) the *situation* or "central event characteristic of the emotion"; and (4) the *function* or the manner in which the emotion supports and furthers certain goals and values of the experiencer (p. 690). Still, while Lindsay-Hartz's study is immensely informative, even she ultimately fails to articulate all the various situational and experiactional constituents of the experience of being-ashamed and how they are interrelated.

In addition, Buss (1980) and Izard (1977) address the conditions for or consequences of shame, but no one describes how any person moves from conditions, through experience, to resolution in an unfolding movement. Various authors give us a slice of the process, but no one gives us the whole picture.

This failure of the literature to articulate the agent's perspective relative to shame as well as to elucidate its lived structure (including its lived temporality) disclosed the need for a systematic and empirical investigation of the lived structure of shame. A study was needed that would enable us to understand the lived meaning of shame as well as to do so in a manner that could be accepted by others as having that level of veridicality characteristic of scientific work.

RESEARCH METHOD

Although my original research project began as an investigation of the experience of shame, it eventually evolved to a study of the experiences of both shame and embarrassment. In retrospect, it is clear that the original choice to investigate the meaning of shame by articulating the lived structures of both being-ashamed *and* being-embarrassed was serendipitous. Without the ability to compare these two cognate phenomena, I would have been unable to truly understand the meaning of either because, as we shall see below, both phenomena entail the experiences of interpersonal exposure and of objectification. Comparing these two with each other, as well as with being-guilty, enabled me to see not only the structures of each, but also how they differed from and how they interacted with each other, that is, how they were related to each other. Nevertheless, while my original study formally encompassed both being-ashamed and being-embarrassed, this chapter focuses primarily on the structure of being-ashamed and on the analyses performed to arrive at its articulation. This section describes the qualitatively analytical method I utilized to comprehend the situated experience of being-ashamed.

In order to capture (1) the experiential dimension of being-ashamed and (2) the temporally unfolding dialectic between the person's experience and behavior, on one hand, and the circumstances and events of the situation, on the other, a formal study was initiated utilizing an *empirical–phenomenological* research method. Besides requiring me to bracket my preconceptions about the phenomenon, it also meant gathering multiple descriptions of actual situated experiences of the phenomenon (the agent perspective) and analyzing them in a rigorous and systematic fashion to arrive at the general structure of the phenomenon. The particular method I used in this study is a personal variation of the qualitative method of analysis elucidated by W. F. Fischer (1974, 1989), Giorgi (1975), and von Eckartsberg (1977). A complete description of my research method, along with illustrations of the various steps, can be found in Vallelonga (1986, pp. 174–281).

Before beginning my formal research project, I conducted two pilot studies. In the first, I solicited descriptions of the "experience of shame" from two subjects and submitted one of the descriptions to a five-step formal analysis that resulted in a "preliminary" characterization of the general structure of being-ashamed. The following is that initial understanding of being-ashamed.

> Being-ashamed means to undergo a perception which suddenly transforms the meaning of how one is behaving in a situation and of who one is in and through that behavior. It means perceiving that one is doing (or not doing) what, according to one's projects (values), one should not and/or must not do (or should and/or must do) and being-guilty. It means also perceiving in what one is doing (or not doing) (and, therefore, in one's guiltiness) that one *is* (or is not) who one, according to one's lived self-projects (thematic or not), one must *not be* (or must be). Being-ashamed, thus, is the sudden perception, in one's behavior, of a discrepancy between who one is (or is not)

and who one must not be (or must be) according to one's self-projects (thematic or unthematic, but usually unthematic). Being-ashamed also means experiencing (as one assumes a retrospective stance) remorse or regret and wishing the past had been different and also, reciprocally (and in a mode of mutual implication), experiencing (as one looks into the future) the call to transform the present either by escaping the situation or by undoing the shamefulness. Since one is already exposed nakedly to oneself, being-ashamed also means experiencing the world and others as mirroring back one's shamefulness to oneself.

When one compares this structure of being-ashamed with the one elucidated by the end of my investigation, one can easily see that my understanding of the meaning of being-ashamed evolved over time, as it should have. It was also clear from the very first analysis, however, that the negative transformation of one's present self that was entailed in being-ashamed called for dialectically correlative negative transformations, on one hand, of one's present world (in being-embarrassed) and, on the other, of one's lived past and one's lived future.

In my next pilot study, I asked 56 community college students to give me descriptions of situations in which they experienced being-ashamed. The question posed to them is the same as that used with the subjects in the formal research project (see below). The following is one of the descriptions received in that pilot study:

> I have almost all my life been ashamed of my weight. For my life, except for a period of one year, I have always been overweight. I know I could have done something to change this, but I guess it was some kind of defense thing for me. I was always ashamed to put on a bathing suit, wear halter tops, etc. Any kind of clothing that shows off too much of me is bad. I really have a skinny-oriented mind. To me, the skinnier the better. And since I have never been this skinny, except for one year of my life, it's really been a hassle. I am ashamed to have anyone touch me because of this fat. I have since decided to do something about this and am on a diet. I am losing weight; and I will continue to stay on my diet until I am, in my mind, skinny enough. Like I said, this problem has an easy solution and I have finally decided to solve it, thus ending my being ashamed.

One of the results of this second pilot study was the discovery that being-ashamed was used sometimes to denote an experience of *intra*personal exposure plus disesteem and sometimes to denote an experience of *inter*personal exposure plus disesteem. That pilot study confirmed for me the belief that shame possesses an inherent ambiguity that called for explanation.

After these pilot studies in which I formulated the question and some of the problems associated with it, I embarked on the formal study. The first phase of the research consisted in *gathering the raw data.* I relied on four subjects (two male and two female) to provide these data. Each subject was asked to describe (in writing and with as much detail as possible): (1) a recent situation in which he or she experienced being-ashamed; (2) how he or she became aware of his or her being-ashamed, as well as what he or she experienced and did in that situation; and (3) how he or she passed through or got over that experience. Subsequent interviews (about an hour in length) were held with the purpose of expanding on the material presented in the written descriptions, filling in gaps, and clarifying ambiguities. The transcripts of these interviews, taken together with the written protocols, constituted the *raw data,* which were then submitted to systematic reflective analysis in the second major phase of the research process.

Two distinct but similar modes of analysis were employed. These modes differed in thoroughness, but each aimed at distilling from the specific situated experiences the general structure of the topic phenomenon (i.e., those patterns of interrelated constituents without which the phenomenon would cease to be what it is).

Mode I of Analysis

The first mode of analysis was used with one of the four subjects and was quite lengthy and thorough. It entailed *five stages* comprising seven levels (A to G). The *first stage* entailed the constitution of the primary data (level A). It involved breaking down the written protocol into manageable meaning units (A1) and weaving into them the materials from the interview transcription (A2). The following is S1's complete written description of her experience of being-ashamed (and embarrassed), along with the division into meaning units.

> /1/ The event which led to an acute feeling of being ashamed on my part occurred recently . . . as I was taking my daughter to nursery school at the YMCA. . . . /2/After I had left my daughter at the school and was in the process of trying to back out of an unusually crowded parking lot, I had the misfortune to accidentally back into the side of another proximately parked car. /3/Although there was a fairly loud crashing sound at the moment of impact, I still was naive enough to hope that I had caused only a slight, barely noticeable dent in the other car. /4/Not taking the time to look, I boldly proceeded to drive away from the scene of the accident. /5/However, to my great chagrin and embarrassment, before I got more than a couple of yards, I noticed a rather obviously distraught middle-aged woman staring rather intently at me from the doorway of the "Y." Alongside her stood the kindly old janitor of the "Y," whom I had known and been friendly with for several years. It was clear to me that both had heard the "crash" from inside the "Y" and that I had been "caught in the act" of not only hitting another car but also of attempting to leave the scene of the accident. My first reaction was one of complete embarrassment and total dismay at "having been caught." /7b/And, at this point, I was still hopeful that I could "save face" by pretending that I didn't know I damaged the lady's car. /6/Within seconds the owner of the other car and the concerned janitor were standing in front of me questioning me as to where I had hit the car, what I had done to it, etc. /7a/ Instead of admitting what I had done and pointing to a fairly long and quite noticeable dent above the car door, I continued to pretend to myself and the others that I didn't think I had done anything. /8/I realize now that, if I let myself admit to damaging the other car, that I would also have to admit to an even greater "sin," that of leaving the scene of the accident. /9/Another factor that increased my embarrassment and led me to try to protect my integrity was that I felt the janitor knew that I was a minister's wife and, although he might be able to accept my behavior if I were an ordinary "Jane Doe," he certainly could not accept it from J. W., the wife of the reverend from . . . church. /10/Finally, when the janitor noticed the dent and pointed it out to the lady owner of the car, the "game was up" /11/but I still tried to minimize it, saying, "That should be easy to fix." She didn't agree with me and proceeded to inquire further as to my name, address and automobile insurance, all-the-time carefully writing everything down. /12/ guess I was somewhat relieved that things were now all "out in the open" and I cooperated fully. /13/However, I believe I was still angry at her for "catching" and embarrassing me. /14/It wasn't until after I got back in my car and proceeded on my way that the full impact of what I had done—and or tried to do—hit me. At this point, the strong feeling of embarrassment that I had had was replaced by an even more uncomfortable feeling of shame and self-loathing. "How could I do such a thing?" I asked myself. How could I, who claim to value honesty so much and who am so critical of others who fail to admit wrongdoing, be so recklessly and carelessly dishonest? /15/As I continued driving to my next appointment (which happened to be a trip to the beauty parlor), the feelings deepened and the more central question of "What kind of a person am I?" became predominant in my being. /16/Concomitant with my feeling of shame was an increasing semi-fear and dread of telling my husband about the accident and how it happened. For I knew that I would have to "confess" to someone, not only what happened (the accident), but how I reacted (my dishonesty). I was definitely concerned with what *he* would think of me too. /17/As I sat in the beauty shop trying to deal with these disturbing feelings, it occurred to me that I must come to grips with this double standard that I obviously cling to: of being harsh with others and easy on myself as far as morality is concerned. Should I try harder to live up to the demands I make of others or should I be less critical of others and more understanding of their humanness even as I am of myself? /18/As I drove home, I pondered over how to tell my husband about this embarrassing and shameful experience. I finally decided that to do it immediately would be the best way—and in the final analysis, the least painful. /19/This I did. My husband responded with a wry smile and a small shake of the head which indicated to me that, although he wasn't pleased with what had happened, he understood that it could happen to anybody. /20/I felt quite relieved and my intense feelings of embarrassment and shame were lessened to a great de-

gree. /21/In general, I find that, when I do something of which I am ashamed, I often seek out someone to talk to who I feel will be understanding. /22/I am unable to forgive myself unless someone else forgives me first. I conclude from this entire experience that I am too other-directed—that I care more about what others think about me than what I think of myself.

Level A2 entailed taking each of the subject's statements from the oral interview and weaving it into the written protocol at the most relevant spot to form a continuously unfolding story of the subject's situated experience. This story constituted the primary data for the subsequent stages of the analysis.

The *second stage* of the analysis consisted of the *first* analytical transformation of the primary data through an *expansive* process of explication and translation (levels B and C). Level B entailed my translating the narrative from the first to the third person and the subject's everyday language into more essential, structural language. In it I also began the process of chronologically ordering the data into the various scenes that together form a coherent, undirectionally unfolding psychological story. Following are excerpts from level B of the analysis of S1's protocol.

S1 gets back into her car and proceeds on her way. As she is driving, the full impact of what she had done or tried to do hits her suddenly and all at once. The strong feeling of embarrassment which she has had is replaced, at this point, by an even more uncomfortable feeling of shame and self-loathing that overwhelms her. She asks herself, "How could I do such a thing? How could I, who claim to value honesty so much and who am so critical of others who fail to admit wrongdoing, be so recklessly and carelessly dishonest?"

Now that she has left the other woman and the janitor and is alone in her car, she feels able to face herself and be honest with herself. She feels freed to face herself because she no longer has to worry about what the others think of her or what their reaction is going to be. She no longer has to protect herself. . . . The reason for the rationalizations is gone. She feels free, now, to come to grips with herself and to admit completely to herself what she had done. . . .

S1 is ashamed of herself on several grounds. First of all, she is ashamed of her boldness in leaving the scene of the accident and not being willing to take responsibility for it and for her actions. Secondly, she is ashamed of being *so* recklessly and carelessly dishonest. . . .

Her recognition of her hypocrisy sweeps over her and she is ashamed. This is experienced as a "moment of truth" when her conscience tells her that she can no longer continue to hide from herself and when she permits herself to recognize what she had done and who she is. . . . Once the process of owning up begins, she realizes that there is no way of stopping it and she now allows it to go forward to the end of admitting completely not only to others but also to herself what she had done and who she is. . . .

S1 attends to the significance of her deeds for revealing who she is as a person. In the moment of being-ashamed, S1 focuses on her self, on her "inner being" and on the "core of her being"—on her "character." In perceiving the self-revelatory character of her deeds, she is radically disillusioned and disgusted with the self she perceives herself to be "really." S1 makes sense of her shameful deeds as revealing that she is not the person that she wants to be and has pretended herself to be. In fact, she sees herself as falling far short of being the kind of person that she has wanted to be. While S1 has always and actually believed herself to be honest, responsible, caring and courageous enough to face the facts and admit when she is wrong, she now painfully discovers herself to be dishonest (hypocritical), irresponsible, recklessly uncaring, and without courage. S1 describes this revelation as a shattering of her self-image. S1 experiences this realization as a moment of total self-awareness (i.e., exposure to herself). She experiences this as a coming in touch with what she "really" is.

Level C continued the process of translating. But more important, in Level C I explicated the subject's text so as to comprehend the described events as fully as possible. This constituted the *hermeneutical* phase of the analytical process. It entailed (1) an explication of what was implicit, (2) a major translation into psychological language (i.e., a further stripping away of details and particulars), (3) a commentary explaining why I was interpreting the text in a particular way, and (4) a series of reflections triggered by the data. This process of expanding the data before condensing it (through elimination of the irrelevant and the redundant) is my own personal variation of the standard enactment of the phenomenological method. Following are excerpts from the level C analysis of S1's protocol:

> Thus, in reflecting back, S1 recognizes several discrepancies on the basis of which she is guilty and ashamed of herself. First of all, S1 realizes that, although she was called (by her value/self-project to be an "honest and responsible person") to stop after the accident and assume liability for the damages caused, she *instead* boldly proceeded to leave the scene of the accident in order not to assume liability for the damages. Furthermore, in the recognition of this transgression of her still-affirmed value and, especially in the recognition of the excessive ("outré") manner (i.e., with boldness) in which she has transgressed that value, S1 recognizes that she *is* the person that she *must not* be according to her lived self-projects. In other words, while S1 desires ("wants") to be an "honest and responsible person" and, therefore, experiences the consequent call ("ought") to stop and assume responsibility for the damages, it is also clear that she *must not* be *boldly* (i.e., *so* unhesitantly and *so* callously) irresponsible.... Furthermore, S1 recognizes, in these further transgressions also, and especially in their excessive quality ("*so* recklessly," "*so* carelessly," "*so* long"), that she *is* the person she *must not* be. Thus, while S1 wants to be an "honest and responsible person," she *must not* be a person who is *so* recklessly and *so* carelessly dishonest for *so* long in doing something which will harm another person....
>
> In other words, S1 painfully recognizes several discrepancies between what she did and what she was called to do by one or more of her values/self-projects (and S1 is guilty in these recognitions). Furthermore, in each of the above painfully-recognized discrepancies and, especially, in a certain excessive quality characteristic of each, S1 even more painfully recognizes a further discrepancy between who she actually is (as revealed in her deeds) and who, according to her already lived values/self-projects, she *must not* be (in this recognition . . . S1 is ashamed of herself). S1, therefore, does attend to the significance of her deeds for revealing who she is as a person....

The *third stage* consisted of the *second* analytical transformation of the primary data through a *condensing* process leading to *situated* structures of the phenomenon (levels D and E). In level D, I began the process of eliminating whatever details were irrelevant to understanding the experience of being-ashamed. I also separated out the overlapping, intertwined, and interconnected experiences of being-ashamed and being-embarrassed, both of which several subjects (like S1) described within the same situation. I also delineated multiple instances of being-ashamed if they were given within a single subject's description.

The function of level E was to identify the situated structure of each instance of being-ashamed described (through a further condensation of level D) and the organization of the constituents within the threefold temporal framework of any affective situation—that is, the "becoming-phase," the "being-phase," and the "getting over-phase." Following are excerpts from the level E analysis of S1's protocol:

> *The Being-Phase* of S1's Second Instance of Being-Ashamed-of-Herself (i.e., the core of the "shameful" situation—the occurrence of S1's being affected in the specific mode of being-ashamed-of-herself).
>
> #1 While S1 experiences the spontaneous impulse to avoid reflecting on her transgressions, she permits herself to do so, despite its unpleasantness, in order to appreciate fully the *personal* significance of her deeds. #2 In reflecting back on or thematizing her being-guilty, S1 not only thematizes the discrepancy between what she had actually done and what she had been called by one of her values to do or not to do, but she also recognizes the excessive ("outré") manner in which she has transgressed her value. #3 At the same time, S1 shifts her focus from the "doing"-dimension to the "being"-dimension (from what she has *done* to who she *is* or *is not*). She sees this "being"-dimension enacted in or revealed by her deeds. (This shift is expressed/enacted in a shift from a focus on verbs to one on adjectives in S1's worded consciousness—e.g., *from* "to stop" or "to leave" *to* being "callous," "irresponsible," "dishonest," "hypocritical"). Thus, *in both* realizing/thematizing the excessive manner in which she has transgressed her still-affirmed value/project *and in* shifting her focus from the "doing"-dimension to the "being"-dimension, #4 S1 painfully recognizes that she *is* the person she *must not* be according to her lived (but not necessarily thematized) self-projects. In other words, S1 painfully recognizes a further discrepancy in

"being," i.e., a discrepancy between who she is (as revealed/enacted in her deeds) and who she *must not* be according to her lived self-projects. In this recognition S1 begins the transformation into being-ashamed.

#7 However, along with the recognition of this failure-in-being (entailed in the discrepancy-in-being), S1 performs a movement of *reduction* whereby she defines herself (either explicitly or implicitly) as being *only* this (already generalized and decontextualized) negative face/self—as being *only* this failure-in-being. This movement of reduction is experienced by S1 as a coming in touch with her "real self" beneath deceptive appearances. #8 Before this abstract, decontextualized negative face/self which S1 had reductively defined herself to be, S1 is disgusted/sickened. She also loathes/hates/contemns this exclusively negative self she has discovered/made herself to be. Thus, S1 spontaneously (and on both a pre-reflective/corporeal and an affective/emotional level) recoils from and rejects as repulsive this negative face/self.

The *fourth stage* consisted of the *third* analytical transformation of the primary data through a further condensing process leading to the formulation of the *general structure* of the phenomenon (level F). This entailed teasing out from the situated structures those circumstances and events that were peculiar to the particular subject's experience and those that were characteristic of all such experiences. It also aimed at interrelating those essential and universal constituents into an unfolding gestalt to form the general structure. This general structure was the goal and final result of the analytical process. Following is a brief selection from that lengthy general structure:

The living of the two movements of generalization and decontextualization (in self-dislike and self-disgust) and of the third movement of reduction (in self-hatred) constitutes a *partial or complete negative transformation of one's present self or personality.* The experience of this negative transformation of one's self constitutes the core of the experience of being-ashamed-of-oneself (in all three modes).

In the light of the above, it is clear that *that-about-which-one* is ashamed (as the phrase, "being-ashamed-of-*oneself,*" literally suggests) is—in all three modes—one's self or identity or being or personality, *as it exists in one's own eyes.* It is one's own *private* self/identity/being/personality, as it were. In addition, that which is *at stake* or in question in all three modes of being-ashamed is the *worth* of this *private self.* Furthermore, essential to all three modes of being-ashamed-of-oneself is the implicit *re-affirmation* or *retention* of the privative self-project which one recognizes is unfulfilled or violated in one's negatively-valued or absolutely-negatively-valued behavior or characteristics.

In simply disvaluing or absolutely negatively valuing a thematized particular behavior or characteristic of one's *"proprium,"* as well as either a particular feature of one's personality or one's entire (reduced) personality, one simultaneously experiences these aspects of one's existence as *to-be-kept-hidden (or secret)-from-Others,* especially from those Others whom one does not want to disesteem one. Such disvalued or absolutely-negatively-valued features of the "doing"-, "having"- and/or "being"-dimensions constitute the *"shameful"* in one's existence. Furthermore, *in* (and as another face of) the negative transformation of these personal realities into lived secrets, one immediately experiences *being-self-consciously-embarrassed* (or being-embarrassed-over-potential-exposure). In other words, correlative to the negative transformation of one's present self, either into a partially deficient self (the self-dislike and self-disgust modes) or into a totally deficient self (the self-hatred mode), one experiences the *spontaneous correlative negative transformation of one's circumambient world into a now potentially* seeing-into and disesteeming world. The experience (i.e., felt perception) of this correlative negative transformation of one's present world *is* one's being-self-consciously-embarrassed; and it is lived as one's being *vulnerable, conspicuous* and, even, *porous.*

The *fifth* and last *stage* was a purely illustrative one (level G). It presented a synoptic view of each subject's entire affective situation, and a sketching out of the temporal unfolding of each subject's affective situation across both dimensions of experiential and chronological time. I performed this step in order to highlight the various dialectics that transpired. I refer the reader to my original study (Vallelonga, 1986) for examples of the level G analysis.

Mode II of Analysis

The second mode of analysis was used with the remaining three subjects. It was an abbreviated analysis and involved only four stages instead of five. It entailed only the constitution of the primary data (level A3) and a single, formal but summary level of analysis (level A4), along with levels F and G.

The second mode of analysis was *not* used to determine the general structure. I felt that the rich data contained in S1's protocol provided enough material to disclose the general structure of being-ashamed. The main function of these additional analyses was instead to demonstrate that I had already achieved *redundancy* in my analysis of S1's protocol and also that the general structure was in fact an essential and universal one. The second mode also had an additional function—to provide a pool of material suitable for illustrating each of the constituents and for sketching out a range of some of the possible variations on the general structure.

RESEARCH RESULTS

1. General Structure of Being-Ashamed

The outcome of the phenomenological analysis described above was the elucidation of the general structure of being-ashamed. In this section, I will present that empirically derived general structure, illustrate it with reference to the protocols, and compare it to the general structures of being-embarrassed and being-guilty (from Vallelonga, 1986).

In the course of this investigation, it became clear that the "being-ashamed" that the subjects had described was more accurately denominated as being-ashamed-*of-oneself.* It is the structure of shame in this sense, therefore, that was elucidated in this project.

As mentioned earlier, one of the results of this study was the articulation of structures that capture the lived temporality of the experience of being-affected in a particular situation. The structures I arrived at are those of *living through certain situations,* and they have a threefold temporal format of (1) "becoming," (2) "being," and (3) "getting-over" (i.e., of waxing, peaking, and waning) that is called for by the unfolding character of the situations themselves. The structures I articulated are thus not those of atemporal, decontextualized moments. They are the *structures of the temporally unfolding dialectical interaction between embodied subjects and their particular situations.*

What was discovered in the course of analyzing the data was that the principal moments in the total affective situation were constituted by dialectically related transformations of past, present, and future self and past, present, and future world. The elucidation of these dialectical relationships among the various negative transformations of self and world across the three dimensions of lived time constituted one of the principal achievements of the study and *provided the foundation for understanding the interrelationships among the various shame affects.*

The analysis of the protocols from both the pilot studies and the formal research disclosed that "shame" is used, with considerable confusion, to denote two categories of phenomena: those in which *inter*personal exposure plus objectification are focal and dominant and those in which *intra*personal exposure plus disvaluation are focal and dominant. The data revealed that being-embarrassed and being-ashamed are, respectively, the chief, but not exclusive, exemplars of each category. Following is an abridged version of the general structure or "whatness" of being-ashamed that I arrived at in this research:

A. *Becoming-Phase*

The becoming-phase of being-ashamed is constituted by the possession or creation of some feature of one's existence which one personally disvalues. One can be ashamed on account of any perceived failure either in moral or nonmoral "doing," on account of some personal characteristic/possession, or on account of what happens to one. The becoming-phase also includes the occurrence of the occasion for one's being-ashamed, i.e., of the circumstance or event which effects the thematization of the discrepancy in "doing" or "having" which in turn precipitates the awareness of the discrepancy in "being" which constitutes the being-phase of being-ashamed.

Subjects in both the pilot and formal studies clearly demonstrated that one can be ashamed on account of almost anything. In other words, one can give negative self-definitional significance to almost anything connected to the self. Thus, one can experience self-disesteem on account of *what one has.* The subjects described being ashamed on account of such things as bad teeth, a large bust, and a poor, working-class background.

One can also be ashamed on account of *what one does or has done.* Subjects talked of being ashamed on account of getting poor or failing grades in school, on account of making love to one girl while engaged to another, on account of getting drunk and getting arrested, and on account of dropping a boyfriend as soon as he got a divorce from his wife. One can see from these examples that one can be ashamed on either moral or nonmoral grounds. In other words, being-guilty is *not* an essential precondition for being-ashamed, as I had prematurely concluded in the first pilot study.

One can also be ashamed on account of *what one is.* Thus, subjects described being ashamed of their ethnic, racial, or religious background, such as being Italian, Polish, Hispanic, black, or Jewish. Others are ashamed of their occupations, that is, for being "only" a janitor, dishwasher, or migrant farmworker. Still others are ashamed of certain characteristics, such as being a "slow learner" or an epileptic.

Likewise, one can be ashamed because of *what happens or is done to one.* Many subjects described being ashamed for having been raped or sexually abused.

Finally, one can also be ashamed because of some negative feature or behavior of *someone else with whom one feels identified* in some way. Thus, subjects described being ashamed on account of a father's alcoholism or of a son's getting arrested.

The results of this study showed clearly that one can be ashamed on account of any aspect of the self. In other words, one can live *any* personally disvalued feature of one's "doing" or "having" dimensions as having *negative self-definitional significance.* Nevertheless, the research also shows that the grounds for being-ashamed tend to cluster into the general category of *failure* (moral or nonmoral), which can be subdivided into instances of badness, weakness, incompetence, inadequacy, or defectiveness.

The becoming-phase of being-ashamed also includes the occurrence of the circumstance or event that permits or precipitates the person's becoming aware of the discrepancy in "being" that is the core of the being-phase of the situated experience of being-ashamed. In S1's case, the becoming-phase of her experience of being-ashamed consisted both in the attempt to leave the scene of the accident (the grounds for her being-ashamed) and in the leaving of the owner of the car and the bystanding janitor (the occasion for her being-ashamed). As S1 left the presence of the owner and the janitor, her embarrassment subsided, along with the need to defend herself against public disesteem. As her embarrassment subsided, she found herself confronted with her own deeds and she experienced a resurgence of her being-guilty and a movement into being-ashamed.

B. Being-Phase

1) The being-phase entails the core events of the affective situation. The central event in being-ashamed occurs in two stages. In the first, one perceives a discrepancy in "doing" or "having," i.e., between what one wants to or must do or have (one's projects) and one's actual deeds or characteristics (the actual circumstances). In the second, one lives this thematized discrepancy in "doing" or "having" as signifying a failure in "being." In other words, one lives these thematized negatively-valued deeds or characteristics/events as having negative definitional significance as to who one is. One lives them as signifying that *one is the negative kind of person one does not want to be* (self-dislike) *or must not be* (self-disgust and self-hatred). Being-ashamed is lived in three modes or degrees: *self-dislike, self-disgust and self-hatred.* In self-dislike the failed project in "being" is lived with a "want-not" motivational valence. In self-disgust and self-hatred, it is lived with a "must-not" motivational valence.

This perception of a failed project in "being" can also be viewed as entailing two or three related movements: (a) a *lived generalization* from a negative valuation of a deed/characteristic/event to a negative valuation of one's personality or self; (b) a *lived decontextualization* in which one focuses on the negative feature and tends to ignore or deny the other positive features of one's self; and (c) (in the case of self-hatred) a *lived reduction* of one's self to complete negativity (i.e., to a totally deficient/worthless self). The perception of the discrepancy in "being" constitutes a partial (self-dislike and self-disgust) or complete (self-hatred) *negative transformation* of one's *present self* or personality.

The being-phase of being-ashamed begins with an *intra*personal exposure that takes place in two stages or steps. In the first, there is an exposure *to oneself* of some negatively valued feature of one's "doing" or "having" dimension (i.e., some behavior or personal characteristic). Thus, S1 starts to be ashamed as she allows herself to see clearly that she did boldly and callously leave the scene of the accident in order to avoid accepting liability for the damages she had caused and that in doing so she had violated her own moral code of behavior.

In the second stage of the intrapersonal exposure (and *as a result of* the exposure in the first stage), one becomes aware of a discrepancy in "being," that is, between the person one does not want to or must not be and the person one believes one is as a result of the just discovered discrepancy in "doing" or "having." One lives the exposed deed or characteristic as signifying that one is the negative kind of person one does not want to or must not be. Thus, S1 clearly moves from a judgment of her behavior to a judgment of her self. As she says, "As I continued driving . . . , the feelings deepened and the more central question of 'what kind of person am I?' became predominant in my being." Further, S1 lives the realization that she "boldly" left the scene of the accident as signifying that she is the "callously irresponsible" and "hypocritical" person she must not be. As she co-constitutes the meaning of these events in this manner, she is ashamed in the manner of self-disgust and self-hatred.

The descriptions I received from my subjects indicated that being-ashamed may be lived in three possible modes: self-dislike, self-disgust, and self-hatred. The differences among these three experiences are determined by (1) whether the motivational investment in the failed privative self-project is either a "want not" (self-dislike) or a "must not" (self-disgust or self-hatred) and (2) whether one reduces oneself to the negative face revealed in the initial perceived discrepancy in "doing" or "having" (self-hatred) or not (self-dislike and self-disgust)—or, in other words, by whether one lives the discrepancy as partially (self-dislike and self-disgust) or totally (self-hatred) definitional of the self. Thus, S1 experiences being-ashamed in the manner of self-dislike in the perception that she is a person who is "irresponsible" and "dishonest." However, she experiences being-ashamed in the manner of self-disgust and, at moments, self-hatred, in the perception that she is a person who is "*so* irresponsible" and "*so callously* dishonest."

It appears that each experience of being-ashamed-of-oneself in the mode of self-disgust or self-hatred entails the experience of an *extreme failure* in "being." The violation of self-projects with such a "must not" motivational valence appears always and intrinsically to entail an *excessive violation* of one's values. Thus, S1 is ashamed in the modes of self-disgust and self-hatred because she is not just irresponsible or dishonest, but is so in an outrageously excessive manner. She says:

> Well, I think my self-image just doesn't permit me to do things like that. And it's shattering to me to know that I'm not really the person that I want to be. That I could fall so short of being the kind of person that I wanted to be.

Although S1 uses the word "wanted" here, the project in "being" that is violated is a "must not" one. While she wants to be a certain kind of person, she must not fall so short of being that kind of person.

Of central importance for understanding the relationship between being-ashamed and being-embarrassed is the recognition that one's living the perception of some negative deed, event, or characteristic as being negatively definitional of one's self constitutes the experience of a *negative transformation* of one's *present self.* As we shall see below, the experience of such a transformation calls dialectically for other correlative transformations that constitute the experience of other cognate shame-phenomena.

> 2) This negative transformation of present self constitutes the core but not the entirety of the experience of being-ashamed. It also includes other constituents. Thus, *dialectically correlative* to this transformation of present self, one undergoes the spontaneous negative transformation of one's present circumambient world into a now potentially seeing-into + disesteeming/rejecting one. This constitutes one's being-self-consciously-embarrassed *in* one's being-ashamed. In this self-consciousness one experiences being vulnerable to interpersonal exposure + disesteem, and therefore experiences a call to live that deficiency in "being" as a secret to-be-kept-hidden-from-Others. Likewise, one also feels both porous/transparent and conspicuous.

Thus, in defining oneself to be this negatively valued (self-dislike) or absolutely negatively valued (self-disgust and self-hatred) self, one undergoes a negative transformation of one's *present self* in the experience of being-ashamed *and* immediately and spontaneously experiences a dialectically correlative negative transformation of one's *present world* into a potentially seeing-into and disesteeming one. This latter transformation is experienced as being-self-consciously-embarrassed in one's being-ashamed. This being-self-consciously-embarrassed is an integral part of the fuller experience of being-ashamed. One of my subjects (S4) captured this immediate correlative transformation of his present world in the following manner:

> I felt chilled and kept saying inside myself: "Oh, God . . . Oh God. . . . How could I be so blind?" I felt real shame—a deep guilt. I was aware of an emotional nakedness, as though everybody could see my shortcomings, my inner feelings of self-disgrace.

Another subject (S3) described the tension in this self-consciousness between feeling as if others can see while knowing that they cannot. She says:

> I felt surely that everybody must know . . . that I am not doing what I am supposed to be doing. . . . I don't think they were [aware]. . . . I guess my own magnitude expanding though I knew it wasn't true . . . It's like surely everybody must know what I've done; and that's stupid.

Those who experience being-ashamed all undergo this sort of "plate-glass feeling" (Laing, 1965, p. 106), in which they feel as though they are porous and transparent and their inner selves are visible to others.

Further, precisely because this feature that has just been exposed to the self is negatively valued, one experiences the immediate call to live it as a "secret," that is, as something to be kept hidden from others. It is the "shameful" that is lived as though it would induce others, at the very least, to disesteem one if it were revealed to them.

It is because of this dialectical self-conscious-embarrassment that being-ashamed-of-oneself is a phenomenon not just of *intra-* but also of *inter*personal exposure plus disesteem. *It is this dialectical relationship between negative transformations of present self and present world that contributes to the ambiguity of being-ashamed and to the confusion between it and the experience of being-embarrassed* (whether over potential or actual exposure). Nevertheless, while being-ashamed entails both intra- and interpersonal exposure, the central and key exposure is still the *intra*personal one:

> 3) The negative transformation of one's present world in being-self-consciously-embarrassed continues as one spontaneously passes over into being-anxious in the face of actual exposure and into being-embarrassed-over-(presumptive)actual-exposure. Thus, one's world almost always continues to get transformed from a potentially rejecting one into a probably rejecting one (anxious) and sometimes into a presumptively actually rejecting one (embarrassed). Being-ashamed appears always to entail the constituent experiences of being-self-consciously-embarrassed and being-anxious in the face of exposure, along with a felt thrust toward being-embarrassed-over-actual-exposure. It sometimes, *but not always*, entails the actual experience of such embarrassment-over-actual-exposure. Thus, the negative transformation of one's present world in being-ashamed calls for a further dialectical negative transformation of both one's *future self* and *future world*. Against the background of the perception that one is and perhaps has always been the negative self one does not want to or must not be, one becomes apprehensive that one will continue to be that negative self. One is thereby anxious in the face of additional future experiences of being-ashamed (future self). Likewise, one becomes apprehensive that Others will confirm one's own negative self-valuation and will enact that valuation in their behavior in several ways (e.g., rejection). One is thereby anxious in the face of such threats from one's future world.

This dialectically required experience of being self-consciously-embarrassed is the *first* step in a process of *progressive* negative transformation of one's world that is triggered in the experience of being-ashamed. The other two steps are: the experience of one's world as *probably* exposing plus disesteeming (being-anxious) and the experience of one's world as *actually* exposing plus disesteeming (being-embarrassed-over-actual-exposure). While this process is not inevitable in its entirety in every full experience of being-ashamed, the movement into being-anxious, at least, appears essential. The movement into being-embarrassed-over-actual-exposure, on the other hand, appears generally to require the physical presence of others. Thus, one can experience the being-self-consciously-embarrassed in being-ashamed over something *without* also experiencing being-embarrassed-over-actual-exposure in regard to the same feature or ground. There is, nevertheless, a felt thrust toward experiencing it if others are physically present.

Thus, in being-ashamed and self-consciously-embarrassed, certain negative eventualities spontaneously arise on the horizon of one's future. The experienced emergence into awareness of these possibilities constitutes a negative transformation of one's *future* (self and world) and the emergence of one's being-anxious in the face of such a negative future. This felt thrust toward being-anxious is almost dialectically *demanded* by the very fact that one's being-ashamed and self-consciously-embarrassed are negative transformations of one's present (self and world, respectively). However, just as being-ashamed and self-consciously-embarrassed constitutes the experience of a negatively transformed *present* (both self and world), so also the anxiousness one moves into spontaneously is an anxiousness in the face of a negatively transformed *future* (both self and world). One is anxious in the face of further instances of being-ashamed (negative transformations of future self), as well as in the face of being ridiculed, shamed, rejected, or otherwise hurt by others (negative transformations of future world).

Thus, S1, in the midst of her being-ashamed for trying to leave the scene of an accident, spontaneously attends to the moment in the near future when she will have to confess to her husband what she has done. In that realization, she is anxious in the face of being-embarrassed-over-actual-exposure to her husband and of being-humiliated by him (negative transformation of future world) and of possibly being further ashamed of and disillusioned in herself (negative transformation of future self). She says:

> Concomitant with my feeling of shame was an increasing semi-fear and dread of telling my husband about the accident. . . . I was definitely concerned with what he would think of me too. . . . I was *afraid* of revealing to him that I was no better than anyone else. . . . *And I was afraid he would say this to me or intensify the feeling that I already had* by what he said—trying to make me feel guilty. . . . I was afraid he would confirm it even more by agreeing with me (emphasis added).

If the conditions are right, the correlative progressive negative transformation of one's present world *culminates* in the experience of being-embarrassed-over-actual-exposure, or humiliated. Generally, this means that if one experiences being-ashamed in the presence of others and if these others show some presumptive sign of disesteem, then one experiences being-embarrassed-over-actual-exposure in regard to that for which one already disesteems oneself. Having already negatively judged oneself, one is prone to construe any ambiguous behavior, circumstance, or event as indicating that others do, in fact, know about one's personal deficiency and are disesteeming one for it. Thus, precisely because one is ashamed, one has a tendency to experience being-embarrassed-over-actual-exposure. Such a being-embarrassed correlative to one's being-ashamed constitutes a further and final negative transformation of one's present world—an advance beyond the negative transformation begun in one's being-self-conscious.

One subject provided a good illustration of being-embarrassed-over- (presumptive) actual-exposure consequent upon and correlative to her being-ashamed. She describes working as a waitress in a resort hotel and being embarrassed when the captain of the waiters yelled at her in front of all the hotel guests. In response to this public censure, she ran out crying and was immediately ashamed for being "so weak" and for not being able to respond effectively to the captain. She also felt even more embarrassed believing that the guests now disesteemed her even further for her "weakness." Her being-embarrassed-over- (presumptive) actual-exposure correlative to her being-ashamed was revealed when she said: "I felt that I could never face the guests again. . . . I was feeling that they were looking down on me for the whole thing."

While it is true that one does not experience being-embarrassed-over-actual-exposure each time one experiences being-ashamed, it is also clear that in each instance of being-ashamed, one does experience being-self-consciously-embarrassed as well as a *thrust toward* being-embarrassed-over-actual-exposure, that is, a *further* negative transformation of one's present world. At minimum, the experience of being-ashamed shapes one's world to such a degree that one is disposed or inclined toward seeing whatever persons are present as *actually* seeing into one's shamefulness and disesteeming one for it. This dialectical thrust toward embarrassment-over-actual-exposure also reveals being-ashamed as a phenomenon of both intra- and interpersonal exposure and disesteem. As indicated above, the confusion in the literature about the nature of "shame" is grounded in this dialectical relationship, within a single experienced situation, between negative transformations of present self (being-ashamed) and present world (being-self-consciously-embarrassed and being-embarrassed-over- [presumptive] actual-exposure).

Although one does not always experience embarrassment over actual exposure correlative to one's being-ashamed, when one does experience it, that very experience confirms and reinforces one's own self-disesteem and increases the felt pain of one's being-ashamed. In fact, they tend to reinforce each other so that the experience of each is intensified:

> 4) In addition, the body participates in the total felt-perception which is being-ashamed. Thus, on a pre-personal level one lives the perceived discrepancy either as literally distasteful (in self-dislike) or as revolting/nauseating (in self-disgust and self-hatred). In both cases the body pre-reflectively enacts its rejection of the perceived deficiency in personal "being." Further, one also experiences a sort of bodily weakness along with a sense of helplessness in response to the perceived failure.

Being-affected is not just a psychic event. Rather, it is a moment in the interaction of a unitary *body-subject* with its "world" (in the broadest sense). The body participates necessarily and meaningfully in the intentional act of being-affected. The intentional act, therefore, is not solely a cognitive event. Being-affected is a matter of a *felt knowing*. The feeling refers both to the affected consciousness (i.e., the painful or pleasant quality of the knowing itself) and to the reverberations or resonations of the perception in or by one's embodied existence. The feeling aspect of being-ashamed thus entails both the painful quality of the felt perception and the body's expression of and response to what is perceived.

The body's involvement in being-ashamed expresses, prereflectively and prepersonally, the core meaning of the lived situation itself. In effect, the body, through its enactment of *distaste* (for self-dislike) and *revulsion/nausea* (for self-disgust and self-hatred), lives out and expresses one's own negative valuation of one's self or personality. The distaste and revulsion that are characteristic of the various types of self-disesteem are bodily and prepersonal enactments of one's *self-rejection*. They are the body's part in the self-disesteem.

S4 clearly reacted bodily to his being-ashamed. He twice used the word "Yuck!" to express his distaste over his shameful behavior. He said: "And in show business, you know the expression—'You step on people to get ahead.' And here this was—Yuck!" He also said the shameful incident "caused this 'Ugh!' ugly feeling in me" and that he "got this terrible sick feeling." In her being-ashamed, Sl described feeling "sort of sickened." Likewise, one of the subjects in my study of embarrassment also described an incident of being-ashamed and declared in the interview:

> I got sick in my stomach. *R:* You mean nauseated? *S:* Nausea, yeah. *R:* You really felt nauseated? *S:* Uh-huh. *R:* As what? *S:* As a realization of what I was doing. Of how I reacted, of being so out of control, of being so violent against them. The realizing what I could have done, you know, made me sick to my stomach. Usually, when I'm ashamed, I start getting sick.

In addition to a feeling of bodily distaste or revulsion, there is also an experience of bodily weakness. One of my subjects said, "I felt weak—just like somebody had just cut me open." The bodily weakness appears to express a certain lack of direction and motivation appropriate to the crisis in self-definition that is entailed in being-ashamed. Being-ashamed always entails some sort of being-demotivated, whether of brief or long duration. The bodily weakness is an aspect of that demotivation. One loses heart, as it were.

Similarly, there is a concomitant feeling of helplessness. S1 declares that in the midst of her feeling "sickened and disgusted," she also undergoes "a feeling of helplessness, of being all alone in this situation—that there was nothing anybody else could do or say that could help [her] out of this."

5) Correlative to the felt-perception that one is the negative person that one does not want to or must not be, one typically undergoes the felt-perception that one is not the ideal person that one still wants to be. This felt-perception of that discrepancy constitutes one's *being-disappointed-in-oneself.* In being-ashamed one focuses on the negative-actuality-pole of the discrepancy in "being"; while in being-disappointed-in-oneself one focuses on the positive-ideality-pole of the very same discrepancy.

Being-disappointed-in-oneself appears to be an isomer, so to speak, of being-ashamed. In both, one perceives and attends to a discrepancy between one's self-project and one's presumptive actual self. However, the difference appears to be that in being-ashamed one focuses on the presumptive *actual negative* self, which constitutes the failure of the self-project one has been living up to now. One does not attend to the continuance of the correlative positive self-project as such, even though it has so far foundered while still being retained. On the other hand, in being disappointed in oneself, one focuses not on the presumptive actual negative self (if that has emerged), but on the *ideal positive* self that one continues to want to be and that one realizes one has thus far failed to be. In both, the negative affect is constituted by a felt perception of the discrepancy between actuality and ideality. But in being-ashamed, one attends to the *negative actuality pole* of the discrepancy, while in being-disappointed-in-oneself, one focuses on the *positive ideality pole* of the very same discrepancy. In the former, one focuses on one's *being* who one does *not* want to be. In the latter, one focuses on *still wanting to be* who one sees one *is not.*

Being-disappointed-in-oneself and being-ashamed are correlative lived generalizations that most often occur together. In being-disappointed-in-oneself, one lives one's deeds/characteristics/experiences as signifying that one is *not* the unambiguously positive self (in part or in whole) that one *continues to want to/have to be.* For example, S1 lives her attempt to leave the scene of an accident as signifying that she is not the honest, responsible, caring person she wants to and must be, and she is disappointed. In being-ashamed, on the other hand, one lives one's deeds/characteristics/experiences as signifying that one *is* the unambiguously negative self (in part or in whole) that one wants *not* to or must *not* be. For example, S1 goes on to live that same event as signifying, not just that she is not the honest, responsible person she must be (which could be the opportunity for the formulation of a truly ambiguous self), but rather that she is the dishonest, hypocritical, uncaring person she must not be. Painful though it may be, being-disappointed-in-oneself is experienced as a species of *loss* (temporary or permanent) of a self-ideal. Being-ashamed is experienced instead as a disaster that befalls one. It is a positively bad event from which one recoils.

While being-disappointed-in-oneself is qualitatively different from being-ashamed, the difference is subtle and most often obscured. It is obscured because both phenomena frequently occur together and because the pain of being-ashamed tends to overshadow that of being-disappointed-in-oneself when they do occur together.

6) In being-ashamed one lives one's present self as unambiguously negative either in part (self-dislike or self-disgust) or in whole (self-hatred). In it one also experiences a dialectical

> thrust towards a correlative negative transformation of one's *past self* and of the ambiguity of that past self. In other words, one is called to live one's behaviors or characteristics as having negative definitional significance not just for who one is now but also for who one *has been in the past and up to now.* If one allows them to have such significance, then one also experiences *being-ashamed-of-one's-past-self.* Further, if this past self had been lived previously, not as ambiguous, but as unambiguously positive (in part or in whole), then the negative transformation of that past self is lived as *being-disillusioned-in-oneself.* Thus, in being-ashamed one is called to live a generalization not just from "doing"/"having" to present "being" but also to *past "being."* All instances of being-ashamed entail at least the felt thrust toward the experience of such *negative transformations of past self.* Further, in these transformations of past self one typically experiences oneself as having been at least *naive* and often as having been a *pretense, fraud or phony.*

Because of the mutual implication of the three dimensions of lived time, when one undergoes the negative transformation of one's present self in being-ashamed, one also experiences a dialectical thrust, not just toward a correlative negative transformation of one's future self and world, but also toward a correlative negative transformation of one's *past self,* that is, toward living one's past self as having been negative as well. In other words, one lives one's behaviors or characteristics or the events at issue as having significance regarding one's understanding/definition, not just of who one is *now,* but also of who *one has been in the past and up to now.* What this means is that one experiences a dialectical thrust toward being-ashamed-of-one's-*past*-self, that is, toward disesteeming who one believes/understands one has been up to now. However, if one's self-understanding in the area at issue has been and is unambiguously positive, then one also experiences being-disillusioned-with-oneself. In certain instances, then, one may experience both being-disillusioned-with-oneself and being-ashamed-of one's *past* and *present* self. (For a discussion of the difference between being-ashamed-of-one's-present-self, being-ashamed-of-one's-past-self, and being-disillusioned-with-oneself, see Vallelonga, 1986, pp. 1116ff.)

Thus, in being-ashamed, one is also inclined to live one's current experiences, deeds, or characteristics as signifying that one has *never* been the self one believed oneself to be and, instead, that one has always been an unambiguously negative self (in one area or in toto).

In being-ashamed-of-one's-past-self and being-disillusioned-with-oneself, one feelingly perceives not only that one is not and has never been the person one wants to or must be, but, more important, that one *is and has always been* the unambiguously negative person (in part or in whole) that one does not want to or must not be. Two of my subjects, S1 and S4, exemplified both the disillusioned loss of the positive self-definition and the ashamed elaboration of the negative self-definition.

Thus, S1's being caught in her attempt to leave the scene of the accident forces her to confront what she has done. As she lives what she has done as having self-definitional significations, she eventually undergoes the experience of being-ashamed-of-her-present-self in the mode of self-hatred and, subsequently, the experience of both being-disillusioned-with-herself and being-ashamed-of-her-past-self. She says:

> Well, I felt completely shattered by the experience. Sort of like my world is coming apart. *My pretentions about myself—that I can't hold onto them any longer.* And sort of like *my self-image is shattered.* . . . That the image that I have of myself as being someone who is honest and responsible, you know, was shattered. I felt like the bottom had dropped out of every thing and that *I wasn't the person I had pretended to be to myself.* . . . I guess a feeling of complete and total helplessness and *nakedness* and total self-awareness—being in touch with *what I really was.* . . . It was sort of a feeling of surrendering to *the truth about myself* (emphasis added).

Thus, S1 lives her recent deeds as signifying that she is not the person she must be (i.e., "honest," "responsible," and "caring"), and she is disappointed-in-herself for that. However, she also lives these deeds as signifying that she has *never* been that unambiguously positive person or that, at least, she has *never* lived these qualities in an unambiguously positive manner, and she is disillusioned-with-herself. S1's case illustrates well that this "shattering" of one's "self-image" is lived as a painful moment of truth, and one never questions that it is "the truth." Precisely because this discovery has the character of a "moment of truth" or of a shattering of illusions (which one has co-constituted, at the minimum), one also experiences oneself typically as having been at least *naive* or, further, a *pretense, phony, sham, or fraud.*

7) As part of this experience of being-ashamed-of-oneself one also experiences the urge/impulse to do something (at this stage, usually vague and unspecified) to ease the pain. One frequently experiences the impulse, *per impossibilem,* to escape from one's own eyes and to disappear.

As in the experience of all negative affects, in the case of being-ashamed one experiences, in the midst of the painfulness of the self-disesteem, an urge or impulse to do something to ease or eliminate the pain. This impulse, in its myriad forms, constitutes the *conative* dimension of the being-affected in the mode of being-ashamed. This conative dimension of the Being-Phase of being-ashamed consists in a *thrust or pull toward some action* (to end the painful intrapersonal exposure and self-disesteem), rather than in the action itself. The action itself, on the other hand, to which one is called to deal with the pain, would constitute the Getting-Over-Phase of the situated experience of being-ashamed.

In every negative affect, including being-ashamed, one experiences the call to do something (at this stage, usually something vague and unspecified)—either to undo the discrepancy the awareness of which is the source of the painful self-disesteem or to escape the awareness of that discrepancy. Several of my subjects illustrated the urge to do something to undo the discrepancy that occasioned the self-disesteem. Thus, S4 experienced the call to do something unspecified. At the moment he was ashamed, he declared: "I couldn't wait until I got out there and did something about it. . . . I wanted to get out there and see that little girl and correct it, to do something about it." Likewise, one of the subjects in my embarrassment study said of the "ugly feeling" he had in being-ashamed and embarrassed: "I wanted to shake it off and I wanted to leave, to get rid of it. . . . I didn't want to really own it. It's uncomfortable." Yet this subject does not know immediately what he is to do to get rid of this uncomfortable feeling.

S4 also experienced the urge to undo the painful awareness and to run from the self-confrontation it portended and demanded. He admitted candidly: "Truly, I wanted to run from my inner thoughts of self-deprecation and shame." He expanded on this by saying that "it's almost like not wanting to go through that therapy with yourself." Nevertheless, he resisted the urge to run from his own "moment of truth."

Sometimes one experiences the urge, as in being-embarrassed-over-actual-exposure, to escape the seeing-into and disesteeming eyes. In being-ashamed, however, it is one's own eyes that are doing the exposing and disesteeming; and one cannot escape one's own awareness that easily. In fact, one does do it through forgetfulness, and this constitutes one of the chief solutions enacted in the Getting-Over-Phase. Still, in the Being-Phase, one sometimes experiences the desire to disappear and, *per impossibilem,* to escape one's own gaze. Another pilot study subject illustrated this impulse when he revealed that he "was so ashamed that . . . I just wanted to crawl into a hole and never come out." The magical nature of this solution aims at undoing both one's own painful self-awareness and any exposure to by-standing others.

8) Similarly, the just-thematized discrepancy in "being" has a power to attract one's attention to it. One typically gets absorbed in this discrepancy and gets temporarily pulled away from one's immediately previous world-of-action and one's then-current and still-pending task-projects in it. In addition, clock time fades away. However, if one continues with one's already initiated task-projects, one typically does so in a *distracted/preoccupied* manner as one is torn, as it were, between attending to the still-pending projects and attending to the unresolved feelingly-perceived discrepancy.

There is a certain flustering that is attached to being-ashamed *by means of* the attendant self-consciousness. Similarly, there is also a certain flustering that adheres to the experience of being-ashamed simply from the power of the painfully perceived discrepancy to attract and hold one's attention. One becomes preoccupied with and fixated on the just-revealed deficiency in one's personality and is thereby distracted from one's current activities. Precisely because of this fixation, one tends to lose track of clock time. One's experiential time moves slowly while clock time, paradoxically, moves quickly. Furthermore, if one manages to continue with one's already initiated task-projects, one typically can do so only in a distracted or preoccupied manner as one feels oneself split and one divides one's attention between the still-pending task-projects and the unresolved feelingly perceived discrepancy in "being."

S3 illustrated the intense preoccupation with the failed self-project and the sort of being-flustered as well as distractedness it engenders, saying:

For two hours I walked around town, not knowing or caring where I was going. . . . The shame caused me to be disoriented in walking and crossing the street and I found myself bumping into people. I was turned inward on myself, facing what I should have done, and what I really was doing. . . . But, see, because I was in my own mind, I was going around in circles. I felt like I was going around in circles . . . with what I shouldn't have done and what I should have done. . . . *R:* I see. O. K. How did you respond to bumping into people? *S:* It didn't wake me up or anything. You know, I just would move away and would just say, "I'm sorry," or something, you know. But it didn't get me out of what I was in. I mean, I was in it.

9) Further, one's future deteriorates even more as one moves into being *unsure* about whether one can avoid being that negative person in the future. This anxiousness constitutes a mid-ground between being-ashamed and *being-depressed.* In it one experiences a call to believe that one will be *unable* to not be that negative person in the future as well. This is a call to being-depressed, which would constitute a final lived generalization—this time from present "doing"/"having" to *future* "being." Further (in self-disgust and, especially, in self-hatred), one's future gets transformed dialectically into an impenetrable negativity which leads one to lose hope. One sees nothing to pull one forward and one loses motivation to go forward. One thereby becomes profoundly discouraged and demoralized. This can progress farther into despair which gives rise to thoughts of death and suicide.

In the realization that one is the person one does not want to be or must not be, one begins to become *unsure* about whether one can cease being that negative self one *continues* to want not to be/must not be. This is the realm of being-anxious in the face of further instances of being-ashamed and in the face of being-depressed. This is a midground between being-ashamed and actually being-depressed. In actually being-depressed, one moves from being unsure about whether one can cease being that negative person to *believing that one cannot stop being* that person, while one *persists* in wanting to/having to not be that person. One moves *from* a present sense of being *able* to not be that negative person despite one's current failure to not be it, *through* a present sense of *unsureness* about whether one can cease being that person, *to* a present sense of being *unable* to not be that person.

Being-depressed thus constitutes the final step in an unholy trinity of lived generalizations. In being-ashamed, one lives a generalization from "doing"/"having" to *present* "being." In being-ashamed-of or disillusioned-with-one's-past-self, one lives a generalization from "doing"/"having" to *past* "being." And in being-depressed, one finally lives a generalization from "doing"/"having" to *future* "being."

The principal lived generalization is the one to present "being"—the experience of self-disesteem (being-ashamed). It appears to be the essential condition for the other two and can occur without the other two. Yet, the more intense the self-disesteem, the stronger is the *felt thrust* to live the other two generalizations as well.

Being-ashamed-of or disillusioned-with-one's-*past-self* expands the negative transformation of oneself across temporal horizons and deepens one's experience of self-alienation. Further, the more complete one's disaffection with one's own self-image, the more likely is it that one will experience oneself as being *unable* to achieve or fulfill one's self-ideal in the future, and the more likely one's being-depressed as well.

Furthermore, the more intense the self-disesteem or the more complete the negative transformation of one's present self (as in self-disgust and, especially, self-hatred), the more complete is the negative transformation of one's *entire* future. In self-hatred, one's future tends to get transformed into an inpenetrable negativity with no access. In this transformation of one's future, one loses all hope. One sees nothing to pull one forward and one loses all motivation to go forward. One becomes profoundly discouraged and demoralized. In fact, the felt impenetrability of one's future and one's demoralization are the same reality.

At that point at which one's future appears impenetrable, life itself begins to appear pointless, and thoughts and desires of suicide (or at least of death) spontaneously emerge. These thoughts or desires reflect the despair that in turn expresses the complete alienation of one's own future in one's self-disgust or self-hatred. In such a moment, one moves from a sense of terrible or complete unworthiness in the present to a sense that one cannot be other than terribly or completely unworthy even in the future.

It is especially likely that thoughts of death or suicide should spring up in the depression flowing from self-disgust or self-hatred. If one sees oneself as profoundly or completely negative and worthless and one sees no hope of becoming otherwise in the future, it would seem reasonable that one would want to escape the intense affective pain and the self-alienation, even if through death. Such a desire is reasonable once one has lived the two quite unreasonable and formally illogical generalizations, first to complete and unambiguous negativity in the present, and then to complete and unambiguous negativity in the future as well.

In being-ashamed, there is a tendency to being-*discouraged* and sometimes even to *despair.* In being-ashamed, one inevitably tends, as we have seen above, to become unsure of whether one can ever cease to be the person one does not want to or must not be (i.e., one becomes anxious in the face of being-depressed). Being-discouraged is the motivational state that goes along with this sense of unsureness about one's future "being." It is another face of one's lack of confidence in one's ability to achieve one's self-project. Since one is unsure about one's ability to succeed, one does not know whether to invest one's effort in the project. One is disinclined to take the risk.

Despair, however, is different. In despair, one does not just have a "sense" that one is unable to achieve one's self-project. *One is sure.* In being-depressed, one goes beyond just being unsure, but one is still not sure. One has a sort of suspicion that one is unable. In despair, one is sure one is unable not to be that person and sees no way of ceasing to be that unambiguously negative person. With that sureness, one has no inclination to invest energy or effort in one's self-project.

None of the subjects in my formal research project exemplified the progression to being-depressed or to despair. S1, however, comes very close. She reaches the point of being unsure about whether she can ever cease to be the person she does not want to or must not be, and she is discouraged momentarily. She certainly experiences tugs in the direction of being-depressed, but she manages to resist them and not make the generalization to future "being" after all. She says:

> Well, I think I almost felt like there was no reason to go on living because, if my real self was just a pretense, you know, and if I couldn't respect my real self, then there was no point in living. . . . I guess a feeling of complete and total helplessness. . . . As I said before, a feeling of helplessness, of being all alone in this situation—that there was nothing anybody else could do or say that could help me out of this . . . self-contempt! . . . I guess there still must have been doubt or I felt that there was still some possibility that I could become the person that I wanted to be.

S1 apparently gets glimpses of being unable not to be that person she must not be as well as of being unable to be the person she must be and in those glimpses comes close to depression and despair. Yet she ultimately resists this negative sureness and retains the possibility that she can still cease to be the negative self she does not want to be/must not be and can still become the person she wants to be/must be. In doing so, she wards off the being-depressed and does not despair. She is discouraged a bit, but she is not demoralized.

C. Getting-Over-Phase

> In the getting-over-phase of being-ashamed, as in that of other affects, one is faced with two principal choices: to *forget* the discrepancy in "being" or to *resolve* it. One can cultivate forgetfulness by using direct or indirect strategies. Because of the absolute disvaluation of the deeds/characteristics and self in self-disgust and self-hatred, however, it is quite difficult to accomplish this distraction and forgetting. One of the indirect strategies involves making excuses for the underlying discrepancy in "doing"/"having," excuses which one does not fully believe (i.e., they are self-deceptive). These excuses enable one to shift blame for one's failure onto Others or the world in general. Likewise, as in other affects, one can also resolve the discrepancy at issue. In being-ashamed this can be accomplished by undoing the negative self-definition or by modifying or giving up one's failed self-project. One can undo the negative self-definition by undoing the founding discrepancy in "doing"/"having" (i.e., by atonement or a firm-purpose-of-amendment or analogous forms of both) or by undoing the lived generalization from "doing"/"having" to "being." It appears difficult, however, to give up the self-projects at issue.

The getting-over-phase is the fulfillment of the *conative* dimension of the being-phase. The calls that were experienced in the being-phase are acted upon in this phase. They are actions aimed at undoing the pain of that felt perception that is the core of the being-phase. *The function of the getting-over-phase is to eliminate the pain of the affected knowing.*

As mentioned above, the heart of the affective experience of being-ashamed is the felt perception of a discrepancy between one's action/characteristics and one's self-projects—a discrepancy between who one apprehends oneself to be and who one does not want to be or must not be. Considering that this is the discrepancy the felt perception of which constitutes one's being-ashamed, there are two basic modes of getting over the experience: (1) by *un-*

doing one's *painful awareness* of that discrepancy in some way or (2) by *undoing* the *discrepancy itself* in some way.

The first alternative is to pursue the *path of forgetfulness*. This course aims not at internal consistency, but at quick relief of the painful being-affected. It ignores the lack of integrity and overlooks the conflict or split within oneself. The second alternative constitutes the *path of resolution,* the course of *reintegration*. It achieves further personal integrity or consistency either by making one's actions consonant with one's moral values and/or self-ideals or by modifying one's moral values and/or self-ideals to have them cohere more closely with one's patterns of behavior and collection of characteristics.

1. The Path of Forgetfulness. Even those who eventually pursue the path of resolution appear to experience an immediate, spontaneous impulse to run from the painful self-awareness and to pursue forgetfulness of one's failure. In fact, almost everyone's first impulse, it appears, is to do precisely that. S4 illustrated this call to run and the felt conflict between running from one's failure in "being," on one hand, and grappling with it to resolve it in a more congruent fashion, on the other. He said:

> Truly, I wanted to run from my inner thoughts of self deprecation and shame. . . . 'Cause it isn't easy for me to admit that I am not what I would like to share with other people. . . . I don't like it when something reminds me that I'm not the self I preach. . . . So I don't want reminders because . . . I suspect that, if I go too deep, I'm going to find a monster that doesn't fit any of my conscious values or anything I appreciate about life. I'm going to find a monster there. . . . I don't want to look. . . . It's almost like not wanting to go through that therapy with yourself.

S4 is afraid to face the discrepancy lest he be radically disillusioned and lose faith in himself. He is anxious that he will never be able to not be who he does not want to be/must not be or be able to be who he wants to be or must be. It is this fear that drives him in the direction of cultivating forgetfulness. It is this fear that we all share to some degree.

Still, while one's immediate impulse is almost always to run from and forget the painful discrepancy, I must say that, at least in the cases of self-disgust and certainly self-hatred, forgetfulness is not easy to achieve. Precisely because, in these two modes of self-disesteem, the failed self-projects have been lived with a "must not" motivational investment and thus are lived intensely and felt deeply, it is not easy to accomplish the desired forgetfulness. One has a difficult time letting go of the focus on the failure, and events tend easily to remind one of it. Even more, what often happens is that one's first responses (as in S1's case) are spontaneously in the direction of cultivating forgetfulness, but one's subsequent responses come to grips with the difficult-to-forget discrepancy. S1, for example, immediately chooses to ignore her shameful awareness that she has not been the careful and responsible person she wants to be. Yet, eventually, this awareness emerges again with intensity, and she chooses to face it and work through it. As she says, that was her "moment of truth" when she "couldn't hide from [herself] any longer."

The forgetfulness that is cultivated is not generally permanent, and it certainly is not complete. Forgetfulness means that one has managed not to attend to the discrepancy until it no longer exerts any call to one on its own. Other events and circumstances, however, can recall the shameful event for one and provoke the reemergence of the discrepancy into one's focal awareness. One does not, as perhaps occurs in amnesia, obliterate from consciousness this discrepancy in "being"; it is only put out of one's thoughts. One still remains vulnerable to attacks of recollection, as it were. Further, even if one manages to forget some specific discrepancy in "being," one does so only incompletely. What occurs in forgetting such discrepancies is that one forgets the concrete grounds or occasion for the self-disesteem, but one is still left with a vague and unthematic sense of being deficient as a person. This unthematic sense of being deficient serves as a ground for one's experiactions and becomes more insidious precisely as a groundal rather than a focal phenomenon.

There appear to be two general modes of cultivating the desired forgetfulness of the painful discrepancy in "being": directly or indirectly. These modes are not mutually exclusive, and one can pursue either or both ways of inducing forgetfulness. The reader is referred to Vallelonga (1986, pp. 1147–1166) for explanations and illustrations of these modes.

2. *The Path of Resolution.* The second way of getting over one's experience of being-ashamed (in all three modes) is the path of resolution, that is, by actively undoing or resolving the feelingly perceived discrepancy between one's self-projects (to not be a certain kind of person) and the current actual events (that one's behavior or characteristics do suggest/reveal one actually to be that certain kind of person). Since the discrepancy entails a divergence between one's projects and the actual events, the only way actively to resolve the discrepancy is to change one of its poles in order to bring both into alignment or congruence with each other. The only choices one has, then, are (a) to give up or modify one's project not to be a certain negative kind of person (i.e., to change oneself or the self-pole of the discrepancy) or (b) to change the fact that one's behaviors or characteristics reveal one to be that certain kind of person (i.e., to change the world or the world-pole of the discrepancy). In other words, the only alternatives one has in resolving the discrepancy at issue are to change either the way one wants things to be or the way they are (from the agent's perspective). Changing the way things are, obviously, can boil down simply to changing the way one perceives things to be.

a. In the first choice (i.e., to change the way one wants things to be), what one is changing is one's self-project or project not to be a certain kind of person. In this mode of resolving the discrepancy, one does not make one's underlying behaviors/characteristics or one's overarching self-definition conform to or fulfill one's self-project, but the other way around. One makes one's self-project conform to, or at least not be discrepant with, one's actual behaviors or characteristics and the self-definition founded on them.

While, on the surface of things, it might appear as easy to change one's self-project as to change the presumptive actual events and one's self-definition, or even easier, such does not appear to be the case. Persons, unless explicitly directed and aided to do otherwise (as in forms of cognitive therapy), appear to hold onto their self-projects in an uncritical fashion. They frequently struggle to live up to their oppressive self-projects or endure the sense of failure rather than question the reasonableness of these constrictive self-projects. There are two ways of changing one's unfulfilled self-projects (i.e., the self-pole of the discrepancy): (i) by further specifying or delimiting one's self-project or (ii) by simply giving up the self-project at issue. (For further explanations and illustrations of these resolution strategies, see Vallelonga, 1986, pp. 1206–1212.)

Suffice it here to give an example of the strategy of giving up one's self-project at issue. Thus, S3 implied that she gave up her project to be a perfect nun. While she was at first ashamed to perceive that she was the kind of nun she felt she must not be (i.e., one who is not always in control of her emotions), she eventually appreciated her failure as an indication of her being "human," which she now positively values. She appeared to give up the project not to be an imperfect nun or to be perfectly in control of her emotions. She accepted her failures as an indication of her humanity and her ability to be more compassionate. She said:

> I was ashamed before . . . , *but the whole thing is kind of endearing in a way.* This is kind of hard to explain. You know—*that I was so human. R:* You see this as a moment of growth too? *S:* Yeah, this whole thing was very growthful. You know, I like knowing that I'm weak *because that's the truth,* you know. But I can recognize it—the experience. *And I think it is the basis of compassion,* you know, and understanding. . . . It's a release. *R:* So there's a lot of positive aspects even to your being imperfect? *S:* Uh-huh, uh-huh. Because this shame did pass (emphasis added).

b. In the second choice (i.e., to change the presumptive actual events), what one is basically changing is the negative, current self-definition that is discrepant with one's self-ideal or self-project. One can change this negative self-definition in two broad ways: (i) by undoing the founding discrepancy in "doing"/"having" by changing either the behavior/characteristic at issue or one's underlying "doing" or "having" project, or (ii) by undoing the lived generalization from this discrepancy to a negative definition of one's self. In these two modes of undoing the discrepancy in "being," one's self-projects remain exactly as they were. One does not change the ideality pole of the discrepancy but its actuality pole, that is, the negative self-definition discrepant with one's ideal self.

The strategies for resolving the discrepancy are numerous, and the reader is directed to Vallelonga (1986, pp. 1166–1212) for examples and illustrations of them. Suffice it here to describe how some subjects resolve the discrepancy founded on a violation of moral values by using actual or analogous "atonement" and a "firm purpose of amendment." S1, for example, does feel ashamed for being careless and irresponsible in not exiting her parking spot slowly and in trying to leave the scene of the accident. She is distracted from this being-ashamed, however, by her getting caught and her being-embarrassed-over-actual-exposure. Toward the end of her interactions with the others, she begins to cooperate and finally assumes financial responsibility for the accident. As she does this, her being-ashamed for being irresponsible is eased and she feels relief. In effect, her assuming financial responsibility undoes the harm done; it constitutes "atonement." She says: "I felt, I think, relieved because now I was doing the right thing. I was cooperating with her and I was giving her my name and address. And, in doing so, I was taking responsibility for what I had done."

S1 adds at the end of her interview that sometimes she seeks symbolic restitution; she seeks out punishment for herself as a means of doing penance and undoing her being-ashamed on moral grounds. She declares: "But sometimes I tell people who I know will not react favorably to something that I have done. And maybe this is the way I have, you know, of being punished—by having someone else criticize me for what I've done." Later on in the same situation, Sl is ashamed for being so hypocritical. Part of the way she resolves the discrepancy in "doing" and "being" is to make a firm purpose of amendment not to be so critical of others. Her answer to the question of the "double standard" is "to be less critical of others and be less critical of myself."

2. How Being-Ashamed Differs from Being-Embarrassed and Being-Guilty

As indicated at the beginning of this chapter, three of the goals of my research project were (1) to elucidate the concrete lived structure of being-ashamed; (2) to clarify the differences between being-ashamed and its cognates, being-embarrassed and being-guilty; and

(3) to uncover the grounds for the persistent ambiguity in the realm of shame. Let us now turn to comparisons of being-ashamed with both being-embarrassed and being-guilty.

a. A Comparison of Being-Ashamed and Being-Embarrassed

The major difference between these two phenomena is that being-embarrassed-over-actual-exposure entails the central experience of *inter*personal exposure and *public* disesteem, while being-ashamed entails the central experience of *intra*personal exposure and *self*-disesteem. In being-embarrassed, one primarily undergoes the negative transformation of one's *present world,* while in being-ashamed one primarily undergoes the negative transformation of one's *present self.* In addition to this central one, other differences were also found.

Thus, in the becoming-phase, the ground for being-embarrassed is a ground that one believes the other disvalues but that one may or may not disvalue oneself. In being-ashamed, the ground is always personally disvalued. Likewise, in being-embarrassed, the grounds tend to cluster in the area of *accidents* and breaches in one's standard public self-presentation (e.g., conspicuousness, impropriety, breaches of privacy); in being-ashamed, they cluster in the area of *failures* in personal "being" (e.g., badness, weakness, incompetence).

In the being-phase, there were other differences besides the central one noted above. In being-embarrassed, one perceives the other as living a generalization from one's "doing"/"having" to one's "being": in being-ashamed, one lives the same generalization about oneself (i.e., one allows one's deeds or characteristics to be negatively definitional of one's identity and one's worth). Likewise, in the former affect, the quality of the motivational investment in the failed project does *not* determine a range of types, while it does in the latter (i.e., self-dislike, self-disgust, and self-hatred). Similarly, on the bodily/behavioral level, one blushes and feels hot in the former one, while one feels a certain physical distaste or revulsion/nausea in the latter. In being-embarrassed, one experiences a call to hide, literally and physically, from the other, while in being-ashamed, one is called to hide figuratively from oneself (i.e., by cultivating forgetfulness). In being-ashamed, one experiences a dialectical thrust toward generalizing from present "doing"/"having" to future "being," that is, to discouragement, depression, and despair, whereas in being-embarrassed, one does not. Likewise, being-embarrassed entails a "metalevel" (embarrassed over being embarrassed), but being-ashamed does not.

In addition, in the getting-over-phase, there are a few differences among many similarities. An example is the fact that one resolves the discrepancy (by changing the world pole) in being-embarrassed by what is called "facework," while in being-ashamed, one does so by what is called "atonement" and a "firm purpose of amendment" (both real and analogous).

The "exposure" type of being-embarrassed and being-ashamed are correlative and cognate phenomena. (For a comparison of the "exposure" and "intrusion" types of being-embarrassed, see Vallelonga, 1986, pp. 695–765.) They are two sides of the same coin of experienced disesteem. Further, there is a dialectical relationship between these two vicissitudes of personal esteem, and the experience of each calls for the experience of the other. In fact, the experience of each can be included as a constituent in the fuller experience of the other. Yet they are not the same event, and they can and do occur without each other. One can be embarrassed-over-actual-exposure without experiencing, immediately or subsequently, any self-disesteem at all. Similarly, one can experience self-disesteem apart from any prior experience of interpersonal disesteem or any subsequent experience of (presumptive) actual interpersonal disesteem. Every experience of self-disesteem, however, calls for and, it appears, *requires* the experience of at least *potential* interpersonal disesteem, that is, being-self-consciously-embarrassed. The dialectical relationship between these two phenomena, along with the equipotentiality of a shame-situation (discussed below), has grounded much of the confusion between these two phenomena.

Before proceeding to the next comparison, the relationship between the two types of exposure plus objectification exemplified by these two phenomena bears some discussion. My research study demonstrated unequivocally that the occurrences of intra- and interpersonal exposure plus objectification (like being-ashamed and being-embarrassed) are *dialectically* related as negative transformations of, respectively, self and world. Just as self and world are mutually horizonal to each other (much like figure and ground), so also are their lived negative transformations. Each one calls for and leads into the experience of the other. It is for this reason that most often the generic label "shame" is used to denote an experience that entails both exposures. Thus, both being-guilty and being-ashamed entail the experience of at least a correlative self-conscious-embarrassment. Likewise, most experiences of *inter*personal exposure tend to lead into other experiences of *intra*personal exposure.

Each of these kinds of dialectically related negative transformations (and exposures) also serves as the central constituent for a group of cognate shame-phenomena. Thus, a negative transformation of present self (or *intra*personal exposure) is the central constituent of being-ashamed, being-guilty, and being-regretful (along with their corresponding attitudes, conscientiousness and shyness). On the other hand, a negative transformation of present world (or *inter*personal exposure) is the central and focal constituent in all types of being-embarrassed, in being-humiliated/shamed, and being-disgraced/dishonored (as well as in their corresponding attitudes, somatic and psychic modesty, tact/discretion, and shyness).

b. A Comparison of Being-Ashamed and Being-Guilty

Both being-ashamed and being-guilty are experiences of *intra*personal exposure and self-disvaluation. Both entail the experience of a negative transformation of one's present self that calls for the correlative negative transformation of one's present world in, at least, being-self-consciously-embarrassed and of one's future self and world in being-anxious. Against the background of these similarities, I can highlight the following differences:

Relative to the becoming-phase of these structures, the ground for being-ashamed can be a deficiency in either the "doing" or "having" dimension and can entail either a moral or a nonmoral failure. For being-guilty, the ground is only a moral failure in the "doing" dimension. In addition, being-guilty requires various enabling maneuvers (i.e., self-deceptive strategies) to accomplish the becoming-phase. Being-ashamed does not.

In the being phase, the primary difference between being-guilty and being-ashamed is that the core constituent of the former consists in the felt perception of a single discrepancy in moral "doing," while the core constituent of the latter consists in the felt-perception of *two* discrepancies—a founding discrepancy in "doing" or "having" and a founded one in "being." Thus, being-ashamed entails allowing one's deeds or characteristics to have negative definitional significance as to one's identity and worth (i.e., a lived generalization), while being-guilty does not. The focus remains on the discrepancy in moral "doing" in being guilty. In addition, there is only one mode or degree of being-guilty, but three of being-ashamed (i.e., self-dislike, self-disgust, and self-hatred). Likewise, being-guilty tends to mobilize one to action, while being-ashamed tends to discourage and demoralize one so that effective action is undermined. Being-guilty thus can lead to personal integrity, while being-ashamed can lead to personal disintegration (e.g., depression, despair).

The only notable difference in the getting-over-phase of both is that in being-ashamed one can also accomplish, not just a real, but also an analogous version of atonement and a firm purpose of amendment (to undo nonmoral discrepancies). Likewise, being-ashamed can be resolved by explicitly undoing the lived generalization from "doing"/"having" to "being."

Further, while Becker (1968) suggested that guilt may be an essential part of shame and Ausubel (1955), on the other hand, indicated that shame is a part of guilt, my research demonstrated that each can be a constituent in the fuller experience of the other. Thus, being-ashamed *may* be part of the being- and getting-over-phases of being-guilty, while being-guilty *may* be part of the becoming-phase of being-ashamed. Thus, being-guilty may be an antecedent to or ground for being-ashamed. One is frequently ashamed on account of moral failures, but one is also ashamed on account of other, nonmoral failures. Likewise, being-ashamed may be a consequent or elaboration of being-guilty. Thus, one's being-guilty frequently, but not always, develops into a being-ashamed.

It is also worth noting that these analyses disclosed that among the shame-phenomena, the dialectic between transformation of self and world was not the only one operative. There is also one among the three dimensions of lived time. Thus, the relationship between being-embarrassed (all types), being-guilty, and being-ashamed, on one hand, and being-anxious, on the other, entailed a dialectic between negative transformations of one's lived present and lived future. Likewise, the relationship between being-ashamed, on one hand, and being-ashamed-of-one's-past-self and being-disillusioned-in-oneself, on the other, entailed a dialectic between one's lived present and one's lived past. Similarly, the relationship between being-guilty and being-ashamed disclosed another dialectic, that is, a quasi-dialectic between the "doing"/"having" dimensions and the "being" dimension.

3. The Equipotentiality of a Shame-Situation

After the essential lived structures of being-ashamed, being-embarrassed, and being-guilty, were articulated, it became clear in my original research that *several shame affects* (or their central felt perceptions of discrepancies) *can and do occur together in dialectically connected clusters within the scope of a single experienced situation.* Thus, no matter whether one is inspecting a description of a situated experience of "shame" or "embarrassment" (to mention two), one is likely to encounter descriptions of any combination of the following moments (among others): the perception of a discrepancy in moral "doing" (being-guilty), the perception of some nonmoral failure (being-regretful), the experience of some degree of disesteem of one's present self (being-ashamed), the experience of some form of public exposure plus objectification (being-embarrassed over actual or potential exposure or intrusion), and the anticipation of various negative eventualities (being anxious). Each of these moments, however, serves as the core constituent of a different shame-affect. That being so, *the entire situation could legitimately be denominated by the name for any of these affective moments,* that is, as, respectively, being-guilty, regretful, ashamed, embarrassed/humiliated/disgraced and anxious in the face of exposure. As is obvious, such *ambiguity or denominational multivalence* has, in fact, produced considerable confusion and controversy.

The point is that there are multiple moments that typically make up a shame-situation. In addition, the shame-situation can be and is denominated differently by different authors and experiencing individuals depending on which moment has the most significance for that person. As we have seen above, the negative transformations of present, past, and future self or world that form the core of the shame-affects can and do occur in *dialectically connected clusters* within the same lived situation. How that situation is experienced and denominated by the one who lives through it depends on which negative transformation stands out as *figural* for that person. Depending on which transformation stands out, the others (along with the other typical constituents of shame-situations) get spontaneously organized around it. Indeed, each situation can be spontaneously organized and experienced/denominated in as many different ways as there are perceived negative transformations of self or world. This state of affairs constitutes the "equipotentiality of a shame-situation."

What this means is that because of the dialectical relationships among the various transformations of self and world across the three dimensions of lived time, there is a certain amount of *ambiguity* that is *inherent* in the domain of shame-phenomena (and perhaps in the domain of other emotional phenomena as well). My empirical–phenomenological investigation of shame has determined which constituents make up the structures of being-ashamed, being-embarrassed, and being-guilty. It has even determined that these various constituents generally occur in most shame-situations and that the specificity of the situation depends on the organization spontaneously given to it by the person living through it. The specificity of the situation is *co-constituted* by the person, not just on the level of the projects he or she is living, but also on the level of what the person selects out perceptually as significant/figural to him or her. What this means is that no researcher can, from the observer perspective alone, examine a situation and declare with certainty that, because of the presence of such and such constituents, the person has experienced or will characterize the situation in a particular or specific way. We can only sketch out a range of possibilities. The actual character of the situation is determined by the person in interaction with the events/circumstances. Like Heisenberg's principle, the "equipotentiality of a shame-situation" declares that there is a certain ineluctable *"indeterminacy"* attached to the shame-domain.

It is this ambiguity and equipotentiality of a shame-situation (founded on the underlying dialectical dimension) that partially explains why different authors may, as indicated at the beginning of this chapter, describe the same constituents yet denominate the experience differently or speak of the same experience and thematize different constituents. It partially explains why some authors consider the experience of self-disesteem a shame-affect in itself, while others consider it a constituent of either being-guilty or being-embarrassed. It also partially explains the confusion in everyday life and in the literature between guilt and shame, and between embarrassment and shame.

CONCLUSION

The research reported and summarized in this chapter has helped to elucidate the essential "whatness" of being-ashamed and how it differs from that of being-embarrassed and being-guilty, and has uncovered what seems to be much of the grounds for the persistent ambiguity in the domain of "shame." In addition, it has demonstrated, contrary to the assertion by Fell (1965, p. 218), that emotional phenomena *are particularly amenable* to explication *solely* by the phenomenological means, that is, by an empirical–phenomenological approach.

REFERENCES

Albers, R. H. (1995). *Shame: A faith perspective.* New York: Haworth.

Ausubel, D. P. (1955). Relationships between shame and guilt in the socializing process. *Psychological Review, 62,* 378–390.

Becker, C. (1968). *An explication of embarrassment and shame within a situation of interpersonal exposure.* Unpublished master's thesis, Duquesne University, Pittsburgh, PA.

Benedict, R. (1946). *The chrysanthemum and the sword: Patterns of Japanese culture.* Boston: Houghton-Mifflin.

Bradshaw, J. E. (1988). *Healing the shame that binds you.* Deerfield Beach, FL: Health Communications.

Broucek, F. J. (1991). *Shame and the self.* New York: Guilford Press.

Buss, A. H. (1980). *Self-consciousness and social anxiety.* San Francisco, CA: Freeman.

Dann, O. T. (1977). A case study of embarrassment. *Journal of the American Psychoanalytic Association, 25,* 453–470.

Darwin, C. R. (1965). *The expression of the emotions in man and animals.* Chicago: University of Chicago Press.

Davitz, J. (1969). *The language of emotions.* New York: Academic Press.

Edwards, D. G. (1976). Shame and pain and "Shut up or I'll really give you something to cry about." *Clinical Social Work Journal, 4,* 3–13.

Emde, R. N., & Oppenheim, D. (1995). Shame, guilt and the Oedipal drama: Developmental considerations concerning morality and the referencing of critical others. In J. P. Tangney & K. W. Fischer (Eds.), *Self-conscious emotions: The psychology of shame, guilt, embarrassment, and pride* (pp. 413–436). New York: Guilford Press.

Fell, J. P. (1965). *Emotion in the thought of Sartre.* New York: Columbia University Press.

Fischer, K. W., & Tangney, J. P. (1995). Self-conscious emotions and the affect revolution: Framework and overview. In J. P. Tangney & K. W. Fischer (Eds.), *Self-conscious emotions: The psychology of shame, guilt, embarrassment, and pride* (pp. 3–22). New York: Guilford Press.

Fischer, W. F. (1974). On the phenomenological mode of researching "being anxious." *Journal of Phenomenological Psychology, 4,* 405–423.

Fischer, W. F. (1989). An empirical–phenomenological investigation of being anxious: An example of the phenomenological approach to emotion. In R. S. Valle & S. Halling (Eds.), *Existential–phenomenological perspectives in psychology: Exploring the breadth of human experience* (pp. 127–136). New York: Plenum Press.

Fossum, M. A., & Mason, M. J. (1986). *Facing shame: Families in recovery.* New York: Norton.

Freud, S. (1953a). Three essays on the theory of sexuality. In J. Strachey (Ed. and Trans.), *The standard edition of the complete psychological works of Sigmund Freud. Volume 7* (pp. 123–243). London: Hogarth.

Freud, S. (1953b). The interpretation of dreams. In J. Strachey (Ed.), *The standard edition of the complete psychological works of Sigmund Freud. Volumes 4 & 5.* London: Hogarth.

Freud, S. (1962). Further remarks on the neuropsychoses of defense. In J. Strachey (Ed.), *The standard edition of the complete psychological works of Sigmund Freud: Volume 3* (pp. 157–185). London: Hogarth.

Friesen, V. I. (1979). On shame and the family. *Family Therapy, 6,* 39–58.

Giorgi, A. (1975). An application of phenomenological method in psychology. In A. Giorgi, C. T. Fischer & E. L. Murray (Eds.), *Duquesne studies in phenomenological psychology: Volume 2* (pp. 82–103). Pittsburgh, PA: Duquesne University Press.

Harper, J. M., & Hoopes, M. H. (1990). *Uncovering shame: An approach integrating individuals and their family systems.* New York: Norton.

Isenberg, A. (1949). Natural pride and natural shame. *Philosophy and Phenomenological Research, 10,* 1–24.

Izard, C. E. (1977). *Human emotions.* New York: Plenum Press.

Kaufman, G. (1974). The meaning of shame: Toward a self-affirming identity. *Journal of Counseling Psychology, 21,* 568–574.

Kaufman, G. (1989). *The psychology of shame: Theory and treatment of shame-based syndromes.* New York: Springer.

Laing, R. D. (1965). *The divided self: An existential study in sanity and madness.* Baltimore: Penguin.

Lewis, H. B. (1971). *Shame and guilt in neurosis.* New York: International Universities Press.

Lewis, M. (1992). *Shame: The exposed self.* New York: Free Press.

Lewis, M. (1995). Embarrassment: The emotion of self-exposure and evaluation. In J. P. Tangney & K. W. Fischer (Eds.), *Self-conscious emotions: The psychology of shame, guilt, embarrassment, and pride* (pp. 198–218). New York: Guilford Press.

Lindsay-Hartz, J. (1984). Contrasting experiences of shame and guilt. *American Behavioral Scientist, 27,* 689–704.

Lindsay-Hartz, J., de Rivera, J., & Mascolo, M. F. (1995). Differentiating guilt and shame and their effects on motivation. In J. P. Tangney & K. W. Fischer (Eds.), *Self-conscious emotions: The psychology of shame, guilt, embarrassment, and pride* (pp. 274–300). New York: Guilford Press.

Lynd, H. M. (1958). *On shame and the search for identity.* New York: Harcourt, Brace.

Mascolo, M. F., & Fischer, K. W. (1995). Developmental transformations in appraisals for pride, shame, and guilt. In J. P. Tangney & K. W. Fischer (Eds.), *Self-conscious emotions: The psychology of shame, guilt, embarrassment, and pride* (pp. 64–113). New York: Guilford Press.

Mead, M. (1950). Some anthropological considerations concerning guilt. In M. L. Reymert (Ed.), *Feelings and emotions* (pp. 362–373). New York: McGraw-Hill.

Merleau-Ponty, M. (1962). *Phenomenology of perception.* London: Routledge & Kegan Paul.

Middleton-Moz, J. (1990). *Shame and guilt: Masters of disguise.* Deerfield Beach, FL: Health Communications.

Miller, S. (1985). *The shame experience.* Hillsdale, NJ: Analytic Press.

Morano, D. V. (1973). *Existential guilt: A phenomenological study.* Atlantic Highlands, NJ: Humanities Press.

Nathanson, D. L. (1992). *Shame and pride: Affect, sex, and the birth of the self.* New York: Norton.

Nichols, M. P. (1991). *No place to hide: Facing shame so we can find self-respect.* New York: Simon & Schuster.

Potter-Efron, R. T. (1989). *Shame, guilt and alcoholism: Treatment issues in clinical practice.* New York: Haworth.

Riezler, K. (1943). Comment on the social psychology of shame. *American Journal of Sociology, 48,* 457–465.

Sattler, J. M. (1965). A theoretical development and clinical investigation of embarrassment. *Genetic Psychology Monographs, 71,* 19–59.

Scheler, M. (1957). Ueber Scham und Schamgefuehl. *Schriften aus dem Nachlass. Gesammelte Werke: Volume 10.* Bern: Verlag Francke.

Schneider, C. D. (1977). *Shame, exposure and privacy.* Boston: Beacon Press.

Shaver, J. H. (1976). *The feeling of shame.* Unpublished doctoral dissertation, Duquesne University, Pittsburgh, PA.

Solomon, R. C. (1976). *The passions.* Garden City, NY: Anchor/Doubleday.

Stierlin, H. (1974). Shame and guilt in family relations: Theoretical and clinical aspects. *Archives of General Psychiatry, 30,* 381–389.

Straus, E. W. (1966). Shame as a historiological problem. In *Phenomenological psychology: Selected papers* (pp. 217–224). New York: Basic Books.

Tangney, J. P., & Fischer, K. W. (Eds.). (1995). *Self-conscious emotions: The psychology of shame, guilt, embarrassment, and pride.* New York: Guilford Press.

Tomkins, S. S. (1963). *Affect/imagery/consciousness. Volume 2: The negative affects.* New York: Springer.

Vallelonga, D. S. (1986). *The lived structures of being-embarrassed and being-ashamed-of-oneself: An empirical phenomenological study.* Unpublished doctoral dissertation, Duquesne University, Pittsburgh, PA.

von Eckartsberg, R. (1977). *Psychological research methods at Duquesne.* Unpublished manuscript, Duquesne University, Pittsburgh, PA.

7

Latin American Women's Experience of Feeling Able to Move Toward and Accomplish a Meaningful and Challenging Goal

Tania Shertock

INTRODUCTION

This chapter presents an existential–phenomenological study of, specifically, Latin American women's experience of feeling able to move toward and accomplish a meaningful and challenging goal. The purpose of the study was threefold: (1) to elucidate the experience being investigated, (2) to add to the literature exploring women's consciousness, and (3) to study the psychology of women who have had this experience, as well as the psychology of women in Latin America.

The direction for this study was derived from an earlier pilot study I completed concerning Latin American women and acculturation in which ten women from ten different Latin American countries were interviewed. I was immediately struck by the fact that in every case the members of this randomly selected group of women were able to significantly improve the quality of their lives since arriving in this country. It became interesting to consider the experience of those women who had been able to excel in Latin America in the sense that they had been able to move toward and accomplish a goal that they considered

Tania Shertock • Institute of Transpersonal Psychology, San Francisco, California 94131.

Phenomenological Inquiry in Psychology: Existential and Transpersonal Dimensions, edited by Ron Valle. Plenum Press, New York, 1998.

meaningful. The study reported in this chapter explored the quality of their experience and the means by which they felt able to empower themselves.

As did J. B. Miller (1991), I considered the definition of power in common usage as concerned with augmenting one's own force by having authority and/or influence as well as controlling and limiting others, as opposed to "the great ability to do, act, or produce" (Gove, 1966). Denmark, Tangri, and McCandless (1978) had pointed out that research on power had largely been focused on male subjects, male personality, and masculine spheres of activity. According to Denmark et al. (1978), there was a dearth of research on power motivation in women, and from their review of the literature, theoretical and experiential research did not seem to take into account any different or unique relationship to power that women might have.

Important theories concerning this subject have emerged from collaborators at the Stone Center for Developmental Studies (Jordan, Kaplan, Miller, Stiver, & Surrey, 1991), who created a series of documents that eventually served as a vehicle to provide alternate descriptions of the traditional system of thought concerning the development of women. For example, they proposed an alternate model of human interaction as "power with" or "power together" or "power emerging from interaction," which contrasts with the more traditional "power over another" model. Surrey (1991) proposes that this more collaborative model of power overrides the active/passive dichotomy of the dominant culture by suggesting that all participants interact in ways to foster connection, in turn enhancing everyone's personal power. The significance of this study lies in its exploration of the process of empowerment in a woman's lived experience.

A REVIEW OF THE LITERATURE: FROM THEORIES OF THE MATRIARCHY TO SCIENTIFIC RESEARCH

This literature review spans many perspectives that address the psychology of women and that are implied in researching the experience that is the central focus of this chapter. A number of studies paint an early picture of a "woman's experience" in studies gathered by scholars who have investigated goddess cultures. Bachofen (1967), Diner (1973), Lerner (1986, 1993), Eisler (1987), and Gimbutas (1974) collectively document the wealth of information derived from archaeological log books and ancient texts on cultures that were variously termed "matriarchal," "matrilineal," or "matrifocal," the existence of which discounts the patriarchal proposition that the secondary roles of women are natural and divinely ordained. It is evident that the current interest of many women in feminine spirituality and the early goddess cultures may serve to inspire them toward a more positive sense of female identity. These studies on the goddess cultures suggest a range of issues involving self-empowerment in women—issues worthy of future research.

Concerning Latin American women, I examined their cultural context, including the historical perspective, their relationship to the family, work, politics, and religion, and stories collected from individual women (Shertock, 1994). It became evident that their understanding and utilization of power were significantly different from those of First World women.

The literature of this area elucidates this relationship between Latin American women and power. For example, Stevens (1973) and Lavrin (1987) contrast *machismo* to the existence of *marianismo*, which has been described in terms of feminine spiritual superiority. Marianismo is a movement within the Roman Catholic church that has as its object the veneration of the Virgin Mary, but its roots are in the ancient mother goddess cults in Europe in which women were venerated as the creators of life. It has been translated into a special form of authority within and outside the home whereby confrontations with men are avoided and there is an understanding of the boundaries of power of each sex. Both Stevens and Lavrin feel, however, that in today's society, the aforementioned value system, which strengthens traditional gender roles and diffuses tensions in male–female relationships, is likely to have undergone changes.

Chaney (1979) has explored how gender roles function outside the home. Her study, entitled *Supermadre,* resulted from a survey she conducted with women in Chile and Peru. She concludes that when women enter politics, they consider their activities as an extension of their family role. Chaney refers to a woman official who described herself as a *supermadre* and reflects on the positive aspects of women's concern with moral issues. She cites Jaquette (1974), who describes this phenomenon in terms of women wielding formidable power in certain spheres and behind closed doors, but laments that it is difficult to document because it is subtle and hidden. Lerner (1986, pp. 4–5) says: "Women are and have been central, not marginal, to the making of society and to the building of civilization . . . they are and always have been actors and agents in history."

Gross and Bingham (1985), for example, have documented a wealth of information that portrays Latin American women who, through the power of collective action, have changed the course of history in their countries. Among many examples are the Madres de Playa de Mayo, who drew world attention to their missing children, the Arpilleras groups in Pinochet's Chile, and a small group of Bolivian coal miners' wives who elicited the support of 1200 people in the service of their cause by going on a hunger strike.

The inner resources that can assist women to accomplish their goals have also been addressed. For example, the Jungian approach to feminine consciousness and psychology is symbolic and, as such, describes feminine forms of activity and understanding within all people, men as well as women, rather than gender-based stereotypes. This approach is important precisely because it adds a dimension not present in this way in other schools of thought. Eminent analyst and professor Ulanov (1971, 1981) explored the feminine principle. Emma Jung (1957) considered the *anima* and the *animus* to be archetypal figures of especially great importance. She contributed to the understanding of the animus, the masculine principle, as it appears in its relation to the individual and to consciousness. She chose to present the anima as an element of being, emphasizing its natural aspect, which she considered the essence of the feminine.

Renowned writer and analyst Singer (1976) adds to this understanding with her work on androgyny, as does Woodman (1982), who offers hope for women who find creative potential in a partnership between their feminine principle, which moves in a spiral motion toward what is meaningful, and the revitalized masculine principle, which has the penetrating power to inseminate and release the creativity of the feminine. Bolen (1984), with her version of a new psychology of women, affirms the diversity of normal variations among women by exploring the various goddess archetypes in order to provide women with additional means of understanding themselves.

According to Walsh (1987), no topic has produced more passion within the field of the psychology of women than Freudian theory. The debates, as chronicled by Walsh, took place between feminists who repudiated the Freudian paradigm (e.g., Greer, 1987) and those who sought a rapprochement between psychoanalysis and feminism. A good example is the controversy that surrounds the psychoanalytical tradition regarding Freud's theories in relation to the development of feminine psychology and gender identity (see Horney, 1987; Klein, 1975; Mahler, Pine, & Bergman, 1975; Thompson, 1987). Various divergent theories, including more sociologically based theories of gender and sex differences are described below. More recent analytical contributors to the theory of women's psychology include Zanardi (1990), who points out that psychoanalysis was the first science to confront the study of sexuality and that it became a means of identifying female sexual repression and its psychological importance to the individual and social development of women. As such, its contribution has been to surpass social determinism by changing the focus from so-called "objective reality" to the subjective realm of intrapsychic processes.

As a founding member of the object relations school, Klein (1990) has contributed to the discourse on feminine psychology through her psychoanalytical work with children. She challenged Freud on his concept of an inferior superego formation in women. Chodorow (1978) made a contribution by exploring the effects of women's mothering, which, she says, "produces asymmetries in the relational experiences of girls and boys as they grow up which account for crucial differences in feminine and masculine personality, and the relational capacities and modes these entail" (p. 169). Chodorow (1989) presents a perspective that moves away from the predominantly sociological point of view of her earlier work toward one that emphasizes object relations theory. It takes into account the multiplicities of gender(ed) experiences.

Gilligan (1982) is renowned as one of a number of researchers who have established a new paradigm for the psychology of women. She explored identity formation and moral development and found that those who operated from this morality of responsibility and care, primarily women, tend to value ongoing dialogue and communication in relation to the problem and seek to develop an understanding of the context for the moral choice. She goes on to propose that "women's sense of integrity appears to be entwined with an ethic of care, so that to see themselves as women is to see themselves in a relationship of connection" (p. 171). Other researchers have investigated gender identity, sex-role stereotyping and adaptability, cross-cultural perspectives, the formation of morality and values, and attributes found in relation to success.

Belenky, Clinchy, Goldberger, and Tarule (1986) built on the work of Gilligan and examined women's ways of knowing based on the intensive interview/case study approach. They describe a number of different perspectives from which "women view reality and draw conclusions about truth, knowledge, and authority" (p. 67):

1. The Way of Silence—Represents an extreme denial of self and dependence on external authority for direction.
2. Received Knowledge—Listening to the voices of others. Women who think of words as central to the knowing process, and who learn by listening, mostly to authorities.
3. Subjective Knowledge A—Truth as private, personal, and subjectively known or intuited. Truth resides within a person and can negate answers the outside world supplies.
4. Subjective Knowledge B—The Quest for Self: Women who have discovered personal authority and truth and have gone from this newly acquired subjectivism to insist on shaping and directing their own world.
5. Procedural Knowledge A—Women who once relied on a mixture of received and subjective knowledge, now look to feeling and intuition for some answers and authorities for others. They later abandon both subjectivism and absolutism for reasoned reflection.
6. Procedural Knowledge B—Separate and Connected Knowing: Connected knowledge *(gnosis)* represents an understanding through a direct, personal acquaintance with an object or person. This view implies acceptance and precludes evaluation, since evaluation puts the object at a distance and quantifies a response to the object that would otherwise remain qualitative.

7. Separate Knowing—Uses standards to evaluate the analysis of something, knowledge being that which implies separation from the object and mastery over it.
8. Constructed Knowing—Integrating the Voices: This way of knowing and viewing the world involves an effort to reclaim the self by attempting to integrate knowledge that they felt intuitively was personally important with knowledge they had learned from others.

Loevinger (1976) offers a model of development of the self in relation to other, a model devised and validated through research on women. Her autonomous stage reflects a balance between the poles of the separation–connection continuum referred to by Gilligan (1982) and Chodorow (1989) and in doing so provides a valuable indicator regarding the psychology of women.

Maccoby and Jacklin (1974) have compiled an extensive review of more than 1200 works covering key areas such as intellect and achievement (perception, learning, memory, intellectual abilities, cognitive styles, achievement motivation, and self-concept) and social behavior (temperament, social approach–avoidance, and power relationships). Their results indicate that measurable and proven sex differences are relatively few.

Cross-cultural research suggests that characteristics that are attached to men and women vary according to culture. This research is important because it established culture as a determinant in the development of gender identity. Margaret Mead's (1949) studies describe societies (the Arapesh, Tchanbuli, and Iatmul) in which expectations for male and female behavior vary quite dramatically, leading her to conclude that cultures emphasize and reinforce behavior according to varying criteria. Barry, Bacon, and Child (1957) used data from ethnographic reports available in anthropological literature about socialization practices in 110 cultures. Their results indicated an overwhelming reinforcement of nurturance, obedience, and responsibility for girls, and achievement and self-reliance for boys. They found, however, that "there are many instances of no detectable sex differences; these facts tend to confirm the cultural rather than directly biological nature of these differences" (p. 332).

Block (1973) discusses some socializing influences as they impinge on the development of sexual identity. She examined the relationship between personal ego maturity and less traditional definitions of sex role by drawing on cross-cultural studies and longitudinal data.

Bem (1975) is recognized for her work equating androgyny with mental health and for creating the Bem Sex Role Inventory. She then shifted her perceptive to what she terms Gender Schema Theory (Bem, 1987). This theory contains aspects of social learning and cognitive developmental theories and proposes that sex typing is a learned phenomenon. As such, it is neither inevitable nor unmodifiable. Cantor and Bernay (1992) researched attributes of women executives in business, government, and education that contributed to their success.

What stands out in these many different studies is that many of the successful women studied perceived obstacles, not as obstacles, but rather as opportunities, and they did not report any serious problems in their formative years. This goes back to the importance of parenting emphasized by Dinnerstein (1976) and Chodorow (1978).

RESEARCH METHOD

In the spirit of phenomenological inquiry, I immersed myself in the literature of existential-phenomenology in order to shape a method that would best suit the nature of my personal process and the experience I wished to research. The result is a blend of styles from Colaizzi (1973, 1978), Giorgi (1985), and van Kaam (1959).

Personal Inquiry

The first step consisted of a radical self-examination that Colaizzi (1973) terms individual phenomenological reflection (IPR). This process involves the researcher's bringing to awareness his or her preconceived notions and biases regarding the experience being investigated so that the researcher is less likely to impose these biases when interpreting the subjects' reports of their experience. When later distilling the written descriptions in order to reveal their inherent meaning, one brackets these preconceived notions or biases; that is, one remains consciously aware of one's preconceptions while reading the descriptions so as not to unconsciously impose these notions on the subjects' descriptions.

Colaizzi asserts that we, as researchers, must incorporate self-reflective knowledge that we obtain from ourselves as conscious subjects in order to really understand the structure of the experiences being described by our fellow human beings.

Selection of Co-researchers

The term *co-researcher* will henceforth be used because it acknowledges the phenomenological perspective that the emergent meaning is co-constituted by the description of the experiences and the interpretive process of the one seeking the prereflective structure of the experience. The co-researchers were selected on the basis of their having had the experience under investigation and their willingness and ability to provide a description of that experience (Polkinghorne, 1989).

The co-researchers were 13 Latin American women who had the experience of feeling able to move toward and accomplish a meaningful and challenging goal. They were asked if in their view they had certain basic attributes as listed by van Kaam (1959): (1) having had the experience under investigation, (2) feeling able to express themselves in written and verbal form, (3) feeling able to express inner feelings without excessive shame and inhibition, (4) being able to sense and express the experiences that accompany these feelings, and (5) feeling a spontaneous interest in the experience. Unlike Colaizzi (1978), who stated that experience with the investigated topic and articulateness sufficed, van Kaam believed that the capacity to provide full and sensitive descriptions required more skills. Consistent with van Kaam's beliefs, co-researchers were informed that the process of reflection on their experience might bring them in contact with certain emotions. In being so informed, they were given the opportunity to consider whether they were prepared and able to participate in the research.

According to Polkinghorne (1989), the selection of co-researchers in phenomenological research differs from the logic of statistical sampling theory in that phenomenologists have as their goal the description of the structure of an experience, not the characteristics of a group that have had the experience. He goes on to say that the purpose of selecting subjects is to generate a full range of variation in the set of descriptions, not to meet statistical requirements for the purposes of generalization. The structure of the experience is derived from a combination of empirical descriptions and additional descriptions generated from imaginative thought experiments. The phenomenological researcher therefore needs to choose an array of individuals who provide a variety of descriptions specific to the experience being explored.

Written Protocols

Van Kaam (1959) devised a means to elicit narrative descriptions. Using his method as a model, the 13 co-researchers were given the following invitation in their native Portuguese language:

1. Select a time and space where you can be alone, uninterrupted and relaxed.
2. Please describe your experience of feeling able to move toward and accomplish a meaningful and challenging goal.
3. Recall a situation or circumstance in your life where you have been able to move toward and accomplish a meaningful and challenging goal.
4. Please describe how you felt at those times.
5. Try to describe those feelings so that someone reading or hearing the report would know just what the experience was like for you. Keep your focus on the experience, not just the situation itself.
6. Please do not stop until you feel that you have described your feelings as completely as possible. Take as long as you would like to complete your description.

Interview

Following completion of the written protocols, the co-researchers scheduled a time for a follow-up "walk through" interview, at which time they were given the following instructions:

> We're going to walk through the experience together just the way you wrote it. I'll read a part of it, then stop, and I would like for you to describe for me anything that comes to your mind about the part of the experience I have just read. You may provide further details about it, describe any thoughts or feelings you have, relate any images or metaphors that may come to your mind; in short, just let yourself go and associate to what I've read to you in whatever way you choose until you feel ready to stop. We'll then move on to the next part.

The purpose of this interview was to give each co-researcher the opportunity to deepen her original description by including anything additional that now came to mind. Once in the "walk-through" situation, however, I soon realized that the interviews needed to be more interactive and interpretive than had originally been planned. The intricacies involved with the differences in language and cultural context offered a larger ground for misinterpretation than was initially recognized. This necessitated my asking specific questions in the moment designed, in the spirit of collaboration, to correct any misperceptions the co-researcher may have had. Throughout this time, I adopted the following stance suggested by Wertz (1984) for conducting interviews with one's co-researchers:

1. Empathic presence to the described situation.
2. Slowing down and patiently dwelling with the interviewee on the details of the description.
3. Magnification, amplification of the details.
4. Turning from objects to immanent meanings.
5. Suspending disbelief and employing intense interest.

Of the 13 interviews, 9 were conducted in Portuguese and 4 in English. An interpreter was present for those conducted in Portuguese as a consultant on an as-needed basis. These interview reports were recorded on tape and then typed verbatim. For those interviews that were conducted in Portuguese, the transcriptions were translated, careful attention being paid to remaining as close as possible to the original meaning and flavor, even when the grammar and syntax were not entirely consistent' with those of English. Each translation was checked by several translators and, finally, carefully rechecked by the researcher to ensure that the original essence of the communication was preserved.

Protocol Analysis

The protocol analysis was partially adapted from Colaizzi (1973) and is outlined below:

1. The protocols were carefully read through in order to acquire a feeling for them.
2. Bracketing my biases, significant statements were then identified, for each protocol, as directly pertaining to the experience being investigated.
3. All the protocols were then reread several times until the significant statements across all protocols began to emerge or suggest themselves as categories or themes. As Colaizzi states, the number of categories that manifest is based solely on the classifications into which the explicit statements naturally group themselves. He goes on to say that the actual protocol statements, and not any derived interpretation of them, determine the number and type of category.
4. The protocols were then reread with the categories in mind to check for the comprehensiveness of the categories. Starting from a preliminary sense of its nature, a deeper and more integral understanding of each category was thus developed.
5. Each categorized statement in each category was then "translated" from its raw form to what are termed "components" for the purpose of communicating their meaning more clearly. Colaizzi describes this step as formulating meaning, by means of creative insight in order to delineate the meaning of each statement.
6. Steps 1–5 were then repeated for the interview data. The data collected in the interview were then compared and integrated with the written protocols.
7. All the written and interview components in all categories were then combined into a series of statements called fundamental descriptions (FDs) by the method of phenomenal study (PS) (Colaizzi, 1973). That is, the researcher focuses on the main phenomenon being studied as it was experienced, and culls out the essence of that particular description. Husserl (in Polkinghorne, 1989, p. 42) elucidates this process as follows:

   > In practice, the process leading to grasping the essential pattern of a structure usually requires a careful working through and imaginative testing of various descriptions of an essence, until the essential elements and their relationship are differentiated from the unessential and particular.

8. For each category, the researcher calculated the percentage of co-researchers whose descriptions contained that particular category (van Kaam, 1959). The percentages provide a more detailed sense of the variety of expression.
9. The fundamental descriptions were then examined in light of the original descriptions to check for their validity. This process of checking, which is utilized at each step until the final structure is distilled, involves a zig-zag procedure described by Polkinghorne (1989), whereby the researcher moves back and forth between the

significant statements, the components, and the fundamental descriptions to ensure that the results accurately reflect the original raw descriptions.

10. Based on Colaizzi (1973), the researcher then reflected on the fundamental descriptions and synthesized them in a way that recognized and held their essence. This process involved tying together and integrating the fundamental descriptions into consistent and systematic general descriptions or fundamental structural themes. Colaizzi describes this process as empirical phenomenological reflection (EPR).

Von Eckartsberg (1986, p. 51) summarizes Colaizzi's reasoning for the distinctiveness of the operational steps in his method:

> The difference between the findings regarding the fundamental description (FD) and the fundamental structure (FS) is seen to arise from a difference in focus on two layers of existence: reflective and prereflective life. Whereas the FD is a description of the subject matter in its reflective dimension which is thematically present to the subject's experience and can thus be directly reported upon request, as an appearance, the FS represents a structural elucidation of the prereflective dimensions which cannot be directly accessed by the subjects but which can be revealed to and elucidated by the reflective explication of the investigator by the act of interpretive reading.

In the study presented herein, the fundamental descriptions are synonymous with the described FD, while the FS is represented by the fundamental structural themes.

RESULTS

The prereflective structure implicit in the original protocols and interview transcriptions emerged as the following 18 categories (it should be noted that the comparison between the written protocols and the interview descriptions did not reveal any new categories):

1. Having a sense of knowing about, or believing in, their capacity and ability to achieve goals that they set for themselves.
2. Identifying needs, goals, and/or desires.
3. Having the ability to conceptualize ideas clearly, to construct an effective plan from the idea, and to organize, mobilize, and stimulate others to help realize the plan.
4. Having a sense that support from the environment (which provides strength to move forward) is derived from either the father, the mother, the husband, friends, teachers, or employees.
5. Feeling able to move forward despite obstacles, even if the task seemed impossible, when entering a previously all-masculine environment.
6. Experiencing work as important and feeling that one has to work very hard. One grows externally and internally, feeling happy about this change.
7. Liking and needing challenge, experiencing challenge as motivating, as empowering, as providing meaning, and as a source of learning.
8. Feeling that the work is necessary for society, and as pioneering, opening opportunities/spaces as yet unavailable to others; feeling the need to contribute or give something back to society, that competitiveness and aggressiveness are not only unnecessary, but counterproductive to accomplishing goals.

9. Feeling satisfied by the result or outcome of accomplishing the goal.
10. Feeling that it is important to work in conjunction with others and that the movement toward the goal has a participatory constituent.
11. Feeling that past education is important.
12. Acknowledging one's sociocultural origins.
13. Seeing the ability of moving toward a goal as a process that is applicable to goals in general.
14. Feeling helped or propelled by unseen forces, God, or divine providence.
15. Feeling an independence.
16. Feeling able to move forward for extended periods of time with one-pointed dynamism, enthusiasm, willpower, courage, and fighting spirit.
17. Acknowledging the importance of balancing family life with professional goals.
18. Having a major life event precipitated by external circumstances provide the impetus to either decide on or change one's life direction.

Fundamental Structural Themes

Following are the fundamental structural themes derived from step 10 of the analysis described in the Protocol Analysis section. Let us look at each of these themes in turn.

1. A Self-Reinforcing Cycle of Confidence

Many of the co-researchers in this study experienced an innate knowledge that they would be able to accomplish the goal that they had set for themselves. This was either a long-standing intuitive awareness about themselves that overshadowed doubts and fears or a sense/knowing that developed during a process whereby the accomplishment of one goal added to the person's confidence in being able to accomplish others. While they were not able to offer a logical explanation, they were aware of this strength that arose from within them. There was also a need or desire to give expression to this awareness that then developed into a life-path decision and subsequent goals.

2. Sociocultural–Political Origins and Events Outside Their Control

When describing their experience, many co-researchers began by characterizing their socioeconomic and cultural background or ascribed significance to it as a constituent of their experience of first feeling willing and then able to move toward significant goals. In many cases, it was the example of one or both parents who inspired them with a value system that placed a high value on work and education combined with an awareness of their limitless potential. This inspiration engendered, in many instances, love and enjoyment of both work and the process of learning derived through education or life lessons.

Another significant constituent was the nature of interpersonal relationships within the family or community that provided support, encouragement, or assistance that the co-researcher was able to translate into feelings of security, self-worth, or motivation to move forward. In some cases, there was a desire to balance reciprocal emotional support with the demands of professional career.

In many instances, circumstances beyond their control forced co-researchers to crossroads in their life at which they felt obliged to reexamine their life direction and make choices about the manner in which they would proceed. These events were either political or personal, and provoked deeply felt life changes.

3. A Vision of Community Purpose and the Stimulation of Lateral Participatory Activity

Co-researchers felt able to envision their ideas very clearly and construct viable plans to implement them. This involved the ability to organize and prioritize and also to motivate and empower others to participate in the plans in order to bring them to fruition. The driving force behind this ability was perceived in direct relation to how necessary the co-researchers felt their plans to be for the society in general. In many cases, co-researchers felt that it was essential for them to feel that their work was of benefit to the larger society.

Concurrent with this was the idea that they were opening spaces and creating new opportunities rather than competing with others. Interdependence was implied throughout the descriptions, including their working relationships with men. While many of their experiences involved working harder than men to prove their competence, their attention was generally on performing their work to the best of their ability and cooperating with the future vision of the project in mind. Many experienced great personal satisfaction from having been able to accomplish their goals, and this satisfaction was increased for some by the knowledge that they had made a difference in the lives of others.

4. Constituents of Their Ability to Continue Movement Toward the Goal

For some, the challenge itself engendered enthusiasm and the requisite strength to continue moving forward. In most cases, the co-researchers were able to move past obstacles, such as the challenges presented by entering into a predominantly male environment or resistance from others to their ideas. In some cases, they converted people to their position by the use of effective communication skills; in other cases, they concentrated their attention on getting the job done well rather than giving the obstacle the power to impact them. For example, some actually said that discrimination exists, but they simply paid no attention to it. In many circumstances, they were able to move forward, tolerating the perception of others that they were operating outside sociocultural norms for women.

Most co-researchers reported being able to face and move through their own weariness, insecurity, and fears in order to maintain their forward momentum. Many felt that once they had made a commitment to a goal, they were able to continue moving forward with perseverance and strong intentionality, without allowing themselves to become distracted until they reached their goal. Some were convinced that it was important to maintain positive emotional and mental states such as passion for the work, spiritedness, optimism, and enthusiasm. While only a few co-researchers mentioned independence specifically, the extent of their achievements is testament to the development of a high level of independence.

Finally, some co-researchers experienced a source of what they termed "spiritual power or strength" that guided them or assisted them when they felt the need for it. From their particular spiritual perspective, they regarded difficult and challenging experiences as preparation for larger humanitarian purposes and were able to perceive certain opportunities as signs from the divine regarding their life purpose.

REFLECTIONS, EXAMPLES, AND INTEGRATION

In many ways, the results of this study substantiate the theories of the Stone Center writers and other proponents of a new psychology of women (reviewed above) by revealing the relational contexts and modes the co-researchers operated within. They also indicate similarities in feminine consciousness that occur cross-culturally, especially the inclination (or drive) to relate/connect, and offer a greater understanding of the collective work of Latin American women and the structures that they have built to accomplish this work.

The Jungian approach emphasized the interaction between the masculine and feminine principles. When considering the meaning of this approach, it seems evident that many of the co-researchers have found a way to balance these principles within themselves and integrate this sense of balance into their relations with men in the outer world. There are, however, certain realities to consider when choosing to work and function in a cultural context that is largely dominated by patriarchal values and standards. These co-researchers expressed a choice to retain what they considered to be feminine attributes and values while moving into male-dominated spheres of influence. One person, a police commissioner, emphasized her humble origins and a desire to remain true to herself, including the notion to remain feminine within a police department that was perceived as arbitrary, brutal, and violent. She described her interpersonal working style as "stimulating others with kind words."

The theme of participation, where women work alongside men, resonated through many of the descriptions and reflected the dynamic side of the feminine described by Ulanov (1971) as an urge to relate, join, reach to, and get involved with concrete things and people, and not to abstract or theorize. One co-researcher attested, for example, that she would not write her speeches or allow them to be written ahead of time because she wished to base what she said on the nature of the personal interaction between herself and the audience. Another co-researcher concluded that political changes can come about only through civilian, grass-roots participatory movements, and she began moving toward the goal of bringing people together in a nonviolent fashion. Another felt that unless there was a high degree of participation and agreement for a project, it probably was not worth pursuing.

Another aspect of this participatory theme emerges in their descriptions of helping to build up and maintain organizational structures designed to promote societal change. This desire appears consistent with J. B. Miller's (1976) criterion of attachment and affiliation. One co-researcher helped to create an organization that would help women learn to come together to develop themselves professionally and personally in a spirit of cooperation. Another helped to start an association similar to the League of Women Voters that would influence politics for the benefit of society at large. A third created an organization that promoted awareness and worked for change on behalf of the physically challenged population.

Another aspect of feminine consciousness described by Ulanov (1971) as feminine comprehension was expressed in the descriptions as a sense of knowing where there is an awareness about something. In this case, it appears that the co-researchers' awareness of their capacity to achieve their goals seemed to arise from within rather than a situation in which they willed or thought out their ideas in a linear fashion. For example, one co-researcher described her process as follows:

> Sometimes I get tired and discouraged. Then something new will happen. Something new shows up and changes everything, opening another door. And I know I am meant to walk through it. So I do.

The Jungian approach also addresses the positive aspects of the masculine principle that, when realized in a woman, provide the ability to choose, discriminate, and find the strength of purpose to move toward a goal. The co-researchers in this study have all found this capacity within themselves, and could also be said to exhibit what Bem (1975) terms *psychological androgyny* in the sense that they display sex-role adaptability by being able to discern a goal and gather the fortitude to move forward and accomplish it.

Hemphill (1989) concludes that an important and potentially healing result of her study on the upward-mobility experience of executive females is that it invites the realization that success is an individual process and not simply gender-related. In the same way, I see the experience of the women in this study as indicative of the kind of experience that will eventually transform the overemphasis that our society currently gives to gender differences. One co-researcher said that after she had successfully integrated herself into a previously all-male domain, the women who came after her were not subjected to the same scrutiny and were judged far more on the basis of their competence.

The importance attributed to their family of origin by the women of this study is central here. Koltuv (1990) posits that the animus is formed out of the experience a woman has with her father or her mother's animus. In other words, if a woman has developed a strong, healthy relationship with her own masculine principle due largely to role models who have been able to accomplish meaningful projects in their lives, there is a stronger possibility that she will accomplish such projects herself. Many of the co-researchers referred to their father or mother as positive role models and inspirations for their life-path decisions. One co-researcher saw her father struggle against poverty, disaster, and a variety of misfortunes to create a substantial livelihood to feed his family with very little more than ingenuity and hard work. Another described how her father improved his situation by working and studying and later became successful in many and varied careers. She herself became an accomplished businesswoman, writer, and sculptor. This same co-researcher described her mother as the landmark of her life and asserted that it would have been very difficult for her to achieve what she did without the support and assistance provided by her mother.

Family dynamics were also addressed as important constituents of the co-researchers' experience. Many of them described parents or a parent who encouraged them to do whatever they chose regardless of the accepted social norms. One respondent was specifically encouraged by her father to become a doctor, a vision so far outside the societal expectations for her sex or class as to almost seem impossible. Another co-researcher believed that her enthusiasm to complete her goals came from her parents, who always encouraged her to do whatever she wanted to do.

Loevinger's (1976) autonomous stage people have the capacity to acknowledge and cope with inner conflict and conflicting needs and duties. They recognize that there are limits to autonomy and that emotional interdependence is inevitable. Many of the co-researchers expressed the importance of emotional interdependence whether it was in the workplace or balancing the demands of family and work and, in doing so, indicated a high level of maturity. Two co-researchers stressed the importance of balancing the giving and receiving of family support in conjunction with work responsibilities. They were also able to acknowledge inner conflict and move through it with their larger humanitarian goal in mind. One co-researcher said:

> In the same way as the consciousness grows, the responsibility increases, gets bigger proportionally, very rapidly. The realization that the difficulties that are given are also more complex brings forth the enormous desire to continue to fight for the common good, to better the quality of life of the population of my country.

Another expressed a great deal of initial despair in relation to her medical school experiences, but was able to move through it because she knew she had something to offer in terms of an alternative to the medical attention that she had received as a patient.

In the sense that many of the co-researchers felt it necessary for their work to be of benefit to society at large, their words reflect the characteristics described by Gilligan (1982). They were able to keep moving toward their goals specifically because they knew that what they were doing was necessary for the community at large. They felt connected to their community at large and needed by it. One woman stated:

> You feel fulfilled. It's the sensation of fulfillment. It's great, it's gratifying because it's not purely and simply that you have done a job but that you have been able to help people. This is important. I have this great need to do people a good turn, to work for the community, to do something that helps the community, to alleviate the suffering and uncertainty, the unhappiness of the people.

This aspect of the relational is evident in the literature on Latin American women. F. Miller (1991), for example, underlines Lerner's (1986) assertion that women's history in the Latin American context is part of the search for social and political justice for all people. This was certainly true for the co-researchers involved with politics. Several indicated that their involvement in politics was not without considerable risk to themselves and their families, but they were able to continue drawing on an inner strength, one aspect of which was a connection to their commitment to their life mission. One reports:

> Before finishing my second year, I abandoned classes to do political action. I was arrested and after my release from jail, I had difficulties to find work in my city, Salvador, so I moved to São Paulo. I arrived here in the beginning of the seventies, the worst phase of the military dictatorship and repression. Despite continuing being loyal to my political ideals, I did not want to run the risk, I thought I had a right to be happy. I had a great challenge in front of me: how to reconcile making money and continue to be loyal to my principles. That's when I discovered children's literature and the necessity of reading in a country of a low cultural level such as Brazil. I think I mastered this challenge very well, not only because I realized a pioneering and important work; I was the first person to be under contract as an editor to edit only children's books, because I realized that with this work I could succeed in earning a salary with which I could live with dignity, without betraying or renouncing principles.

Another woman described intense and very realistic fears that she had for the safety of her family and herself. Yet, she persevered and felt empowered by knowledge that she was working for the good of the city, a process that, in her own words, was irresistible. She says:

> Then—something will happen. Something new shows up and changes everything, opening another door. And I know I am meant to walk through it. So I do. And sometimes I'm scared. Sometimes there is hostility, the very real danger of physical aggression. Sometimes I fear for my family and their safety, and mine, and I feel quite helpless—and from way down deep inside, I feel the strength return. From somewhere.

The experiences expressed by women who engaged in political activity in this study are poignant examples of themes running through the literature on Latin American women. The concept of marianismo referred to by Stevens (1973), for example, is a significant distinguishing feature of Latin American feminine consciousness as it describes one of the ways the feminine currently functions in Latin America, and because of its connection to the ancient matriarchies. In this sense, there is a connection to the literature concerning the return to goddess consciousness, a consciousness that was always there but hidden and not spoken of for its own protection.

In the area of political activity, an important distinction between Latin American and North American women needs to be emphasized. Jaquette (1976) invites us to an expanded notion of female political participation via her examination of Latin American women's participation in groups and the roles they have played in informal networks of communication and influence. Throughout the literature, it is suggested that the political participation of Latin American women is not always overt and obvious but is nevertheless powerful. Latin American women's particular style of influence is fertile ground for research with vast interdisciplinary implications, including sociology, political science, and Latin American feminine consciousness.

The study presented in this chapter has shown that there are similarities in feminine consciousness that occur cross-culturally, especially the drive to connect, and interdependent modes of operating. In addition to this, however, as Angeles Arrien (personal communication) has stated, the feminine in Latin America has multiple ways of connecting to achieve political and social ends. As evidenced in both past literature and in this study, there is a much greater tendency for Latin American women to work together in a group than is generally seen in the political participation of North American women. I concur with Arrien that North American women have a greater tendency to imitate the male model of doing it alone, whereas Latin American women have a greater understanding of collective work. Throughout my study, women emphasized the need to generate cooperation. They view the achievement of goals as accomplished by the fostering of collective motivation.

Chaney's (1979) concept of supermadre sheds light on the means by which a woman can enter into a male-dominated domain and accomplish change. The co-researchers who worked in the political arena expressed a desire to work for education and the creation of a good community, using empathy, intuition, and cooperation (which Chaney describes as predominantly female characteristics). One co-researcher started her political career with the intention of helping disabled individuals, but eventually began to focus on helping all those on the margins of society and assisting them to become more integrated into the society. She describes her attention to the inner, human, psychological, and moral aspects of the isolated people of her constituency.

Another important theme that emerged from this study was the inclination to open new spaces and create new opportunities. The co-researchers described this impetus in terms of creating new areas of endeavor and addressing needs formally not addressed by the society, rather than competing on the grounds already established by others. Although the literature does not make specific reference to this aspect or theme, there are many examples of women who used unconventional means to move toward their political goals, including the women who protested in the Plaza de Mayo in Buenos Aries, the participants of the Arpilleras groups in Chile, the Bolivian coal miners' wives, and the many others who used other forms of parapolitical behavior to achieve their goals (see Gross & Bingham, 1985).

McManus (1991) describes the Mothers of Courage, or Madres, in Argentina and the nonviolent methods they used to combat the military junta. McManus (1991, pp. 96–97) describes what he believes to be the true legacy of the Madres:

> . . . simple truths, stated in measured tones, carrying great moral authority. The Madres pricked the conscience of many in Argentina and abroad. They became the conscience of their people. And in so doing, they saved many from degradation and despair. They lit the candle in honor of human dignity and protected it with their lives. . . . Asked to look back and explain the significance of their struggle, Juanita Pargament (Madres treasurer) replies, "To have learned that when a moral witness uplifts a country, truly it is a great merit. And our country has achieved respect here and abroad precisely for this."

The findings in my study correspond to Hemphill's (1989) results in that the co-researchers found that their ability to overcome obstacles and barriers fortified their sense of competence. They transcended roadblocks through various methods, concentrating on the task at hand and not allowing themselves to be distracted. While they did not deny the existence of these obstacles, they refused to allow them to be a significant problem. One co-researcher said:

> I never felt discriminated against, but I was discriminated against. It has always been like this. There is not a woman in this country that does not suffer discrimination, but I never felt it, never.

This statement points to the attitude that the co-researchers held toward challenges and the process of overcoming obstacles. Some relished and enjoyed the challenge itself and even felt the need to continue to find challenges. Others felt that the challenges were part of their life path's lessons and, as such, prepared them for important work. It appears that having a positive attitude toward obstacles and seeing them as opportunities is an important constituent of the experience of feeling able to move toward a difficult goal. Cantor and Bernay (1992) were surprised to find that their participants saw obstacles as opportunities, concluding (p. 282): "It's not that there are no obstacles to women's success in our culture, but the psychological 'comfort zones' provided by these women's early lives buffeted them from the impact of the obstacles."

Some co-researchers reported an event or series of events that interrupted the flow of their lives and played a part in shifting their perspectives. Two examples involved severe physical trauma, and one involved arrest and imprisonment. After this event, they made decisions about the course their lives would take and developed a strong conviction that they would be able to reach meaningful goals. The tragic life events seemed to empower them to move forward.

This study stands in contrast to much of the literature on Latin American women in that it validates the individual woman's struggle to overcome internal and external obstacles to make her life meaningful without projecting blame. There is a remarkable lack of focus on oppression despite the acknowledgment that it exists. One respondent, a police commissioner, reports:

> I'm happy. I'm at peace with myself, have no traumas, no frustrations, in whatever I do I'm happy. This is the important thing. All the difficulties I encountered in my life; this is what happens to all people. In my case I could never tell you about the suffering that I had to endure to arrive at this point in time. No, I'm not saying that I never ever suffered, but I struggle and fight and have no time to suffer. I really don't have time for that. . . . It [discrimination] did happen, but I never really felt it.

This passage illustrates the perspective that successful people view obstacles differently from unsuccessful people. All these women, to one degree or another, personify feminine values in practice and in doing so provide eminent role models for those privileged to come into contact with them. They have been able to maintain the integrity of their feminine value system while utilizing the masculine principle within themselves to move out into the world and in order to materialize goals that benefit the larger society. This study has revealed aspects of both the behavioral and psychological dimensions of this masculine–feminine integration. The transpersonal–spiritual implications of this intrapersonal process offer a promising area for future phenomenological research.

REFERENCES

Bachofen, J. J. (1967). *Myth, religion, and mother right: Selected writings of J. J. Bachofen.* Princeton, NJ: Princeton University Press.

Barry, H., Bacon, M., & Child, I. (1957). A cross-cultural survey of some sex differences in socialization. *Journal of Abnormal and Social Psychology, 55,* 327–332.

Belenky, M. F., Clinchy, B. M., Goldberger, N. R., & Tarule, J. M. (1986). *Women's ways of knowing: The development of self, voice, and mind.* New York: Basic Books.

Bem, S. L. (1975). Sex role adaptability: One consequence of psychological androgyny. *Journal of Personality and Social Psychology, 31*(4), 634–643.

Bem, S. L. (1987). Gender schema theory and its implications for child development: Raising gender-aschematic children in a gender-schematic society. In M. R. Walsh (Ed.), *The psychology of women: Ongoing debates* (pp. 226–245). New Haven, CT: Yale University Press.

Block, J. (1973). Conceptions of sex role: Some cross-cultural and longitudinal perspectives. *American Psychologist, 28*(6), 512–526.

Bolen, J. S. (1984). *Goddesses in everywoman: A new psychology of women.* New York: Harper Colophon. *Latin America.* Philadelphia, PA: New Society Publishers. pp. 48–62.

Cantor, D. W., & Bernay, T. (1992). *Women in power: The secrets of leadership.* Boston: Houghton Mifflin.

Chaney, E. M. (1979). *Supermadre: Women in politics in Latin America.* Austin: University of Texas Press.

Chodorow, N. (1978). *The reproduction of mothering: Psychoanalysis and the sociology of gender.* Berkeley: University of California Press.

Chodorow, N. (1989). *Feminism and psychoanalytic theory.* New Haven, CT: Yale University Press.

Colaizzi, P. F. (1973). *Reflection and research in psychology: A phenomenological study of learning.* Dubuque, IA: Kendall-Hunt.

Colaizzi, P. F. (1978). Psychological research as the phenomenologist views it. In R. S. Valle & M. King (Eds.), *Existential-phenomenological alternatives for psychology* (pp. 58–62). New York: Oxford University Press.

Denmark, F. L., Tangri, S. S., & McCandless, S. (1978). Affiliation, achievement, and power: A new look. In J. Sherman & F. L. Denmark (Eds.), *The psychology of women: Future directions of research* (pp. 393–460). New York: Psychological Dimensions.

Diner, H. (1973). *Mothers and amazons: The first feminine history of culture.* New York: Doubleday Anchor Books.

Dinnerstein, D. (1976). *The mermaid and the minotaur: Sexual arrangements and the human malaise.* New York: Harper Perennial.

Eisler, R. (1987). *The chalice, the blade: Our history, our future.* San Francisco: Harper.

Gilligan, C. (1982). *In a different voice: Psychological theory and women's development.* Cambridge, MA: Harvard University Press.

Gimbutas, M. (1974). *The goddesses and gods of old Europe: Myths and cult images.* Berkeley: University of California Press.

Giorgi, A. (Ed.) (1985). *Phenomenology and psychological research.* Pittsburgh, PA: Duquesne University Press.

Gove, P. B. (Ed.). (1966). *Webster's third new international dictionary.* Springfield, MA: G. & C. Merriam.

Greer, G. (1987). The female eunuch. In M. R. Walsh (Ed.), *The psychology of women: Ongoing debates.* New Haven, CT: Yale University Press.

Gross, S. H., & Bingham, M. W. (1985). *Women in Latin America: The 20th century.* St. Louis, MO: Glendhurst Publications.

Hemphill, H. A. (1989). *The upward mobility experience of executive females who have attained senior management positions in United States corporations in the mid 1980's: A phenomenological study.* Unpublished doctoral dissertation, Saybrook Institute, San Francisco, CA.
Horney, K. (1987). The flight from womanhood. In M. R. Walsh (Ed.), *The psychology of women: Ongoing debates.* New Haven, CT: Yale University Press.
Jaquette, J. (Ed.). (1974). *Women in politics.* New York: Wiley.
Jaquette, J. (1976). Female political participation in Latin America. In J. Nash & H. I. Safa, (Eds.), *Sex and class in Latin America* (pp. 221–244). New York: Praeger Publishers.
Jordan, J. V., Kaplan, A. G., Miller, J. B., Stiver, I. P., & Surrey, J. L. (1991). *Women's growth in connection: Writings from the Stone Center.* New York: Guilford Press.
Jung, E. (1957). *Animus and anima.* Dallas, TX: Spring Publications.
Klein, M. (1975). The effects of early anxiety situations on the sexual development of the girl. In *Psychoanalysis of children* (pp. 194–239). New York: Free Press.
Klein, M. (1990). The Oedipus complex in the light of early anxieties. In C. Zanardi (Ed.), *Essential papers on the psychology of women* (pp. 65–87). New York: New York University Press.
Koltuv, V. S. (1990). *Weaving woman: Essays in feminine psychology from the notebooks of a Jungian analyst.* York Beach, ME: Nicholas Hayes.
Lavrin, A. (1987). Women, the family, and social change in Latin America. *World Affairs, 2*, 108–128.
Lerner, G. (1986). *The creation of patriarchy.* Oxford: Oxford University Press.
Lerner, G. (1993). *The creation of feminist consciousness: From the middle ages to eighteen-seventy.* Oxford: Oxford University Press.
Loevinger, J. (1976). *Ego development.* San Francisco: Jossey-Bass Publishers.
Maccoby, E., & Jacklin, C. (1974). *The psychology of sex differences.* Palo Alto, CA: Stanford University Press.
Mahler, M., Pine, F., & Bergman, A. (1975). *The psychological birth of the human infant: Symbiosis and individuation.* New York: Basic Books.
McManus, P. (1991). Argentina's mothers of courage. In P. McManus & G. Schlabach, (Eds.), *Relentless persistence: Nonviolent action in Latin America.* Philadelphia: New Society Publishers.
Mead, M. (1949). *Male and female.* New York: William Morrow.
Miller, F. (1991). *Latin American women and the search for social justice.* Hanover, NH: University Press of New England.
Miller, J. B. (1976). *Toward a new psychology of women.* Boston: Beacon Press.
Miller, J. B. (1991). Women and power. In J. V. Jordan, A. G. Kaplan, J. B. Miller, I. P. Stiver, & J. L. Surrey (Eds.), *Women's growth in connection* (pp. 197–205). New York: Guilford Press.
Polkinghorne, D. E. (1989). Phenomenological research methods. In R. S. Valle & S. Halling (Eds:), *Existential–phenomenological perspectives in psychology: Exploring the breadth of human experience* (pp. 41–60). New York: Plenum Press.
Shertock, T. (1994). An existential–phenomenological study of woman's experience of feeling able to move toward and accomplish a meaningful and challenging goal in a Latin American cultural context. Unpublished doctoral dissertation, Institute of Transpersonal Psychology, San Francisco.
Singer, J. (1976). *Androgyny: The opposites within.* Boston: Sigo Press.
Stevens, E. P. (1973). Marianismo: The other face of machismo. In A. Pescatello (Ed.), *Female and male in Latin America: Essays.* (pp. 90–101). Pittsburgh, PA: University of Pittsburgh Press.
Surrey, J. L. (1991). Relationship and empowerment. In J. L. Jordan, A. G. Kaplan, J. B. Miller, I. P. Stiver, & J. L. Surrey, *Women's growth in connection* (pp. 162–180). New York: Guilford Press.
Thompson, H. (1987). The mental traits of sex: An experimental investigation of the normal mind in men and women. In M. R. Walsh (Ed.), *The psychology of women: Ongoing debates* (p. 2). New Haven, CT: Yale University Press.
Ulanov, A. B. (1971). *The feminine in Jungian psychology and in Christian theology.* Evanston, IL: Northwestern University Press.
Ulanov, A. B. (1981). *Receiving woman: Studies in the psychology and theology of the feminine.* Philadelphia: Westminster Press.
van Kaam, A. (1959). Phenomenal analysis: Exemplified by a study of the experience of "really feeling understood." *Journal of Individual Psychology, 15*(1), 66–72.
von Eckartsberg, R. (1986). Life-world experience: *Essential–phenomenological research approaches in psychology.* Washington, DC: Center for Advanced Research in Phenomenology; University Press of America.
Walsh, M. R. (Ed.). (1987). *The psychology of women: Ongoing debates.* New Haven, CT: Yale University Press.
Wertz, F. J. (1984). Procedures in phenomenological research and the question of validity. In C. Aanstoos (Ed.), *Exploring the lived world: Readings in phenomenological psychology* (pp. 29-48). Carrollton: West Georgia College.
Woodman, M. (1982). *Addiction to perfection: The still unravished bride.* Toronto, Ontario, Canada: Inner City Books.
Zanardi, C. (Ed.). (1990). *Essential papers on the psychology of women.* New York: New York University Press.

8

The Intentionality of Psychodiagnostic Seeing

A Phenomenological Investigation of Clinical Impression Formation

Scott D. Churchill

A phenomenologically based psychology seeks to investigate the phenomena of human experience as manifested in concrete situations. One of the specific interests of phenomenological psychology is how consciousness is involved in structuring the world of everyday life: What is going on such that *there is*[1] this or that experience? How is experience *given* to us, both as lived and as known? The "givenness" of experience refers to both the active and the passive ways in which the experiencing person becomes aware of something. Such an interest is characteristic of the phenomenologist's focus upon the "constitution" of experience. Previous philosophical investigations of constitution (Husserl, 1973, 1982, 1989) have given us the concept of "intentionality," which refers to the structural (i.e., dynamic, dialectical, and irreducible) relationship between the perceiver and the perceived. An intentional analysis of experience is an examination of the ways in which the perceiver is present to an object such that there is a thematic grasp of the object as having a particular quality or meaning. Phenomenological philosophy raises the question of givenness or constitution at

[1]Here I am making implicit reference to Heidegger's (1968) later thinking regarding the origins of being: the *es gibt* (literally, "it gives," but more typically translated as "there is"). Heidegger invites us to reflect upon the "it" that "gives" experience. In his earlier work, Heidegger (1962) locates being in *Da-sein* ("being-there"), that is, the (human) being that is the "there" (*"da"*) where meaning arises.

Scott D. Churchill • Department of Psychology, University of Dallas, Irving, Texas 75062.

Phenomenological Inquiry in Psychology: Existential and Transpersonal Dimensions, edited by Ron Valle. Plenum Press, New York, 1998.

the transcendental[2] level, that is, within the realm of "pure" subjectivity. In contrast, an existential–phenomenological psychology at the empirical level takes as its domain the intentionality of human subjects in specific situations.

In a study I conducted some years ago, the question of how experience is given or constituted was addressed in the situation of psychodiagnostic interviewing (Churchill, 1984a, b). Although texts on psychodiagnostic assessment are filled with discussions of the various instruments, techniques, standard procedures, and interview formats to be adopted by assessors, there is a notable absence of research on the psychologist's own experience within the psychodiagnostic interview. My interest was to articulate a phenomenological understanding of the psychologist's experience during a clinical assessment interview for the purpose of better understanding how impressions of personality are co-constituted by the psychologist's unique style of attentive presence. Although practicing psychologists generally acknowledge that they themselves are "instruments" of psychodiagnosis, one finds little in the clinical research literature clarifying concretely *how,* in fact, the psychologist is involved both actively and passively as a perceiving agent in the task of psychodiagnosis. One reason for this absence of research might be that "the art of clinical examination comes from attitudes and qualities that are neither obtained nor easily detected by scientific procedures: the clinician's awareness of people and human needs" (Feinstein, 1967 [quoted in Enelow & Swisher, 1979, p. 7]). Indeed, empirical research on clinical inference and clinical judgment has focused largely on data processing and decision making, while the lived experience of the diagnostician is dismissed as nonaccessible to investigation.

The problem I faced was thus a methodological one. Spiegelberg (1975, p. 237) has pointed out that in studying any form of person perception, "there seems to be no escape from claiming some kind of priority to the individual consciousness in which the we-phenomenon takes shape." Erikson (1958, p. 67) likewise observed that "indeed, there is no choice but to put [the clinician's] subjectivity in the center of an inquiry into evidence and inference in clinical work." Holt (1971, p. 11) has acknowledged that there are "informal" processes of observing and listening to the client that do "not generally operate by means of a highly conscious, rational, explicit drawing of inferences"; the actual forming of a clinical impression is a spontaneous event that "typically happens outside the spotlight of our fullest (or *focal*) awareness." In this sense, "informal assessment" might be seen as an example of what Polanyi (1966) referred to as "tacit knowing." Some of the behavioral cues from the assessee, as well as experiential phenomena on the side of the clinician, are "known" to the clinician only to the extent that they tacitly signify something else that does become thematic, namely, the personality of the subject. The informal processes at work in assessment would include everyday modes of person perception such as empathy, intuition,

[2]The term "*transcendental*" refers here to the a priori characteristics of consciousness that exist apart from the contingencies of its engagement in this or that "transcendent" (or "objective") reality. Husserl uses the adjective "transcendental" in reference to consciousness when the latter is viewed through the purifying reflection of the phenomenological reduction; the adjective "transcendent" refers to those *objective* realities that remain "outside" one's focus after performing the reduction. "Transcendent" as used here should therefore not be confused with its spiritual meaning within the discourse of transpersonal psychologists. Outside the phenomenological tradition, "transcendent" is often used to refer to a sacred dimension. If such a dimension were to be found in Husserl, it would more properly belong to what he calls the "transcendental realm." Commenting in 1925 upon the teachings of Gautama Buddha, Husserl made the following remark: "Indeed, the purest essence of Indian religiosity . . . is, I would say, not so much 'transcendent' as 'transcendental'" (cited in Hanna, 1996, p. 367). Later, in *The Crisis of European Sciences and Transcendental Phenomenology,* Husserl (1970b, p. 137) would remark that the transcendental attitude of the phenomenologist was perhaps "destined in essence to effect, at first, a complete personal transformation, comparable in the beginning to a religious conversion."

snap judgments, stereotyping, attribution, and projection. There is also the psychologist's "biographical presence" (C. T. Fischer, 1978, p. 214), which refers to the multitude of ways that one's personal and professional history is co-constitutive of one's "clinical" impressions. It is this "subjective" side of the assessment process that researchers of clinical judgment seek to systematize in the interest of making clinical intuition more reliable: "As soon as the methods of science are applied to an informal, everyday process, it begins to turn into a discipline . . ." (Holt, 1971, p. 43).

Psychologists attempting to investigate the clinician's "subjectivity," however, have complained that the clinician's thinking is too "impressionistic" (Marks, Seeman, & Haller, 1974), "complex" (Lanyon, 1972), or too "private, quasi-rational and non-repeatable" (Hammond, 1966) to be studied empirically. Thus, some researchers would be deterred from a phenomenological approach to studying clinical impression formation because it appears that psychologists' judgments are a "function of a process that they cannot trace" (Hammond, 1966, p. 18). The result of this disparaging stance toward the clinician's capacity for self-examination is a body of research that fails to illuminate in a comprehensive way how clinicians are "subjectively" involved in the assessment process.

What the research on clinical judgment *has* contributed is an evaluation of: (1) the relative utility of various kinds of information used for making clinical decisions, (2) the relative efficiency of different methods of processing information, and (3) the relative value of possible decisions regarding treatment. Researchers of clinical judgment have, in effect, abstracted clinical judgment from the rest of the clinician's experience and treated it as an isolated information-processing activity. Instructive clinical texts continue to dwell almost exclusively on the tasks of "formal" assessment—namely, testing and data interpretation (e.g., Aiken, 1996; Cramer, 1996; Sperry, 1995) or "what to look for" during the clinical interview (e.g., McWilliams, 1994; Morrison, 1995)—while alluding only obliquely to the fundamental task of the assessor to elicit and perceive the data in the first place. The irony is that clinical inference is thought to be more "empirical" when it is based just upon objective test data—that is, when it is *not* grounded in the clinician's experience. To the extent that the psychologist's own involvement within the assessment process is clearly a "function of covert processes" (Sullivan, 1954, p. 54), these processes ought to be brought to light so that their influence can be properly evaluated.

Returning to the methodological issue raised earlier, the question for me was how to make the clinician's subjectivity accessible. Based on an inspiration from Husserl's (1973, 1982, 1989) own investigations into the intentionality of experience, I conceived my research as an intentional analysis of the psychologist's *acts of consciousness* insofar as they are directed toward understanding a person in the context of psychodiagnostic assessment. The research was thus a "return to beginnings" in the psychologist's experience—an attempt to bring into focus the way that clinical impressions originate in the psychologist's illuminating presence to the client.

THE RESEARCH TOPIC: DESIGNATING THE PHENOMENON

The process of forming clinical impressions relates to many cognate phenomena studied in both social and clinical contexts: person perception, person cognition, social perception, interpersonal experience, impression formation, snap judgments, attribution, attention, implicit personality theory, naive psychology, clinical inference, clinical judgment, and clinical attitude. The difficulty in naming my phenomenon is that the research topic fits no one of these designations exclusively but at the same time encompasses all of them.

The expression "psychodiagnostic presence" might be used in a general way to refer to the phenomenon, but the problem with this designation is the nebulous way in which the term "presence" has been used in phenomenological discourse. "Presence" is certainly involved in a fundamental way in the process of forming clinical impression, but a more determinate expression is needed—one that explicitly acknowledges the perceptual dimensions of the psychologist's experience. Alfred Schutz's (1962) discussion of "cognitive styles" provides a framework for approaching this general kind of phenomenon, but the name is not quite right, as the work of psychodiagnosis calls upon thinking, feeling, imagining, sensing, and intuiting all at once; "cognitive style" is thus too limited a way of referring to the phenomenon. Psychodiagnostic "attitude" might work, but then there is the problem of the more common association of the term with social values and beliefs, rather than to "attitude" understood phenomenologically as a constitutive stance of consciousness. Foucault's (1973) use of the common French expression *regard* is tempting, but the problem then becomes how to translate this term into English. (His translator's choice of the more uncommon English equivalent "gaze" invites a possible misconstruction of the phenomenon as nothing more than a casual glance.)

I eventually decided that the expression *seeing* in its common, everyday usage was well suited as a way of referring to the kind of phenomenon under investigation. "Seeing" preserves the fullness of the phenomenon being sought, in its face-to-face, perceptual, contemplative, embodied nature. It is a term commonly used to refer to understanding ("Do you see what I mean?") and imagination ("seeing in the mind's eye") as well as perception. The idea of perspectivity is implicit, and the term "seeing" is also used to express the processes of learning, discovering, admitting something to one's presence, experiencing, inquiring, and reflecting. My choice of the term here is also inspired by Castaneda's (1971) usage, insofar as for him *seeing* refers to an embodied worldview, an alert style of presence to one's situation, and, most essentially, a seeing-through the surface appearance of things. All things considered, "seeing" is a comprehensive and meaningful way of referring to the psychologist's conscious activity during a diagnostic interview.

Let us now turn to a clarification of what is meant by a seeing that is "psychodiagnostic." For the purpose of this study, the term *psychodiagnosis* is used to refer in a generic sense to the process through which a psychologist comes to know the personality of an individual. Etymologically, the term means "to have seen through to the psyche." The perceptual basis for psychodiagnosis can be lost, however, when we think of diagnosis as a species of "knowledge" (the latter being a term derived from the Greek *gnosis*). I have for this reason chosen the expression "psychodiagnostic seeing" as a way of emphasizing the perceptual nature of the phenomenon being investigated. The term "psychognosis" was used in 1874 by Brentano (1973) to refer to a descriptive psychology employing the method of "inner perception" as a means of describing one's *own* intentional acts. With the advent of psychoanalysis as a method of observing *other* people's psychological lives, the term "psychodiagnosis" became employed to refer to a process of seeing that could penetrate "through" (*dia*) to the depths of another's psychological life. Thus rooted in the tradition of depth psychology, psychodiagnosis came to be understood as a process through which the psychologist comes to "know" the character structure of the individual. As such, *"psychodiagnosis* may be regarded as a portmanteau word signifying the reconciliation of *psy-*

chodynamics and *diagnosis*" (Rosenzweig, 1949, p. 3). Allport's (1937) understanding of "psychodiagnostics" as a science of "character reading" that sees through expression to the "inner structural consistency" that comprises the whole of the personality is consistent with the notion of psychodiagnostic seeing used in this research.

Allport (1937, pp. 500–501) writes: "The fact that one perceives a personality at first contact, not by fragments pieced together with painful slowness, but with swift 'intuition,' . . . is a matter of considerable theoretical importance." This observation is important precisely because it brings us back to the perceptual origins of clinical impressions. Allport suggests that an empirical understanding of the psychologist's experience of the expressive behavior of the client is necessary to lay the foundations of "psychodiagnostics," which he refers to as "a modern scientific extension of the ancient arts of physiognomy" (p. 497) and the "most complex branch of psychology" (p. 496). He tells us, in short, that psychodiagnostics consists of the discovery of the significance of expression for understanding the "inner qualities of personality" (p. 484). Character reading is not, however, a simple matter of correlating various expressive features with separate traits; rather, "the feature acts as a 'point of condensation'. . . for judgments which are so elusive and complex that their basis cannot be fully recognized" (p. 505). Allport's remarks regarding the psychodiagnostic process of "reading" expressive behavior are of value to us in two ways. First, he informs us of a special way in which clinical impressions are initially grounded in the perceptual experience of the clinician. Specifically, the embodied expression of the client already signifies, or directs our clinical perception to, the "who" of the person being studied. Second, he tells us that clinical impressions are further developed through a process of "judging" that is so difficult to circumscribe that an understanding of what goes into the makeup of a clinical impression (with respect to both *what* is perceived and *how* it is perceived) escapes even the psychologist who forms the impression. These observations helped to shape the focus of the current research by directing our attention to the implicit constituting activities that are lived rather than known by the psychologist.

Empathy

One such constitutive activity is the psychologist's "empathic perception" of the client's embodied expression. The notion of empathy has, in fact, been used by many psychologists in describing the clinician's presence to a client (Allport, 1937; Hamlyn, 1974; Rogers, 1951; Sarbin, Taft, & Bailey, 1960; Schafer, 1983; Taft, 1955; Tagiuri & Petrullo, 1958; Wiens, 1976). Originally, this term was used to suggest "the imitative assumption of the postures and facial expressions of other people" (Allport, 1937, p. 530). In the context of assessment, it is usually said that the clinician "puts himself in his client's place" (Combs & Snygg, 1959, p. 254), "uses his imagination" (Hamlyn, 1974), or "listens with the third ear" (Reik, 1948) to understand the client's experience. Empathy can be described as "the process by which one person is able to imaginatively place himself in another's role and situation in order to understand the other's feelings, point of view, attitudes, and tendencies to act in a given situation" (Gorden, 1969, pp. 18–19). Theodore Lipps (1903), in his presentation on the concept of *Einfühlung* (which was subsequently translated into English as "empathy" by Wundt's student, Titchener), suggested that it is the kinesthetic experience of assuming the postures and facial expressions of others that is the very source of our knowledge of other people. According to Lipps, slight motor movements are induced in the observer by another's presence, and the experiential qualities that are evoked by these movements are perceived in the other person *over there*. "The patient's expressive movements involuntarily bring about *an imitation* in our own organism" (Reich, 1972, p. 362). We sense in and through our own bodies the intentions and affects that animate the other, and we simultaneously understand our tacit experience as significative of the other's expression.

What is important about the concept of empathy for the current research is that it raises the question of whether the forming of clinical impressions is to be understood as a perception followed by a judgment, or rather as the singular intuitive experience of a perceiving, imaging, embodied subject. Proponents of the "naive" view prefer to describe person perception as a direct intuition of meaningful qualities, whereas the more commonly held psychological view holds that person perception is a function of inference. These two views are in fact derivations of two opposing epistemological traditions.

Inference

"Inference" theorists follow the lead of the British empiricists and associationists, who purport that impressions of other people are received passively and then organized according to laws of association. Accordingly, it is the latter process of organization that makes sense out of what was originally without sense. Locke's "association of ideas" becomes, in modern psychological parlance, "inference." In the context of person perception, it refers to "a cognitive process in which characteristics of a general class are *attributed to* an individual *taken as an instance of* that class" (Sarbin et al., 1960, p. 5 [emphasis added]). To the extent that the characteristic is attributed by means of association of the initial sense datum with information held in the perceiver's memory, the perceived characteristic cannot be said to exist independently of the perceiver's consciousness (a point with which intuitionists would not quarrel). But the characteristic is then understood to be the end product of a chain of cognitive events, the mechanism of which would account for the act of attribution (hence the trend in assessment literature to supplant the clinician with the computer [beginning with Holtzmann, 1966]). It is with this reduction of perceived characteristics to a causal network of cognitive processes that the opposing epistemological trend takes issue.

Intuition

The "intuition" approach can be traced back to Leibniz, a 17th-century contemporary of Locke, for whom perception could not be understood in terms of mechanical transactions. Kant, Brentano, Dilthey, Husserl, James, and Koehler all follow in this tradition, which today continues to rival the Anglo-American empiricist influence in cognitive theory. The Continental school, so named because of its origins on the European continent, takes a "hermeneutic–dialectic" (Radnitzky, 1970) approach to human phenomena, an approach that enables one to see the individual as the active source of his or her acts (Allport, 1955). With respect to perceptual acts, meaning is understood to be achieved through acts of direct intuition (i.e., consciousness unmediated by logical operations), rather than inference. That is, meaning is not a matter of "information processing," but of embodied perception. In the case of person perception, empathic intuition rather than cognitive association is understood to be the basis for knowing others.

Even though Husserl (1970a) privileges that species of intuition directed toward one's own "inner" experience as "the ideal of adequation" (Levinas, 1973, p. 136), he still allows for an act of intuition directed toward the other's experience. Following Husserl (1982, 1989), Levinas (1973, p. 150) describes how intuition is a primordial means for discovering others: "A phenomenological intuition of the life of others, a reflection by *Einfühlung* opens the field of transcendental intersubjectivity." Thus, in the phenomenological tradition, empathy and intuition are related as species to genus, and one can properly refer to an "empathic intuition" whereby one encounters the psychological life of another (see Husserl, 1989, pp. 170–178 & 239–247).

One of the major controversies in the person perception literature has been a debate over whether or not (or, at least, to what extent) our perceptions of others are mediated by cognitive processes, namely, inferences or attributions. In general, the existing research into person perception can be described as efforts to determine what goes into the makeup of our understanding of others. Some investigators have focused on the perceptual determinants of our understanding, namely, *what* we react to (From, 1971); others focus on *how* we react (Jones & Davis, 1965; Mischel, 1974; Tagiuri & Petrullo, 1958). The study reported in this chapter is interested in "what" is perceived only insofar as this is revelatory of "how" the perceiver perceives. Unlike traditional empirical studies, which seek to establish cognitive mechanisms to *explain* the "how" of person perception, this study seeks to *understand* the "how" by illuminating the constitutive perceptual framework (*Einstellung*) of the perceiver.

THE RESEARCH SITUATION: LOCATING THE PHENOMENON

The question of access to a research phenomenon is a question of finding concrete evidence by means of which to make one's discoveries. Heidegger (1962) distinguishes three components of any inquiry: that which is asked about (the *Gefragtes*), that which is interrogated (the *Befragtes*), and that which is to be found out by the asking (the *Erfragte*). In psychological research, these three aspects would correspond to the research question (i.e., the phenomenon under investigation), the data (i.e., the experience or situation that serves as a location for the phenomenon), and the results. The question of access to the research phenomenon (the *Gefragtes*) is a question of finding concrete evidence (the *Befragtes*) by means of which to make one's discoveries (the *Erfragte*). Empirically based phenomenological research can thus be described as an investigation *of* a situation selected for study *about* the phenomenon of interest. My research phenomenon (which I eventually referred to as "psychodiagnostic seeing," but which at the outset of my study remained nameless, or only vaguely alluded to with expressions such as "clinical intuition") had to be made accessible in such a way that the individual conscious experience of the psychologist could become thematic. This amounted first of all to locating the phenomenon in a situation that could be approached empirically and that could manifest the phenomenon as prominently as possible. Having considered therapeutic and even everyday situations as possible contexts for a study of psychologists' acts of impression formation, I decided upon the assessment interview as the situation that provided the best means of access for this study.

Access to the phenomenon, however, required more than a situation to investigate. It required an approach, a way of focusing on the situation in order to allow the phenomenon to come into view. It is precisely upon this *approach* that I shall focus in the next part of this chapter. First, I will summarize here the procedure that I used for obtaining empirical evidence of the phenomenon of psychodiagnostic seeing [the full statement of this procedure is presented in Churchill (1984b, pp. 70–94)]. Two male clinical psychologists volunteered to participate in the study; both worked in institutional settings in which assessment was a regular part of the psycholgist's work. Moreover, both were involved in training psychology interns in assessment interviewing. My data generating activities consisted of the following:

1. Preliminary interview. The purpose of this step was to collect background data to be used by the researcher in subsequent analyses of the protocol data. The general issues that were raised included: (1) how does the subject see the function of psychodiagnosis; (2) what is the role of theory in his assessment of others; and (3) what motivates him to do this kind of work. (Note: The masculine gender is used exclusively here because of the gender of the research subjects.) Such background information was considered to be essential insofar as "the 'sedimented' structure of the individual's experience is the condition for the subsequent interpretation of all new events and activities" (Schutz, 1962, p. xxviii). The data collected at this stage of the research were transcribed and put aside for subsequent collating with the other data.

2. Preassessment protocol. Immediately prior to the assessment, the subject was asked: "How are you going into the assessment? What are your referral questions and what impressions about the client do you already have as you go into the interview?" The purpose of this research step was to discover *prior to* the assessment interview, and prior to the subject's reflection on the interview, what specific questions and impressions were, in fact, "perceptual horizons" of the subject's subsequent perception of the client.

3. Interview phase of a psychodiagnostic assessment. This was the concrete research situation itself, lived by the psychologist acting as a research subject, and videotaped for review during subsequent research interviews. In the assessment, the research subject (the psychologist) encountered his patient and together they went through what, for the psychologist, was a "standard" psychodiagnostic interview—one that was representative of his psychodiagnostic work.

4. Postassessment protocol. Immediately following the assessment, the psychologist was handed these written instructions:

> Would you please describe, as fully as you can, what you just experienced during the interview phase of your interaction with the client. I am interested in the course of events as they took place *for you* as you moved through the interview, including those momentary experiences and associations that might otherwise not seem directly relevant to assessment as such. I would like for your account to have provided an overview of the entire interview, including your entrance into testing, and detailed accounts of some of the different kinds of experience and activity that you went through. Thank you for your help.

Together with the preassessment protocols, the subjects' responses here constituted the empirical data for this research and thus the basis for further elaboration and clarification of the psychologists' experience.

5. Reading and posing questions to the protocols. The protocol data were read and reread in order to gain a preliminary sense of the content and organization of the subject's description. Through gaining a familiarization with the whole, I was able to grasp the contextual significance of individual statements and to facilitate the eventual clarification of this significance by posing questions to the subject in the follow-up interview. The questions arose out of my encountering the protocol statements in terms of *how* the psychologist was living the psychodiagnostic interview during the moments being described.

6. Follow-up interviews. These were the most important research interviews insofar as they were directed toward the elaboration and clarification of the experience to which the protocol data referred. The transcripts of these interviews were used as primary data for reflection by the researcher. The interviewing process consisted of a collaborative dialogue between the researcher and subject, in which the subject was asked to read aloud from the protocols while I interjected questions to facilitate clarification of issues relevant to the research. During these interviews, the videotape of the assessment was ready at hand to facilitate the subject's recollection of his experience. (It was especially helpful in enabling me to direct the psychologist's attention to his own activities and appearance on tape.)

7. Collating the data. The purpose of this research step was to assemble a collated text out of all the data collected from the subject. This consisted of juxtaposing statements that spoke to a common theme from different places in the various interviews to form paragraphs of data, and arranging the paragraphs into a somewhat chronological sequence.

The research thus consisted of videotaping and collecting descriptions of the experience of diagnostic interviewing from two male clinical psychologists, and then proceeding with a modified version of Giorgi's (1975) method of protocol analysis in order to make explicit the "structure" and "style" of psychodiagnostic seeing. Using the data base of protocols and transcribed interviews obtained from the psychologists both before and after an actual diagnostic assessment, I proceeded to analyze their experience phenomenologically to reveal the structure of psychodiagnostic seeing in terms of its motivational and preconceptual horizons as well as the modes of attention and explication exhibited by the psychologists.

THE RESEARCH APPROACH: "TO THE AFFAIRS THEMSELVES"

In the traditional "empirical" approach to perception, both the perceiver and the perceived are viewed from an external point of view as things with objectifiable properties. In studying perception, one looks to the properties of the respective objects involved and attempts to trace a correspondence between the properties of the perceived and the properties of consciousness that are activated by the perceived. Consciousness itself is taken here to be just another aspect of the objectified body, something ultimately reducible to material events happening in the brain. Thus, to study my perception of a tree, for example, I might look to the properties of its leaves that make them appear green, then to the properties of my sensory apparatus that allow me to see green, and attempt to "fine-tune" my understanding of this relationship according to the laws of physics and physiology. If consciousness itself has any place in all of this, it is merely as the implicit starting point of the inquiry (i.e., to study the perception of green leaves requires that I first be conscious of the tree as green and on that basis set out to investigate my perception).

In contrast to this positivist approach, which seeks material facts and marginalizes consciousness as an epiphenomenal starting point for "hard" scientific research, there is an alternative paradigm for the study of consciousness that comes out of the phenomenological tradition. As Brentano (1973) suggested, the realm of the psychological is different from the realm of the physiological on the basis of there being given to us through "inner perception" [what Dilthey (1977b) and Heidegger (1962), in essence, called *Verstehen*] a whole world that is of a different nature from what is revealed through "external perception." Memory and imagination—along with thought, intuition, and feeling—are among the processes of inner perception that are interwoven in the fabric of our experience, within which all the powers of psyche function together as a whole (Dilthey, 1977a). What this meant to Husserl (who was Brentano's student) was that a newly understood correspondence between the perceiver and the perceived could be established through a rigorously scientific philosophy he eventually called *constitutive phenomenology.* Husserl's project, in essence, was this: The correspondence he sought was between (1) the features, not of the thing *in itself,* but of the thing *as perceived by the experiencing subject,* and (2) the essential characteristics of consciousness that allow things to be perceived as having such qualities as are perceived. That is, if we take the perception of a "tree," Husserl's interest was *not* what identifiable qualities of a given object correspond to the designation "tree," but rather what essential characteristics of consciousness make possible the perception of this object *as* a "tree." Thus, for instance, consciousness would have to be able to recognize a relation between "form" and "content" (to use familiar Platonic categories) in order to look at this particular object and say "I see a tree." If I see a tree, it is because I am capable of seeing the novel in terms of the familiar; it is because I carry with me to every perceptual encounter a host of typifications that illuminate the objects of the world for me insofar as they are seen through the lens of my language system.

A constitutive phenomenology, however, goes far beyond the study of language itself as a horizon of experience. At a more personal level, it enters the deeper psychological world of feelings and desire, as well (Husserl, 1970b, p. 317):

> Here the research interest is directed toward human beings as persons . . . what is in question is not the world as it actually is but the particular world which is valid for the person . . . the question is how they, as persons, comport themselves in action and passion—how they are motivated to their specifically personal acts of perception, of remembering, of thinking, of valuing. . . .

To give a simple example, borrowed from Sartre's (1948) analysis of the "sour grapes" experience: I perceive these grapes hanging over my head as delicious, and their deliciousness is given to me in my sudden experience of desire for them. One might say that my desire for the grapes and their desirability are intentional correlatives. The grapes hanging out of reach are constituted in their deliciousness not on the basis of any scientific evidence concerning their chemical makeup, but rather as a function of my desire for them. The truth of this can be demonstrated very simply: A moment later, when I discover that I cannot reach them, I turn away in disgust: "The grapes are sour anyway." Is it on the basis of the chemical makeup of the grapes that I ascertain their deliciousness, or even their sourness? No. These perceptions are a function of something else, namely, the illuminating presence of a consciousness that intends, first, a desirous relationship to the grapes and, at the next moment, a disgusted relationship. The motivation for this change in attitude is still another story, one that would bring us more fully into an existential psychoanalysis of the scene. For our purposes here, however, we have at least established a context for understanding Husserl's undertaking: A relationship can be established between the *object* of perception (not as it *is* but as it *is perceived*) and the *act* of perception that *sees* (and thus *seizes upon* or *grasps the meaning of*) the object. Husserl called the act of consciousness the *noesis* and the intentional object (i.e., the object as meant, in the profile in which it is illuminated) the *noema*.

With his own "noetic–noematic" investigations into the constitution of experience, Husserl (1973, 1989) provides a general framework that can guide empirical research focusing on particular noematic interests, such as my own interest in a psychologist's perceptual experience of a person. From a phenomenological point of view, an inquiry into the experiential origins of psychodiagnostic impressions will lead us to an understanding of the attitude or perceptual style of the psychodiagnostician.

According to Husserl (1970b), we can adopt a variety of attitudes or orientations toward the world within the "natural standpoint" of everyday life. A change in "attitude" is a change in one's mode of involvement with the world, although we never really abandon our natural experiential perspective toward objects of perception, even when we assume scientific or professional roles or attitudes (Husserl, 1982, p. 55). One such specialized attitude, "to take our own case, is the psychological point of view in which the glance is directed upon [everyday] experience" (Husserl, 1962, p. 150). Within this general psychological orientation, Husserl (1970b, p. 318) designated an even more specialized standpoint, a "personalistic attitude" that is characterized by an "interest in personal habitualities and characteristics," that is, an interest in what is unique and particular to the individual person. This "personalistic viewpoint" is essentially the attitude of the psychodiagnostician, namely, a theoretical and practical interest aimed toward making thematic statements about an individual's experience and behavior that goes beyond a general interest in human behavior and motives. But this interest turns out to be just one part within the whole perceptual "style" of the psychodiagnostician. In seeking to understand what constitutes a psychological *way of seeing,* one must necessarily look to the perceptual "horizons" or contexts of psychological explications. For Husserl, these contexts ultimately point back to the same prescientific realm of perceptual receptivity for all theoretical or practical standpoints.

INTENTIONALITY OF PERCEPTUAL EXPERIENCE

Having indicated that an intentional analysis of perceptual experience yields an explication of "noematic" and "noetic" horizons of experience, I would like to clarify and elaborate the meaning of these terms. *Noematic horizons* constitute the thematic field of perception, namely, the ever-widening contexts of meaning within which the theme (*noema*) presents itself in a given act of experience. Such horizons belonging to the perceptual object include the spatial, social, biographical, and/or historical contexts within which the object is actually or virtually perceived. By a shift of attitude, one thematic field may supersede another, and yet the theme remains the same, as when, for example, we first consider a person with respect to his or her personal development and then with regard to a situation in which one finds oneself involved (Gurwitsch, 1964, p. 322). Thus, *noematic horizons, as they might appear to a psychodiagnostician, would pertain to the situations within which a person being evaluated is observed:* the family life described to us, the group therapy in which the person interacts with others, the game room where one reveals one's locomotor skills and practical intelligence, or the clinical interview in which one reveals one's transferences and defenses.

Noetic horizons, in contrast, refer to those constitutive dimensions of perceptual presence that are "anticipatory" by nature and that include the motivational and conceptual contexts of the perceiver, or what Heidegger (1962, p. 192) refers to more generally as the "fore-structure of understanding." Here, the connection between separate moments of perception is given a basis in the intentionality of the perceiver's frame of reference: Anticipations make possible subsequent acts of recognition by pointing the rays of the attentive regard in particular directions, thus connecting a present perception with the future unfolding of profiles. Gurwitsch (1964) writes:

> With the actual perception are interwoven expectancies of future perceptions of such a nature that what appears through them will adhere to certain typical delineations. . . . The future perceptions are anticipated to fill in the structural pattern of the inner horizons of the present perception, which is specified only as to its typical style and general outline. . . . . (p. 281) These anticipations are of potential perceptions which, when actualized, will render visible that unseen at the moment (p. 293).

To speak of noetic horizons of perceptual experience is to understand that there is presupposed in any given act of perception "a *field* of pregivenness, from which a particular stands out and, so to speak, 'excites us' to perception and perceptual contemplation" (Husserl, 1973, pp. 71–72). This field necessarily belongs to the particular perceiver, insofar as what is noetically pregiven always refers back to the consciousness in which perceptual adumbrations take place: "It is a field of prominences *for me,* toward which my perception is oriented" (p. 72). What is "pregiven" as the noetic field of perception is the "foreconception" of the perceiver, which in turn directs one's attention ("foresight") toward particular features of a perceptual object (Heidegger, 1962, p. 191).

Husserl's (1973) noetic analysis of the structures of receptive experience (which serves as the ground for all acts of judgment, including those of the psychologist) clarifies the constitutive role of attention in the forming of an impression: "In general, *attention* is a *tending of the ego toward an intentional object,* toward a unity which 'appears' continuously in the change of the modes of its givenness and which belongs to the essential structure of a specific act of the ego. . . ." (p. 80). To the extent that attention is part of the essential structure of perceptual acts, any understanding of styles of perceiving would have to disclose the attentive modes or "cognitive styles" (Schutz, 1962, p. 230) of which various species of perception are the intentional correlates. Collectively, *the noetic horizons operative in psychodiagnostic seeing belong to the clinician; they consist of the interests, theoretical knowledge, and attentive style that are peculiar to the individual clinician.*

PERFORMING AN INTENTIONAL ANALYSIS

> The phenomenological method par excellence is the method of "intentional analysis." (Gurwitsch, 1964, p. 292)

In performing an intentional (or *noetic–noematic*) analysis, one proceeds by "beginning with a 'sense' already elaborated in an object that has a unity and permanence before the mind and then separating out the multiple intendings that intersect in this 'sense'" (Ricoeur, 1967, p. 36). This is, in fact, the approach that I took in my own research on clinical intuition. I was interested in articulating the "intentionality," that is, the structural relationship between perceiver and perceived, revealed in the moments of my subjects' experience that were described in their protocols and interviews.

Data

To demonstrate such a focus, I will offer a reflection on the first impression that one of my research subjects (a clinical psychologist) had of his client. The following statement is taken from the original protocol and follow-up interviews with this clinical psychologist, who volunteered to participate in this study by describing his initial impressions of a client referred to him for evaluation (Churchill, 1984b, p. 344):

> . . . On the phone his voice seemed rather businesslike, on the cold, mistrustful side. [R: Can you recall what gave you that impression? Was it something *that* he said or was it more in *how* he said it?] It wasn't anything to do with *what* he said. It was very cut and dry in terms of content. I said, "Can you come in during the day?" and he said, "No, it's hard because I work." It was totally in the way that he came across on the phone. Well, you know, there's all kinds of ways people can come across. Some people come across nervous and confused, other people come across needy and dependent. And he just sounded very distant—almost as though I was talking to an insurance salesman or someone. Very businesslike. [R: Was it that you had trouble feeling a closeness or an empathy for him, or that you were picking up in his voice his being distant from you?] It has nothing to do with empathy, because I wasn't trying to empathize with him. It was a professionalism. And later on I found out when I met with him that he's a professional counselor himself. So I think what was coming across on the phone was a professional voice, which, one of the reasons it stood out, was that I'm not used to it here.

Idiographic Analysis

On the basis of these self-report data, I performed a reflection guided by the principles stated earlier. My goal here was to articulate the "intentionality" revealed in the moments of the psychologist's experience that were described within each paragraph of data. The question that I formally posed to the data was: *What does this moment of the psychologist's experience reveal about his meaning-constituting presence as a clinician?* This question placed my focus squarely on the perceiving as distinguishable from (and ultimately irreducible to, even while being inextricably bound to) the perceived. It was precisely the latency of the *act* of perceiving within the perceptual moment that I was seeking to uncover.

As Gurwitsch (1964, p. 232) indicated, a constitutive analysis proceeds first with a focus upon the noetic horizons of a single noema:

> Here we are studying a certain phase of the process rather than the process itself as a whole in its progressive development. . . . We confine ourselves to the study of a single perception occurring at a certain phase of the perceptual process. Our task is to give as complete a description and analysis as possible of the single perception and its noematic correlate, namely, the appearing object exactly as it presents itself through the single perception. We ignore here any changes occurring when the perception under discussion is integrated into the unity of one sustained perceptual process.

Gurwitsch clearly states here the *focus* of a noetic analysis, but we still need to indicate the *manner* of such an analysis. For Merleau-Ponty (1962, p. 258), "To experience a structure is not to receive it into oneself passively: it is to live it, to take it up, assume it and discover its immanent significance." Similarly, Dilthey (1977b) spoke of an act of "reexperiencing" (*Nacherleben*) that would involve "the *transferring* of one's own self into a given set (*Inbegriff*) of expressions of life" (p. 132). In this transposition of oneself into the psychic life of another, there is a "fully sympathetic reliving . . . along the line of the events themselves" (p. 133). What is essential is that the researcher be capable of "co-performing" the subject's intentional acts: "In empathy I participate in the other's positing" (Husserl, 1989, p. 177). To "posit" is to take a stand in relation to something, to "position" oneself in such a way as to illuminate certain meanings within one's situation. Thus, in empathy, one participates in the other's unique positioning of himself or herself within a situation. While maintaining one's own position as researcher, one gradually allows oneself to feel one's way into the other's experience.

Beginning with my taking up or "reexperiencing" of the psychologist's explication of a particular profile or aspect of his client, I thus proceeded to reflect back to the intentional horizons that would be the necessary contexts for such an act of thematization. My analysis thus proceeds from "what" was seen to "how" it was seen, that is, to what constituted the *"seeing"* as such. The work of the intentional analysis became an effort on my part to focus on the very *acts of seeing* through which the psychologist was able to understand his client, but which necessarily remained invisible to the psychologist as long as he remained focused (even during our research interview!) on the client. Indeed, it was the "how" of the subjects' experiences that they were not able to articulate for me during the research interviews. Whenever I asked a subject *"How* did the client appear to you as such and such?", the subject would typically reply in terms of *what* he experienced. For example, when I asked the psychologist how he came to perceive the client as "businesslike, on the cold, mistrustful side," he replied: ". . . he just sounded very distant—almost as though I was talking to an insurance salesman. . . ." (Churchill, 1984b, p. 318). What the subjects were present to were the vicissitudes of what they saw in their clients. *How* they were seeing the client, that is, how *they* were *seeing* the client, remained veiled from sight. It was precisely my job to thematize what the psychologist himself could not see: his own act of seeing.

Here, then, is a sample of my analysis:

> For Dr. S., the initial imaging of his client occurred during a phone conversation. What first stood out was the style of the expression: . . . he just sounded very distant—almost as though I was talking to an insurance salesman. . . . Very businesslike. . . . one of the reasons it stood out was that I'm not used to it here." This reveals a spontaneous perceptual tendency toward the unusual and, at the same time, a tendency toward explicating the unusual in terms of the familiar. In describing his client's voice as "businesslike," Dr. S was hearing his client as speaking in the more general manner that belongs to businessmen and professionals; that is, Dr. S recognized a *general style* of expression which the client's particular manner was *like* (thus revealing an anticipatory horizon at work in Dr. S's experience). More concretely, the client's manner of speaking was "like" Dr. S's prior phone conversations with insurance salesmen. To experience the new voice as "businesslike" implies that Dr. S's past experiences were part of the context of his listening and that he was seeing a similarity between the voice he was listening to and a kind or type of voice he had encountered in the past. In describing the voice in terms of this type, Dr. S was using the familiar to understand the unfamiliar. That is, in order to have articulated his perception of the client's voice as businesslike, Dr. S must first have "constituted" the meaning of the voice in terms of what was already familiar to him through past experience. To perceive the client as a "kind" of person requires familiarity with kinds of people: prior knowledge or acquaintance with kinds of people against which to compare a present perception. For Dr. S, this horizon of his own past experiences becomes the basis for subsequent acts of typification. This implies that the horizon of past experiences, while essentially latent in any perceptual act, becomes functional only if motivated by a genuine interest in the present object of perception. The activation of such anticipatory horizons in Dr. S's experience of his present client is constituted, then, by his psychodiagnostic interest in the client. The horizon of past experiences includes not only Dr. S's histories with other clients, but his general acquaintance with kinds of individuals.
>
> Also, in light of Dr. S's self-described gentle, receptive style of moving-toward others, it appears that hearing a businesslike voice became a salient feature of his perception to the extent that it stood out against his own preferred style of sociality. Dr. S's perception of his client's voice was further adumbrated by his attunement to the affective quality of the voice as "cold, mistrustful," which in turn led to the thematization of the client as interpersonally "distant." [An imaginative variation of this would be a "warm" feeling regarding the client being further adumbrated in the psychologist's experience as a sense of interpersonal closeness."] He consequently experienced the

client, on the basis of the cold affective tonality of his voice, as being "mistrustful." So, under his psychodiagnostic gaze, there was a spontaneous explicative movement from the typical and physiognomic features to the characterization of the client in terms of affective and interpersonal dynamics—in this instance, a dynamic which became uniquely thematic for this psychologist in the context of his own social style. In addition, he heard the client's slow speech over the phone as the speech of someone who was somewhat depressed. Dr. S used this term descriptively (rather than diagnostically) to refer to the "slowed-down" quality of the client's disposition as perceived in the physiognomy of his speech, that is, in its slowness. A variation of this would be physiognomically rapid speech being experienced by the psychologist as revealing excitement. The movement from physiognomy to disposition is itself an empathic elaboration on the part of the perceiver. Although grounded in the psychologist's prepredicative receptive experience, the image of the client *as* depressed was already a movement toward explication, i.e., toward a predicative act of interpretation.

Eidetic Reduction

What I have described up to this point is an application of Husserl's method of noetic–noematic analysis at the idiographic level. The movement from individual to general levels of meaning is undertaken in service of the phenomenological interest in universal structures of experience, and is accomplished by means of the two essential (and mutually implicative) "steps" of phenomenological analysis as defined by Husserl (1973, pp. 340ff): (1) an "eidetic reduction" whereby what is idiosyncratic to a particular experience is eliminated and (2) a subsequent procedure whereby invariant qualities are confirmed by producing "free imaginative variations" of the experience under investigation. The essence of these research steps consisted of regarding the particular moments of experience (from which my structural thematizations had been derived) as empirical variations of a general phenomenon that could take on other possible variations.

The research question posed at this point was: What *kinds* of conscious phenomena have been represented in the data? Under the eidetic reduction, the experiences that formed the basis for the original idiographic analyses were no longer of concern with respect to their status as actual facts (i.e., as representative of *specific* experiences), but were transformed in my own imagination into *possibilities* of psychodiagnostic experience lending themselves to "free variations." Their empirical status as arbitrary examples of an *eidos* were taken as a point of departure for envisioning a multiplicity of variations through which a unity of sense became evident. Concretely, I would read a passage from my analysis of one subject's experience, then pause and consider what other ways a psychologist might experience the "same kind" of moment during an assessment. A sense of what was the same about the imagined variations and the original empirical example was then articulated as a general theme, that is, a theme common to any and all variations. What I sought was the invariant form without which the empirical examples and free variations would not be possible, that is, without which they could not be grasped as examples of a *kind* of experience. The *eidos* or general structural theme was that sense that was congruent to all the variations.

On the basis of the intentional analysis of the psychologist's *first impressions* presented above (supplemented by empirical and imagined variations), I was able to begin my thematization of *psychodiagnostic seeing* as a perceptual style. The following is a brief excerpt from the complete analysis (which can be found in Churchill, 1984b):

> Initially and at times throughout the interview, the psychologist allows an impression of the client to form, not as a result of deliberate contemplation, but rather as a spontaneous adumbration of images. The initial imaging of the client under the receptive regard moves rather spontaneously

toward making explications about the psychological meaning of what has been perceived, even if these explications are not yet formally integrated into a diagnostic statement.

The psychologist's "first impressions" are essentially based upon a prepredicative experience of the client that is primarily empathic in nature—that is, a receptive, affective attunement with the individual in which the psychologist's acts of seeing are lived as a tendency toward the vivid presence of the client's experience in the psychologist's own experience. This tendency to "feel" a sense of the client is preparatory for the psychologist's making explications about the client. The psychologist's empathic "feeling into" the experience of the client is lived pre-thematically: only after the encounter does the psychologist look back and recognize a moment of his experience as one where he was "feeling a sense of" the client's experience, or imitating the client's behavior. Such moments of co-experiencing give way to a passive genesis of images which subsequently become languaged as explications about the client. In this way, the physiognomic features that were originally striking to the psychologist are subsequently taken up in an act of predication whereby the "sense" or meaning of those features becomes thematic. This sense is what is meant by a "first impression," but also any impression to the extent that it is grounded in the perceiver's perceptual experience. It is always a predicative (i.e., explicative) act of the psychologist, motivated by psychodiagnostic interests, that constitutes prepredicative (empathic, sensory) experience as "information."

Knowledge that the psychologist has about people in general, either from informal acquaintance or formal training, is a horizon of psychodiagnostic seeing: past experiences constitute the thematic field of the psychologist's perception of the client. There is a spontaneous, nondeliberate, associative relating of the person presently being seen by the psychologist to persons seen in the past. Understanding the present client involves the awakening of already familiar understandings about similar kinds of people, without which the client would remain unfamiliar. The awakening of sedimented understandings makes possible recognition and anticipation, each of which is a part of the structure of psychodiagnostic seeing. Concretely, the psychologist finds himself informed by the different histories he has had with different kinds of people, in a way that is motivated distinctively by the task of psychodiagnosis.

ORGANIZING THE RESULTS

In formulating a general structural description out of the individual analyses, it was necessary to find points of convergence around which the findings at the individual level could be brought together. A major challenge was deciding how to structure the results: how to present the phenomenon of psychodiagnostic seeing in an organized way. I was eventually struck by the way that my research phenomenon was different from others that had been the subject of empirical–phenomenological investigations. Instead of focusing on a *kind of experience* such as jealousy, disappointment, anxiety, decision-making, anger, frustration, being pleased with oneself, and so on, I was focusing on a *constitutive attitude,* a perceptual stance or *way of seeing.* It is important, then, that we discuss the various ways that one might go about the task of constructing a results section that would be adequate to one's particular research phenomenon.

Noematic Explication: Constituent and Narrative Models

Analysis of typical human experiences gives heavy emphasis to the process and "noematics" of the subject's experience, to the unfolding of "what" happened psychologically. At least two different but related models for composing empirical–phenomenological explications have been developed at Duquesne University. The first, which we might call "constituent analysis," began with van Kaam's (1959) phenomenal study of the experience of "really feeling understood." His method was "empirical" insofar as he collected self-report data. His results were "descriptive," though not quite "eidetic," since he did not employ Husserl's essential procedure of free imaginative variation; rather, he limited his analyses to the empirical variations presented in the data. Van Kaam's procedure was developed by Giorgi (1971, 1975, 1985) and Colaizzi (1973, 1978) into a truly empirical–*phenomenological* method reaching beyond the *phenomenal* to the *structural* ("intentional") features underlying psychological processes such as learning and cognition. In the Giorgi–Colaizzi approach, constituents are related to each other as moments within a structural whole.

Another approach (with at least two different "styles") developed out of the first; here, more emphasis is placed on the emergence of a *psychological story*. W. F. Fischer's research on being-anxious (1974), being-emotional (1978), and self-deception (1985) traces a particular sequence of psychological events, and the descriptive results are formulated as "scenes" or "moments" that interconnect both meaningfully and temporally. Von Eckartsberg's hermeneutical (1986) and ecological (1993) approaches to stream-of-consciousness research likewise culminate in a storying of the essential features of the uniquely personal experience under investigation. What Fischer and von Eckartsberg have in common is the mapping of human experience onto an existential matrix: With Fischer, there is an implicit presence of Heidegger's (1962) "care-structure" in his languaging of his subject's experience; in von Eckartsberg, one finds Schutz's (1967) social phenomenology as well as Gurwitsch's (1964) field of consciousness (among other frameworks drawn from Heidegger, Scheler, Binswanger, and Rosenstock-Huessy) as backdrops against which "experiaction" falls into relief (von Eckartsberg, 1971, 1986). In fact, both paradigms—constituent (or theme) analysis and story (or narrative) analysis—are compatible with each other and can be brought together in an effective and mutually implicating way, as illustrated in Chapter 3 in this volume.

Noetic–Noematic Explication: A Configural–Structural Model

When, as in this investigation, the analysis is focused upon a *way of seeing* (and not on the actual "formation" of an impression), it is not appropriate (or really possible) to arrange the structural description chronologically as a process. Imagined variations on the chronology of psychodiagnostic interviews, for instance, reveal that the actual sequence of events within a particular instance of forming a clinical impression would not be an invariant feature of the phenomenon in general. We still find a multidimensional intentionality that needs to be unraveled analytically so that the various constituents can be grasped and articulated. With a perceptual style, however, it is not a matter of establishing a temporal logic, but rather of articulating the interweaving of the perceptual themes and horizons that appear through the intentional analysis. The "structure" or interweaving appears as a configuration, as in the relationship of figure to ground (except that this relationship exists in the perceiving rather than in the perceived). Thus, for instance, past experiences become the horizon or "ground" of the psychologist's perceiving acts.

I therefore developed a *configural* (as opposed to a *chronological*) structure of psychodiagnostic seeing to articulate the various frames or layers of the psychologist's experience. My general structural description was organized around three general structural aspects of psychodiagnostic seeing that were discerned by reading through the individual meaning units (within the collated text of data) and asking: What *kind* of constituent (or "horizon") of psychodiagnostic seeing is being described here? The various "kinds" of constituents were then regrouped into more embracing categories until the most general level was obtained. The general level was reached when I had grouped all of the constituents of psychodiagnostic seeing into three broad dimensions that were irreducible to each other, while being structurally intertwined.

The general structures of seeing that emerged in the final analysis were the psychologist's (1) constitutive projects and interests, (2) modes of attention, and (3) modes of explication. These three aspects of psychodiagnostic seeing would be general structural aspects of any kinds of seeing that yield meaningful impressions. In other words, they are dimensions that would have to be interrogated and explicated in order to arrive at a structural understanding of *any* perceptual experience (or "intuition" in the strict sense) that results in a thematization of the perceptual object. Although the three aspects are themselves universals of seeing, it is the particular way in which these are found to be concretized in psychodiagnostic assessment that makes the results of my research distinctive of psychodiagnosis.

Initially, I attempted to "step back" as far as possible in viewing the psychologist as a person engaged in a psychodiagnostic encounter. I strove to identify the "ultimate" psychological horizons of psychodiagnostic seeing. The question I posed to the findings was: What have I discovered in the data that speaks to the most general (trans-situational) horizons of the psychologist's experience during the assessment interview? In what way is psychodiagnostic seeing an expression of more general projects in the lives of the subjects? This, then, became the first constitutive horizon of psychodiagnostic seeing presented in the results, namely, the *motivational context.* To arrive at the "next" horizon, I simply reduced the range of my questioning from the realm of general life projects to specifically psychological projects. An example would be the psychologist's interest in the realm of "the psychological" as constitutive of psychodiagnostic relevancy across any or all assessment interview situations. A further reduction of scope thematized the specific interests of the psychologist going into this particular assessment interview: What preconceptions and questions did he bring with him to the encounter? Existentially, these horizons move from general life projects that are a more ongoing part of the psychologists' lives to the projects that are more immediate and transient (i.e., situation-specific) in nature.

Having delineated the constitutive projects and interests of the psychodiagnostician, I next focused upon what those interests were constitutive of, namely, the attentive regard of the psychologist during the psychodiagnostic interview. In the corresponding section of the general structural description, I first presented the psychologist's modes of attention to the client's self-presentation; then I described the psychologist's attention to the task of establishing rapport with the client; finally, I focused on the psychologist's self-reflective awareness during the interview (see the sample of results presented below).

The third area of results corresponds to the psychologist's thematic understanding of the client. My explication here was made possible by a shift in focus from the aforementioned motivational horizons and style of presence belonging to the psychologist to the psychologist's descriptive, diagnostic, and interpretive thematizations: What kinds of explications were revealed in the data?

In fact, all the layers of this general structural framework are "woven together" in experience at the same time; a chronological structure was therefore simply not appropriate. The metaphor of layers is used here to indicate that the different constitutive horizons of an act of seeing are differentiated not by their place in a sequence of dynamic events, but by their relatively prethematic status in the perceiver's experience. Each horizon of seeing is a horizon precisely because it is not itself "seen." And yet some horizons are closer to becoming thematic than others; for example, the referral questions that are constitutive of what the psychologist is looking for are prethematic, to be sure, but eventually become thematized in the psychologist's description of the patient. On the other hand, personal motives for being a psychologist are not as likely to become thematic for the psychologist in the course of perceiving and describing the patient. Such motives are still constitutive horizons and can be described metaphorically as a different layer of experience than the referral questions. In this sense, one can say that individual moments of perceptual experience are "layered" by the various horizons that constitute the psychologist's involvement in and co-creation of the perceptual encounter.

In summary, the solution opted for in organizing the results was to divide the general structural description into three sections, each emphasizing one of the general aspects of psychodiagnostic seeing, while allowing each to demonstrate the interrelatedness of the structure as a whole. Since each section was written from the point of view of the whole, the reader will find a latent presence of the whole structure in the description of each of the particular constituents.

RESULTS

In order to communicate the vivid sense of the phenomenon that was revealed at the individual level, the results were formulated as an "illustrated general structural description." The illustrations that follow the general structural statements in the results presented below are drawn from the data that served as the basis for eidetic reduction and free variation. Thus, the presentation of results follows an approach "in which a structure, a system of order, is derived from the instances, and which serves to unify them as parts of a whole" (Dilthey, 1977b, p. 138).[3]

A General Structural Description of Psychodiagnostic Seeing

In the psychologist's presence to the client's self-presentation, there is revealed a dialectical relationship between presence and understanding: There is an interest in a kind of information that the psychologist can both recognize and facilitate. The psychodiagnostician's interest *is* how he is looking. Simultaneously, the psychologist listens and looks for a particular kind of self-disclosure and attends to facilitating the client's engaging in this kind of self-disclosure. In listening and looking for a certain kind of information, the psychologist already has in mind a general sense of what constitutes such information. It is this preconception of what he is looking for that constitutes what will stand out as relevant. In attending to facilitation, the psychologist is attending to the interpersonal dynamics of the encounter through which he can get the person to reach the "level" or "kind" of self-disclosure sought after. Both "listening for" and "attending to facilitating" serve the psychologist's reaching toward a kind of understanding. Psychodiagnostic seeing is a grasping, or reaching toward, constituted by the stance of the particular psychologist who is doing the reaching. That is, in the psychologist's "reaching toward," there is revealed a "from which"

[3]Because of space limitations, only selections from the second area of findings (pertaining to "the attentive regard of the psychologist") are presented here. The complete results are available in Churchill (1984b, pp. 95–152).

the psychologist reaches, namely, the embodied, biographical, factical, finite (and ultimately *human*) perspective of the psychologist. Psychodiagnostic seeing is not just a recording or considering of information *given* by the client; it is an active seeing that is at the same time a grasping that *looks up,* that *seeks,* that *constitutes* precisely what about the client's self-givenness is to be considered information in the first place. The self-givenness of the client does not become "information" for the psychologist unless he understands it in such a way that it is relevant to his psychodiagnostic interests. Virtually everything about the client's self-givenness that is relevant has been constituted as such by the psychologist's particular style of seeing.

Passive Receptivity: Apperceptive and Empathic Modes of Presence

Although aimed toward the goal of making categorical judgments or "predications" about the client, psychodiagnostic seeing is lived originally as a prepredicative or receptive regard for the client through which the psychologist finds himself or herself "struck" by perceptually given features that are physiognomic (physical characteristics, expressions, dispositions). The perceptual salience of such features is a function of the psychologist's own style of presence to others, which attunes his or her gaze to features that carry a meaning relative to that style. Here the psychologist experiences the forming of impressions in a receptive mode: "just listening to," or "feeling a sense of," the client. This mode is a passive observing that follows the lead of the client. In hearing and feeling the client's speech while observing the client's bodily presence, the psychologist is open without being totally passive, and interested without being directive. The openness of the psychologist is characterized by an attunedness for the unusual or unexpected. At such moments, the person "constitutes himself" in a way that is experienced by the psychologist as not under his direct control. It is the person who is "forming an impression" for the psychologist.

Illustration. S2 was first struck by his client's physical appearance, and he saw this appearance as reflecting a "mode of life" consistent with his preconception of the client: The client's physical delicacy (very slight, very blond, very white) were congruent with the delicate (hesitant) attitude he exhibited toward the appointment, as manifested by his "antsy" stance in the waiting room. Unusual facial features (a small face and bug eyes) were perceived as revealing a "strange" personal quality. (S himself has dark hair, a heavy beard, and robust build, and wears glasses that diminish the appearance of his own eyes.) P's unusually large eyes and lips, which gave him the appearance of an alien, constituted for S2 the anticipation of a strangeness to be encountered in P's experience and behavior. What is strange, alien, unusual, is precisely that which does not present itself as familiar or typical. Similarly, in S2's first impression over the phone, what stood out initially was the style of expression: ". . . his voice seemed rather businesslike. . . ." ". . . one of the reasons it stood out was that I'm not used to it here." What became salient for S2 here was something he was not used to in the context of his work with patients—a style of "professionalism" over the phone: ". . . he just sounded very distant—almost as though I was talking to an insurance salesman. . . . Very businesslike." It was the unexpected quality of Ps voice that was first languaged by S2 in terms of a style of expression that was already familiar to him but in a different context. In his attentiveness to that which was out of the ordinary, there is revealed a style of attention that distinguishes psychodiagnostic seeing from the everyday "selective inattention" of the natural attitude. S1 went even further in stating that he was always looking for something out of the ordinary. So, while there is a spontaneous contemplative tendency toward the familiar (e.g., in terms of which typifications are produced), there is a perceptual tendency toward the unusual that is typical and distinctive of psychodiagnostic seeing.

S2's attitude of openness, embodied in his "evenly hovering" presence, was directed toward developing a global impression of the client's experience. In forming this impression, S2 was not calculatedly thinking of a diagnostic label; rather, he stayed at the level of the client's self-description and attempted to get a feel for it: "Gee, this guy's really been depressed: he hasn't been sleeping." S2's attentive regard is revealed here as a tendency toward the concrete, vivid presence of the client's experience in S2's own experience of P, and is understood by S2 as a function of his own personal style of interpersonal presence ("gentle," "reflective, quiet," and "just being pretty open") "poking through" his professional presence. S2 entered the interview "just listening to" the client's description of his situated feelings, allowing a "first-level" meaning to be constituted by the client's own description. As he allowed the client to form an impression in his experience, S2 was not explicitly thinking about it. As the impression began to form, S2 noticed himself "feeling easy and comfortable." Later, he found himself interested in the client's reaching even greater depths in his self-disclosure. There is, then, a feeling of satisfaction that accompanies S2's interested receptivity to the client. S2 feels easeful and comfortable with respect to the satisfaction of his interest in the client.

Active Seeking: The Psychologist's Illuminating Presence

Psychodiagnostic seeing, although at times a more passive feeling, observing, listening *to,* is at other times a more active listening *for* that is a function of the foresight of the psychologist. This foresight (i.e., the directive influence of the psychologist's preconceptions and questions concerning the client) is lived as an active meaning-constituting presence to the client's self-presentation. In this mode of attending to the client, there is an active seeking that is motivated by the aims of assessment and guided by the psychologist's preconceptions of psychological typicalities and by the referral questions and information. Part of what makes psychodiagnostic seeing a praxis is precisely this interest in developing a certain kind of articulated understanding. The psychologist's imagination is called into play as a means of filling out initial impressions in a tentative way so that even in the absence of concrete expectations, the psychologist's foresight is constituted by imagined possibilities. In seeking, there can be at one extreme a closed-mindedness that shuts out from sight everything except what is being sought. In this case, seeking becomes a blinding of oneself to the fulsomeness of the client's self-presentation. A more open seeking is possible when the psychologist holds in view his or her anticipations of how the client might appear, while keeping those anticipations more horizontal. In any case, seeking reveals the psychologist's presence as a tendency toward thematic understanding. That is, it is on the basis of anticipations that the psychologist contemplates the client and thus not only receives images (presentations from the client that enter the psychologist's sensory experience), but also moves toward an articulation of the sense of those images. This articulation consists of a thematic penetration to the "inner horizons," or personality, of the person being perceived. The psychologist is not content to merely observe what is apparently self-evident about the client's self-presentation, but seeks to "see *through*" [*dia*-gnosis] the apparent to the psychological phenomena [*psycho*-diagnosis] that lie waiting to be brought to the surface by a more penetrating gaze. This "looking for" is at once an act of knowing [*gnosis*]. It stands between the signifier (the client's appearance) and the signified (the client's personality), and sees the latter "through" the former. Thus, psychodiagnostic seeing is not only a question of simple observation (i.e., the reception of the simple self-presentation of the client in his or her altering modes of appearance) but also, and essentially, a question of constituting and deciphering a relation of signifier to signified, or a relation of the visible to a hidden level of visibility that is "in-visible." While present to the relationship of signifier to signified, the psychologist does not simply observe, but actually constitutes this relationship on the basis of his or her psychological foreknowledge.

Illustration. In light of his psychoanalytical background, S1 was "looking primarily for historical information," as well as seeking to find out whether his client's intellectual defenses were masking a psychotic process. This latter interest became a horizon against which S1's perceptions were understood; that is, presently perceived overt behaviors were grasped as possibly being the appearance of some covert process. In this active mode of seeking, the preconceptions and questions of the psychologist become the conditions of possibility for whatever is to be understood about the client. For S1, the penetrating psychoanalytical nature of his gaze made possible his viewing the client's interview behavior as the surface of a depth and, at the same time, as significative of that depth. This signification is not entirely self-evident; otherwise, anyone present to the client would observe his psychological depths. A psychodiagnostician is called in on the case precisely because of his special powers of perception. This power does not consist of any magic, but rather refers to the psychologist's constitutive presence as an instrument of diagnosis. For example, S1's seeking "how in touch with his unconscious process he is apt to be" was constituted by his "basic assumption" that some of the individual's behavior "gets stored in something called the unconscious."

S2's perceptual activity was not only a passive listening; he also saw himself consciously "looking for threads"—that is, discerning the essential elements of his client's experience—"and raising them to that level" of understanding that was constituted by S2's own theoretical stock of knowledge. The latter provided S2 with the psychological concepts that generally inform his understanding of individuals, that is, that enable him to "go back to the person and say, 'I think *this* is what's happening.'" S2's work thus consisted of moving beyond the level of concreteness that was characteristic of his client's self-description. To accomplish this, he viewed the events described "as an example of some psychological thing which ties in with all kinds of other things." This was clearly an essential and distinctive constituent of his psychodiagnostic seeing, namely, that he was explicitly present to his preconceptions of psychological phenomena that provided his gaze with a kind of "foresight" or way of seeing *in-to* the psychological meaning of the client's lived events. S2 reported that at one point when he asked the client if he felt depressed,

> his expression changed, become more grim, sad, almost despairing, and I got a feeling of, this guy is much sadder than he's putting forth to me.... I remember thinking to myself, "Wow! This guy is really in the depths" . . . his whole posture changed, and his face just grew much sadder, and he

> mentioned feeling uncomfortable, and for me, I thought, "Great!" I was really glad when that happened because it felt like there was some emotion in it, and when that happened I felt like, "Good, now we're getting some depth." . . . It was like I hit a raw nerve—and I was getting some results.

S2 was clearly not happy that his client was depressed; rather, he was pleased that the client had expressed his depression. It was not even the content of the client's expression here, but rather the kind of expression, toward which S2 most fundamentally directed his attention. Given his personal interest as a psychologist in depths of expression, S2 was actively on the lookout for this in his client's self-expression, and was pleased when he finally saw that his client was "in the depths" because he was finally "getting some results." Here it is clear that getting to the depths of the client's experience was the aim of S2's work. We find here, as in the case of empathic receptivity, a feeling of satisfaction that occurs when the perceptual interest finds fulfillment in the client's self-presentation.

Similarly, when S1 found himself less distracted by the videocamera and thus able to direct "more of my attention to listening to the client," he said to himself, "Thank God!" He was pleased that at least his most general psychodiagnostic interest (i.e., to direct his attention to the client) was being fulfilled. Later, toward the end of the interview, S1 found himself getting tense and anxious because he had "not fully gotten the kind of history" he had wanted. Here again, the psychologist appears to be affected by either the fulfillment or the disappointment of his interest, which also implies the success or failure of his project. At the end of the interview, S1 was "satisfied that the mission had been accomplished, that I had gotten what I had gone in there for," and "was feeling better about the client's situation—more hopeful." Here we find three different interests fulfilled: (1) the task interest, namely the project of accomplishing a self-chosen goal; (2) the psychological interest, namely, seizing upon information from the client relevant to S1's psychological interests; and (3) a prognostic interest with a distinctively altruistic character, namely, an interest in the client's future welfare.

The striving toward fulfillment of interests during the course of psychodiagnostic seeing reveals something essential about this mode of seeing. It is not, even in its more passive moments, anything like the casual regard of a man sitting on a park bench with freedom to turn his attention toward anything at will. Nor is it, in its more active moments, like the intense and discerning attention of the person driving an automobile through a snowstorm. The attentive regard of the psychodiagnostician is not characterized by the capriciousness of the man on the park bench because, regardless of the particular direction of the gaze and the particular style of "turning toward" characteristic of the psychologist, there is always *a striving toward the contemplation of the "self" of the client,* a striving that seeks to remain focused on the client as thematic object. Thus, psychodiagnostic seeking involves an active turning away from other possible thematic objects. Neither is the psychodiagnostic gaze characterized by the kind of "depth" of the attention of the automobile driver, because the former is a reflective gaze, a seeing of meanings that hover around one central perceptual object, rather than an intense scanning of the surface of an ever-changing perceptual field. (The psychologist's attention to the individual's *personal* depths is, indeed, what gave birth to the expression "depth psychology" in the early part of this century.) These two characteristics of the attentive regard of the psychodiagnostician are distinctive and essential: It is a turning toward one perceptual object and, most important, it is an interest in the explication of "psychological" profiles or aspects or meanings of this object.

The Self-Reflective Experience of the Psychologist

The anticipating, imaging, feeling, perceiving, remembering, and task-facilitating activities of the psychologist are all directed toward a contemplative explication of the client's psychological life, otherwise known as a "clinical impression." All of these contemplative acts, directed toward the client as thematic object, are "unreflective" to the extent that the psychologist's attention, or "positional consciousness," is directed away from his own acts, which remain unthematized as the noetic *means* of his or her noematic syntheses. There are, however, moments of the psychologist's experience during the course of psychodiagnostic seeing when the interest is directed toward the psychologist's own experience, toward his or her own feelings, behaviors, thoughts, and intentions, taken as thematic object.

Self-Monitoring Presence. Psychodiagnostic seeing is lived simultaneously as an attention to the client and a self-monitoring presence to the psychologist's own feelings and expressions. This self-awareness is called into play in two ways. First, the psychologist's own affective experience is understood as revelatory of the client's psychological life. Concretely, the psychologist's own gut level reactions to the client are not only taken notice of, but also seen as signals to something that the client is doing. Second, the psychologist controls expression of these reactions in the interest of maintaining rapport with the client. The psychologist is aware of aspects of his or her own experience that are visible, as in the specific awareness that the act of psychodiagnostic seeing (e.g., the penetrating gaze) is itself seen by, and thereby affects, the client. Even beyond this, there is a control of the psychologist's perceptual tendencies. Specifically, the psychologist tries to guard against distractions of any kind, whether they be personal concerns, extraneous interferences, or even the psychologist's own curiosity. In general, distractions are the correlates of an intending gaze that, for the moment, is not motivated by psychodiagnostic interests. It is the self-monitoring alertness of the psychologist that redirects the psychologist's attention to the task-related interests of the interview. Self-monitoring presence is thus distinctive of psychodiagnostic seeing insofar as it constitutes a very specific form of self-experience consisting of observation and control of one's own perceptions, feelings, and actions.

Self-Consciousness and Role-Awareness. The psychologist lives his experience of the psychodiagnostic assessment as a kind of performance that can be perceived as others: He is aware of aspects of his own experience that are visible and invisible, aware of how his colleagues might judge his performance if they were looking, and aware of the impact of his presence upon the client. The experience of being seen, whether actual or virtual, can make thematic for the psychologist an awareness of living a role, a self-consciousness about who one is (and how one measures up) in the psychodiagnostic interview situation. Although role-awareness can become a distraction from the psychodiagnostic task, for the most part the psychologist lives the role of diagnostician prethematically. Moments when the psychologist's role shifts can occasion a thematic awareness of what one is up to. In assessment, this awareness is revealed particularly in the transition from the interview phase into testing, which is experienced by the psychologist as a modification of both his role and the atmosphere of the encounter. The atmosphere becomes more task-oriented with more of a "work flavor," more formal and serious. The psychologist experiences a shift in his role from "interested friend" to "technician."

Critical Reflection. Distinctive of psychodiagnostic seeing is the psychologist's critical presence to the explications made regarding the client's psychological life, whether they originate in the client's understanding or the psychologist's. To the extent that the client's own thematizations are understood to be revelatory of the psychological realm that is of interest to the particular psychologist, there is a positing of interest in these thematizations (which constitute the client's "story"). Although the psychologist's empathic presence to the client is characterized by an implicit belief in the validity of the client's statements for the client, there occurs a disruption of this belief when the psychologist becomes aware of a discrepancy between the client's description of his or her experience and what the psychologist perceives as self-evident on the basis of the client's behavior and expressive gestures. In the psychologist's critical evaluation of the client's use of psychological or otherwise sophisticated language to describe his or her experience, there is revealed a bracketing or stripping away of intellectualizations and a positing of belief in both the client's naive experiential descriptions and the psychologist's own perceptions of the client. This reveals the psychodiagnostic gaze as a directing of attention toward the client's experience as lived; this attention, in turn, is lived by the psychologist as a suspension of belief in accuracy with respect to the subject's description of his experience. The psychologist not only "checks" the client's self-report against his or her own perceptions, but also brings his or her own explications under the same scrutiny. Aware that wishes, anticipations, and preconceptions can result in self-fulfilling prophesies, the psychologist critically reflects on his or her impressions in an effort to purge them of any such preconceptions not validated by the client's self-givenness. The psychologist's personal sense of the validity of explications (i.e., the adequacy of languaging) is constituted by the psychologist's preconceptions within the realm of "the psychological." Although the achievement of such an adequacy of description (e.g., in psychological report writing) is limited by or relative to the psychologist's capacity for critical reflection, such reflection appears as an essential constituent of psychodiagnostic seeing.

THEORETICAL AND PRACTICAL IMPLICATIONS

I first discuss implications of my findings with respect to two pairs of distinctions that have been made in the literature: (1) "informal" versus "formal" modes of assessment and (2) "inference" versus "intuition" theories of person perception. Following this discussion, I suggest some implications for the training of clinical psychologists.

Informal versus Formal Assessment

A topic of interest to clinical and social psychologists alike is the relationship between everyday and clinical styles of impression formation. The clinical interest is to better understand the informal (everyday) processes of person perception in order to facilitate a more disciplined employment of these acts in clinical contexts. In this section, I discuss how clinical psychologists have typically understood the relation of everyday to clinical styles of perception and offer another approach to this issue in view of my findings. My position is that the assessment literature[4] has generally failed to adequately address the issue of incorporating

[4]Notable exceptions to this criticism would include Allport (1937), Sullivan (1954), Dana (1982), and C. T. Fischer (1994).

everyday processes of person perception into psychodiagnosis due to a misconception of the relationship between the two.

The distinction made in the clinical literature between formal (clinical) and informal (everyday) assessment suggests that formal assessment is more disciplined than informal assessment. It follows that to incorporate informal modes of person perception into clinical assessment would consist of taking up the former in a more disciplined way than in everyday situations. Holt (1971, p. 43) writes: "As soon as the methods of science are applied to an informal, everyday process, it begins to turn into a discipline. perhaps eventually into a technology, and even a science." The question I am raising here is whether the activities that fall under the rubric of "formal assessment" are, in fact, a scientific reappropriation of informal impression-formation activities.

Holt suggests that pedestrian modes of person perception can be formalized by applying scientific methods to them. He means that scientific investigation would establish the validity of everyday modes of person perception so that they could then be legitimately brought into the framework of formal diagnosis. The question is: How would these informal modes be incorporated into a "technology" of formal assessment? Would they remain as informal natural processes, or would they become changed and possibly artificial as a result? Holt (1971) appears to be ambivalent on this point. On one hand, he suggests that "in *discovering* psychological principles . . . and in applying them to everyday life, we should not hesitate to make use of empathy and to rely on our feelings" (p. 13). Thus, we can use empathy, feelings, and intuition to discover that empathy, feeling, and intuition are human processes involved in person perception. Furthermore, we are free to apply these processes to everyday life situations. On the other hand, Holt states that in order to apply these processes to the clinical situation, they must be "systematized."

To systematize person perception would consist of first *researching* it and then, on the basis of having opened one's eyes more fully to its dimensions, *practicing* this everyday process in the clinical situation. This is essentially what Sullivan (1954) meant when he described the clinician as, ideally, an "expert in interpersonal relations." Although calling for a systematic improvement of the clinician's social skills, Holt (1971) ultimately places more emphasis on what appears to be a regimentation of information-assessment activities. His own chapters on formal assessment do not present a clinical application of principles of person perception derived from empirical research. Rather, he discusses such conventional topics as structured interviews, free association, objective tests, and projective techniques. The "discipline" of formal assessment is thus to be found in the structuring of interviews and in the formality of tests; the assessors own powers of understanding are not brought into the discussion. Instead of simply allowing for a degree of informality within assessment, Holt ends up trying to formalize an essentially informal process and, in doing so, replaces the informal process with technique.

In my research, I found that the subjects were cognizant of their experience of their clients in a way that I would not expect to find in the average person. This leads me to believe that what makes clinical assessment "formal" is not the perceptual processes involved but how these processes are taken up—namely, more deliberately. Rather than seeing the distinction between formal and informal assessment as corresponding to different kinds of impression-formation activities, which then have to be reconciled in the practice of assessment, I would suggest that we view these activities along the lines of James's (1950, p. 221) distinction between "knowledge of acquaintance" and "knowledge-about." That is, it would seem that the "informal" and "formal" categories refer not so much to types of assessment as to a relative degree of clarity with respect to how one is structuring one's perceptions. This relativity exists in both everyday and clinical contexts and is a function of the perceiver's reflective alertness to his or her own perceptual process.

What is distinctive of psychodiagnostic seeing is that the perceiver is relatively more alert to all that is happening, both "formally" and "informally" (in Holt's sense), than in everyday perception. The clinical alertness of my subjects was evident, for example, in their wondering if they were getting enough information, in their awareness of which of their own idiosyncratic interests tended to distract them from the task of assessment, and in their reflection upon the significance of their being affected by the client or of the fact that they were mirroring each other's gestures during the interview. Their rather spontaneous consciousness within the interview situation can be viewed as an informal "knowledge of acquaintance" of the client that is no less a part of psychodiagnostic seeing than the "formal," that is, deliberate, attempts to scrutinize these "informal" impressions. It is the self-scrutiny of the psychologist that facilitates development of the more precise "knowledge-about" the client that is characteristic of clinical impressions. In this respect, I would agree that the difference between pedestrian and psychodiagnostic styles of impression formation lies not in the perceptual processes utilized, but in "the degree of validity sought and the definiteness of the inferences made" (Sarbin, Taft, & Bailey, 1960, pp. 16–17).

Another way in which my findings speak to the idea of a formal/informal distinction in assessment is in the elaboration of types of attention and explication. Holt's (1971) global differentiation can be delineated further through the results of this study. For example, the sections on "passive receptivity" and "active seeking" clarify, on the level of the psychologist's presence to the client, how in fact both the psychologist's informal and formal perceptions are constituted by task-related interests. Likewise, the section on spontaneous and deliberate modes of explication clarifies, with respect to the psychologist's impressions of the client, the genetic relationship between simple modes of explication (of the "knowledge of acquaintance" variety) and more complex, integrative modes (of the "knowledge-about" variety). The formal/informal distinction can be further elaborated with respect to the psychologist's task-facilitating presence. Both subjects reported a more personable, social (informal) style of comportment toward the client during the interview phase, as opposed to a more technical, task-oriented (formal) style in the testing phase. This, in turn, sheds light on why Holt emphasizes the "test" features of assessment in his presentation of "formal assessment." We often think of formality with respect to customs and manners, so it is in the psychologist's comportment toward the client that the distinction becomes obvious. By analyzing psychodiagnostic seeing with respect to attentive regard and formation of impressions, this study has helped to clarify the relation of "everyday" to "clinical" modes of perception.

Inference Theory versus Intuition Theory of Impression Formation

There has been a long-standing theoretical debate over the process by which we form an impression of another person. On one hand, there is the view that the subjectivity of the other is not revealed directly in perception, but is inferred by means of a distinct act of cognition. On the other hand, there is the view that sensing and judging occur simultaneously in an immediate act of intuition, so that an impression of the other is formed perceptually without the mediation of inference. The resolution of this controversy appears, in light of my findings, to be more a matter of "both/and" than "either/or."

Two modes of attentive regard were discovered. The first, *passive receptivity,* would correspond to what is generally referred to as acts of "intuition." The second, *active seeking,* would be involved in any act of "inference" whereby the perceiver seeks to discover something beyond what is immediately evident. For example, S1 was interested in finding whether his client's intellectual brightness was masking a psychotic process such as schizophrenia or depression. In raising the question of psychosis, there was certainly an act of inference, in the sense that S1 wondered whether the client's brightness was a sign (or mask) of a hidden process. In making hypotheses, the psychologist does use inference. The issue is whether the psychologist, in confirming a hypothesis, proceeds only by inference. The data indicate that there is also the possibility of a direct intuition. S2 observed that his client appeared to be depressed, although not in a diagnostic sense. Staying at the level of description, S2 articulated how he saw the client's depression in the latter's posture, tone of voice, and complaint of being unable to sleep. He did not infer depression; for him, it was given perceptually in the client's self-presentation.

One reason that inference theorists deny intuition is that their approach to perception is based upon a disparaging view of the body. On one hand, the other's body isn't "enough" to convince us of his or her conscious agency; on the other hand, the perceiver's body is nothing more than a conveyer of information—a mechanical means of representation, not an intelligent presence. If we understand that "all perceivers have bodies and are oriented to embodied existence" (Giorgi, 1976, p. 6), then our approach to the perception of others is quite different. Rather than attempting to isolate the intervening cognitive variables of impression formation (see Arthur, 1966; Hammond, Hursch, & Todd, 1964; Holtzman, 1966; Lanyon, 1972), we would approach the latter as an act of the integrated body-subject (Merleau-Ponty, 1962). Acknowledging the problem that decision-making technologists face in attempting to replace clinical intuition with data processing, Holt (1971, p. 12) writes: "The objective ideal of completely mechanized assessment will be impossible until computers can be taught to feel, to judge, and to care about people." Technological thinking, with its striving for efficiency, loses sight of human intuition in its "one-track" search for fast and reliable formulas of data processing. It treats the psychologist as a mind, not as a person. It cannot replace the experience of intuition, because only persons can feel and thereby intuit; pure minds would be able only to infer, insofar as they would be programmed to move from one piece of information to another. As science continues to investigate the cognitive activity of the clinician, one might ask: "What becomes of the man—not of the brain but of the man. . . . What becomes of the face-to-face, the meeting, the seeing, the forming of the idea . . . ?" (Heidegger, 1968, p. 42).

The problem is that research aimed toward improving the reliability and efficiency of clinical decision making often results in a theory and practice of diagnostic assessment that loses sight of its human context. Part of that context is the human way in which the psychologist perceives the client. My subjects were hard-pressed to analyze their impressions into an orderly sequence of perceptual and cognitive acts. Rather, they described their impressions as having a "sudden" quality to them, while acknowledging that there was a lot going on all at once in their experience. This "all-at-once" quality should not be dismissed merely as a function of the subjects' poor powers of introspection, as has been done in the past by cognitive psychologists. Rather, such self-reports should be taken as a clue to how impressions are in fact formed. Clinicians might learn how to form more meaningful impressions if they become more open to their humanness instead of thinking like machines. Another part of this human context is, of course, the client. The psychologist must first be oriented toward the client in his or her humanness in order to be able to form an adequate impression. A computer is capable neither of assuming a human posture toward others nor of being oriented toward others in their humanness, because no program is capable of representing the activities of a body subject.

Psychodiagnostic seeing, as revealed in this study, is more a matter of direct intuiting than of processing data. To be sure, these results do indicate that there are moments when the psychologist does form hypotheses based on inference. But these hypotheses are auxiliary to the process of seeing, not the basis for it. The implication is that instead of inference and intuition being left to stand as two opposing theoretical understandings of impression formation, they can be understood as two different modes of person perception, two modes that can best be understood as parts of a whole that are experientially integrated within the impression-formation process.

Informal processes of assessment such as empathy, imagination, and intuition have been presented above as essential constituents of the psychologist's experience of psychodiagnostic seeing. What follows are some implications for the practice of clinical assessment that emerged from a further exploration of these phenomena.

Imaging and Empathic Contemplation in the Formation of Clinical Impressions

The assessment literature dwells on clinical judgment and information processing while paying very little attention to the psychologist's use of images in diagnosis. The findings of this study reveal that images play an essential role in the genesis of clinical impressions. Psychodiagnostic knowledge is achieved by the psychologist's acts of "predication," which build upon images to produce explications that are revelatory of the psychological life of the client. Images are produced spontaneously as correlatives of the psychologist's receptive regard. Although the term "image" is sometimes used to denote pictorial representation in the imagination, I am using the term here as it is customarily used to connote the correlative of both perceptual and imaginary acts. Furthermore, as with terms such as "gaze" and "seeing," there is no visual bias intended by the term "image." Rather, in what Merleau-Ponty (1962) calls "a natural attitude of vision" (p. 227), that is, a common, naive understanding of the visual and thus of visual language, there is revealed "a 'primary layer' of sense experience which precedes its division among the separate senses" (p. 227). Similarly, in speaking of "images" I am referring to what the psychologist experiences in a feeling, sensing, listening, observing presence to the client, which is in fact the way the term was used in context by my research subjects.

The psychologist's imaging of the client, from simple apprehension of physiognomy to the more contemplative "filling out" of first impressions, is initiated by the client's entering into the perceptual field of the psychologist. For example, S2's experience of his client's voice over the phone was not a deliberate contemplation but rather a spontaneous adumbration of images, "just sort of noting [the client's] tone of voice, his expressions, his capacity to share personal information." This psychologist experienced the client forming the impressions *for* him. Initially, he was not thinking in terms of classification; rather, his presence was more empathic:

> I'm trying to get some kind of global impression of what's going on with him. I don't sit there thinking, "Hmm, unipolar depression." What I think is, "Gee, this guy's really been depressed, he hasn't been sleeping," and I might wonder to myself if he's having nightmares, I wonder . . . if he can't fall asleep or wakes up early. I try to fill that in. . . .

The wondering here was the filling out of the initial image by S2's empathic taking up, or feeling into (*einfühlen*), the fact that the client had not been sleeping well. Wondering appears as a way of setting up anticipations about the client. That is, S2 utilized his imagination in a disciplined way to open up possibilities for subsequent confirmation. Here, the movement into imagination is grounded in and motivated by S2's concern for and contemplative presence to his client. He momentarily turns away from direct observation in order to return better informed.

S2 also said that he would "try to fill in" the particular nature of the client's sleep disturbance as he further developed his clinical impression. The filling in of the impression could be either explicitly imagined or implicitly anticipated through the constitutive presence of S2's foreknowledge of possibilities of sleep disturbances. That is, foreknowledge can either motivate the filling out of the initial image in the psychologist's imagination or remain implicit as an anticipatory horizon for his subsequent act of recognition in which possibilities imagined earlier find fulfillment (i.e., are re-cognized) in the perceptual reality.

It is upon these simple acts of explication that the more complex explications are founded. There is an abundance of literature on the psychologist's acts of typification, prediction, and interpretation, which comprise the explications arrived at by means of a more deliberate striving. Such findings during an assessment are the intentional correlatives of a regard that seeks. I have given equal emphasis in the general structural description to the spontaneous images that are the correlatives of an empathic, receptive regard because these images form the basis or noematic nucleus for all subsequent perceptions and explications that are integrated into the clinical impression. The initial imaging is an essential part of the "art" of assessment even if it is an informal procedure that has only a tacit presence in the assessment findings. Psychologists prefer to speak of their "formal" findings, and so the literature on assessment emphasizes prediction and diagnostic classification.

The implication I draw from my findings regarding the genetic importance of initial imaging is that clinical psychologists should explore more rigorously in their own experience the development of first impressions, which is a domain of research usually reserved for social psychologists. Informal assessment activities such as empathic and imaginative contemplation indeed must come under the psychologist's self-scrutiny if they are to become a disciplined part of psychodiagnostic procedure. Trainees in assessment might begin with a long period devoted to critical discussion of how they form first impressions. Person-perception research could be consulted to make them more aware of sources of error as well as their own untapped resources. The trainee might strive to become more attuned, for example, to the potentially rich source of psychodiagnostic insight to be found in the more spontaneous upsurge of images during the encounter with a client. The finding that there are active and passive modes of attention and impression formation suggests that psychologists might be trained to develop their aptitude for each style, rather than just accepting that "this is my style—this is the way I am." While allowing for stylistic differences, namely, tendencies toward either the more passive "receptive" mode or the more active "seeking" mode, training programs might try to enhance psychologists' skills in their weaker modes in addition to working with their strengths. Both modes can be effective diagnostic listening skills, similar to the way that introversion and extraversion (as "experience types" on the Rorschach) are both effective styles of coping with reality. Although psychologists certainly need to "be themselves," they also have a responsibility to call upon all their psychological skills to form the most complete picture of their clients that they can.

CONCLUDING REMARKS

To the extent that psychodiagnostic seeing is a "function of covert processes" (Sullivan, 1954, p. 54), the question of "knowing how you are looking" becomes a difficult one both for the clinician and for the researcher who attempts to bring these processes to light. This study follows a phenomenological approach in attempting to articulate the often implicit perceptual and cognitive acts involved in the forming of clinical impressions. It is hoped that the reader will have gained some useful insight into both the process of intentional analysis and that of psychodiagnostic seeing.

REFERENCES

Aiken, L. R. (1996). *Personality assessment: Methods and practices.* Seattle, WA: Hogrefe & Huber.

Allport, G. W. (1937). *Personality: A psychological interpretation.* New York: Henry Holt.

Allport, G. W. (1955). *Becoming: Basic considerations for a psychology of personality.* New York: Yale University Press.

Arthur, A. Z. (1966). A decision-making approach to psychological assessment in the clinic. *Journal of Consulting Psychology, 30,* 435–438.

Brentano, F. (1973). *Psychology from an empirical standpoint.* New York: Humanities Press.

Castaneda, C. (1971). *A separate reality: Further conversations with Don Juan.* New York: Simon & Schuster.

Churchill, S. D. (1984a). Forming clinical impressions: A phenomenological study of psychodiagnostic seeing. In C. M. Aanstoos (Ed.), *West Georgia College Studies in the Social Sciences: Volume XXIII* (pp. 67–84). Carrollton: West Georgia College. (Reprinted under the title *Exploring the lived-world: Readings in phenomenological psychology.* Carrollton, GA: Eidos Press.)

Churchill, S. D. (1984b). *Psychodiagnostic seeing: A phenomenological investigation of the psychologist's experience during the interview phase of a clinical assessment.* Unpublished doctoral dissertation, Duquesne University, Pittsburgh, PA.

Colaizzi, P. F. (1973). *Reflection and research in psychology.* Dubuque, IA: Kendell/Hunt.

Colaizzi, P. F. (1978). Psychological research as the phenomenologist views it. In R. S. Valle & M. King (Eds.), *Existential–phenomenological alternatives for psychology* (pp. 48–71). New York: Oxford University Press.

Combs, A. W., & Snygg, D. (1959). *Individual behavior: A perceptual approach to behavior.* New York: Harper and Bros.

Cramer, P. (1996). *Storytelling, narrative, and the Thematic Apperception Test.* New York: Guilford Press.

Dana, R. H. (1982). *A human science model for personality assessment with projective techniques.* Springfield, IL: Charales C. Thomas.

Dilthey, W. (1977a). Ideas concerning a descriptive and analytical psychology (1894). In *Descriptive psychology and historical understanding* (pp. 21–120). The Hague: Martinus Nijhoff.

Dilthey, W. (1977b). The understanding of other persons and their expressions of life. In *Descriptive psychology and historical understanding* (pp. 121–144). The Hague: Martinus Nijhoff.

Enelow, A. J., & Swisher, S. W. (1979). *Interviewing and patient care.* New York: Oxford University Press.

Erikson, E. H. (1958). The nature of clinical evidence. *Daedalus, 87,* 65–87.

Feinstein, A. R. (1967). *Clinical judgment.* Baltimore: Williams & Wilkins.

Fischer, C. T. (1978). Personality and assessment. In R. Valle & M. King (Eds.), *Existential–phenomenological alternatives for psychology* (pp. 203–231). New York: Oxford University Press.

Fischer, C. T. (1994). *Individualizing psychological assessment.* Hillsdale, NJ: Lawrence Erlbaum. (Reissue of 1985 Brooks/Cole text.)

Fischer, W. F. (1974). On the phenomenological mode of researching "being anxious." *Journal of Phenomenological Psychology, 4,* 405–423.

Fischer, W. F. (1978). An empirical–phenomenological investigation of being-anxious: An example of the meanings of being-emotional. In R. S. Valle & M. King (Eds.), *Existential–phenomenological alternatives for psychology* (pp.166–181). New York: Oxford University Press.

Fischer, W. F. (1985). Self-deception: An existential–phenomenological investigation into its essential meanings. In A. Giorgi (Ed.), *Phenomenology and psychological research* (pp. 118–154). Pittsburgh, PA: Duquesne University Press.

Foucault, M. (1973). The birth of the clinic: *An archaeology of medical perception.* New York: Vintage Books.

From, F. (1971). *Perception of other people.* New York: Columbia University Press.

Giorgi, A. (1975). An application of phenomenological method in psychology. In A. Giorgi, C. Fischer, & E. Murray (Eds.), *Duquesne studies in phenomenological psychology: Volume II* (pp. 82–103). Pittsburgh, PA: Duquesne University Press.

Giorgi, A. (1976). *The implications of "the primacy of perception" for the science of psychology.* Paper presented at the Symposium on Perception, Evidence, and Truth held at the State University of New York, Binghamton.

Giorgi, A. (1985). *Phenomenology and psychological research.* Pittsburgh, PA: Duquesne University Press.

Gorden, R. L. (1969). *Interviewing: Strategy, techniques, and tactics.* Homewood, IL: Dorsey.

Gurwitsch, A. (1964). *The field of consciousness.* Pittsburgh, PA: Duquesne University Press.

Hamlyn, D. W, (1974). Person-perception and our understanding of others. In T. Mischel (Ed.), *Understanding other persons.* Totowa, NJ: Rowan & Littlefield.

Hammond, K. R. (1966). Probabalistic functioning and the clinical method. In E. I. Megargee (Ed.), *Research in clinical assessment.* New York: Harper.

Hammond, K. R., Hursch, C. J., & Todd, F. J. (1964). Analyzing the components of clinical influence. *Psychological Review 71,* 438–456.

Hanna, F.J. (1996). Husserl on the teachings of the Buddha. *Humanistic Psychologist, 23*(3), 365–372.

Heidegger, M. (1962). *Being and time.* New York: Harper & Row.

Heidegger, M. (1968). *What is called thinking?* New York: Harper & Row.

Holt, R. R. (1971). *Assessing personality.* New York: Harcourt Brace Jovanovich.

Holtzman, W. H. (1966). Can the computer supplant the clinician? In E. I. Megargee (Ed.), *Research in clinical assessment.* New York: Harper.

Husserl, E. (1962). *Ideas: General introduction to pure phenomenology.* New York: Collier Books.

Husserl, E. (1970a). *The idea of phenomenology.* The Hague: Martinus Nijhoff.

Husserl, E. (1970b). *The crisis of European sciences and transcendental phenomenology.* Evanston, IL: Northwestern University Press.

Husserl, E. (1973). *Experience and judgment.* Evanston, IL: Northwestern University Press.

Husserl, E. (1982). *Ideas pertaining to a pure phenomenology and a phenomenological philosophy—First book: General introduction to a pure phenomenology.* Boston: Martinus Nijhoff.

Husserl, E. (1989). *Ideas pertaining to a pure phenomenology and to a phenomenological philosophy—Second book: Studies in the phenomenology of constitution.* Boston: Kluwer.

James, W. (1950). *Principles of psychology: Volume I.* New York: Dover.

Jones, E. E., & Davis, K. E. (1965). From acts to dispositions: The attribution process in person perception. In L. Berkowitz (Ed.), *Advances in experimental social psychology: Vol. 2.* New York: Academic Press.

Lanyon, R. I. (1972). Technological approach to the improvement of decision making in mental health services. *Journal of Consulting and Clinical Psychology,* 39, 43–48.

Levinas, E. (1973). *The theory of intuition in Husserl's phenomenology.* Evanston, IL: Northwestern University Press.

Lipps, T. (1903). *Leitfaden der Psychologie [Textbook of psychology].* Leipzig: Engelmann.

Marks, P. A., Seeman, W., & Haller, D. L. (1974). *The actuarial use of the MMPI with adolescents and adults.* Baltimore: Williams & Wilkins.

McWilliams, N. (1994). *Psychoanalytic diagnosis: Understanding personality structure in the clinical process.* New York: Guilford Press.

Merleau-Ponty, M. (1962). *Phenomenology of perception.* London: Routledge & Kegan Paul.

Merleau-Ponty, M. (1964). *Signs.* Evanston, IL: Northwestern University Press.

Merleau-Ponty, M. (1968). *The visible and the invisible.* Evanston, IL: Northwestern University Press.

Mischel, T. (Ed.) (1974). *Understanding other persons.* Totowa, NJ: Rowan & Littlefield.

Morrison, J. (1995). The first interview: *Revised for DSM-IV.* New York: Guilford Press.

Polanyi. M. (1966). *The tacit dimension.* New York: Doubleday.

Radnitzky, G. (1970). *Contemporary schools of metascience.* New York: Humanities Press.

Reich, W. (1972). *Character analysis.* New York: Simon & Schuster.

Reik, T. (1948). *Listening with the third ear.* New York: Farrar, Strauss.
Ricoeur, P. (1967). *Husserl: An analysis of his phenomenology.* Evanston, IL: Northwestern University Press.
Rogers, C. (1951). *Client-centered therapy.* Boston: Houghton-Mifflin.
Rosenzweig, S. (1949). *Psychodiagnosis: An introduction to the integration of tests in dynamic clinical practice.* New York: Grune & Stratton.
Sarbin, T. R., Taft, R., & Bailey, D. E. (1960). *Clinical inference and cognitive theory.* New York: Holt, Rinehart & Winston.
Sartre, J.-P. (1948). *The emotions: Outline for a theory.* New York: Philosophical Library.
Schafer, R. (1983). *The analytic attitude.* New York: Basic Books.
Schutz, A. (1962). Scheler's theory of intersubjectivity and the general thesis of the alter ego. In *Collected papers I: The problem of social reality* (pp. 150–179). The Hague: Martinus Nijhoff.
Schutz, A. (1966). Edmund Husserl's ideas, Volume II. In *Collected papers III: Studies in phenomenological philosophy* (pp. 15–39). The Hague: Martinus Nijhoff.
Schutz, A. (1967). *The phenomenology of the social world.* Evanston, IL: Northwestern University Press.
Sperry, L. (1995). *Handbook of diagnosis and treatment of the DSM-IV personality disorders.* New York: Brunner/Mazel.
Spiegelberg, H. (1975). *Doing phenomenology.* The Hague: Martinus Nijhoff.
Sullivan, H. S. (1954). *The psychiatric interview.* New York: Norton.
Taft, R. (1955). The ability to judge people. *Psychological Bulletin. 52,* 1–23.
Tagiuri, R., & Petrullo, L. (Eds.). (1958). *Person perception and interpersonal behavior.* Stanford, CA: Stanford University Press.
van Kaam, A. (1959). Phenomenal analysis: Exemplified by a study of the experience of "really feeling understood." *Journal of Individual Psychology, 15,* 66–72.
von Eckartsberg, R. (1971). On experiential methodology. In A. Giorgi, W. F. Fischer, & R. von Eckartsberg (Eds.), *Duquesne Studies in phenomenological psychology: Volume I* (pp. 66–79). Pittsburgh, PA: Duquesne University Press.
von Eckartsberg, R. (1986). *Life-world experience: Existential–phenomenological research approaches in psychology.* Washington, DC: Center for Advanced Research in Phenomenology; University Press of America.
von Eckartsberg, R. (1993). The person's psychocosm and the stream of consciousness. *Methods: A Journal for Human Science.* Annual Edition, 5–28.
Wiens, A. (1976). Psychological assessment. In I. Weiner (Ed.), *Clinical methods in psychology.* New York: Wiley.

9

Dissociative Women's Experiences of Self-Cutting

Faith A. Robinson

INTRODUCTION

This research was developed to discover the essence of the phenomenon of nonsuicidal, self-cutting behavior among highly dissociative persons. Because of their own personal fears grounded in a lack of understanding, therapists, emergency room personnel, and crisis intervention workers often back away from or are ill prepared to help nonsuicidal self-cutters who have dissociative disorders. Subsequently, professional ignorance about self-cutting often leads self-cutters to withdrawal, isolation, and increased shame and guilt. Many of the women who participated in this study had not revealed their self-cutting behaviors to anyone except their therapist; often, their therapists had not asked for details about their cutting behaviors. In fact, I heard three of these women say, "No one ever asked me about this."

During this research, a complex interplay between fears, needs, trust/distrust, and guilt became apparent, and I could see how the complexity of emotions would inhibit openness in therapy. For such a clinically relevant behavior to be undisclosed complicates therapeutic dynamics and adversely affects an individual's prognosis. One of the positive outcomes from this research is the opening of communication channels about this very complex and secretive behavior.

I chose the phenomenological method of research to answer important clinical questions and provide insight into the forces operating within self-cutting behaviors. Rather than trying to answer the first obvious question of "why," I wanted to know something of the meaning of the experience from the views of the cutters themselves. I felt that assisting

Faith A. Robinson • 26485 Carmel Rancho Boulevard, Suite 6, Carmel, California 93923.

Phenomenological Inquiry in Psychology: Existential and Transpersonal Dimensions, edited by Ron Valle. Plenum Press, New York, 1998.

professionals to develop the art of comprehending *the essence* of this potentially dangerous behavior could bring about greater understanding, patience, and compassion and, ultimately, be most helpful to the cutters themselves.

The participants in this study met the criteria for Multiple Personality Disorder according to the diagnostic criteria defined in the *Diagnostic and Statistical Manual of Mental Disorders—Third Edition, Revised* (DSM-III-R) (American Psychiatric Association [APA], 1987). Most references will be to the diagnostic category in use at the time of the research (Multiple Personality Disorder [MPD]), since this research occurred prior to the publication of the *Diagnostic and Statistical Manual of Mental Disorders—Fourth Edition* (DSM-IV) (APA, 1994), which expanded the criteria and renamed the category Dissociative Identity Disorder (DID).

CURRENT PERSPECTIVES ON DISSOCIATION AND SELF-MUTILATION

Recognition and Prevalence of Multiple Personality Disorder (MPD)

Although once viewed as a "rare" disorder, MPD is currently being recognized and studied with increasing frequency. MPD is a "complicated clinical disorder" that involves "fragmentation of self and the transformation of identity" (Ross, 1989, p. 9). A major review of the history of MPD cites 33 reported cases from 1901 to 1944, 14 from 1944 to 1969, and 50 personally known to one researcher (Greaves, 1980). Braun (1986) estimates that the number of identified and reported MPD cases went from 500 in 1979 to 5000 in 1986.

This number may be low considering the magnitude of research currently reported. For example, an epidemiological survey of 454 Canadian adults from Winnipeg's general population found a 1% prevalence of MPD and 10% prevalence of some type of dissociative disorder (Ross, 1991). A symptom severity study of 166 MPD and 57 Dissociative Disorder Not Otherwise Specified (DDNOS) patients led Ross to conclude that pathological dissociation may be as common as anxiety, mood, and substance abuse disorders (Ross et al., 1992). A total of 50 clinical MPD patients in Indiana were studied by Coons, Bowman, and Milstein (1988) to assess etiological factors in MPD. A 1992 Trenton State Prison study of 49 inmates found that 60% of the population met the criteria for dissociative disorders (34% MPD, 26% DDNOS) (Culiner, 1993).

Complexity of MPD Cases

The increased reporting of the number of MPD cases comes along with an increased reporting of the complexity and severity of cases. The earliest reports involved only a few alter personalities. Kluft (1984) studied 33 MPD patients who had an average of 13.9 alters. In 1988, his practice contained 26 patients who had been under clinical observation for 3 or more years and who had over 25 alters each. Cases now have been reported with as many as 100 alters (Kluft, 1988).

Childhood Exposure to Violence and Abuse

A look at the literature reveals strong evidence for histories of physical and/or sexual childhood abuse preceding self-mutilative behaviors (Carol, Schaffer, Spensley, & Abramowitz:, 1980; Green, 1967, 1978; Grunebaum & Klerman, 1967; Roy, 1978). Descriptions of "some or excessive" violence in the home during childhood have been disproportionately reported. Self-mutilating patients report not being allowed to express anger in the home (Caroll et al., 1980) and recall childhood memories of excessive physical punishment (Roy, 1978).

Trauma plays a significant role in the etiology of MPD (Coons, 1980; Greaves, 1980). Histories of such persons reveal a high incidence of incest trauma (the most commonly reported trauma) (Blume, 1986; Putnam, 1986). Sexual abuse is a debilitating event that causes children to fear punishment, family breakup, disturbance of "their" world, and death if they divulge the "secret." In addition, survivors are often forbidden to express anger; the unexpressed anger must therefore be held inside until it becomes a deep rage (Blume, 1986).

Dissociation as a Childhood Defense Mechanism

The development of the capacity to dissociate during childhood is now seen as a defense mechanism that is a psychological, biological adaptation to a set of repeated childhood traumas (Putnam, 1989). Depersonalization is also a powerful form of defense. Research shows a strong connection between depersonalization and self-mutilative behavior. Mounting anxiety leads to depersonalization, overwhelming emotion, and loss of contact with reality just prior to self-mutilation (Rosenthal, Rinzler, Walsh, & Klausner, 1972). In fact, there is extensive evidence that many cutters report no pain or an altered perception of pain during the process of cutting (Bach-y-Rita, 1974; Grunebaum & Klerman, 1967; Rosenthal et al., 1972). Some see cutting behaviors as responses to depersonalized states and to desires to return to reality, rather than as direct responses to inner rage (Pitman, 1990).

In summary, several factors are apparent. First, dissociative disorders are clearly associated with early childhood traumas in the form of emotional, physical, and/or sexual (usually incestuous) abuse. Second, a common childhood defense against these traumas is depersonalization. Third, there is an association between depersonalization and self-abuse in the form of self-mutilation. This evidence strongly suggests that client self-mutilative behavior should lead clinicians to a serious consideration of the possibility of a dissociative disorder.

Self-Mutilation Associated with Borderline Personality Disorder

Unfortunately, self-mutilation is more clearly connected in the literature to the diagnostic category of Borderline Personality Disorder. I think some of the reasons for this are: (1) MPD was a mysterious, little-understood, and underreported disorder until the 1980s; (2) dissociative disorders are very complex, and too few clinicians are seeking education in this clinical area; and (3) self-mutilative behavior is a major symptom of Borderline Personality Disorder (APA, 1987, 1994). Together, I think these factors have impeded accurate diagnoses and treatment for may self-mutilators who suffer from dissociative disorders.

Looking at the diagnostic criteria for MPD, the DSM-III-R suggested that MPD persons were often misdiagnosed as having *only* Borderline Personality Disorder because of mood swings, poor self-image, and disruptive interpersonal behaviors (no reference is specifically made here to self-mutilative behaviors). A single reference to self-mutilation is found under possible complications: "Suicide attempts, self-mutilation, externally directed violence (including child abuse, assault, or rape), and Psychoactive Substance Dependence

Disorders are possible complications" (APA, 1987, p. 271). The updated version of the DSM has a more accurate description of the disorder with no cross-reference to Borderline Personality Disorder and more information linking self-mutilative acts with dissociative disorders (APA, 1987, 1994).

In Walsh and Rosen's (1988) *Self-Mutilation: Theory, Research and Treatment,* chapters are dedicated to several clinical populations, including adolescents, borderlines, psychotics, and retarded or autistic populations. There is no index reference to dissociative disorders in this 273-page book.

In Favazza's (1993) *Bodies Under Siege: Self-Mutilation in Culture and Psychiatry* (a 270-page book on self-mutilation), there are several index notations for Borderline Personality Disorder and one single index notation (referencing one single paragraph) to the dissociative disorder MPD. Favazza describes MPD as a "rare diagnosis" and self-cutting as "brutal acts" performed by alter egos. Included in this single paragraph is Bliss's (1980) excellent point that women will direct rage toward themselves, while men will direct rage toward others. Favazza concludes with Bliss's reference to "strange assaultive and lethal crimes" committed by possible MPD males and a passing reference to Putnam's (1986) study comparing "internal homicide" to acts of self-mutilation (p. 135).

Despite sparse and misrepresentative offerings in these two books on self-mutilation, if one looks at books specifically addressing dissociative disorders, one will find references on self-destructive behaviors and ample support for the connection between dissociative disorders and self-mutilative behavior (Kluft & Fine, 1993; Loewenstein, 1991; Putnam, 1989; Ross, 1989; Ross et al., 1989). Blume's (1986) work with sexual abuse survivors reveals elements common to both MPD patients and cutters, including depersonalization, alienation from one's body, and self-mutilative acts. Putnam (1986) found self-mutilative behaviors in at least 33% of MPD patients. In yet another study, self-mutilation was a common occurrence among 48% of 100 dissociative clinic patients (Coons & Milstein, 1990).

As mentioned above, there is strong evidence linking dissociative disorders to severe, recurrent traumas. The abuse suffered by the more severely disturbed dissociative patients is believed to be more sadistic and bizarre. Severe abuse histories and complex personality systems increase the likelihood of self-destructive behavior (Ross, 1989).

To thoroughly understand dissociative disorders in today's climate, one should become aware of related, controversial issues. Such issues include the validity of recalled memories and the actuality of reported ritual abuse. Both of these issues are pertinent to the data collected in this research. I will present some information here to help the reader be somewhat aware of the issues raised for each of these controversial subjects.

Reliability of Reported Memories

First, let us look at the concept of reliability of memory. One of the major controversies surrounding reports of alleged abuse involves the credibility of reports of the survivors' memories. All memory is false to some degree. Memory depends on existing mental schemata and integrates experience in a schema based on previous experience. A well-functioning hippocampus is required for the mind to adapt and categorize data properly. Experiences during states of intense emotion follow different neural pathways of memory consolidation. A century ago, Pierre Janet suggested that "vehement emotions" accompanying traumatic experiences interfere with information processing and recategorization at a verbal and symbolic level (van der Kolk, 1993).

That a memory might be distorted or false does not mean a person is intentionally lying. Some details of the memory may be authentic and other details distorted. Therapists need to consider the intensity of presenting behaviors relative to a patient's usual presentation and inquire about further details to evaluate the validity of a report. Van Benschoten (1990, p. 28) states: "The literal truth is intricately and inextricably woven together with threads of misperception, suggestion, illusion, dissociation, and induced trance phenomena, to form the complex web which becomes the survivor's memories." She provides several examples of studies in which memories of traumatic events were radically changed.

A now-famous study (Loftus, 1993) of implanting "created" memories of being five years old and lost in a shopping mall reveals the suggestibility of the mind and the unreliability of memory. After 2 weeks of suggestion, the young man in Loftus's study related vivid details that were continually expanding; these details included the balding head and glasses of the fictitious man who had rescued him (p. 532).

Ritual Abuse

Next, let us look at the controversial topic of ritual abuse. Clinical cases of ritual abuse are being documented, including clinical accounts that contain emotionally shocking and almost unbelievable information. The term "ritualized abuse" was initially used to categorize a survivor's experiences of satanic abuse memories. Pazder's (1980) definition, cited in Kahaner (1988, p. 201), is "repeated physical, emotional, mental, and spiritual assaults combined with a systematic use of symbols, ceremonies, and machinations designed and orchestrated to attain malevolent effects."

A broader definition of ritual abuse was printed in *Survivorship*, a publication for survivors of ritual abuse, torture, and mind control (Star Dancer, 1993, p. 4):

> Ritual Abuse is *any* repeated, systematic mistreatment perpetrated in the name of an ideology or dogma. This abuse may be mental, physical, emotional, spiritual or sexual and frequently combines all of these types of abuse in an attempt to condition every aspect of the victim's humanity toward that ideology. Ritual abuse is often practiced in the context of religious or pseudo-religious ceremonies, but is supported through chronic instances of individual, non-ceremonial abusive acts where doctrine is incorporated by the perpetrator as a rationalization, either deliberately or spontaneously (because the perpetrator is him or herself a victim of the mind control system).

The literature reveals that involvement with some type of allegiance to or worship of Satan in rituals is being reported in clinical settings. This allegiance to Satan is thought to be a means of justifying the activities performed (Van Benschoten, 1990). It is important, however, to carefully distinguish between ritual abuse that is satanic in nature (Satanic Ritual Abuse [SRA]) and ritual abuse that is not satanic in nature (StarDancer, 1993; Van Benschoten, 1990).

StarDancer (1993, p.4) lists subtypes of extrafamilial and intrafamilial organizations in which abusive practitioners are known to perform ritual abuse including sectarian (e.g., Satanic, Wiccan, Children of God, Dionysian), nonsectarian (e.g., organized crime, neo-Nazi, Scientology, pyramid schemes, EST), culturistic (e.g., Aboriginal, Ku Klux Klan, Santería, Vodun), multiperpetrator idiosyncratic or charismatic (e.g., Jonestown), single-perpetrator idiosyncratic (e.g., Richard Ramirez, Jeffrey Dahmer, Leonard Lake), and pantheistic occult (loosely connected networks in which elite perpetrators use knowledge of metaphysical systems in which archetypal figures represent aspects of Godhead or Primal Powers and prey upon members' vulnerabilities).

According to Van Benschoten (1990, p. 26), "The practice of deliberate abuse and neglect by many totalist groups on religious, pseudo-religious, and other ideological grounds has since been well documented." Systematic harm within cultic groups raises fears of "former members, law enforcement agencies, child welfare organizations, psychotherapists, and the medical profession" (p. 26).

Ritual abuse survivors seeking therapeutic support are helping us gain understanding of the modus operandi of cults. Neswald, Gould, and Graham-Costain (1991, p. 47) address this in their discussion of the mind-control programming observed in SRA:

> Concurrently, the etiological underpinnings and treatment demands of these special patients are being unraveled and understood as never before. As a result, it is becoming increasingly clear that perhaps the most demanding treatment aspects of such cases concern the problems posed by what is known as "cult programming."
>
> So called cult "programs" are really no more than conditioned stimulus–response sequences—consistent with basic learning theory. Such conditioning is achieved through a large variety of sophisticated and sadistic mind control strategies involving the combined application of physical pain, double-bind coercion, psychological terror, and split brain stimulation. All programs are stimulus-sensate triggered. Thus, programs may be enacted (triggered) via auditory, visual, tactile, olfactory and/or gustatory modalities. Classical, operant, and observational/modeling paradigms all are utilized by the cults and their "programmers."

This information on these two controversies associated with the diagnosis and treatment of dissociative disorders is presented to emphasize the importance of professionals' keeping an open and critical mind about all sides of these controversial issues. This is only a sampling, however, and should not be considered inclusive of all points of view. My personal recommendation is to be well-read and open-minded, but to take a middle-of-the-road position that considers all possibilities without endorsing any extremes.

PHENOMENOLOGICAL METHODOLOGY

Uniqueness of the Experience Being Studied and Associated Obstacles

The uniqueness of this topic and of this clinical population raised several considerations to be addressed in the research design phase. Ethical concerns were initially discussed with the research and human subjects review committee. Advice was sought from outside professionals, an MPD cutter, and an expert on state and American Psychological Association ethics.

A delicate balance was required between the goals of research and the goals of clinical psychology in order to capture the essence of the self-cutting experience. Of foremost consideration was a concern for subjects' personal safety and welfare. Mechanisms for enhancing subject safety were therefore incorporated in the research design. First, this study was designed to study *nonsuicidal* cutting to avoid the dangers inherent in asking suicidal clients about intense emotional experiences. Then, to minimize risks, only dissociative clients who had been diagnosed by their therapists as having MPD and who had experienced nonsuicidal cutting were sought as subjects in this study. These subjects were considered to be functioning at higher levels of stability and had well-developed therapeutic support systems.

Referrals were solicited from licensed professionals (nine psychologists and two MFCCs), who signed a referral form stating that they had a professional, therapist–client relationship with their MPD client who had engaged in nonsuicidal cutting. They felt that the client was sufficiently stable to participate without potentially detrimental effects. Further, these therapists were available for emergencies during data collection times and scheduled a therapy session with each client immediately after their research interviews.

Another important consideration was whether or not to attempt to identify different "personalities" for each referred client in the interest of securing a protocol from each of their subpersonalities. After considering the added complexity in the data collection process, the difficulty (if not impossibility) of accessing all of a subject's personalities, and the potential interference caused in the subject's therapeutic integrative progress, a decision was made to collect a single protocol from each subject without any attempt to address subparts of the subject's personality system. It was decided that each subject would have complete control over the type, amount, and source of data provided and that the researcher's role would be as nonintrusive as possible.

Methodological Steps

I felt the topic of study was worthy of careful consideration and handling. I also was sensitive to the painful data that would likely emerge from this study, and wanted the subjects who participated to know that their participation was treated with the greatest respect and seriousness. I designed each research step carefully to maximize the validity of the results.

Self-Reflection

The initial step involved my own self-reflection to articulate presuppositions with regard to self-cutting and dissociative subjects. This process of awareness, identification, and articulation of my biases decreased the likelihood that my personal thoughts would later interfere with my analysis of the subjects' data. I identified five general ideas that addressed preexisting thoughts and feelings about the experience being studied. These ideas were that self-cutting involved: (1) an internal state of absorption and disconnection from the outer world, (2) a temporary state of release, (3) some experience of pleasure, (4) an addictive process, and (5) the development of tolerance.

Subject Selection

Subjects were 11 females between 29 and 44 years of age who lived in diverse geographic areas. Five were married, three were single, and three were divorced. Two were high school graduates, three had bachelor's degrees, two had master's degrees, and four had advanced degrees and/or trained specialties within the medical field. Five received financial support from disability, one from her family, and five from self-employment.

Initial Data Collection

The initial data collection was highly structured and noninteractive. A subject was handed written instructions that included the following paragraph:

> Please recall an experience of self-cutting. Please describe how you felt during this experience. Describe your feelings just as they were so someone hearing or reading your report would

> know exactly what the experience was like for you. Be sure to describe the experience itself and not just the content of the situation. Please do not stop until you feel you have described your feelings as completely as possible. Take as long as you would like to complete your response.

The initial data collection typically took 60–90 minutes to complete. Unique to this study, sometimes different subpersonalities within the personality system presented data for the same protocol, each from its own perspective. These different subpersonalities were discerned by handwriting styles, physical appearance changes, and/or content shifts.

Follow-up

Follow-up interviews consisted of a structured "walk-through" part and a nonstructured "dialogue" part. During the "walk-through," the initial description of the self-cutting experience was read aloud phrase by phrase, each time allowing the subject an opportunity to clarify or enhance her original description. Discussion in this process was minimal and highly structured in order to keep the data "pure" and unbiased. The original description plus the "walk-through" enhancements represented the "protocol" that would later be analyzed.

From the onset of data collection, it became apparent that additional dialogue would be needed to fully understand the subjects' descriptions of their self-cutting experiences. Words and phrases were sometimes symbolic or cryptic. References to complex interactions within themselves and interactions between the subjects and persons or events in their private lives made understanding the data a more challenging task than with a nonclinical population.

During the dialogue phase, I asked unstructured and impromptu questions in order to better comprehend unclear phrases or concepts. With this more personal dialogue, the subjects' personal stories became "alive," allowing me to frame their descriptions in more accurate contexts. While these latter data were not used directly in the phenomenological analysis, they did provide an essential function by clarifying the writers' intentions in using specific words and phrases.

Data Analysis

Because of the tremendous richness of the data, extra care was taken during the data analysis stage to not "lose" any of the essence in the process of distillation as I moved from the raw data successively down toward the essential constituents. Embedded in the process was a painstaking carefulness to allow the process to unfold naturally and slowly. Because of the intensity of my own particularly painful emotions, it was necessary to walk away from the data after each step of the process to allow time and distance to reestablish my personal mental, emotional, and spiritual balance. I believe that this systematic distancing kept my own biases at a minimum as the distillation process continued.

Reading the Protocols. Much time was saved by using a computer for storing, working with, and printing data. This provided benefits to the project because of the amount of data involved. For example, organization and management of data were easier, specific components could be located more easily, and working with table formats allowed me to focus more on the actual data without being distracted by a manual process.

Each protocol was encountered initially through several readings of its contents. As each protocol was collected, it was entered into a computer file. In follow-up interviews, protocols were read aloud to their writers in the "walk-through" and "dialogue" phases. The walk-through enhancements were added to the computer file by using brackets and bold type in order to distinguish these words from the original data. The dialogue questions and answers were kept separate from the protocols. After all protocols had been collected, they were privately read aloud in a single sitting to capture a sense of the "wholeness" of the collective experiences.

Discriminating Meaning Units. The first step of phenomenological analysis was to discriminate "meaning units." Each protocol was slowly read to discern natural breaks in psychological meaning pertinent to the experience as a whole. According to Giorgi (1985), meaning units do not exist as such, but are instead perceived transitions relative to the psychological sensitivity of the researcher; hence, it is important to recognize that these "meaning units exist only in relation to the attitude and set of the researcher (p. 15)." This being the case, the final interview's dialogue process between researcher and subject was invaluable in aligning the "attitude and set" of the researcher with that of the subject as she described her self-cutting experience.

Discovering the Psychological Relevance. Next, redundancies and irrelevancies (ideas unrelated to the experience of self-cutting) were removed. The remaining meaning units were printed and cut into individual slips of paper to facilitate rearrangement by "relatedness."

Still working with individual protocols, related meaning units were grouped to find similar items, such as pain and lack of pain, physiological sensations, and any reference to a quality associated with blood. The process at this point involved Colaizzi's (1978, p. 59) "precarious leap," whereby the researcher moves from the subject's words 'to his or her own words, thereby transporting the researcher from an experience-specific to a psychological context.

At this point, the data exhibited great complexity. In addition to emerging themes, interconnections between themes were becoming apparent. Their appearance necessitated occasional shifts of data from one identified emerging theme to another. When the refinement of emerging themes was completed for each of the 11 protocols, the psychologically relevant themes were printed and cut into slips of individual themes. A different color was used for each protocol as a means of identifying the source protocol for any given theme.

At this point, there began a new phase in which analytical processes combined with intuitive, feeling processes to intuitively sort these several hundred colored pieces of paper into categories of similar meanings. This process led to the identification of 58 common themes referred to as "constituents."

It was sometimes difficult to determine which of two constituents was more prominently identified within a theme. For example, the interconnections in "Cutter feels pain subside as she watches blood drip" relate both to the feeling of pain and to her experience of seeing blood. In this particular case, the theme was placed with the constituent of "Pain" and a notation was made under "Pain" for a secondary reference to "Blood." This process identified a high number of constituents and complexity of interconnections. The meaning units could easily have become lost in the volume and complexity of the data. Also, the relatedness between constituents seemed "essential to the comprehension of the common elements of the experience for the co-researchers as a group" (Robinson, 1994, p. 51).

To establish a focal point in this complexity and interconnected data, a weighting technique was developed to measure the number of references made to each constituent, the number of cross-references between a constituent and other constituents, and the number of subjects who referenced each constituent. Ultimately, a factored Table of Constituents, Items & Subjects, and Cross-References was created to list each constituent in its respective order of relative importance. Readers interested in specific details of this analysis or in the actual table are referred to page 49 of my original dissertation manuscript (Robinson, 1994).

At this point, the quantification and analysis process required a deliberate effort from me as researcher to move once again into the qualitative nature of the data. Reprocessing the initial 58 constituents brought further reduction to 32 constituents that could not be reduced any further.

Descriptive Statements. In organizing the 32 constituents for the final descriptive statements, several elements were considered: the data collection interviews, the constituents as seen through the eyes of the subjects, the factored table, and whether the constituents related meaningfully to the beginning, middle, or end of the self-cutting experience. These were composed into an all-inclusive detailed general essential structure.

Then, each of the 11 protocols was encapsulated into a one-page, context-specific summary that captures the personal story of the particular woman (situated essential structures). The stage was now set for the final synthesis: reducing the individual, personalized data and the psychologically relevant data to a one-page composite. This final summary (general essential structure) defines the "essence" of the self-cutting experience.

PHENOMENOLOGICAL RESULTS

Essential Attributes of Self-Cutting (General Essential Structure)

Table 1 lists the final 32 essential constituents of the act of self-cutting in the dissociative population. Not all of these constituents pertain to every case, but the 32 in sum are representative of the composite constituents of the independent experiences of 11 women.

General Essential Structure

The general essential structure describes my best view of the experience of cutting after this exhaustive handling of the data collected (Robinson, 1994, p. 86):

> Emotional states prior to cutting include feelings of being trapped, inner emotional chaos and noise, intolerable emotional pain, intense anger, and feelings of separation from significant others. Cutters are wanting a release of tension, a relief from their enduring pain, and a quieting of the chaos and noise. They reach a place where it feels as if they will explode or die or where they are feeling out of control and desperate. It feels as if physical pain could take away their emotional pain.
>
> Alters emerge who seem responsible for the cutting. The actual cutting takes on a compulsive, habit-like, automated process. The cutter may or may not have knowledge of plans to cut beforehand.
>
> Cutters shift into an altered mental state. They find a pleasurable place where they are feeling in total control of their life. Their pain subsides. As the intensity climbs, they discover relief and

Table 1. Final 32 Constituents

Altered states of consciousness	Postcutting emotional states
Blood	Precutting emotions
Compulsive quality	Preplanning
Control	Protective quality
Cutters' attitudes toward having emotions	Punishment
Death/suicide	Relationship with one's therapist
Desire to have "it" stop	Relationships with others
Fear	Release function of cutting
Feelings during cutting	Scars
Feelings toward cutting	Self-image
Getting enough	Sexuality
Inner quiet	Spiritual connection/love
Loss	Stopping cutting
Memories of trauma and rituals	Temporary relief
Pain	Triggers
Pleasure	Who does the cutting?

inner quiet, peace, comfort, nurturing, love, and feelings of safety. This mental state is a highly-pleasurable aspect of cutting, one that they wish to prolong.

The sight and feel of flowing blood has much symbolism for cutters. It represents life, the possibility of death, release of "things trapped" in the body, and survival.

Child alters find means of expression through cutting since they are afraid to communicate for fear of punishment. Visible scars become a sign to the outer world of their pain, fear, and needs.

Cutters experience much fear and shame after cutting. Their good feelings and relief are always temporary. They feel sad knowing cutting brings their greatest comfort and meets their love and nurturing needs.

Many cutters have been ritually abused. Ritually-abused cutters are often programmed to not experience feelings. It is likely they have punishing alters who punish them for feeling, disclosing secrets, experiencing pleasure, and remembering abusive incidents. These cutters may experience a protective quality in cutting that keeps their punishers away. It is also likely that ritual-abuse survivors have been programmed to equate blood, pain and violence with sexual urges and satisfaction.

Complexity of the Experience—Complexity of the Results

The results of this research reflect the complexity of the individuals and experiences being studied. From the original stories of these 11 women, to the initial 58 constituents, and then to the final 32 constituents, one can see that self-cutting is not a simple behavior to understand. The complexity of the research data is found not only in the quantity of significant elements but also in the intricacies of cross-weavings of connections between the elements.

The complexity of the personal histories of these women opened my eyes to the potential of human beings to act in cruel ways toward other human beings, even their own family members. The complex, systematized manners of abuse in some cases were shocking. Even more complex were the intricate, survival-seeking mechanisms within the psyches of the abuse survivors.

Ritual Abuse: An Unexpected Discovery. In the developmental phase of the research, a consultant who had a practice including many dissociative patients expressed his interest in comparing patients who had been ritually abused to those who had not been ritually abused. I explained my intentions to have a "simple" study, since this was my first research project. Truthfully, since this was my first encounter with the notion of "ritual abuse," my inclination was to run quickly and far. Fully unaware, I was headed toward exactly this kind of discovery. (I was at a conference when I was introduced to the woman who became the study's first subject. Her first words were, "I'm a 13th-generation SRA survivor. Do you still want me in your study?" I asked her if there was any potential danger to either herself or me if she were to participate. She answered, "No." This was the beginning of a remarkable journey of discovery.)

As the protocols were being collected, what started to emerge was a qualitative difference among some of the protocols. This sense of difference emerged particularly as the protocols were being entered into the computer or read aloud. The qualitative difference that I was sensing had something to do with "complexity" and "depth" of meaning. After all the protocols were received and I began the phenomenological process, it struck me that the protocols written by the seven women who reported some type of ritual abuse all appeared more complex in many different ways, including literary style, phrasing, hidden meanings, symbolism, length, and switching between alters.

After the descriptive statements had been written, I went back to the colored slips of paper from which I could identify the protocol behind each element. I created a table of elements referenced by each of the protocols. The results were interesting, to say the least. Of the original 58 constituents, 31% were not referenced by those women who had not been ritually abused.

From these results, we can see a significant difference between the act of self-cutting of a ritually abused survivor and that of a non-ritually abused survivor. The difference is apparent in the complexity of their personal experience, the complexity of their memory storage processing, externally reinforced distortions of the concept of good versus bad, the complexity of their internal coping mechanisms, the depth of fear structures, and their degree of development of masking emotions.

Clinical Significance of This Phenomenological Inquiry

This is a new finding in the field of psychology. It imposes on clinicians a new responsibility to be aware of and sensitive to a fine distinction in historical abuse backgrounds, and to recognize differences in treatment approaches and treatment outcomes for these different groups.

A Rare Glimpse into Private Experience. There is no doubt that this study offers a very privileged view into a private and secretive phenomenon. Self-cutting is an incredibly private experience surrounded by much pain, guilt, shame, embarrassment, secrecy, distrust, fear, and self-hate.

Looking back, it is somewhat miraculous that this research could be done at all. Without a doubt, many obstacles were overcome in the process, as moments of trust and faith were required in order to establish the necessary contacts and rapport needed every step of the way. First, the "human subjects review committee" had to be persuaded that this study held merit and could be done safely. The success of this research required vision and energy from the research committee and the 11 referring, supportive therapists. It also required great courage from the 11 women who shared one of their most intimate, embarrassing, and private experiences. Trust had to be earned from these women who were not accustomed to trusting.

I was deeply moved by their personal commitment. These women were at risk of punishment if the perpetrators of their abuse were to discover their disclosures. Some had to protect themselves from their own internal perpetrators. I wondered why they would want to endure such pain, reopen their emotional wounds, and place themselves at risk. One woman wrote, "I'm willing to do this so others won't have to hurt as much." Another said, "Knowledge is power."

Understanding and Empathy. In doing this research, I never intended to answer the question, "Why do people cut themselves?" My purpose was to describe the experience so all of us might begin to understand its dynamics and essence. The women who shared their experiences wanted to be heard and understood, not only for themselves but also for others who hide their experiences in secrecy and shame. The general essential structure provides a context within which to frame such experiences.

Additionally, the hundred-plus pages of interviews provide a rich source of background material to help clinicians gain understanding and insight into 11 personal histories leading to self-cutting experiences. Readers who are willing to undertake a reading of the dialogue data can draw their own conclusions regarding answers to many questions (e.g., "Why do people cut themselves?").

I will never be able to fully understand the life histories of the women in this study. As I read and reread their stories, I personally glimpsed more and more of what they had experienced. It is my belief that only in understanding an experience are we able to truly empathize with the experiencer.

Clinical Approaches. More and more clients are coming forward with symptoms common to the dissociative disorders that need to be evaluated and diagnosed. Clinicians working with such clients have a responsibility to keep themselves up to date on dissociative disorders by attending conferences, following the rapidly increasing amount of literature, and establishing peer contacts with professionals who have experience with dissociative disorders and self-cutting. Dissociative cutters, because of their abuse histories, are naturally suspicious and very perceptive of and sensitive to other people. They want their therapist to ask questions and understand, but become resentful of having to "teach" the helping professions, in general, about aspects of dissociation, abuse, and cutting behavior.

This brings us to some important questions. For example, can therapists properly diagnose and treat the cases they encounter? The possibility still exists that many persons diagnosed as Borderline Personality Disorder are undiagnosed dissociative disorder cases. This is a delicate matter, since therapists need to exercise caution not to overdiagnose dissociative disorders. When one considers the intricacies of clinical attitudes toward both these diagnostic categories (borderline versus dissociative), then one must further wonder what kinds of treatment are being rendered and whether the applied treatment in each case is appropriate. Treatment methods should be in harmony with recommended approaches established by today's experts in the field.

One of the major discoveries in this research is the importance of the therapeutic relationship. The cutting behavior further fragments the world of dissociative clients, as it isolates them from those who cannot see their pain or understand their needs. In most cases, therapists were the only external source of support mentioned by the participants in this study. In most cases, the emotions after cutting included apprehension and fear of what the therapist would say and do.

Another question that arises is whether the therapist's personal discomfort with self-cutting behaviors interferes with his or her ability to discuss these issues with the client. Discussing the cutting incident with their therapist seemed to be an important part of the process for the 11 women in this study. It brought empathy, nurturing, insight, and closure to an intense, painful incident. Writing about and discussing the self-cutting experience, in general, appeared to have some measurable healing effects. In fact, four of the 11 participants have reported that the intense healing benefits received from participation in this research have extinguished their need for further self-cutting incidents.

Inquiring about what personal meaning the cutting experience holds for a client offers great potential benefit. Euphoric states of a spiritual quality, for example, might be a link in cases of serious injury. A client might cut to reexperience this pleasure and not want to withdraw, unconsciously or consciously, thus prolonging the bleeding or deepening the cutting to intensify the pleasure. The most difficult decision a therapist faces is determining the client's need for medical care subsequent to cutting. The women in this study indicated that they wanted their therapist to "trust" them to know when their cuts needed medical attention. The therapist faces immense liability and responsibility concerns, on one hand, and risks losing the bonds of the therapeutic alliance, on the other.

A third question is whether therapists have an understanding of related controversial issues. We have evidence from victims' reports corroborated by law enforcement files and photos, for example, that ritual abuse occurs. There is an extensive psychological literature on recent research in this area. In the dialogues of the 11 women who participated in this particular study, seven described experiences of ritual abuse. The complexity of the internal psychological structures of those who report histories of ritual abuse requires specialized understanding and skill in treatment.

A final area of concern involves scope of practice and limitations. Are therapists able to recognize their own limitations, are they willing to refer cases that are beyond the scope of their practice, and do they know how to find such referral sources? In a conference presentation on *Trauma, Beliefs, and Recovery,* Connors (1993) stated:

> Patients reporting a history of ritual abuse describe countless profoundly abusive experiences. Looking beyond the sensationalism and horror of their memories, experienced clinicians recognize the underlying dynamics of extensive and sophisticated mind control. The outcome of alleged mind control experiences is the creation of patients with pervasive cognitive distortions that impact their beliefs, self-image, and world view. Further, these distortions are profoundly resistant to change.

This is consistent with clinical studies on memory distortion that report how the details of memories of traumatic events are often imprecise (Loftus, 1993). Clinical accounts of mind control and programming tactics tell us that trickery, illusions, and drugs are often employed to alter persons' perceptions (Connors, 1993; Young, 1993; Young, Sachs, Braun, & Watkins, 1990). These findings highlight the importance and validity of a victim's emotional experience and the need for therapeutic support.

PERSONAL GROWTH FROM PHENOMENOLOGICAL INQUIRY

I started out designing this research project because I had met a woman who had recently cut herself and who wanted to talk about her experience with me. I was surprised at the intensity of my emotional response to her behavior. My extreme emotional reaction and her subsequent communicative "shutdown" are what prompted this research.

Self-cutting was an "unknown" to me. In searching for understanding of this one unknown, I encountered another, the concept of ritual abuse, that was both disarming and powerful. It was also an "unknown" for me, but one of imagined greater power. I made an initial conscious decision to categorize it as irrelevant and to avoid it. At first, I was fearless because I was naive. I had no clue as to the extent of facts I had yet to encounter. As the process of the research unfolded, elements of ritual abuse appeared from beginning to end. I could no longer ignore it. That which we do not understand frightens us. Somewhere in this process, I realized that my fears held the power to keep me in ignorance and that they also held the key to expanded awareness. I remembered the words of one of the women in this study, "Knowledge is power."

I have spoken about the courage of the 11 women in this study, but have not mentioned my own process in this regard. As my awareness grew, I must confess there were moments when I had to face my own fears and disbeliefs. It was the courage and conviction of these 11 women that helped me through this incredible journey. In horror and shock, I focused on understanding the data I was analyzing. The elements of ritual abuse were the hardest to face. It is easy to deny, ignore, walk away from the uncomfortable. I learned that it takes courage to work through emotional discomfort as we encounter painful possibilities.

I would expect few people to immerse themselves in these areas to the extent required for this research. But for those who are willing to encounter their own fears, the data are available if they want to reach further. The people who have read these accounts say they will never again be the same because, through reading, they have been transformed by feeling the subjects' experiences of pain. As a researcher, I lived with these experiences for almost a year, facing the pain over and over again as I worked with the subjects and their data. I will never be able to return to the state of naiveté that I held prior to this research.

All of us can resort to denial to protect ourselves from the uncomfortable. Therapists are vulnerable to their own "self-protective incredulity" of the "extraordinarily sadistic and prolonged experiences of satanic ritual abuse" (Van Benschoten, 1990). Van Benschoten (1990, p. 25) cautions against taking an extremist position in either direction: "Denial protects us from the intolerable realization of man's capacity for brutality. Over determination protects us from becoming complacent. . . . To realize the danger in not taking patients' accounts of satanic abuse seriously, one only has to consider instances in which reports of atrocities were initially denied and later found to be true."

CONCLUSION

Dissociative Identity Disorder (formerly Multiple Personality Disorder) is still viewed by many health care professionals as a mysterious, rare disorder despite large amounts of literature documenting its prevalence. Self-mutilative behavior is also viewed as a rare, mysterious behavior, partly because of the degree of secrecy and shame surrounding its occurrences and the scarcity of public information. This research illuminates dissociative disorders and self-cutting behaviors. It also highlights the varying degrees of complexity within the scope of both phenomena relative to the intensity of imagined or real childhood abuses. In light of these discoveries, the results of this research may expand our awareness of the realities depicted within current psychological literature and research, and deepen our personal and professional acceptance of these psychological phenomena.

REFERENCES

American Psychiatric Association. (1987). *Diagnostic and statistical manual of mental disorders* (3rd ed., revised). Washington, DC: American Psychiatric Association.

American Psychiatric Association. (1994). *Diagnostic and statistical manual of mental disorders* (4th ed.). Washington, DC: American Psychiatric Association.

Bach-y-Rita, G. (1974). Habitual violence and self-mutilation. *American Journal of Psychiatry, 131*(9), 1018–1020.

Bliss, E. L. (1980). Multiple personalities. *Archives of General Psychiatry, 37,* 1388–1399.

Blume, E. S. (1986). *The walking wounded: Post-incest syndrome. Sex Information and Education Council of the U.S., 15*(1), 5–7.

Braun, B. G. (1986). Issues in the psychotherapy of multiple personality disorder. In B. G. Braun (Ed.), *Treatment of multiple personality disorder.* Washington, DC: American Psychiatric Press.

Carol, J., Schaffer, C., Spensley, J., & Abramowitz, S. (1980). Family experiences of self-mutilating patients. *American Journal of Psychiatry, 137(7)*, 852–853.

Colaizzi, P. F. (1978). Psychological research as the phenomenologist views it. In R.S. Valle & M. King (Eds.), *Existential–phenomenological alternatives for psychology* (pp. 48–71). New York: Oxford University Press.

Connors, K. (1993). *Trauma, beliefs and recovery.* Paper presented at the 10th International Conference on Multiple Personality & Dissociative States, Chicago.

Coons, P. M. (1980). Multiple personality: Diagnostic considerations. *Journal of Clinical Psychiatry, 41,* 330–336.

Coons, P. M., Bowman, E. S., & Milstein, V. (1988). Multiple personality disorder: A clinical investigation of 50 cases. *Journal of Nervous and Mental Disease, 176*(9), 519–527.

Coons, P. M., & Milstein, V. (1990). Self-mutilation associated with dissociative disorders. *Dissociation, III*(2), 81–87.

Culiner, T. (1993). *The neglected dissociative population: Convicted felons.* Paper presented at the 10th International Conference on Multiple Personality and Dissociative States, Chicago.

Favazza, A. R. (1993). *Bodies under siege: Self mutilation in culture and psychiatry.* Baltimore: Johns Hopkins University Press.

Giorgi, A. (1985). Sketch of a psychological phenomenological method. In A. Giorgi (Ed.), *Phenomenology and psychological research* (pp. 1–21). Pittsburgh, PA: Duquesne University Press.

Greaves, G. (1980). Multiple personality: 165 years after Mary Reynolds. *Journal of Nervous and Mental Disease, 168,* 577–596.

Green, A. H. (1967). Self-mutilation in schizophrenic children. *Archives of General Psychiatry, 17,* 234–244.

Green, A. H. (1978). Self-destructive behavior in battered children. *American Journal of Psychiatry, 135*(5), 579–582.

Grunebaum, H. U., & Klerman, G. L. (1967). Wrist slashing. *American Journal of Psychiatry, 124*(4), 527–534.

Kahaner, L. (1988). *Cults that kill.* New York: Warner Books.

Kluft, R. P. (1984). Treatment of multiple personality disorder: A study of 33 cases. *Psychiatric Clinics of North America, 7,* 9–29.

Kluft, R. P. (1988). The phenomenology and treatment of extremely complex multiple personality disorder. *Dissociation, 1*(1), 47–58.

Kluft, R. P., & Fine, C. G. (Eds.). (1993). *Clinical perspectives on multiple personality disorder.* Washington, DC: American Psychiatric Press.

Loewenstein, R. J. (Ed.). (1991). *Multiple personality disorder* (3rd ed.). Philadelphia: W. B. Saunders.

Loftus, E. F. (1993). The reality of repressed memories. *American Psychologist, 48*(5), 518–537.

Neswald, D. W., Gould, C., & Graham-Costain, V. (1991). Common "programs" observed in survivors of satanic ritualistic abuse. *California Therapist,* 47–50.

Pitman, R. (1990). Self-mutilation in combat-related PTSD. *American Journal of Psychiatry, 147*(1), 123–124.

Putnam, F. W. (1986). The scientific investigation of multiple personality disorder. In J. M. Quen (Ed.), *Split minds/split brains*. New York: New York University Press.

Putnam, F. W. (1989). *Diagnosis and treatment of multiple personality disorder.* New York: Guilford Press.

Robinson, F. A. (1994). *An existential–phenomenological investigation of the experience of selfcutting in subjects with multiple personality disorder.* Doctoral dissertation, California Institute of Integral Studies, 1994. *Dissertation Abstracts International, 55–07*, 3025B.

Rosenthal, R. J., Rinzler, C., Walsh, R., & Klausner, E. (1972). Wrist-cutting syndrome: The meaning of a gesture. *American Journal of Psychiatry, 128*, 1363–1368.

Ross, C. A. (1989). *Multiple personality disorder: Diagnosis, clinical features, and treatment.* New York: Wiley.

Ross, C. A. (1991). Epidemiology of multiple personality disorder and dissociation. *Psychiatric Clinics of North America, 14*, 503–517.

Ross, C. A., Anderson, G., Fraser, G. A., Reagor, P., Bjornson, L., & Miller, S. D. (1992). Differentiating multiple personality disorder and dissociative disorder not otherwise specified. *Dissociation, V*(2), 89–90.

Ross, C. A., Heber, S., Norton, G. R., & Anderson, G. (1989). Differences between multiple personality disorder and other diagnostic groups on structured interview. *Journal of Nervous and Mental Disease, 177*(8), 487–491.

Roy, A. (1978). Self-mutilation. *British Journal of Medical Psychology, 51*, 201–203.

Smith, M., & Pazder, L. (1980). *Michelle remembers.* New York: Congdon & Lattes.

StarDancer, C. (1993). Sibylline shackles: Mind control in the context of ritual abuse. *Survivorship*, 4–8.

Van Benschoten, S. C. (1990). Multiple personality disorder and satanic ritual abuse: The issue of credibility. *Dissociation, III*(1), 22–29.

van der Kolk, B. A. (1993). *The intrusive past: The flexibility of memory and the engraving of trauma.* Paper presented at the 10th International Conference on Multiple Personality and Dissociative States, Chicago.

Walsh, B. W., & Rosen, P. M. (1988). *Self mutilation: Theory, research and treatment.* New York: Guilford Press.

Young, W. C. (1993). Sadistic ritual abuse: An overview in detection and management. *Family Violence and Abusive Relationships, 20*(2), 447–458.

Young, W. C., Sachs, R. G., Braun, B. G., & Watkins, R. T. (1990). Patients reporting ritual abuse in childhood: A clinical syndrome. Report of 37 cases. *Child Abuse and Neglect, 15*, 181–189.

10

Psychology of Forgiveness
Implications for Psychotherapy

Jan O. Rowe and Steen Halling

We live in a world fraught with hostility and alienation at both interpersonal and societal levels. The media barrage us with reports of ethnic wars growing out of centuries of hatred and distrust, political terrorism, random violence in the streets, domestic violence, child abuse of various sorts—the list seems endless. Clearly, there is a tremendous amount of hurt in our lives—both personal and cultural. What is the impact of injury on our lives? How can we heal from it?

These were questions we and our fellow researchers had over a decade ago when we embarked on the study of forgiveness. Partly because of the complexity of the phenomenon, we decided to focus on forgiveness in everyday life as opposed to more dramatic social or political contexts. Even so, forgiveness turned out to be a challenging topic to research. For example, while people were willing to share their experience of forgiveness on deep and personal levels, their doing so often involved a great deal of emotional pain.

On another level, conducting this research was not only a difficult but also a groundbreaking enterprise, because in Western culture forgiveness is a word with religious connotations, and many people are uncomfortable with it as a topic of dialogue. Moreover, forgiveness is spoken of more often as an abstract ideal than as central to one's experience. We have come to wonder whether this confusion concerning the experience of forgiveness may not be an expression of specific contemporary cultural values that run counter to the attitudes necessary for forgiveness: openness to oneself and others, to the metaphorical or mysterious in living, and to mercy. Instead, justice has become synonymous with punishment, mercy with weakness, strength with power over others. Given this situation, it is no

Jan O. Rowe and Steen Halling • Department of Psychology, Seattle University, Seattle, Washington 98122.

Phenomenological Inquiry in Psychology: Existential and Transpersonal Dimensions, edited by Ron Valle. Plenum Press, New York, 1998.

wonder that people are flocking to various mental health practitioners with chronic guilt, shame, resentment, dis-ease, and feelings of estrangement.

While the cultural attitude toward forgiveness may make forgiveness a topic that is uncomfortable and unfamiliar for many people, through our research we have come to see how powerful it is when it does occur. In terms of their actual experience, people consistently reported a sense of liberation, reconnection, and hope about the future. Given such benefits, the question becomes: How can we facilitate the movement toward forgiveness? This question, we believe, is particularly important for those who work with those in psychic pain, in particular, psychotherapists.

This chapter gives a brief overview of the literature, describes the qualitative method we employed in the research, and outlines two dimensions of the phenomenon—forgiving another and forgiving oneself. Finally, we will address the implications of our research for the practice of psychotherapy and sitting with people who have deep-seated hurt.

REVIEW OF SELECTED PSYCHOLOGICAL LITERATURE

Until recently, it appeared that psychology had treated the topic of forgiveness with benign neglect (Halling, 1979). In the last decade or so, however, there has been an increased interest in forgiveness as evidenced by a beginning tradition of empirical research, a number of doctoral dissertations, and a burgeoning literature that presents strategies for facilitating forgiveness in psychotherapy clients.

The neglect of forgiveness in psychology can be attributed in part to some of the social and cultural factors that we have suggested are responsible for its marginalization within our society generally. In addition, as a discipline that is highly self-conscious about its scientific status, psychology has traditionally shied away from phenomena that are closely related to theology and religion. Further, psychology, insofar as it has been conceptualized as a natural science, has placed priority on its method, that is, the experimental method. It has then avoided topics that cannot be easily studied by this method (Giorgi, 1970), and a topic as profound but resistant to simple definition and direct observation as forgiveness clearly falls within this category.

In the discussion that follows, we focus primarily on psychological literature that is empirical in focus, that is, based on interviews, observations, or work with patients. We want to provide an overall portrait of the way forgiveness is being studied on the basis of a review of selected literature and on that basis demonstrate that there is a need for phenomenological investigations of forgiveness. Our basic point is that despite the increased attention given to this topic, there are still very few systematic studies of the actual experience of forgiveness.

Since the time of Freud, the case study has been an important vehicle for advancing psychological understanding. Hunter (1978), who writes from a psychodynamic perspective, considers paranoid reactions and forgiveness as two dramatically different responses to psychological injury. Going beyond the specifics of the case he presents, he speculates about how developmental factors are related to the capacity for forgiveness and describes stages that patients go through insofar as they move from blame and anger toward forgiveness in therapy. Martyn (1977) attempts to integrate psychoanalytical concepts regarding personality structure with theological concepts about forgiveness by considering the situation of an abused child who is in play therapy. However, given the lack of a more comprehensive framework from which to reflect on these two specialized approaches with their distinct assumptions and technical terms, this integration is not readily accomplished. On the other hand, Close (1970) describes in everyday language the movement toward forgiveness of a young woman who had been sexually abused. In this context, he contends that a movement beyond blame and recrimination requires that the injured person look at his or her own responsibility.

There is no question that the case study method contributes to our knowledge of issues such as forgiveness. Yet case studies are limited in that they often use the situation of a particular client primarily for illustrating a certain preconceived theory. Further, the struggle with forgiveness in therapy may take a different form and direction than in everyday life. For example, in therapy, obstacles to forgiveness may be resolved due to the systematic intervention of an attentive and empathetic professional.

Insofar as clinicians have come to believe that forgiveness can heal hurts and overcome resentment, it is to be expected that some, at least, will attempt to develop specific techniques or procedures to facilitate the movement toward forgiveness. One of the earliest clinicians to do so was Fitzgibbons (1986), who regards forgiveness as a way to overcome anger and rage, especially in adolescent males. On the basis of his clinical interventions, Fitzgibbons distinguishes three kinds of forgiveness: cognitive, emotional, and spiritual. He takes clients through a number of specific steps, such as analyzing the origin of the pain or hurt, reenacting the painful situation, and persuading them to make a deliberate cognitive decision to forgive. It is not clear from Fitzgibbons's article, however, what the theoretical and empirical basis is for either the interventions or the conception of forgiveness that he proposes.

Worthington and DiBlasio (1990) have written guidelines for helping couples with troubled relationships move toward granting and seeking forgiveness. They discuss at length the preparation for what they call the "forgiveness session" (e.g., they tell clients not to expect too much and encourage them to focus on how they have hurt the other rather than on how the other has hurt them) and provide criteria far evaluating whether clients have the capacity for such a session. They clearly do not follow procedures mechanically; instead, they offer a basic approach that requires clinical judgment in its implementation.

A variety of studies on promoting forgiveness have been developed at the University of Wisconsin under the leadership of Robert Enright. In a theoretical article, Enright, Gassin, and Wu (1992) examine a variety of writings on forgiveness—religious, philosophical, and psychological—and formulate developmentally based (especially drawing upon Kohlberg) psychological models of how people think about and go about forgiving others. A number of related studies have been carried out to test specific interventions to facilitate forgiveness with various populations such as incest survivors (Freedman & Enright, 1995) and adolescents (Enright, Santos, & Al-Mabuk, 1989).

Attempts have been made to study the issue of forgiveness using traditional research methodology. For example, there are two relatively early experimental studies (Gahagan & Tedeschi, 1968; Tedeschi, Hiester, & Gahagan, 1969). Both investigations involve a prisoner's dilemma game situation. Within this experimental context, forgiveness is operationally defined as the giving of a cooperative response by a subject after his or her opponent has made a competitive response to a prior cooperative response. Although a simple change in behavior, along this line, may be associated with forgiveness, this phenomenon cannot reasonably be defined so narrowly because the process entails a fundamental shift in attitude. Such a shift can by no means be fully explicated from an observer stance, because neither the inner meaning of the act of forgiveness nor the significance of the process is directly "visible."

More recently, McCullough and Worthington (1994), also working from within a natural science psychological perspective, have evaluated claims that clinicians have made for the positive consequences of forgiveness. These consequences allegedly include better relationships, diminishment of resentment and anger, and improved physical and mental health. McCullough and Worthington (1994) argue that the studies they reviewed lack control groups, fail to specify the treatments involved, do not use standardized measures, and so on, and they conclude, "there is not enough data to conclude that forgiving has any clear physical or psychological benefits" (p. 5). It may well be that some clinicians not only fail to substantiate their claims but also have come to see forgiveness as a general solution to many of life's problems. Patton (1985), who writes from a pastoral counseling perspective, has raised questions about the belief that "forgiveness is good for you and that you ought to become forgiving as soon as possible" (p. 124).

But the McCullough and Worthington review raises a more fundamental question: How does one understand the nature of forgiveness? Our concern is that many of their own assumptions about forgiveness are unexamined and erroneous. For example, they refer to forgiveness as "a religious behavior" (McCullough & Worthington, 1994, p. 4), as something that a person grants; they also refer to it as a "promising therapeutic tool" (p. 4), and assume that counselors, especially those who identify themselves as Christian, ought to encourage clients who have been hurt to forgive their offender. In other words, they have assumed, without any examination of the actual experience of forgiveness, that it is a "variable" with effects that can readily be measured and that it is facilitated by specific kinds of interventions, especially persuasion. In this context, the French philosopher Gabriel Marcel's (1991) warning against humanity's excessive reliance on technique and calculation—and with it our disregard of anything that does not depend on us or is not under our control—seems to be very much to the point.

So far, we have commented on the lack of a systematic experiential basis for much of the theorizing about forgiveness and the development of interventions to bring it about. There are, however, a few studies (in addition to our own) that look at descriptions of the movement toward forgiveness on a descriptive basis. Rooney (1989) has done a phenomenological study of how five patients came to find forgiveness through their participation in intensive individual psychotherapy (this was not psychotherapy designed explicitly to foster forgiveness). This study is distinctive in that it is the point of view of the person who forgives, not that of the psychotherapist, that is being considered. For these patients, what was especially important in psychotherapy was "a confessional exchange between patient and therapist, in which the patient experiences a continuation of acceptance by the therapist in the context of the mutual recognition of the patient's wrongdoing" (Rooney, 1989, p. ii). The outcome of forgiveness included a restoration of the belief that one was worthy of belonging, a sense of physical and emotional cleansing, and a more compassionate stance toward both one's own wrongdoings and those of others.

Flanigan (1992) has carried out a much larger-scale study based on interviews with 70 people who had forgiven "the unforgivable." These people were recruited through newspaper advertisements. For Flanigan, "unforgivable" injuries are those that were inflicted by persons close to the individual (e.g., a parent or spouse), that involved a betrayal, and that were deeply wounding and destructive of a person's sense of morality.

We can only agree with Flanigan that persons who have forgiven such injuries have much to teach the rest of us. However, her book is written in the form of a step-by-step manual for persons seeking to forgive, and thus one does not get a complete sense of the subjects' experience. In addition, Flanigan interprets the stories from a cognitive and social psychological perspective, which, in our view, does not do justice to the richness of people's lives.

THE DIALOGAL METHOD

As we have already indicated, this chapter is based on two studies, one on "forgiving another" (Rowe et al., 1989) and another on "self-forgiveness" (Bauer et al., 1992). Here, we want to give a brief overview of the method used in these studies. Along with other phenomenological researchers (e.g., Giorgi, 1970), we believe that the method ought to be appropriate to the content that one studies and that one cannot adequately evaluate and interpret research findings without an understanding of how these findings were arrived at. One of the distinctive features of our investigations is that they led us to develop a new research method even though that had not been our explicit intention.

Each of the studies was carried out by ourselves and four to six graduate students from the master's program in psychology at Seattle University. In approaching the topic of forgiving an other, we found that this phenomenon, which is fundamentally interpersonal, could be studied most appropriately using a method characterized by open and ongoing conversation. This conversation or dialogue took place on two levels: among the researchers and between the researchers and the phenomenon; hence the name *dialogal*. The dialogal method differs significantly from other phenomenological methods in its process, although not in its aim. In contrast to most other studies, faithfulness to the data is fostered through open dialogue among the researchers in relationship to the data rather than through adherence to a set of explicitly spelled-out procedures.

On one hand, this approach to research developed spontaneously as we struggled as a group to find an appropriate way to study forgiveness. On the other hand, we were also influenced, both theoretically and practically, by our prior reading of thinkers such as Buber (e.g., 1958, 1965), Gadamer (1975), Jaspers (1970), Palmer (1983), Polka (1986), and Strasser (1969), all of whom emphasize how understanding and interpretation arise out of dialogue. Encouraged by the success of our first study (Rowe et al., 1989), we decided to follow a similar approach when we studied self-forgiveness (Bauer et al., 1992).

During the early stage of each of the studies, the researchers shared their initial impressions and conceptions about the phenomenon in question, whether forgiving another or forgiving self, and reviewed and summarized pertinent literature. They wrote personal descriptions about their experience of the phenomenon, and discussed these descriptions, allowing for the recognition of specific assumptions that additional descriptions might bring into question. Then the structure of the research was discussed, tentative procedures were mapped out, and the key research question was formulated. For the first study, this question was:

> Can you tell us about the time during an important relationship when something happened such that forgiving the other became an issue?

The question was deliberately phrased in such a way as to not only gather descriptions of the completed process but also allow interviews with those in the midst of the process. Similarly, for the second study, the question was:

> Can you describe a time in your life when self-forgiveness became an issue?

Each member of the research groups posed the research question to a research contributor (or subject), and after the first interview was transcribed, there was a second interview to allow for clarification and further elaboration. Subsequently, each interviewer and another researcher would write narrative summaries for each contributor. The analysis phase of the research followed as the entire research group discussed individual interviews and summaries and questioned the dyad that had written the summaries, in search of themes that were specific to the particular description. Themes in individual accounts having been identified, the next step was to compare narratives to identify common themes. Gradually, a basic understanding of the phenomenon emerged.

Much of our understanding of the dialogal group process came from the work of Michael Leifer, who was a member of both research groups and who audiotaped the group discussions throughout the research process of the first study. On the basis of reflections on these recordings, and on his reading of Gadamer (1975), Leifer (1986) characterized the discussions as having three levels: preliminary, transitional, and fundamental. Generally speaking, the movement from preliminary to fundamental dialogue is a movement from a relatively abstract and disjointed discussion of a topic to a much more experientially grounded conversation. At this level, the researchers become increasingly attentive to what others are saying, and increasingly mindful of the phenomenon as something actually present in the group through the narratives and transcripts as well as through the researchers' attunement to its experiential reality in their own lives.

Space limitations prevent us from elaborating on the nature of the dialogal method, but interested readers can consult Rowe et al. (1989), Halling and Leifer (1991), and Halling, Kunz, and Rowe (1994) for more detailed discussions of the implementation and theoretical justification of this approach.

THE NATURE OF FORGIVING ANOTHER

As stated earlier, we recognized almost from the beginning of our work together that forgiveness is a complex phenomenon with many dimensions. For the sake of manageability, we limited our initial research to the study of forgiveness in an interpersonal context. (Later, we undertook the investigation of "self-forgiveness"—a topic that proved to be somewhat more elusive. This section gives an overview of "forgiving another," and the following section addresses "self-forgiveness.")

The descriptions we gathered, both from ourselves and from others, included ones that described the completion of forgiveness as well as ones from people in the midst of the process. In the following interpretation, based on these descriptions, we attempt to identify the qualities and stages of the process of responding to harm and coping with injury. These "stages" are far from sharply delineated or easily defined. By speaking of stages, we are addressing the seeming evidence and dominance of specific kinds of experiences at certain points in the process.

From these descriptions, it was evident that the process begins when one perceives oneself as harmed by another and comes to a resolution insofar as it ends in a psychological, if not face-to-face, reconciliation with the one who was perceived as hurtful. There are two basic levels to this process. First, forgiving another is most immediately experienced as interpersonal; it occurs within the context of a relationship involving another who has deeply affected one in a hurtful way. Second, and perhaps more profoundly, the experience of forgiving another also has qualities that transcend one's relationship with that person and opens one up to oneself and the world in new ways. It is more than a letting go—it is also a new beginning. The specific nature of these qualities, which become apparent only toward the end of the process, led us to describe the experience as being spiritual or transpersonal as well as interpersonal. It is noteworthy that our research group did not start out with a "religious" or "spiritual" agenda.

The need for forgiveness arises when someone has acted in such a way as to fundamentally disrupt the wholeness and integrity of one's life. Initially, on a deep, almost organic level, there is a tearing of the fabric of one's life, one's world. The injury that involves forgiving another is one that violates the person's sense of self. The unfolding of one's life and identity is impeded or terminated. The future, as it was anticipated before the event, is irrevocably changed; a particular future is experienced as lost altogether, destroyed. A more general future, one beyond "injury," is simply not there for oneself as a possibility, except insofar as particularly engaging activities or situations take one away, momentarily, from the recollection of the hurtful event. When one does recollect, the hurt, pain, and loss of future reemerge at the center of one's life. Thus, the injurious event and relationship are somehow central or pivotal to the network of one's identity in such a way that the disruption impinges upon one's "only world," one's "only meaningful identity as perceived at that time."

In the face of the realization of the hurt, this disruption is profoundly felt. One feels uprooted, "off center." In the words of one woman whose lover's infidelity was revealed to her by a woman friend: "As [the friend] talked, my throat became dry and restricted. It was suddenly extremely cold in my apartment and I began to shiver. I was stunned and unaware of how to react." Relationships to the world and others at this point are characterized by distance and dis-ease, and most dramatically so with respect to the injuring person. The distance or sense of dis-ease remains in relation to that person even after connections with friends and objects in the familiar world have been reestablished—when, for example, familiar streets no longer seem foreign and forbidding. However, the deeper levels of meaning of the disruption to one's sense of self are typically not yet articulated (or conscious). These levels will unfold later.

On a lived level, one experiences the injury as a blow inflicted by the other. There is the conviction that the other's behavior was aimed at oneself, that one was the target of the other's demeaning or intentionally unjust and damaging behavior. At the very least, one believes the injury to have been avoidable had the other person been sensitive to and respectful of oneself. One man said, "If she hadn't known how I felt about it, then it wouldn't have made any difference." He believed that she acted in conscious disregard of what mattered to him.

Oftentimes, an acknowledgment of responsibility, an apology from the other person, is thought to be necessary for healing: The man quoted above said, "She could acknowledge that her position is costly to me. She could apologize, not for her decision, but for how it affects me. That'd be nice. I would like that." Underlying the wish for an apology is often a wish for the other to be different from the person as experienced. Many times, we seem to believe that the situation can change only if the other does.

The ongoing experience of hurt entails a preoccupation with the injury. At the time, one is likely to assume that the other's actions were the simple and sufficient reasons for the hurt and disruption in one's life. Typically, it is only later that one starts to look at the deeper implications of "the injury" in relationship to one's sense of self. The hurtful interaction is remembered as the transition point between a comfortable and familiar sense of the world and an existence that is disturbing and uprooted. The following is a particularly vivid description of this transition:

> The next morning I felt a slow hideous obsession creeping into [me]. I felt it taking over my life. I felt fear and then the fear turned into a cold terrifying anger. . . . I cried and screamed at the injustice of it. . . . It wasn't fair. Why? Why? Why? I asked myself . . . what had I done to deserve this? My questions remained unanswered, and I became angrier.

The initial hurt is often accompanied by anger; in other cases, anger becomes an issue later. It is important, however, that anger and blame be allowed to be experienced. Genuine forgiveness cannot take place if there is a disavowal of some vital aspect of one's own experience and of the relationship to the other person. In some cases, it may be especially hard for a person to allow anger to emerge—if the other is someone one depends on, if one has a habit of blaming oneself, if one believes that it is "bad" to be angry at someone one is close to.

Along with anger, there is frequently a desire for revenge or retribution. These fantasies carry the promise of some sort of partial balancing of an injustice; they provide, however artificially, a "future" of sorts. Most important, they offer a future in which one is no longer a victim, but the victimizer. The possibility of forgiving the other seems unlikely at this point, and the anger may be perceived as extending indefinitely. One middle-aged man seemed to be in this place: "My mother is a stubborn, bigoted, disappointing woman. I don't see how I can forgive." This quote also provides a good example of the perspective that views one's own reaction as simply a function of the other person behavior.

If the other person and the relationship are valued, and if one is troubled by one's own obsession with the hurt, thoughts of revenge are likely to become interspersed with a wish for reconciliation. Although thoughts of the other as blameworthy may still predominate, increasingly there are moments of questioning oneself. So we may ask: Did I misconstrue the intentions of the other? Did I do something to contribute to the problem? During this phase, the enmeshment the hurt has brought about and the self-referential perspective begin to unravel. One begins to catch glimpses of the other in terms apart from the immediate relationship. Preoccupation with the other's wrongdoing begins to be pierced by guesses at explanations for his or her behavior that make it more "understandable" or "acceptable." And there is a dawning awareness that one is somehow helping to keep alive the feelings of discomfort in relation to the other person. Yet exactly how, and therefore how to cease doing so, are not clear. One woman wondered:

> I see the obstacle in front of me but I can't seem to move it: How do I forgive her without her showing me she knows how much pain I've experienced? How do I forgive and not forget so I can go on? How do I rid myself of the selfish demand that she acknowledge my pain?

Aside from concerns about restoring the relationship or the growing desire to feel peaceful rather than haunted by what happened, one may also be moved by an inner obligation to forgive; additionally, there may be a sense of guilt about being angry with the other. But one is unable to simply let go of the hurt and recriminations. At this time in the process, some critical form of healing has not taken place, and there is a moving back and forth such that one might speak of being caught between what seems like irreconcilable opposites: holding on to hurt and anger that create distance and accepting the relationship as it is at present by somehow letting go of the past.

Letting go, although consciously preferable or at least an "ought," does not as yet really feel possible. There is a sense of clinging to the hurt and anger, which is to be distinguished from earlier phases of more spontaneous hurt and anger. This clinging appears to have the function, partly at least, of keeping oneself away from the other while staying engaged with what might have been. As distancing implies, mistrust is often a pervasive theme. This phase may be experienced as an impasse and one feels trapped. One man said, "I did not like the anger and rage I felt, but I also did not know how to leave behind the hurt."

To achieve resolution, one may "try" to forgive, may even say one has, only to find the old pain, anger, and confusion returning. One woman described such an attempt thus:

> I wrote her . . . that I (forgave) her, . . . you see I know that not forgiving her would only destroy myself. . . . By going through the motions I hoped to *feel* forgiveness. But I continue to hang on. Perhaps it is because I feel forgiving her would mean I would have to forget what she did to me. I don't want to forget because if I do, it may happen again.

There may also be some awareness that clinging to the hurt and anger may serve to move one away from specific "inner" experiences such as grief. This grief may concern both the loss of what was and/or could have been, and, on a deeper level, the loss of a particular way of viewing one's self and the world. The latter loss is the deeper metaphorical level of meaning that is not yet entirely clear. One woman, after forgiving her father for years of hurt, said: "[I] am left . . . with a deep sadness for me, for my dad, for all of us who keep ourselves separate out of hurt and fear" She went on to consider, "The avoidance of this . . . sadness may be one reason why [I] resisted forgiving."

During this time, there also may be moments when one feels freed from hanging onto the injury. These times, however, are fleeting and cannot be willed. One man said:

> My hurt and anger vanished as I thought about [her]. . . . I felt healed. . . . This experience was deeply moving, but I would hesitate to call it dramatic. The next day . . . I was back to my previous state, and yet I knew that something was possible even though I had no idea how to get "back there."

The resolution, in the form of forgiveness, appears to come to us in an unexpected context, often at an unexpected moment. And yet, as one is surprised by the resolution, it becomes apparent that at some level it was sought, one was willing to forgive and open to this possibility. It seems that this willingness is crucial for forgiveness to occur. Not imagining how he might forgive his mother, one man said: "I don't see how I can forgive anything. Maybe it's because I'm stubborn or maybe I've talked myself into not being able to back down."

Experientially, however, the moment of forgiveness appears to be the moment of recognition that forgiveness has *already* occurred. Rather than being aware of changing, one realizes that one *has* changed, one *has* forgiven the other. Forgiveness comes as a revelation and is often viewed as a gift. One woman reported:

> I proceeded to call him and apologize for the letter he was going to receive and in the same breath I said I forgave him. When I said this I was taken by surprise. It had in a sense come out of nowhere.

It is important to note that there may be a series of revelations; that is, one may forgive a number of aspects of a relationship independently, or all the injury may be forgiven at once. Whether forgiveness is piecemeal or all at once may depend upon whether the injury was a discrete event or a more complicated series of happenings, as well as upon the intensity and significance of the hurt.

Previous thoughts about what conditions make it possible to forgive (e.g., if the other were to apologize) turn out not to fit the reality as experienced. The focus has been on what "the other should do" and less on what *one* needs to do in order to overcome the injury. Even when apologies were forthcoming, they did not typically enable people to forgive; likewise, people forgive even without acknowledgment on the part of the wrongdoers. In a parallel vein, while the immediate experience of the hurt is very conscious, it seems doubtful that there was clarity to the broader, deeper meaning of the injury. One seems long in coming to a realization of what significance the wrongdoing had in terms of one's life as a whole. As previously noted, the focus was on the wounding rather than on the underlying meaning of the injury.

The critical dimension of forgiving is that one experiences a shift in one's understanding of and relationship to the other person, one's self, and the world. The implications of the original situation are cast in a new light: The hurt is no longer merely an injury that another has inflicted, and that therefore acts as a barrier, but instead becomes appropriated as pain shared with other human beings. In some sense, it is disengaged from the "injuring" person or at least no longer solely referential to that person. One man described this awareness: "[I now felt her] as another human being who was struggling and who basically did not mean me any harm." There is an experience of reclaiming oneself, which at the same time involves a shift into a larger perspective. No longer does one see oneself in a relationship of victim and victimizer: One is freed from the status of being the object of another's actions and so is able to return to oneself. No longer is there only one possible connection with the other person. There are alternatives where before there were none, and this new vision reinstates choice into one's life. A sense of responsibility for one's life and relationships is recovered. After forgiving her father, one woman said:

> My life immediately began to change. After spending almost six years in a profession which I did not enjoy, but had entered to gain my father's approval, I decided to return to school to study what I loved. By opening my mind to forgiveness, I was able to open my heart, and the transformation affected my life.

After forgiving a family member for sexual abuse, one person stated, "For the first time in my life I feel free." Another person said: "I realize that forgiveness has set me free. Free to continue my life, free to exist without pain and anger, and free to love again." The vision of newness is so compelling, so like a gift of grace, one will not choose other than to move gratefully into it. The future—an immediate sense of being on the verge of new beginnings—is again available where before it was not; the past, while neither forgotten nor rationalized away, is no longer a haunting, heavy, and troubling issue.

At the level of lived experience, there is a release of tension, yet this release is one in which one's active participation is acknowledged on some level, although perhaps most clearly in retrospect; thus, people frequently speak of being able to "let go" of anger, hurt, and recriminations. One experiences a restoration of a sense of wholeness and of inner direction and an opening up to perceiving how other people and situations are in their own right, as distinct and separate from one's own needs and desires: "I stopped trying to pigeonhole her into a ready-made mold." One has an attitude of openness to the other; as one person said: "I feel more relaxed and can look her in the eye, where before I couldn't." On a reflective level, one sees the other as having acted in a way human beings do, out of his or her own needs and perceptions; there may even be the recognition that what he or she did is something one has done or could well do: "Forgiving came with acknowledging that we aren't perfect." One understands the other person, and oneself, in a new and fuller way.

The experience of forgiveness is one of radically opening to the world and others, as well as to the person who hurt oneself. There is a sense of arriving home after a long journey and the world is welcoming, so well remembered and yet transformed. One woman wrote: "I knew at last that home was where I was. The past was no longer menacing . . . the future was no longer foreboding. . . . [I] was no longer adrift in a sea of chaos but at the helm in a world that welcomed me. I wept for joy." Others emerge as persons separate from oneself, and yet one's connection with them is more tangible than before. There is a clarity about one's relationship to self and others. There is a sense of relatedness and freedom that did not exist before.

It is because of the transforming nature of forgiveness, coupled with the experience that this transformation involves more than one's own will, that we are suggesting there is a spiritual dimension to forgiveness. More specifically, as we have already indicated, forgiveness comes as a gift or a "revelation," and it involves coming to a deeper sense of connection to oneself, to others, and, in some cases, to something beyond oneself. There is a movement of transcendence, that is, an unanticipated and yet welcomed opening up to the new and an experience of being freed from burdens and restrictions. These aspects were also evident, as we shall see, in the experience of "self-forgiveness."

THE NATURE OF "SELF-FORGIVENESS"[1]

During our study of forgiving another, we began realizing that this phenomenon is intimately related to forgiving oneself. We suspected that they might be two sides of the same coin, so we decided to turn our focus to forgiving oneself.

From the descriptions we have collected, we have come to understand that self-forgiveness becomes an issue, although not necessarily articulated, as a result of an event, such as divorce or suicide of a loved one, that leaves one acutely aware of being estranged from self and others. Generally, this awareness is accompanied by a judgment that one is fundamentally a bad person. This experience is so intense it pervades one's existence, and the embodied belief is that nothing will ever change; the future seems dark and foreboding.

In general, the structure of self-forgiveness, or more accurately that of "experiencing forgiveness for oneself," involves a shift from fundamental estrangement to being-at-home with one's self in the world. This at-homeness involves a change in one's identity, which simultaneously feels very new and very familiar, as though one were recognizing for the first time someone who has always been there: That which one has avoided accepting fully about oneself, for example, the capacity to be enraged or hurtful, is now acknowledged as part of who one is. One moves from an attitude of judgment to embracing who one is. This shift in

[1]This section is based on an article by Bauer et al. (1992) published in the *Journal of Religion and Health*. We are grateful to the journal and the publisher, Human Sciences Press, for permission to include this material here.

identity grows out of the larger meaning the given incident has for one's life: Whereas the initial distress is experienced in the context of a specific occasion or "wrongdoing," at some point there is an awareness that one is in need of forgiveness for merely being human. There is a clarity about oneself and one's place in the world, a sense of connectedness and freedom in the face of the future. The journey is an arduous one requiring both an openness to the mysterious in living and a faith, even in the face of seeming hopelessness, that things can change. What follows is a more detailed account of the structure of self-forgiveness that involves experiencing one's brokenness and estrangement, moving toward healing, and being at home in the world.

Self-forgiveness is a pervasive and ongoing process. It begins when one is no longer able to avoid or deny an increasing awareness that something is fundamentally wrong with one's self or one's life. This awakening may occur in an obvious way, following a specific "crisis" or catastrophic event, such as was the case for one woman whose son had committed suicide.

Alternatively, self-forgiveness may follow a series of difficult and profound changes in a person's life, changes that may appear quite ordinary to outsiders, but are deeply felt by the person experiencing them. Often, it was only in hindsight that the persons interviewed were able to see a particular point in their lives as the start of a process ultimately leading them toward self-forgiveness:

> . . . before I was able to forgive my mother the issue of self-forgiveness never arose because I didn't think I had done anything wrong, because I had just put it all on her. Then when I was willing to forgive her for what she had done, I was able to see my side a little bit. As I opened up to her point of view I also opened up to what was happening to me in a different way. And I saw that, yeah, I was involved with drugs, I was mean to her, I was cruel. . . . So that brought me to feeling . . . remorseful for what happened and my involvement in it.

At this point, one experiences a sense of "brokenness," an estrangement from self and others that is deeply painful. This disconnection is often accompanied by intense feelings of self-recrimination as one replays the situation in one's mind, wracked with confusion, guilt, anxiety, and despair. One's faults and fallibilities can no longer be denied or contained. One feels agonizingly vulnerable, naked before self and others.

The closer one gets to realizing how much one has hurt oneself or has been hurtful to others, the more one's sense of being "bad" or "wrong" intensifies. Often one becomes preoccupied with the very wrongness of the precipitating event itself. One fears one's weakness will be discovered and desperately tries to fix the situation by oneself.

There is a struggle in the midst of a deep sense of remorse. Emptiness, sadness, and intense loneliness may emerge, alternating with cynicism and anger. Self-recrimination often takes the form of "beating oneself up." At this point, one is not sure whether things will ever change and one fears becoming stuck: never recovering from devastation, never moving into healing, but remaining isolated, bitter, and cynical. One woman who suffered a psychotic break after realizing her intense rage at her husband spoke of how hopeless life seemed to her during this period: "It was like I entered into the darkest part of creation and really couldn't see much purpose in being alive."

The overall movement of self-forgiveness can be described as one from estrangement to feeling at home, from darkness to light, from deception and denial to honesty and acknowledgment. This movement is not smooth or linear; it involves a great deal of struggle and vacillation between acceptance and harsh judgment, but certain aspects of it can be highlighted and described.

It is particularly important to experience some kind of loving acceptance from others, especially of those parts of ourselves that we find disturbing: our anger, hatred, inadequacy, mistakes, ignorance, hurtfulness, alienation, irresponsibility. Such relationships are found with therapists, priests or pastors, friends, family; in fact, it may not be too strong a statement to say that what is misleadingly called "self-forgiveness" always takes place in the context of some variation of loving relationships with others. One woman described her friends' support after the suicide of her son: While being supportive of her, they respected and accepted her feeling of extreme vulnerability without being intrusive. Another described a moment that was particularly important in her healing process:

> . . . the sacramental moment was when I confessed to being angry at my husband . . . and the priest said it's not a sin to be angry. And I remember looking up at him just totally astounded that this was not a sin. . . . As I looked at the priest . . . [i]t was as though I was looking into the eyes of love.

The nonjudgmental acceptance of another priest led her to refer to him as an "oasis in the desert." Generally, there is an anchor of some sort—perhaps one's children or one's work—that helps one to keep going.

One woman recounted a powerful experience of healing through religious symbols during her pregnancy. The anchor in the process was the priest who guided her through a meditation exercise that resulted in her experiencing Christ's love and forgiveness in a compelling and surprising way. The priest's follow-up through regular phone calls to her after this healing validated her experience of being accepted and helped her through periods when self-critical thoughts started to push aside the memory of the acceptance she had experienced. And it was most important to her that this person, who had come to know about the thoughts and feelings she had concealed from others, did not look at her differently.

Another important aspect of the self-forgiveness process is a faith or determination that the pain of experiencing one's brokenness will lead to healing, that is, a faith that there is something on the "other side." This faith keeps one "hanging in there" at times when nothing else offers hope of change. In one woman's words: ". . . still within me there was, no, by God, I am going to stick this out . . . I am gonna do this, I don't care what there is. So even though the darkness was there, there was still that determination inside of me."

One man, who identified part of his self-forgiveness process as reconnecting with parts of himself that had been "cut off" since he was a child, talked about the "stuckness" he had felt in trying to find his life's work. He described his determination: ". . . you have to be able to make that affirmation, to be able to say 'O.K., I'm going to go ahead and face failure . . . and forgive myself.' "

For another man, the faith that is integral to forgiving oneself gives direction and shape to one's life. Faith also opens us to the importance of the metaphorical or symbolic level of experience. This openness acknowledges that there is something larger than or beyond our immediate experience that helps to make sense of it. It also acknowledges our own human limitations about what we can and cannot control, can and cannot know; openness also creates a sense of wonder about our lives and the world. In this regard, one person described how encountering the work of Joseph Campbell affected her: ". . . that was very healing for me because I could look at that, the experience of mine, and say, Well, it looks like an initiatory experience of a shaman. And that helped me to express it in a way that was much more peaceful."

For the woman whose son had committed suicide two years earlier, a dream was a powerful avenue of transformation. In the dream, she was in a large house with some friends and her son, who left after lunch to go to the bathroom. Suddenly, she heard a terrible scream from her son and found him sitting on the toilet with blood gushing out of him. She knew immediately that he was dying, as did he, and there was nothing she could do, so they held each other during his last moments. This dream marked a profound turning point in her healing; after it, "life looked like itself again." She was able to reconnect with her experience and with others in a new way.

Sometimes, the transformative experience is simultaneously physical and metaphorical. In the context of a guided meditation at a retreat, the woman who was pregnant was encouraged to invite Jesus to come into the closed room in her mind where she believed all the things for which she had not forgiven herself were kept. As she did so, she realized that the room was musty and empty rather than full of horrible things. At that point, she felt Jesus' presence depart, leaving a void within her. She felt the child turn inside as though responding to the void, and she became very concerned, thinking, "Now I am really in trouble." But with the encouragement of the priest, she continued the meditation. Then, much to her surprise, she had the experience of being filled, and she felt forgiven and spoken to as an individual.

We were continually reminded that the self-forgiveness process is one that includes moments of "grace," like the ones described previously, coming unexpectedly as a gift, almost out of nowhere. It is faith that allows one to be open to these moments, but they are experienced at the time as being not of one's own making, but as coming to one. One person described the experience in this way:

> It had been raining that day, there was a beautiful rainbow that stretched from one side of that valley to the other. So I just pulled up on the rest area up on top of the hill there and just parked the car and just took it all in. It was as though that was God's covenant with me that yes, all the struggles we've gone through are OK and you're going to make it. . . .

In all of this openness, there is a kind of letting go—letting go of one's old identity, expectations, and beliefs, especially the belief that one can heal oneself. This may be combined with a sense that "life is too short" to hang on to old grudges, to punish oneself. "Letting go," however, is not an intellectual, conscious act that we can engage in at will. One man pointed to this distinction: "People think they have forgiven themselves or somebody else when they've just figured out or they've understood why they did what they did, but that's not forgiveness. Understanding is in the head and forgiveness is a surrender of the heart."

An important aspect of the self-forgiveness process is experiencing the grief that comes with letting go: grieving for what might have been, feeling regret for what was. In one man's words: "Not feeling sorry *for* yourself, in that self-deprecating mode, but feeling sorry *towards* yourself, really kind of giving up to yourself that open, like, 'I'm sorry' and allowing yourself to be sorry and to forgive yourself . . . then you can move on."

One also comes to a new understanding of responsibility. Where before one was primarily in a denying or blaming stance toward oneself, now there is the honest acknowledgment of one's participation in the event. One man, who had been rageful toward his mother, described this change as follows: "When I recognized . . . that I was feeling the need to forgive [his mother], I was feeling the possibility of starting to see my own shortcomings, my own irresponsibility, how I was involved in the situation, that I wasn't just a victim." This awareness of one's responsibility frees one to move into a more accepting relationship with oneself.

At some point, one realizes that one has experienced forgiveness. What one has previously negated or tried to change in oneself is now accepted as a part of who one is. This acceptance leads to a new relationship with oneself that has the quality of "being at home," a sense of ease about oneself and one's place in the world. This was a gradual, subtle change for most of the people we interviewed. It was not a case of "Before I was a stranger in a strange land" and "Now I'm home for good." There was instead a growing sense of ease about one's identity and a lessening of self-recrimination and anguish over one's relationship to the world. For one person, forgiving meant "loving myself, including my mistakes or hurting of other people, things like that . . . [it] seems very peaceful and empowering."

There is a shift of focus to a meta-perspective that can embrace all aspects of one's self. One is no longer defined solely by incidents or feelings of being "wrong" or "bad." Rather, there is more of a sense of balance and movement. It is not that one never feels bad or wrong, but that these feelings do not pervade the entire fabric of one's life. The woman who had become psychotic in the face of her anger talked about darkness giving way to longer, more frequent periods of light:

> . . . it was as though a light broke in the darkness on a more regular basis after I began to talk to [her priest]. . . . It's almost like it's the other side of where I was, where the light seldom broke into the darkness, and now the darkness doesn't break into the lightness [as often]. . . . [The darkness] is not as frightening, it doesn't seem like it's forever. . . .

This quality of "at-homeness" is fundamentally an acceptance of one's humanness, often described as a kind of integration or reclaiming. As Buckley (1971, p. 207) has written so aptly in his discussion of the phenomenology of at-homeness, "A genuine openness to reality is clearly a primary requisite for feeling at home." Owning parts of oneself that were previously denied or split off gradually awakens a sense that one's "dark side" is less threatening. One feels ordinary, neither saint nor devil. One becomes more honest with oneself and, in doing so, also judges others and oneself less harshly. One woman described this feeling as "a very gentle resolve"; others pointed to an awareness that judgment had ceased to be the primary way of relating to the world.

Full acceptance of one's humanness involves an awareness of one's connection with others and the world. Life may go on more or less as usual, but there is a deepened, intimate sense of involvement. Words such as "reconciliation" and "belonging" were often used to describe this quality of forgiveness toward oneself. One no longer has to betray one's true self, or the darker aspect of oneself, in order to feel in community with others. One person spoke of being a "part of humanity without losing anything, except a false self . . . I'm not afraid to look other people in the face and tell them who I am."

With renewed trust and acceptance of one's place in the world, one becomes more fully integrated into the ebb and flow of day-to-day living. Life tends to be analyzed less and participated in more spontaneously and freely. One man stated:

> . . . by becoming more of a pattern of the way that I relate to myself [forgiveness] becomes a little bit easier, it becomes more frequent, and maybe even a little less self-conscious. You know, after a while of swimming you don't have to think of the motions. You just do it. . . .

Simultaneously, one experiences one's own separateness. One woman spoke of a deepened sense of individuality as she healed: "I was able to just value that self-care, that affirmation of claiming boundaries, just holding my own. . . ." In reference to her ex-husband, she said: ". . . he just very seldom comes to mind . . . I thought, 'Well that's wonderful, I really am a separate person now.'"

This at-homeness is accompanied by peace of mind, a sense of unity, a feeling that life is fundamentally right and needs no correction. As one group member wrote:

> . . . then there is some other place, as yet nameless, which embraces all, giving everything a connection, and giving [me] a place of belonging. It may be a place of effortless power, of warmth and light which dispels the darkness and chill of the night, at once deeply familiar and entirely new.

THE IMPLICATIONS OF A PHENOMENOLOGICAL UNDERSTANDING OF FORGIVENESS FOR PSYCHOTHERAPY

We live in an age when many people experience a strong sense of alienation from community and themselves. This alienation is often accompanied by feelings of guilt, shame, anxiety, hostility, and depression. For some people, this situation is reflected in severe dysfunction in their living. Others may experience a general sense of malaise and a lack of meaning in their lives. It is not unusual for people, in trying to remedy their dis-ease, to seek out the service of professionals who are concerned with psychological health. While not always evident or articulated, forgiveness of self and others is often an important issue for them. This section will briefly outline the authors' assumptions regarding the process of psychotherapy, the place of forgiveness in a hypothetical therapeutic process, and, finally, the role of the therapist in facilitating movement toward forgiveness.

The authors bring the following assumptions to their understanding of psychotherapy:

1. Healing and change take place in the context of a profoundly interpersonal relationship composed of someone who is willing—hesitantly perhaps—to make himself or herself vulnerable and an open, responsible, respectful therapist who is willing to commit to this endeavor in service of the patient.

2. We have a fundamental faith in people's individual processes and that of psychotherapy when it involves two well-intentioned people who meet the criteria enumerated above. This faith is imperative when the next step in the process is unclear and/or frightening.

3. The relationship develops over time and evolves out of the interaction—both spoken and tacit—between the psychotherapist and the patient. Each therapeutic relationship and process is unique and a reflection of the persons of the two people involved.

4. We never know what is best for the other. This does not mean we are passive receptors: We bring our life experiences, personal psychotherapy, and training to bear in response to patients and their living; we are mindful, however, of our limitations as human beings and psychotherapists.

5. We assume that when someone chooses psychotherapy there is a desire for things to be different even if exactly how this will occur is not clear either to the patient or the psychotherapist and the route is not immediately obvious.

6. We believe people are essentially interpersonal beings. When there is a disturbance in their capacity to enjoy and participate in community, there is a disturbance in their ability to enjoy themselves.

People come to psychotherapy for a variety of reasons, for example, depression, insomnia, poor interpersonal relationships, feeling empty. Basically their life is not the way they want it to be, and they are in some degree of distress, often profound distress. For many people, seeking help from a professional therapist still carries a stigma, meaning one is a failure, unable to handle one's own problems and maybe even "crazy." There is also anxiety about being seen ("seen through") by the psychotherapist. So people often turn to psychotherapy as a last resort, when all else has failed. They often come somewhat embarrassed and anxious—unsure of what to expect. But they know one thing: They want things to be different. They want something to be corrected.

The response of the psychotherapist is crucial at the early stage of the process. A posture that communicates "I hear you and I can sit with you" begins to provide a foundation of hope that things might improve. This early phase necessarily involves establishing a relationship (the relationship is always central, but early on it is explicitly so as two strangers meet). The psychotherapist is interested in who this patient is, what brings the patient to treatment, what it has been for the patient to be alive, what daily life is like, what the patient sees/hopes for the future, and so forth. Is this someone to whom I can be of help? Simultaneously, the patient is assessing whether the psychotherapist can be relied on, trusted, and is helpful and competent. Out of this develops a relationship with a quality unlike any other. If the psychotherapist and patient decide they want to work together, the patient begins a process of self-exploration.

In this process, it often becomes clear that the patient believes deeply that there are certain ways of being that are necessary to be loved and other ways of being that will exact punishment. For example, some people learn to be caretakers and cannot allow themselves to be cared for, others learn that they have to be perfect to be acceptable. Some are convinced that they have to be tough or they will be taken advantage of, others that they must always be positive and never angry or critical. The problem with these ideal self-images is that no one can live up to them satisfactorily and they limit the full array of human experience. So people end up feeling badly about themselves and quite alienated from their deepest aspirations, feelings, and potentials. But the self-image is such a strong definition of what people believe they should be like that they hope in psychotherapy to finally realize that image. For example, the tough guy wants to become tougher or the perfectionist wants to be more perfect.

This is a critical point in the psychotherapy process: One's self definition is called into question, and this is frightening. If the patient is committed to his growth and feels sufficiently supported in the therapy relationship, he may embark on a process of becoming more familiar with himself. This is often a difficult and painful time, often involving shame and guilt. For example, the person who always takes care of others may feel very ashamed about his own deep desire to be dependent. Clearly, this is an area in which self-forgiveness is an issue.

This process also involves facing feelings and thoughts about parental figures who taught one how to be in order to be loved and may have been abusive in various ways. This realization is often slow in coming. For example, it is not untypical for people to come into therapy painting a rosy picture of their childhood or at least minimizing the difficulties of their growing-up experience. We often believe that our parents were justified in what they did and said because they were crucial to our psychological survival. This belief is often carried into adulthood, and we are reluctant to consider that they might have been wrong and hurtful (however well-meaning).

But as the patient begins to realize that she has been given a very limited definition of who she can be to be accepted, she often becomes very angry toward those who imposed this definition on her. Suddenly, the parents who were seen through rose-colored glasses are seen in all their failings. The process now becomes one of allowing the client to experience the disappointment, the anger, the sadness, the grief that come with realizing that someone very important has let one down. This is a period when the patient focuses on who she is in her life and also faces who her parents were/are. Slowly, the patient begins to realize that her task is not to become more perfect, more tough, or more positive, but to accept herself as being utterly unique and ordinary at the same time. And as the patient becomes more accepting or forgiving toward herself, parents also seem more ordinary—neither saints nor Satans, but people who had their failings and limitations but also loved in the only ways they knew how.

As the psychotherapy comes to an end, people report many things: feeling lighter, more loving toward self and others, freer, having more choices, more able to maneuver the storms of life, calmer, less conflicted emotionally. They are no longer rigidly tied to the past, and the future offers choices and opportunities. There is at once a sense of self-possession, of independence, and of being a part of the human community. It is this embracing of one's own humanity and the humanity of others that reflects the deeper meaning of forgiveness.

The role of the psychotherapist in facilitating the movement toward forgiveness requires the same qualities that working through other psychological issues requires: acceptance, patience, sensitivity, and knowledge of human experience. These qualities are not mutually exclusive but obviously interrelated.

In his study of forgiveness in psychotherapy, Rooney (1989) found that the most significant factor for clients was the accepting presence of the psychotherapist. Many people come to psychotherapy with deep-seated feelings of pain, guilt, shame, and anger about who they are, how they have behaved, and how they have been treated. At some level, they assume that when they are really known they will be rejected. The fear is that the psychotherapist will be repulsed by who they really are or overwhelmed by the intensity of their feelings. This expectation reflects how they are in relationship to themselves and how they have experienced others. The key is the genuine attention by the therapist to the experience of the patient. This is more important than any specific verbal exchange. It is very powerful for a patient when the psychotherapist holds his experience, be it anger, self-loathing, confusion, fantasies, or fear, without being punitive. To be in relationship with someone who is not rejecting is disarming to someone who expects such a reaction. The patient is thus called upon to reconsider via the therapist his relationship to self and to others.

Second, the psychotherapist needs to be patient. Each person's process of healing has its own course and needs to be honored. To offer an interpretation, such as "You need to forgive your parents" or "You're being unduly harsh on yourself," may reflect what eventually becomes obvious but if ill-timed interferes with genuine movement. The patient might feel unheard and/or attempt to proceed in a certain direction in order to please the psychotherapist.

Forgiveness comes, but cannot be willed. The psychotherapist needs to be careful to allow the process to unfold and not attempt to rush it. This is true at each point. For example, when a patient is wavering between feeling kindly toward herself and hopeless, or between seeing parents as flawed human beings and malicious people, the psychotherapist may be wise to either sit with the wavering or make some comment to the effect that "it's important not to go too fast." This again reassures the patient that who she is at that moment is acceptable and her own experience is appropriate for the work she is doing.

Third, the psychotherapist must be sensitive to the import for the patient of what is being expressed. When the client is revealing strong emotion or "irrational" thoughts, the psychotherapist may be tempted to minimize or correct the patient's experience, as a way perhaps to protect the psychotherapist from the intensity of the affect and/or to help alleviate the patient's suffering. For example, when someone is blaming himself, it is not helpful, initially, to try to be reassuring or talk him out of it. The patient needs to feel what he is experiencing and have that be accepted. Later, there will be time to explore the meaning of the self-criticalness.

On the other hand, sometimes people continue to express a certain affect—for example, anger—when it doesn't quite ring true. Sometimes this occurs when deeper feelings, such as sadness, are lurking under the surface. Sometimes it happens out of habit. A person so used to feeling a certain way may have difficulty recognizing a decrease in intensity or the presence of some other emotion. It is helpful if the psychotherapist can discern the authenticity and depth of what is being expressed. This enables patients to embrace their own experience more freely as they become more aware of themselves and their relationship to others.

A knowledge of human experience is also useful when helping people come to a fuller and richer understanding of themselves and others. A therapist can be very helpful by offering a different perspective than the patient may have. For example, when someone is feeling murderous, suggesting that such feelings make sense in light of what has happened can help make the feeling less threatening. Knowing about the human condition, particularly forgiveness in this case, gives the therapist insight into what may be happening. For example, gentle confrontation of a patient when she seems "unable to move" (an observation that requires careful discernment), such as "What would life be like if your mother [father, spouse, children] never changed?" or "Can you imagine what life would be like if you weren't angry at ____________ [or yourself]," can be helpful. Such well-timed inquiries provide the patient with the notion that there are other possibilities, that there may be deeper, unapparent reasons for resisting change, and that one's fate is not sealed.

When forgiveness is an issue for people, an injury, an injustice, whether the result of one's own behavior or that of another, is the focus of one's consciousness. Yet in forgiving either self or another, there is a deeper meaning that has significant ramifications for how one experiences life. As discussed earlier, when one has been deeply hurt by another, one is suddenly faced with questions about one's value and place in the world. Similarly, when self-forgiveness is an issue, one is confronted with attitudes and beliefs about one's self and one's failures. Out of struggling with these deeper questions of identity, one comes to a fuller, richer, and more complete embracing of self and others as human beings.

Part of the process that is crucial for forgiveness to occur is the open acknowledgment and examination of the injustice as experienced. As we have seen, this openness to the injury and its consequences create room to move past the hurt. This attitude of genuine regard for experience is at the heart of depth psychotherapy. It is not surprising, therefore, that the two processes—forgiveness and psychotherapy—are harmonious and that the acceptance, patience, and sensitivity of the psychotherapist are ideal facilitators of forgiveness. Further, the better the psychotherapist understands the nature of injury and forgiveness, the more comfortable he or she will be as "witness" to the process, even when forgiveness is never explicitly on the agenda.

REFERENCES

Bauer, L., Duffy, J., Fountain, E., Halling, S., Holzer, M., Jones, E., Leifer, M., & Rowe, J. O. (1992). Exploring self-forgiveness. *Journal of Religion and Health, 31*(2), 149–160.

Buber, M. (1958). *I and thou.* New York: Scribner.

Buber, M. (1965). Distance and relation. In M. Friedman (Ed.), *The knowledge of man: Philosophy of the inter-human* (pp. 59–88). New York: Harper & Row.

Buckley, F.M. (1971). An approach to a phenomenology of at-homeness. In A. Giorgi, W. F. Fischer, & R. von Eckartsberg (Eds.). *Duguesne studies in phenomenological psychology, Volume I* (pp. 196–211). Pittsburgh, PA: Duquesne University Press.

Close, H. (1970). Forgiveness and responsibility: *A case study. Pastoral Psychology, 21*(205), 19–26.

Enright, R. D., Gassin, E. A., & Wu, C. (1992). Forgiveness: A developmental view. *Journal of Moral Education, 2,* 99–114.

Enright, R. D., Santos, M. J. D., & Al-Mabuk, R. (1989). The adolescent as forgiver. *Journal of Adolescence, 12,* 95–110.

Fitzgibbons, R. P. (1986). The cognitive and emotive uses of forgiveness in the treatment of anger. *Psychotherapy, 23*(4), 629–633.

Flanigan, B. (1992). *Forgiving the unforgivable.* New York: Macmillan.

Freedman, S. R., & Enright, R. D. (1995). *Forgiveness education with incest survivors.* Paper presented at the National Conference on Forgiveness, University of Wisconsin, Madison, March.

Gadamer, H. (1975). *Truth and method.* New York: Seabury.

Gahagan, J. P., & Tedeschi, J. T. (1968). Strategy and credibility of promises in the prisoner's dilemma game. *Journal of Conflict Resolution, 12*(2), 224–234.

Giorgi, A. (1970). *Psychology as a human science.* New York: Harper & Row.

Halling, S. (1979). Eugene O'Neill's understanding of forgiveness. In A. Giorgi, R. Knowles, & D.L. Smith (Eds.), *Duquesne studies in phenomenological psychology: Volume III* (pp. 193–208). Pittsburgh, PA: Duquesne University Press.

Halling, S. Kunz, G., & Rowe, J.O. (1994). The contributions of dialogal psychology to phenomenological research. *Journal of Humanistic Psychology, 34*(1), 109–131.

Halling, S., & Leifer, M. (1991). The theory and practice of dialogal research. *Journal of Phenomenological Psychology, 22*(1), 1–15.

Hunter, R. C. A. (1978). Forgiveness, retaliation, and paranoid reactions. *Canadian Psychiatric Association Journal, 23,* 167–173.

Jaspers, K. (1970). *Philosophy: Volume 3.* Chicago: University of Chicago Press.

Leifer, M. (1986). *The dialogal method in phenomenological research.* Unpublished manuscripts, Seattle University, Seattle, WA.

Marcel, G. (1991). *The philosophy of existentialism.* New York: Citadel Press.

Martyn, D. W. (1977). A child and Adam: A parable of two ages. *Journal of Religion and Health, 16*(4), 275–287.

McCullough, M. E., & Worthington, E. L., Jr. (1994). Models of interpersonal forgiveness and their applications to counseling: Review and critique. *Counseling and Values, 39,* 2–14.

Palmer, P. (1983). *To know as we are known.* New York: Harper & Row.

Patton, J. (1985). *Is human forgiveness possible?* Nashville: Abingdon Press.

Polka, B. (1986). *The dialectic of biblical critique: Interpretation and existence.* New York: St. Martin's.

Rooney, A. J. (1989). Finding forgiveness through psychotherapy: *An empirical phenomenological investigation.* Unpublished doctoral dissertation, Georgia State University, Atlanta.

Rowe, J. O., Halling, S., Davies, E., Leifer, M., Powers, D., & van Bronkhorst, J. (1989). The psychology of forgiving another: A dialogal research approach. In R. S. Valle & S. Halling (Eds.), *Existential–phenomenological perspectives in psychology: Exploring the breadth of human experience* (pp. 233–244). New York: Plenum Press.

Strasser, S. (1969). *The idea of dialogal phenomenology.* Pittsburgh, PA: Duquesne University Press.

Tedeschi, J. T., Hiester, J. T., & Gahagan, J. P. (1969). Trust and the prisoner's dilemma game. *Journal of Social Psychology, 79*(1), 43–50.

Worthington, E.L., & DiBlasio, F.A. (1990). Promoting mutual forgiveness within the fractured relationship. *Psychotherapy, 27*(2), 219–233.

11

Women's Psychospiritual Paths Before, During, and After Finding It Difficult to Pray to a Male God

Kathleen Mulrenin

SHUG: Tell me what your God look like Celie.
CELIE: Okay . . . He big and old and tall and graybearded and white. Sort of bluish-gray [eyes] Cool. Big though. White lashes.
SHUG: Ain't no way to read the Bible and not think God white. . . . When I found out I thought God was white, and a man, I lost interest. (Walker, 1982, pp. 201–202)

In this brief excerpt from a dialogue, the novelist Alice Walker (1982) captures an important shift in perspective that one woman, Shug, has already undertaken and is gently leading another woman, Celie, to consider. Shug's perspective is that of a black woman who is aware of the use of religious images in a specific cultural context, namely, the image of a white, male God, in a white, male-dominated society. Research in cultural anthropology (Eisler, 1987; Lerner, 1986; Stone, 1976) and writings by black (Cone, 1970; Jones, 1987) and feminist theologians (Geller, 1983; Ronan, Taussig, & Cady, 1986; Spretnak, 1982) suggest that changes in images of the divine often reflect changes in individual and cultural psychologies. Jones (1987, p. ix) writes: "Especially the concept of God, when it is alien, can hold one, mind and body, in bondage to a distorted understanding of one's own true selfhood, one's own humanity, and in the case of Black people, the meaning of one's own blackness."

Kathleen Mulrenin • Butler Hospital, Providence, Rhode Island 02906.

Phenomenological Inquiry in Psychology: Existential and Transpersonal Dimensions, edited by Ron Valle. Plenum Press, New York, 1998.

In my research, women of different Judeo-Christian religious traditions were asked to describe a time in which they found it difficult to pray to a masculine image of the divine[1] and how, if at all, they resolved it. In contrast to Valle's description of spiritual phenomena (Valle, 1989, 1995; Chapter 12 in this volume), my research is not an investigation of unitive or self-transcendent experiences. Instead, I examine what happens when praying to God, the Father, Jesus, or some other male image of a divine fails to bring about a desired effect—when there is disharmony rather than resonance in the relationship. In examining explicitly problematic, "gendered" spiritual relationships, we might better understand what it means essentially, for some women, to be women, to be women in relation to men, to male images of the divine, and to Judeo-Christian religious communities.

This starting point places my research somewhere on the banks of everyday experiences, offering a few planks in the building of a bridge to span the existential and the transpersonal dimensions of phenomenological psychology. Like other existentially oriented researchers, I examine experiences rooted in everyday life, specifically, the lives of women raised in Judeo-Christian traditions. My work remains on the ground of existential experience, aware of, but not necessarily leaping into, the flow of transcendent experiences. I study the constellation of relationships, events, feelings, and thoughts that bring women to realize their struggles with their religious images. In this study, I examine those paths leading from childhood beliefs in masculine images of the divine to the "awakening of consciousness" to oneself as a woman, one's God as male, and one's culture as patriarchally organized.

In this chapter, I offer a concise overview of research in psychospirituality, including women's psychospiritual development. I also summarize the work of philosophers, theologians, and social scientists who place current theological and psychological questions concerning gender and spirituality within a broader historical perspective. Following this review, I discuss the phenomenological research design I employed. Finally, I present and discuss my qualitative findings on women's psychospiritual paths before, during, and after they find it difficult to pray to a masculine image of the divine.

THEORIES OF PSYCHOSPIRITUAL DEVELOPMENT

The theorists I review in this section take differing approaches to understanding psychospiritual development. Allport (1950) focused not only on how individuals image the divine, but also on how they develop systems of beliefs that concur with that image. Fowler (1981) attended more to the cognitive dimensions of beliefs systems than to specific images of the divine. He relied substantially on Kohlberg's (1968, 1969, 1974), Piaget's (1962, 1967), and Erikson's (1963, 1968) models of development as vital components of his own thought on faith development. After reviewing these "stage" theorists, I summarize the works of Rizutto (1979), an object relations theorist, and Randour (1987), whose research deals exclusively with women's psychospiritual development. I include also in my review

[1] To broaden the scope of possible responses, I chose to use the word *divine* rather than *God* in my research question. From pilot studies, I learned that to use the phrase "masculine image of the divine" left respondents free to describe any male images that they found difficult rather than restricting them specifically to a male God or, simply, God. For example, one woman wrote of her image of a "cherub-cheeked infant Jesus."

the provocative thought of van Kaam (1975), a phenomenological psychologist and prolific writer in the area of formative spirituality, and Goldenberg (1993), a professor in the psychology of religion. Given these many varied approaches, it is important to mention a commonality regarding images of the divine. According to the research of Nelsen, Cheek, and Au (1985), Pagels (1976), Roof and Roof (1984), and Vergote and Tamayo (1980), the dominant images of the divine are masculine in our Western culture. Furthermore, when alternative images are employed, women are more likely than are men to use these different images. The review of the literature of psychospiritual development as well as the new light my findings shed on psychospiritual development should help us understand how women might come to question the dominant cultural androcentric images of the divine and their resolutions to those questions.

Stage Theories of Psychospiritual Development

According to Allport (1950), a psychologist and early writer on the psychology of religion, the divine is perceived in direct relation to the needs of the individual. In regard to psychospiritual development, Allport (1950) wrote that the divine is represented: (1) as omnipotent when the attribute of power is important to the individual; (2) as a source of strength and security when affection is desired; (3) as omniscient when knowledge and guidance are sought; (4) as peace and comfort when solace is sought; and (5) as redeeming when absolution is needed. Allport (1950) also provided a schema of the difference between mature and immature religious sentiments. An immature image of the divine stresses the egocentric needs of the individual and is based primarily on magical thinking and wish fulfillment. In such imaging, the emphasis is on self-gratification; the person remains unreflective and inconsistent. For example, God is like Celie's old, white man, a Santa Claus–like figure. For Allport (1950), such an image does not provide the individual with an integrating or unifying philosophy. It is this process of integration that distinguishes a mature from an immature religious belief. He further elaborated these criteria to include an interest in values that transcend bodily needs and an ability to reflect and gather insights from one's own life.

A cursory overview of Fowler's (1981) stages of faith highlights the similarity of his conceptualization of faith development to the work of Allport (1950). For example, in the first stage, the *Intuitive–Projective Stage,* Fowler (1981) described faith as expressed in the rudimentary use of language, in imitation, and in fantasy. In the second stage, the *Mythic–Literal Faith,* because the child's cognitive skills also have grown, he or she begins to hear and speak the family story, drama, and myth. The world begins to cohere with reference to the family unit. In the *Synthetic–Conventional Stage,* a child learns an ideology, supported by external authorities, anchored in the larger social sphere of family, friends, work, and school. The locus of authority is outside oneself. Most children recognize authority as resting with parents, teachers, and clergy. Generally, social custom supports the child's perception of the authority of others. However, in order to move to the fourth stage of faith development, the *Individuative–Reflective Faith,* the locus of authority must shift to oneself, to the development of what Fowler (1981) termed the *executive ego.* The fifth stage involves a widening of one's own worldview to embrace what is paradoxical to one's view (*Conjunctive Stage*). In the fifth stage, individuals know dialogically. This kind of knowing involves affirming both poles of a dialectic. Simply put, one sees both/and rather than either/or. One asks a question and knows that in asking the question, one already knows the path of the answer. Fowler (1981) likened this dialogue to the emergence of unconscious material to conscious awareness. He also compared his understanding of dialogic knowing to Buber's (1937) mutual relating of the *I–Thou* relation.

The sixth stage involves a willingness to overcome the paradoxes of life and faith through self-sacrifice (*Universalizing Stage*). Universalizers willingly live their lives in sacrifice for a greater good. Fowler (1981, p. 200) wrote of these individuals: "Their heedlessness to self-preservation and the vividness of their taste and feel for the transcendent moral and religious actuality give their actions and words an extraordinary and often unpredictable quality." When one lives a life of self-sacrifice and unpredictability, inevitably tensions and conflicts arise. Biblical figures are frequently depicted as struggling with the conflict between the desires of "human nature" and the call to sacrifice self for others. According to Fowler (1981), universalizers are often filled with such turmoil, conflict, and struggle. The locus of authority at this stage of faith development is "in a personal judgment informed by the experiences and truths of previous stages, purified of egoic striving, and linked by disciplined intuition to the principle of being" (Fowler, 1981, p. 245). In this final stage, authority is based on a finely tuned intuition developed over years of attentiveness to curbing self-centered interests and in intimate agreement with guiding spiritual, ethical, principles. In the theories of both Allport (1950) and Fowler (1981), mature spiritualities reflect the shift of authority from an external locus to oneself, the development of universal guiding principles, and the sacrifice of self, especially bodily needs. Having considered the more cognitive approaches to psychospiritual development, let us turn to an object relations approach offered by Rizutto (1979).

Object Relations Theory and Spiritual Development

Drawing from the insights of object relation theorists (Sandler, 1960; Schafer, 1968; Winnicott, 1965, 1971, 1975), Rizutto (1979) provides a psychoanalytical analysis of the development of images of the divine or what she calls "God representations." She identified the use of God representations as in the service of development or as impediments to growth. Rizutto (1979) also articulated the elaboration and transformation of these representations as an individual matures and adapts to changes in life circumstances. Citing clinical examples, she characterized the God image as directly experienced and rudimentarily created early in the child's development (Rizutto, 1979, p. 188): "The mirroring components of the God representation find their first experience in eye contact, early nursing, and maternal personal participation in the act of mirroring."

Rizutto (1979) understood God representations to be formed from early representations of parents and, in some cases, of siblings. These images, though initially inspired by the images and interactions with the parents and others, exceed the child's experience of others. The divine image becomes the creation of the child—offspring of the child's fears and desires. Thus, the child's representation of the divine may be more punitive than the disciplining real father or more nurturing than the compassionate real mother. Developmentally, these representations shift as the child grows. In the words of Rizutto (1979, p. 90):

> In the course of development each individual produces an idiosyncratic and highly personalized representation of God derived from his object relations, his evolving self-representations, and his environmental systems of belief. Once formed, that complex representation cannot be made to disappear; it can only be repressed, transformed, or used.

Rizutto (1979) neither proposed a specific developmental theory different for each gender nor considered in any depth the cultural influences on the development of God representation. She did call attention, however, to Freud's lack of attentiveness to women's psychospiritual development as distinct from men's psychospiritual development. Rizutto (1979, p. 42) suggested that Freud's theory offered "no explanation for the God representation in women except 'cross inheritance.'"

Women's Psychospiritual Experiences

The psychologist Randour (1987) does offer a qualitative analysis of women's spiritual experience. Because of the foundational importance of her work, I will quote liberally from her text. Randour's (1987) study included interviews of 94 women about their spiritual experiences. Randour (1987, p. 27) asked women to "describe what was, for you, an important or memorable spiritual experience . . . describe the situation and how you felt in the situation as completely as possible." Randour's qualitative research approach yielded descriptive categories or subgroups that she identified as *emerging women, selves in conflict, resolute women,* and *mystical women.* Emerging women described their spiritual experiences as part of a process of self-development, of discovering themselves, their beauty, and power. "Not one of them speaks of God the Father" (Randour, 1987, p. 44). These women located authority within themselves, and many of the women were cognizant of feminist theology.

Selves in conflict were women under the age of 21. They described their spiritual experiences as struggles with the divine, as trying to achieve a balance between self-assertion and a traditional understanding of sin. Unlike the emerging women, the selves in conflict were less at home with self-assertion and pride. They struggled to balance a sense of self-authority and power with relationship with others (the divine, family, friends).

Resolute women consisted of "older women," though Randour (1987, p. 48) gave no ages:

> These women, resolute in their understanding of their spiritual task, do not waver between seeking to satisfy self and seeking to satisfy others. Talking in the tones of traditional religious voice, they strive for submission to a higher authority, not self-enhancement. Accepting the traditional Christian conception of sin for them means that self-assertion endangers their relationship to God and that acquiescence enhances it.

Mystical women experience themselves as one with the divine. Women in this group subsumed their will in the will of the divine. Unlike resolute women, mystical women did not experience themselves as handing over their power or authority. Rather, this experience of the divine or the sacred is one of openness, of allowing the sacred within oneself. One woman described her experience of the sacred within (Randour, 1987, p. 52) as

> buried in a darkness in which I could not see, a vacuum in which I could not hear, in a spaceless-timeless void—I "saw" God, I "heard" God, and I was "in" God. . . . I was "wrapped" in God and carried a sphere of light, warmth, and indescribable peace around me.

Other research in psychospiritual development (Jackson & Coursey, 1988; Pargament et al., 1988) supports these mystical women's notion of collaboratively relating to the divine. In a collaborative relationship, authority is located both within oneself and within the divine. Randour's (1987) research provides a schema that we might reference as we look closely at the transformations of experiences that occur in women's psychospiritual development.

Gender Conflicts with Male Gods: Evidence of an Immature Spirituality?

In this section, I examine the work of van Kaam, a noted theorist and practitioner of formative spirituality and a founding member of the faculty of phenomenological psychology at Duquesne University. In particular, I succinctly summarize his comments relating psychological developmental issues to spiritual conflicts with male Gods.

Van Kaam (1975, p. 124) cited an explicit example of children who felt anxiety over the image of God the Father because "their mothers used to instill fear in their children, threatening to hand the child over to father if the child misbehaved." Van Kaam (1975, p. 112) understood an individual's conflict concerning God's gender as a quasi-spiritual phenomenon rooted in the "infraconsciousness mode" of the body-self rather than the "supraconsciousness mode" of the spirit-self. Infraconsciousness is restricted to the body and mundane human interaction. Supraconsciousness involves an openness to a relationship with God that by definition transcends everyday concerns. According to van Kaam (1975, p. 110), the conflict over God's gender is an inauthentic conflict that recedes in importance as one develops a more mature, adult spirituality. A mature spirituality embodies an "openness to what is beyond us, to the whole of all that is, the Holy."

Is van Kaam's (1975) basic premise of splitting the gendered body from the spiritual self really a more mature spirituality? Allport (1950) and Fowler (1981) might agree, since both of their psychospiritual developmental theories saw bodily needs and existence as secondary to higher ideals and principles. As I discuss in more detail below, van Kaam's (1975) dichotomous thinking is certainly well within the tradition of Western psychology and spirituality. If it is the case that maturity involves a relinquishing of concerns with the body, then women who find their male God's gender problematic would be considered immature in their psychological and spiritual development. And yet, rather than assuming maturity or immaturity on the basis of a particular sociocultural understanding of spirituality, we must ask what role that culture has in the development and maintenance of these male images. Furthermore, how do women experience the possible interplay of culture and religious symbols in their psychological and spiritual development? What meaning is there in the statistical fact that masculine images of the divine are more likely to be called into question by women? What are women telling us when they say they find it difficult to pray to masculine images of the divine? How shall we understand the body, the gendered body, in psychospiritual development for women? To respond to these questions with reference to previous research, I will review research in the psychology of religion as well as writings of theologians and mystics in the Judeo-Christian tradition.

EMBODIED OR TRANSCENDENT SPIRITUALITIES

Underhill (1961), the preeminent scholar of mysticism, carefully described two different traditions of mysticism that are distinguished by the ways in which individuals experience their corporeality. One tradition, the Neoplatonist, stems from the writings of Plato. Within this tradition, the divine is experienced as utterly transcendent. One experiences emanations or attributes of the divine, like the shadows on Plato's cave wall. Dionysius, a Greek philosopher, the Cabalists (Jewish medieval mystics), and Christian ecstatic mystics, notably St. John of the Cross, are historical examples of this kind of mysticism. Underhill

(1961, p. 98) suggests: "Obviously, if this theory of the Absolute be accepted, the path of the soul's ascent to union with the divine must be literally a transcendence: a journey upward and outward," out of the body. In the other tradition, the theory of immanence, which W. James endorsed as that of a "healthy mind" (Underhill, 1961, p. 99), "the quest of the Absolute is no long journey, but a realization of something which is implicit in the self and in the universe: an opening of the eyes of the soul upon the Reality in which it is bathed. For them the earth is literally 'crammed into heaven.'" Underhill cautioned, however, against seeing these two traditions as mutually exclusive. For example, St. John of the Cross (1946) wrote stirring sensual metaphors of union with the divine and also admonished his readers, "Wheretofore, he that will place great reliance upon bodily senses will never become a very spiritual person" (p. 46). Frequently, mystics refer to their experiences in explicitly sensual or sexual tones. These metaphors suggest a passionate desire for union, intimacy, and relationship that seems to be best described through sexual or sensual terms. What if we took these phrases more literally? What if spiritual desire included embodied experience?

Eisler, a cultural anthropologist, has written extensively in her most recent book (Eisler, 1996) on the evolution of our current Eastern and Western religious cultures and their changing understanding of sex, pleasure, and pain. Of the Indian Tantric yoga tradition, Eisler (1996, p. 27) describes pleasure and sex as integral to spiritual practice: "The central rite is *maithuna* or sexual union. The purpose of this rite is to awaken the *kundalini* or divine energy, which is often explicitly identified with *shakti,* the creative power of the Goddess." However, Eisler (1996, p. 147) suggests that within the Tantric tradition there is, by way of idealizing, a distancing of the embodied female from the path of enlightenment: "The female sexual energy is described from the male perspective, with the woman playing an instrumental—and in that sense secondary, indeed peripheral—role to the male's spiritual enlightenment through sex." Nevertheless, as Feuerstein (1989, p. 253) points out, Tantric yoga "introduced a battery of means that hitherto had been excluded from the spiritual repertoire of mainstream Hinduism, notably Goddess-worship and sexuality."

Boyarin (1993), a religious scholar, also has undertaken a study of sex and the body in the Judeo-Christian traditions and their precedent religions. Boyarin (1993, p. 5) suggests that "for rabbinic Jews,"[2] "the human being was defined as a body—animated, to be sure, by a soul—while Hellenistic Jews (such as Philo) and (at least many Greek-speaking) Christians (such as Paul), the essence of a human being is a soul housed in a body." For the midrashic, or rabbinic, Jews (Boyarin, 1993, p. 65): "The same drive that in the study-house will lead a man to study Torah will in bed lead him to have intercourse with his wife, and this is the very same drive that will lead him into sin when he is alone with a woman to whom he is not married. The passion is one." Spiritual and sexual desire are, one might say, genetically identical. For the Hellenistic Jews and Christians, however, the body (matter) became linked with the feminine (Eve) and the spirit or soul (spirit) became associated with the masculine (Adam). This dichotomy led to whole theologies and philosophies founded on a mind–body split as well as misogynistic principles. For example, Boyarin (1993, p. 58) quotes the Jewish philosopher Maimoindes:

> How extraordinary is what *Solomon* said in his wisdom when likening matter *to a married harlot,* for matter is in no way found without form and is consequently always like a *married*

[2]Boyarin (1993, p. 2) explains that "the term *rabbinic Judaism* refers not to the Judaism practiced by Rabbis but to the Judaism practiced by Rabbis and by those who considered the Rabbis their spiritual authority."

> *woman* who is never separated from a *man* and is never *free*. However, notwithstanding her being a *married woman*, she never ceases to seek for another man to substitute for her husband, and she deceives and draws him on in every way until he obtains from her what her husband used to obtain.

Centuries later, employing similar logic, the Christian medieval theologian Aquinas wrote, "Man is above woman, as Christ is above man" (cited in de Beauvoir, 1952, p. 110). What developed from the Hellenistic tradition of emanational transcendence was a hierarchy not only with regard to the individual's relation to his or her own body but also with regard to gender relations, with men seen as primary over women. This discussion brings us back to women's developing spiritualities. In finding it difficult to pray to a masculine image of the divine, women are attending to gender, to the body. Will this embodied context and experience necessarily involve a more immanent rather than a transcendent experience of the divine? We shall see, but before presenting my findings, I outline the writings of Goldenberg (1993), a professor of the psychology of religion at the University of Ottawa, who analyzed Western philosophical, religious, and psychological theory from a feminist and engendered perspective.

Goldenberg: Resurrecting the Body

Goldenberg (1993) centered her analysis around a feminist interpretation of the male's fear of castration. According to Goldenberg (1993), the fear of castration is an artifact of the boy's loss of connection to the mother through the severing of the umbilical cord. Radically disconnected from the mother, the boy develops a fear of the maternal and of his own aggressive desire to return to the mother. The fear of castration is embedded in the original experience of dismemberment.

But what has this psychoanalytical theorizing to do with women's difficulty with masculine images of the divine? Goldenberg (1993, p. 211) described the religious consequence of the original traumatic experience of separation from the mother through the severing of the umbilical cord as the development of a belief in a *transcendent* divine:

> Transcendence is a wish for something beyond the body, beyond time, and beyond specific relationships to life. Such a notion of perfect safety involves negation of this world and is probably motivated by a characteristically (but not exclusively) male fear of being merged with matter. Theologians envision salvation as up, out, and beyond, and call this hoped-for state of dissociation the ultimate reality. . . . In the course of Christian history, men have much preferred to promote [an] image of suffering and agony instead of any image of love and playfulness such as that of Mary with her baby.

Goldenberg (1993) saw the Platonic foundation of Western theologies, philosophies, and most psychologies as providing the philosophical ground for the split between the body and the spirit. Rather than seeing Platonic Forms (Spirit) as the place of ultimate reality, Goldenberg argued that in the beginning, there was in fact *not* the dis-embodied heavenly Word but the earthly Flesh. She grounded her thea-ology in the body, the human body. In this regard, she also turned to object relations and more Freudian-oriented psychoanalytical theory rather than, for example, to the archetypal, Platonic ideas of Jung (1969a, b, 1980). Writes Goldenberg (1993, p. 207):

> As women become more and more central to all branches of Western thought, the belief in the transcendent entities will become less and less tenable. This is true, I suggest, because the tran-

scendent, the immaterial, and the metaphysical is actually the embodied, the physical, and the female. It is the exclusion of things female from philosophy, psychology, and theology which has allowed these disciplines to construct notions of: abstract presence which creates the world and then controls it.

Continuing her discussion, Goldenberg suggested that a bodily and femininely based (though not exclusively a femininely experienced) "transcendence" might be experienced as an intersubjective, embodied phenomenon, in which connection and life rather than separation and death are the key. Goldenberg's analysis undermines the classic Hellenistic Judeo-Christian religious traditions and in fact also calls into question basic Western philosophical assumptions. Despite Goldenberg's (1993) provocative thought, it remains to be seen whether in women's spiritual experiences there is in fact an embodied, life-oriented experience. Her writing itself may be more wish than carnal reality. I shall review my empirical research of women's psychospiritual development, my method, and findings. Finally, in my discussion, I will reconsider in light of my findings the writings of Goldenberg and others.

METHOD

Subjects

Five women, ranging from 24 to 46 years of age, volunteered to participate in this study. Two of the women were white, Roman Catholic, and married with children. One woman was West Indian, an Anglican ministry student, and a single, celibate, lesbian mother. Another woman of Lebanese descent was raised in an Orthodox Jewish family and identified herself as a lesbian in a relationship. The fifth woman had been raised Roman Catholic and as a child also attended a Methodist church. She joined a nondenominational Christian charismatic church months before she entered the study. She identified herself as a lesbian.

Procedure

The beginning chapters of this book set out the scope and foundations of phenomenological inquiry in psychology and present the case that phenomenological empirical methods allow for a rigorous study of human experience as experienced. These chapters also show how qualitative findings are structurally and psychologically informative, often complementing the methods of our quantitatively oriented research cousins. In adopting a phenomenological research approach, like other phenomenological researchers, I assume neither one particular theoretical psychospiritual stance nor a position of "objectivity." By definition, the study of human experience is the study of subjective experience. Thus, rather than arriving at "objective," ahistorical findings, by using a phenomenological method, my research reveals a subjective, historically contextualized understanding of women's psychospiritual development. With a phenomenological psychological approach, I have the tools to examine the essential *subjectivity* of women's experiences.

The phenomenological–dialogal method I developed was a variation on methods developed by Fischer (1971, 1988), Fischer and Wertz (1979), Giorgi (1970, 1985) Halling and Leifer (1991), Kunz, Clingaman, Kortsep, Kugler, and Park (1987), and von Eckartsberg

(1971, 1986). I used a group format in which five women participated in three sequential four-hour group discussions. Each of the first three group discussions became progressively more analytical and structurally oriented, while concurrently building increasing levels of trust and personal reflection in the group. A fourth group discussion, at the conclusion of the project, provided an opportunity for additional dialogue and for a follow-up on the women's continued development.

Prior to the first group discussion, I had received from the participants or co-researchers their original written accounts of a situation in which they found it difficult to pray to a masculine image of the divine and how, if at all, they resolved it. I reviewed these accounts or protocols, inserted questions in the text where I wanted them to elaborate on their experiences, and sent these protocols with my questions back to the women for them to review and to consider their responses prior to group discussion I. In the first group discussion, they read their protocols aloud, including in these accounts their responses to my questions. In this meeting, they listened attentively to each others' experiences. In the second group discussion, we reviewed together my preliminary analysis of the data to the level of narratives with highlighted themes. We continued to clarify aspects of their individual experiences, identified and discussed general themes, and imagined other situations in which this experience might occur. In the third group discussion, we reviewed the general structure that I had formulated from all of the data, including their individual protocols, the narratives, and the group discussions. I asked the participants to review the general structure for understandability and for internal and external validity. I asked them to consider the following questions: Did this general structure articulate the essential constituents and structure of their individual experiences? Did the structure elucidate the experience as a whole and any variation that we could imagine of that experience? For our fourth and final group discussion, we met to celebrate the completion of the project and to follow up on how they had continued to grow and develop in the 2 1/2 years since I had begun to work with them.

Below are two *original* protocols describing experiences of finding it difficult to pray to masculine images of the divine. These accounts do not include information gained in the group discussions; nevertheless, they are representative of both the kinds of accounts written by the women in this study and the basic nature of the experience of finding it difficult, as women, to pray to a masculine image of the divine.

RESULTS

Claire's Original Protocol[3]

About five years ago, Sacred Heart Church, in the afternoon, a side altar, dominated by a cross, with Jesus, the man, hanging, well built, only a loincloth for covering. (There is, in the background mural, Mother Seton, in white, virginal, young woman, on her knees, supplicating a Priest. As usual, I was depressed.)

Kneeling, and I allowed to fully form in my thoughts, and knowing that I found the image of Jesus, the man, interfering. I began to wonder, think about what was underneath the loincloth—a visual image of penis—Jesus as man. Did Jesus get aroused? Was his penis limp in death? What

[3]Claire's writing style is a composite of prose and poetry. Except where clarification seemed necessary, I chose not to edit her protocol.

was the nature of man/Jesus to woman, Claire? How did sex enter into prayer—was it a factor? Who was this Jesus I prayed to? Could I pray to an alive, virile Jesus? Did I want to? Bottom line, I think was [that] I allowed the image in—[I allowed in] the fullness of this body and its multiple ramifications to me as a praying woman. The upshot—the Presence of the Spirit? The Presence of God—laughter and a pigeon pooping on the image. Normalizing the icon, my thoughts—a balm, a blessing, an opening. "It is perfectly OK—this line of thought, this inquiry, this interference and at the same time, because this is not it, at all." This experience basically opened the door for me to play with image—Mother/Father Mystery God. Go back and forth.

Miriam's Original Protocol

The last time I chose to attend services at the Orthodox synagogue to which my family belonged was the first time I remember a clear inability to pray to a male deity—namely God. So many other memories from my youth have disappeared, and yet this one is very clear.

I was wearing one of my better shuul-going dresses, a black smock-type affair with a lace collar and a print of red and pink roses scattered over it. It was a perfect Fall dress, which was appropriate since it was Yom Kippur, the Jewish day of atonement which comes at the end of the Jewish Year, somewhere around late September/early October. Yom Kippur is perhaps the most sacred of the Jewish holidays (it is one of the "High Holidays"); Jews fast beginning at sundown the night before, and spend most of the day in shuul praying for good fortune in the coming year and asking forgiveness for mistakes made in the past.

I was especially excited this Yom Kippur as I had decided to fast the way the adults did, meaning no food *and* no water. (I figured that since I had been Bas Mitzvahed about a year earlier, I was, according to Jewish tradition, a "woman.") To me this symbolized an attempt to move closer to a faith about which I had always had a high degree of ambivalence. I was known in my Hebrew School as one who frequently interrupted class with questions such as "How do we *know* there is a God" or "Why do we hate the Arabs so much?" I always looked forward to Passover, and yet the plague idea bothered me a great deal. I loved Purim and yet ran into trouble when I wanted to dress up as Haman. (Purim is the Jewish child's Halloween; everyone dresses up as a character in one of the famous stories of the Jews' liberation from oppression. Girls almost always dress as Ester, the womon[4] [sic] who saved her people, and boys as Haman, the evil oppressor-for-the-King.) I engaged in a continual tug-of-war with, Judaism, never quite able to make a place for myself within its boundaries.

However, on some level I knew that although Judaism didn't necessarily *call* me, I needed a spiritual connection in my life. Since Christianity had never seemed to me to be anything more than a flimsy excuse for a religion, and also because I was not aware of any other real options, Judaism was my only choice. So I took it, considering Yom Kippur to be an appropriate time for a new beginning.

I walked to Shuul that day feeling very pious—I remember entering this feeling of piety as if I had folded myself into a robe. It felt very comfortable and safe: very *familiar.* Once in the building, I walked upstairs to the main room where services were being held. I opened the door on the left (men and womyn sit separately from each other in Orthodox congregations: at Share Torah womyn sat on the left side of a glass wall, and men on the right) and walked up the aisle to my seat. I was alone in my row; my mother and sister had chosen not to come to services, and I don't remember where my father was—probably somewhere on the other side of the wall.

I picked up the siddur (prayer book) and tried to figure out what prayer they were on. Failing to do that, I decided to just flip through the book and read the prayers—I felt that the sounds of davening (praying) voices all around me were inspirational enough to create a holy space, and would seep into my consciousness in some way. I argued as long as I was reading something relevant, I would still be sharing the experience.

So I read. The siddur quickly proved to be rather boring so I switched to a copy of the Old Testament which I found lying on the bench next to me. Turning to Deuteronomy (why, I don't know), I again read. It was at this time that things began to go wrong.

All of a sudden my eyes were swimming in Lords and Gods and hes and hims. Every name was a man's name. Every experience was a man's experience. Feeling more and more uncomfortable, I

[4]Miriam used the term *womon* for "woman" and *womyn* for "women" to distance these feminine referents from the masculine root *man*. The plural "womyn" is also found in the *New Shorter Oxford English Dictionary* (Brown, 1993) as a nonstandard alternative to the word "women." I adopted these terms in describing one of the adult experientially based spiritualities, using the phrase "womon-centered." For a feminist philosophical discussion of these terms, see Hoagland (1988).

raised my eyes to look around me, and my discomfort turned to mild anger as I realized, *really realized,* that here I was *on the womyn's side.* The womyn around me were chattering and gossiping, mostly sitting on their benches, while the men on the other side of the glass were up and swaying, heads bent, davening up a storm. What the hell was going on? "Either I should be over there, or that damn barrier should come down," was pretty much exactly what I thought. As crossing over was prohibited, and the wall was there to stay, I decided to try to find a more agreeable section of the bible in which I could lose myself. Here came the final blow: I couldn't *find* a more agreeable section. Throughout the entire book, it was *him him him.* And a connection was made in an unmistakable, very specific way. *This* god was a *man.* Pure and simple. It was clear to me, in that moment, that this book was *not* using male pronouns to talk about God as a matter of convenience, as I had been taught; *this* book was telling me that God = man. Further, there was no room for *womon.* I had enough of a sense of my self as female by this time that I clearly felt rejected, scorned and rendered invisible by this male god.

As a last ditch effort, I called up my previously (vaguely) comforting father-image which I had created for myself as God and it only angered me further. I wasted no more time. I stood up and walked out. Within a week I told my parents I wouldn't be returning. Within a few years I found the book "The Moon and the Virgin" which introduced me to the concept of the feminine spirituality. This book truly called to me from the shelf of a used bookstore; it put words to the awe I felt at the sight of the full moon.

Since then I have followed a winding path to the Goddess, and I have explored varied nature-based religions as ways of bringing myself closer to Her. A decade later my father-image has been erased from my mind and I see my deity as female. I feel truly blessed to have been able to break my chains to a male god at such an early age, and thank, the Goddess daily for this gift.

General Structure

From qualitative analysis of these accounts and the three others in the study, and from analysis of information gained during the group discussions, I formulated a general structure of the experience of finding it difficult to pray to a masculine image of the divine. In the general structure, aspects of women's psychospiritual paths before, during, and after finding it difficult to pray to masculine images of the divine are highlighted. I illustrate the essential elements of women's pathmaking, that is, how in childhood the beginnings of their paths, their spiritual beliefs, are influenced by their families and religious communities, and how these beginnings are then inextricably linked later to women's choices of direction and their adult experientially based spiritualities.

The experience of finding it difficult to pray to a masculine image of the divine for the five women of this study, raised in the Judeo-Christian tradition, arose after years of involvement in and active questioning of their religious traditions and communities. The experience involved three phases. The first phase, *growing up in a religious tradition,* included learning androcentric religious images, questioning traditional religious tenets and practices, and learning that females are seen as "less than" males. This phase also included the experience of domination by male authorities. These women, as girls, understood themselves as different from others and developed an empathic awareness of specific socially marginalized groups. These foundational experiences and understandings provided the ground from which later questioning and growth emerged.

The second phase, *awakening* to the gender of their Gods and the gradual realization of the personal implications of the recognition of their Gods' male gender, began in a specific historical context, either in prayer alone or with others, either in a traditional place of worship or in another environment. At a time of profound emotional and interpersonal turmoil, these women sought solace, guidance, or a meaningful connection with their Gods. Failing to receive the attentiveness or comfort that they sought from their Gods, they became aware of the maleness of their divine images. Their explicit awareness of their Gods' maleness and their gradual or sudden implicit or explicit awareness of the social construction of their images of the divine, as well as their awareness of their own femaleness, mark the *turning point* in their relationships to their Gods and to themselves.

The quality of these relational shifts in the awakening process and the affect they experienced and perceived in their Gods delineated two distinct paths of awakening. The first path, *relating mutually,* was characterized by one woman's shift in the quality of the relationship with the divine from a childlike request for assistance to two adults relating reciprocally. Affectively, this woman experienced a shift from feeling fearful and depressed to feeling a sense of openness and humor. On the second awakening path, *relating angrily and willfully,* four of the women, after recognizing their Gods' gender and experiencing them as contrary to their expectations, renewed their efforts to make their Gods respond as they wished. This unrequited second effort was followed by feelings of shame or of being flawed, to which these women responded by angrily withdrawing from the religious context and from the relationship with their Gods. They angrily refused further engagement with their traditional religious communities and images and/or disengaged from all spiritual contexts and relationships. Having separated from their traditions, these four women found themselves facing a choice of two other paths: the path that continued away from religion and spirituality and the path that reconnected with religion and/or spirituality. The three women who chose the second path, like the woman who related mutually, developed their own adult spiritualities.

In the third phase, *forming an experientially based adult spirituality,* four of the five women continued to develop spiritual relationships in their lives, but on their own terms. Three evolving spiritualities were based on the women's embodied experiences of relating mutually to their Gods/Goddesses/sacred processes and on imaginary variation: (1) developing a "womon-centered" spirituality, exclusive of male images; (2) allowing a family of images, with male/female and gender-free images; and (3) creating or adopting a gender-free spirituality, including entire relinquishment of a belief in the divine. Their experiences of embodied self-authority and self-acceptance also highlighted a transformation of childhood experiences of "feeling less than males," "different," or "dominated by male authorities." The adult experientially based spiritualities reflected important relationships in the women's lives (including, for some women, relationships to males) and were characterized by specific qualities of mutuality, namely, choice of one's image of the divine based on one's own embodied life experiences, bringing one's whole self into relationship emotionally, physically, sensually, and spiritually, and reciprocity within diverse communicative styles.

For all five women in this study, the experience of finding it difficult to pray to a masculine image of the divine was essentially a recognition that their childhood images of God were too constrictive for their growing sense of themselves as adults. As they developed an embodied whole sense of self, their images of the divine also expanded in both image and relational capacities. The evolving process of self–other understandings led them to see that they will never again believe in God the Father or Jesus in the same manner that they once did.

EXISTENTIAL–PHENOMENOLOGICAL REFLECTIONS AND DISCUSSION

In this section, we will consider closely the paths of women's psychospiritual journeys before, during, and after finding it difficult to pray to a male God. We will see how traditional theories both help and hinder our understanding of women's psychospiritual experiences.

Growing Up in a Religious Tradition

For all the women in my research, their psychospiritual paths began in childhood. (I had explicitly recruited women who were "raised in" Judeo-Christian traditions.) Religious beliefs were focal to their families' sense of family. Each woman belonged to a family, and the family belonged to a religious community. They were "cradle" Anglicans, Catholics, or Jews. As is probably a common experience for most children, they were not given a choice in childhood among many religious traditions or images of the divine. For these women, the absence of choice about religious beliefs early in life would figure prominently in the development of their adult spiritualities. Their childhood worlds were relationally complex and yet were also simply dichotomous. The complexity of their world included intimate and conflicted relationships with parents, siblings, and grandparents, among others. There were, however, those who were given more power and privilege at birth and those who were given less authority and power from birth. The women, as girls, learned that men are more empowered by the world than are women. As Pam, a participant, said, "Being a female was a mark against me when I was born."

Masculine Images of the Divine, Male Authorities, and the Meaning of Being a Girl

As Allport (1950) and Fowler (1981) suggested, childhood images of the divine are most often anthropomorphized and "magical." One's early images are consistent with developing systems of beliefs that initially place authority outside oneself in God, parents, and other socially recognized authorities. In my research, despite the attempts by women, as girls, to question traditional religious tenets, images, and practices, generally they accepted the predominantly male images of the divine that were given to them, as well as the gender role stereotypes within their religious practices. Similar to the aforecited insights of black (Cone, 1970; Jones, 1987) and feminist theologians (Geller, 1983; Ronon et al., 1986; Spretnak, 1982), the women in my research developed restricted or thwarted senses of themselves consistent with cultural and familial teachings and values about women in society. The male God was the supreme authority in a patriarchal world, that is, a world that was perceived and presented as hierarchically organized with primarily males holding positions of authority, power, and privilege. As Miriam described her childhood image of the divine:

> That father image was totally out of a book, an old, white, man-in-the-sky, long flowing beard, flowing white robe, like the Sistine Chapel, total Dad image. . . . it was almost exactly, "He'll take care of everything. It's in his hands."

From my research, I cannot assert that obeying external religious authorities is exclusively a gender-related aspect of these women's psychospiritual development. However, submitting to religious authority while also developing in a context in which, as a girl, one was aware of oneself as "being less than" or perceived as "less than" males impeded the women's developing sense of self-authority. The women in my study recalled having perceived their grandmothers, mothers, other women, or themselves as submissive, compliant, or valued less than males in their own home, in their religious community, or in other social contexts. Added to this dimension of seeing oneself as "less than males" was religious authorities' outright rejection of the women's assertion of critical thought, even when they were not being held in comparison with male others. These religious authorities were not exclusively males; female religious authorities also dismissed or refused to respectfully acknowledge the budding critical thinking of the young girls. The women's overt expression of their individual perspectives as children on religious tenets and stories were thwarted. As an example of this dynamic, here is a quote from Claire:

> I was in the second or third grade asking a question and it was probably one of the most interesting questions I ever asked. I remember asking it in class and the question was shocking to this Catholic school nun that I said it. It was about Jesus and the story they were telling us. I think it was really asked in innocence. I think I was real interested and the response I got was, you just don't do that in this class.

In summary, being a girl carried specific meanings for women who would later find it difficult to pray to masculine images of the divine. Growing up in their Judeo-Christian familial traditions suggested to these women as girls that to be female meant to be more restricted and have less opportunities than males. Being a child meant not to question religious authority. Examples of exclusions based solely on gender included not being permitted to be on the bema (altar) at the synagogue or to speak the prayers aloud, not being permitted to serve as acolytes, not being permitted to play cricket, and not envisioning oneself as the hero in children's stories because the heroes were all males. In addition, each of the women, as girls, experienced at least one experience, though typically multiple experiences, of older males misusing their authority by physically, sexually, or emotionally harming them.

A Fragmented Self: Being Sexual or Being Spiritual

As others thwarted their curious, active, questioning involvement in the world and discriminated against them as women, the women in my study, as girls, felt increasingly "different," "fragmented," or "split off." They experienced themselves as fragmented in two distinct ways. First, they split off their reflective selves from their active selves, experiencing an incongruity between their thoughts and their actions. As we saw in the example cited above from Claire's experience in the second grade, they learned to refrain from expressing their dissenting or curious thoughts about their religious beliefs. Second, when they experienced sexual desires or an awareness of their sexual interests in religious or spiritual relational contexts, they also felt a sense of bodily fragmentation, of bodily shutting down.

These specific findings are not necessarily gender-related. In other words, it could well be that boys might experience a fragmentation in self-experience, a splitting of their thoughtful, inquisitive selves from their actively, questioning selves. Furthermore, boys might also develop a psychological split of their spiritual selves from their sexual selves. Future research will have to clarify whether this splitting is in fact gender-related. Nevertheless, these two essential constituents of fragmentation—(1) the dissenting, curious, and inquisitive self as distinct from an active, social self and (2) a polarization of one's spiritual self and one's sexual self—rang true through all religious contexts in the women's protocols. To be spiritual in childhood, and often well into adulthood, means to be passive and accepting, not attending to oneself in one's wholeness, cognitively, sensually, and spiritually. Alison, who struggled from childhood to reconcile her homosexuality with her sense of herself as a spiritual person, described the sexual/spiritual conflict most poignantly:

> That's why it's hard for me to pray. I just get a block. I can't do it. I don't know what to do about it. I'm not sure if I'm allowed to because of my sexual preference, [if] I'm allowed to do that,

to pray. I'm so scared right now. I'm not sure I'm allowed to pray or even how to do it. This situation is hard because I am a spiritual person.

Early in their lives, the women in my research learned to split off their experience of themselves as sexual beings from their experiences of themselves as spiritual beings. Both Allport (1950) and Fowler (1981) held that "self sacrifice" and the neglect of bodily needs in the service of a higher principle or cause were evidence of a more mature spirituality. Listening to the experience of women, we learn that this splitting off of dimensions of one's experiences brought with it a feeling of fragmentation, of an uncomfortable distance from one's own bodily experience. Perhaps these experiences might help to foster the "higher principles" in the more "mature" spiritually developed individuals of which Allport (1950) and Fowler (1981) have written. For example, in my study, from these disempowered and fragmented experiences of self, women developed a sense of empathy for others. They learned to broaden their own perspective of the experience of being thwarted to include other marginalized groups who also suffered discrimination. Claire gave a telling example of her own experience of both discriminating against and empathizing with another marginalized person:

I remember very early being aware of others. I grew up in the South and racism was out there and visible. I can remember, I must have been four years old, and we did these skits. Somebody blackened my face. I was going to do something, dancing, perform in some way. One of the Black women, who was at the youth center, was watching the performance. I realized I was just appalled that I was doing what I was doing. I can remember going to her and apologizing. I remember going to this woman and asking her to forgive me. This woman's response was that it was okay. I was always aware of this stuff.

This finding of women's tendency to develop empathy for others based on their own experience of disempowerment helps us to understand how it is that women develop empathy for others, but at what cost? According to my research, the cost in childhood is that of seeing themselves, or of believing that others perceive them, as "less than" males. This foundational childhood experience of the imbalance of power between the genders lies at the heart of the turning point in praying to a male God.

Turning Point: In a Context of Interpersonal Turmoil, Seeking Solace or Guidance from One's God

The women in my research found it difficult to pray to a masculine image of the divine when they approached prayer with a profound desire to connect with this Other in a time of great turmoil. In prayer in this moment, they sought solace, guidance, forgiveness, or acceptance from a Father or Jesus, believing He should and would offer His kindness or guidance. Prayer was an attempt on their part to have their emotional turmoil addressed or redressed. The women turned to their powerful, male Gods for assistance. As we reviewed above, Allport (1950) and Fowler (1981) suggested that early in psychospiritual development, one locates authority outside oneself. Allport (1950) astutely pointed out that most adults retain this childhood belief in external spiritual authorities. For some women in my research, this experience of the divine as external and "magical" did continue well into their adulthood. All of the women in my research at the time of finding it difficult to pray to a masculine image of the divine, whether in adolescence or adulthood, approached their Gods as they had in childhood; they petitioned Him as an external authority for assistance. They awaited His helpful response.

Their prayer concerned interpersonal turmoil in their lives. One woman recounted a profound loneliness and a desire to connect with her God and her religious community in a meaningful way as a teenager. Another described the need for comfort following a sexual assault. Another sought comfort following a threatened physical assault, which evoked traumatic memories of childhood. A fourth woman desired guidance after the emergence of memories of childhood sexual assaults and prior to revealing her lesbianism to her mother. The fifth woman sought acceptance following a rape, the loss of a close friend, who accepted her lesbianism, and the rejection of her church community because of her sexual orientation.

In prayer, God's maleness became thematic and problematic. As Claire wrote: "What was the nature of man/Jesus to woman, Claire?" His maleness obstructed their desires to be comforted, to be accepted, to feel connected. The male image became known to them through the use of masculine pronouns in prayer, through stories pertaining primarily to the experiences of men, through the names Father and Lord, through the bodily image of a male deity, or through the presence (sometimes in the role of intermediaries) of male clergy. Recognizing God's maleness meant recognizing that God's maleness was constructed within a particular social context in which men were more privileged than were women. As Miriam succinctly wrote, "God = man."

The awareness of the social construction of the gender of the male God made explicit the previously unacknowledged imbalance of power in the relationship. Where once they prayed to God the Father/Jesus to make things all right, now they recognized that He was a male and not responding to their petitions as they had expected. Equally important as the recognition of the social construction of God's male gender was the recognition of the meaning of being female. The explicit or implicit understanding of the relation of gender and power, and the instantiation of this power relation in one's childhood image of the divine, marked the turning point in women's difficulty with male images. We will see in the spiritualities some of the women eventually created that their *new* God/dess/sacred processes would share authority with them rather than maintain the imbalance of power present in their childhood beliefs in male Gods. Their experientially based spiritual relationships enjoyed collaboration rather than one-sided obedience, submission, and petition. Before elaborating on their experientially based spiritualities, let us first consider in more detail the paths that led to their developing new spiritualities or, for some of the women, forfeiting all spiritualities.

Relating Mutually or Relating Angrily and Willfully

Once the women recognized the social construction of their male Gods and the significance of their being female in a male-privileged social milieu, their paths diverged. Claire's protocol is illustrative of the path of relating mutually. Miriam's protocol is an example of relating angrily and willfully.

At the first fork in the road of women's paths in finding it difficult to pray to a masculine image of the divine, surprisingly, it was not God's gender that dictated whether or not the relationship would continue. Rather, it was the experienced closeness of one's male God to one's lived world. If the women experienced God the Father/Jesus as interested in her *as a woman,* then the relationship grew in mutuality and in ways previously unforeseen by the women themselves. For Claire and for women like her, the first step in developing a relationship of mutuality was to become interested in God's "humanity." It was Claire who first "leveled the playing field" of the perceived power in her relationship with her God. Claire, the woman, disrobed Jesus, the man. Claire imaged Jesus in his naked humanity. Claire acted and, by the very nature of her act, assumed power as a woman. Her power was not a power over her male God, but an empowered stance with her God, in mutual attraction. It is important to note that for Claire and for women like her, imaging the divine in alternative ways included imaging her *male* God in ways that she had not previously experienced Him.

How could Claire have acted from this empowered position so suddenly when other women did not? Two interrelated psychological constituents made this shift possible. First, a woman must attend to her embodied experience—literally attune to her senses and herself as a woman. Second, through this newly embodied presence in a spiritual relational context, she must cognitively broaden her understanding of her relational experience of herself and her divine Other. For example, Claire consciously attended to the detail of the maleness of her God and to herself as a woman. She also attended to the wider context in which her experience unfolded. For Claire, this wider context included the experience of seeing a bird defecate on a statue of Jesus. In that moment, the Jesus image her religious community, her family, and she had constructed collapsed. By humorously and openly attending to her own and Jesus' detailed *embodiment,* that is, by literally *seeing* and experiencing what bad been previously ignored (Jesus's virility, her femininity, the revealing of images as mere images), Claire widened her relational perspective. By being in her body, by attuning to the possibility of sexual arousal in a spiritual relationship, she broadened her own and her divine Other's relational, including affective, capacity. She experienced herself and the Other as sensual, humorous, respectful, passionate, and open . . . as all things and experiences.

In contrast to the openness and good humor on the path of relating mutually, women who embarked on another path felt angry and willful toward their Gods. These women experienced their Gods as disinterested, condescending, and distant. Unlike Claire, the angrily and willfully relating women did not experience an affirmation of their femininity in relation to their male Gods. At first, in response to their Gods' indifference, the women tried to further appease them. Initially, they maintained the divine–human, male–female, powerful–powerless imbalance that was prominent in the structure of their spiritual relationships in childhood. Only after years of withdrawal and self-exploration were they able to move beyond the angry withdrawal to a more mutual and empowering relationship to the divine. However, one of the women in my study stayed on this path of angry withdrawal, separated from spiritual relationships and religious contexts, skeptical that she would ever have a spiritual relationship again. This choice of refusal to engage spiritually is an essential possibility for women who have difficulty praying to a masculine image of the divine. For other women, whose leave from spirituality was only temporary, the possibility of a mutual relationship with the divine began when they were invited back to spirituality by other women. These invitations (sometimes in the form of books) included alternative images of the divine accentuating explicitly feminine experiences.

Experientially Based Adult Spiritualities

The essential constituents of an adult spirituality for women who have had difficulty praying to a masculine image of the divine are a profound self-acceptance as a woman and an understanding of spirituality and religion on their *own* terms, according to their *own* life experiences. In contrast to asking for assistance from a God/Goddess or a spiritual process located solely outside the self, these women developed a sense of the divine or of a spiritual process within themselves. In childhood, each woman had felt different from others, outside the mainstream. In practicing their adult experientially based spiritualities, the quality of their self-acceptance emphasized an appreciation of themselves as a woman. This enjoyment of their femininity often explicitly related to the healing of a traumatic childhood experience (the abuse of male authority) or limiting self-perceptions. In addition, they explicitly highlighted a welcoming of their sexualities. Fundamentally, these women came to a different understanding of themselves, their engendered, embodied experiences, their own sexual desires and spiritual relationships. The experiences of the divine within and the freedom to embrace their sexualities carried with them a sense of personal wholeness, a coming together of previously fragmented senses of self.

The Role of Choice, Embodied Gender, and Reciprocity in Developing Spiritual Relationships of Mutuality

Mutuality in adult spiritual relationships was characterized by three prominent features. First, the women *chose* their divine images on the basis of their own embodied life experiences and relationships. Second, they brought their whole selves to the relationship, including their *sensual, sexual, embodied femininity* and authority. Third, their spiritual relationship emphasized *reciprocity* but within diverse communicative styles, such as playfulness, conversation, and eroticism, I will discuss in more depth each of these characteristics of mutuality within adult experientially based spiritualities.

Mutuality in the context of these spiritual relationships involved choice. Following difficulty with their childhood male images of the divine, the women, as adults, embraced images that best represented their experiences of themselves and relationships of primary importance to them. Adult spiritualities involved being agents in their spiritual relationships. In contrast to childhood, in which images were given to them, in adulthood, they were consciously involved in the development of their divine images and of beliefs congruous with these images. They chose and created their images from their broader life experiences with their Gods/Goddesses/sacred processes, lovers, friends, family, and nature, and with their abstractions about life itself.

For the women in my research—indeed, it would seem, for many who, raised in traditional Judeo-Christian religions, choose images of the divine based on their own embodied experience—the act of choosing one's divine image is no small psychological or spiritual task. This shift of authority from outside oneself to within oneself is similar to the findings on psychospiritual development of Allport (1950) and Fowler (1981), but different in at least one important way. In my findings, what Fowler (1981) termed the "executive ego" is an *embodied* executive ego. Through the crisis of women finding it difficult to pray to a male image of the divine, self-authority became not only a conscious choice but also an embodied consciousness, an *act* of self-authority. The exercising of an embodied authority calls into question the Hellenistic Judeo-Christian tradition of a body/spirit split (Boyarin, 1993; Underhill, 1961). Embodied feminine authority within a spiritual context also addresses the gendered role models in religious traditions of women as less than men (Boyarin, 1993; Goldenberg, 1993). The fall of God the Father as transcendent, distant, and without gender happened as women became aware of their own bodies and of the "divinity" of their own bodies. All of the women in my study included feminine images of the divine in their adult spiritualities, though their divine images were not always exclusively feminine.

Experiencing themselves as embodied women, while being spiritual, meant attending to themselves as sensual and sexual beings. They accepted being lustful and experiencing bodily pleasure, while concurrently affirming their being spiritual. As Claire described: "I can be prayerful in sexuality with my husband now. In a way that I could not have before this experience. That was shut off." Unlike the Hellenistic Judeo-Christian tradition and similar to the Indian Tantric tradition, sexuality is welcomed, in fact incarnated, in the relational domain of the spiritual.

Finally, the experience of mutuality in their adult spiritualities assumed different communicative styles that all involved reciprocity. For Claire, relating to her male God with its erotic overtones involved a reciprocal attraction of her God toward her. Claire's account also resembled mystical experiences described by Teresa of Avila (Ramge, 1963), Fowler (1981), and Randour (1987). Moreover, Fowler's (1981) *Universalizing Stage* involved an individual's capacity to endorse seemingly opposite poles of a dialectic. He likened one's attitude in this stage to that of the acceptance of emerging unconscious material, a knowing and not knowing that is ever opening oneself to paradoxical truths. The paradoxical quality of this relationship also holds true for the women in my study. Paradox within a context of mutuality or reciprocity characterizes Randour's (1987) *mystical women*'s sense of being one with and yet distinct from the divine. For the women in Randour's research, this oneness was *not* a handing over of one's authority to God. Likewise in my research, there is paradox within the oneness, as Claire wrote:

> I no longer understand what other is, no longer have solid images, rather the possibility of all images, all concepts, all thoughts. . . . We are all god.

Relating mutually with their new images of the divine leveled the previous hierarchically structured spiritual relationships and essentially contributed to the women's experience of self-authority, a sense of being whole, of being the ground of their own existence.

Van Kaam's (1975) assertion that the experience of finding it difficult to pray to a masculine image of the divine is a phenomenon characteristic of an immature spirituality, of the "infraconsciousness" mode of the "body-self," is descriptively false when compared to my data (van Kaam, 1975, p. 112). It rings false when we consider maturity as based on a "spirit-self" or "supraconsciousness" disembodied from the infraconsciousness of the body-self (van Kaam, 1975, p. 112). Women, like the mystical women in Randour's (1987) research, who lived their lives undisturbed by the gender of God the Father, are not more mature or immature than women who, after finding it difficult to pray to a masculine image of the divine, relate to more diverse images. I believe the psychospiritual maturity of a woman might be better understood by considering, among other characteristics, her acceptance of herself (as an embodied, spiritual, sexual, relational authority) and the quality of her relationship to the divine. Only as *some* women transform the obstacle of God the Father or Jesus into broader or other (including feminine) images do they step toward further self-growth, authority, and acceptance. More recent research on spiritual relationships concurs with this possibility of spiritual relationships as embodied and/or collaborative (Fox, 1981; Jackson & Coursey, 1988; Pargament et al., 1988).

I will now briefly outline the three subtypes of adult spiritualities the women developed. These spiritualities are understood to be evolving and fluid, changing with the advent of new life experiences and self-understandings.

Forging a Womon-Centered Spirituality

All of the women who developed adult spiritualities included feminine images of the divine in their new spiritualities. In the womon-centered spirituality, however, these divine images were *exclusively* feminine. Negatively judgmental of anything masculine, womon-centered spirituality evolved in the experience of two of the women in my research. Their spiritualities reflected their important life relationships as lesbians. In Miriam's words:

> I associate with the Goddess a natural way of living. I don't associate a patriarchal God with walking in the woods at all. To my perception that's not where that happens. It doesn't happen there but the femaleness happens there. I really do make that kind of black and white distinction. It hasn't gone beyond gender. It's very womon, womon, womon. I can't include men anywhere. All nature has the female gender.

Like the other adult experientially based spiritualities, the womon-centered spirituality emphasized choice based on the experience of being women, that is, being embodied, gendered persons. This kind of spirituality drew its images from women's life experiences, of themselves as women, of other women, of nature, and other dimensions of life. Finally, similarly to the other adult experientially based spiritualities, the womon-centered spirituality was characterized by a self–other process of mutuality, in which the women felt themselves to be full partners in the relationship.

Allowing a Family of Images

Those women in my research who maintained masculine images of the divine did so only by broadening their original experiences of their male Gods and by including other diverse images of the divine, like Mother/Father, Sister/Brother, Mystery/Absurdity. Allowing a diversity of images more accurately represented the relationships and experiences that were significant to these women. These included relationships with their parents, their spouses/partners, their siblings, children, and friends. This broadening of images also reflected their experiences of life itself—absurdity, surprise, mystery. For the women of this type of spirituality, questioning their relationships with their male Gods did not necessitate a "letting go" of that relationship. Rather, they found ways to broaden their relational spheres to include new understandings of familiar male images, while also adding entirely new feminine and abstract relationships.

Maintaining male images of the divine in no way forfeited the essential constituents of choice, of self-acceptance of one's whole sensual, feminine, spiritual being, and of reciprocity. As Claire's protocol exemplified, one might become aware of one's own feminine gender and embodied authority in relation to a male image of the divine. Claire felt her sense of self-affirmation and self-will in this regard no less than did those women who imaged feminine images of the divine. What shifted for Claire with respect to her male image of the divine was her sense of choice. She chose this new male image among many other images newly available to her. She also rooted this image in her sense of herself as a woman. Furthermore, this newly expanded male image was "closer" to her, sensually real, "really real" and present to her in a way that was mutually respectful and affirming of her sense of herself as a woman.

Creating or Adopting a Gender-Free Spirituality

A third alternative to androcentric images of the divine offers a spirituality free of anthropomorphic images and therefore also free of gender. This finding was based on data in my research and on existing literature. In keeping with the formulated constituents of adult experientially based spiritualities, this kind of spirituality includes the three prominent characteristics of mutuality: (1) choice; (2) being present with one's whole self, sensually, sexually, spiritually, and relationally; and (3) experiencing a reciprocity within a diversity of communicative styles. The "images" of this spirituality, however, are not of gendered beings but of what I have termed *sacred processes.* Claire's use of the word "mystery" or "absurdity" is a reflection of dynamic processes in life in which one may participate, to which one might relate, without identifying specific gendered beings to whom one is relating. Zen Buddhism might be a spirituality to which women would turn following difficulty praying to a masculine image of the divine—although from the data collected in this study, it would be essential that the practice of Zen for some women include the characteristics of choice, of an affirmation of self, body, mind, and spirit, and a sense of reciprocity within the sacred process. Creationism, as Fox (1981) envisions it, rather than as the fundamentalist Christians see it, would also be an example of a modern spirituality that is not necessarily gender-focused. Fox (1981) includes an affirmation of the embodied, sensual–spiritual self in his understanding of spirituality.

SUMMARY

As the data suggest, the experience of finding it difficult to pray to a masculine image of the divine is essentially a phenomenon of a woman's developing an awareness of the sociocultural significance of gender and its impact on her psychospiritual development. As children, these women, raised in Judeo-Christian traditions, believed in a male God who would take care of them or who would respond authoritatively to their pleas for assistance. As children, they understood themselves as females to be seen as "less than males." Through the experience of turning to their male Gods in a time of interpersonal turmoil, and experiencing these Gods as not responding to them in the manner in which they had hoped, they were opened to new understandings of the gender of their male Gods and the sociocultural significance of gender. Through the process of finding it difficult to pray to a masculine image of the divine, the women in this study developed more affirming senses of self and self-authorities rooted in their experience and acceptance of themselves as *women.* Most of the women in the study continued to develop spiritual relationships beyond their experience of difficulty with their male Gods. In their new spiritualities, they defined their spiritual images and relationships on their own terms, based on their own relational experiences. The shift in spiritual authority from an external, male authority to a mutually empowering relationship with a being (male or female) or sacred process (without gender) was essential to this phenomenon. The shift in authority from an external locus to an internal locus has been documented in previous studies on psychospiritual development. What this study also highlights, however, is a kind of spirituality overlooked or underemphasized not only in psychological research but also within the studies of religion and spirituality, that is, the experience and importance of embodied spiritualities. Traditionally, affirmation of gender, or more primordially, of the sensual, sexual body, within spiritual contexts has been discouraged or even denigrated. In the structural analysis of the experience of women who have difficulty praying to masculine images of the divine, in addition to developing a deeper understanding of women's psychospiritual development, we have offered some rudimentary contributions to understanding a shift from a spirituality of the transcendent to one of immanence and embodiment.

REFERENCES

Allport, G. W (1950). *The individual and his religion.* New York: Macmillan.

Boyarin, D. (1993). *Carnal Israel: Reading sex in Talmudic culture.* Berkeley: University of California.

Brown, L. (1993). *The new shorter Oxford English dictionary: Volume 2.* Oxford, England: Oxford University Press.

Buber, M. (1937). *I and Thou.* Edinburgh, Scotland: T &T Clark.

Cone, J. H. (1970). *A black theology of liberation.* Philadelphia: J. B. Lippencott.

de Beauvoir, S.(1952). *The second sex.* New York: A. Knopf.

Eisler, R (1987). *The chalice and the blade.* San Francisco: Harper & Row.

Eisler, R. (1996). *Sacred pleasure: Sex, myth, and the politics of the body—new paths to power and love.* San Francisco: Harper.

Erikson, E. H. (1963). *Childhood and society.* New York: W. W. Norton.

Erikson, E. H. (1968). *Identity, youth, and crisis.* New York: W. W. Norton.

Feuerstein, G. (1989). *Yoga: The technology of ecstasy.* Los Angeles: Tarcher.

Fischer, C. T. (1971). Toward the structure of privacy: Implications for psychological assessment. In A. Giorgi, W. Fischer, & R. von Eckartsberg (Eds.), *Duquesne studies in phenomenological psychology: Volume I* (pp. 149–163). Pittsburgh, PA: Duquesne University Press.

Fischer, C.T. (1988). *Ethical dimensions of qualitative research: Is there a moral to the story?* Paper presented at the Seventh International Human Science Research Conference, Seattle, WA., June.

Fischer, C.T., & Wertz, F. J. (1979). Empirical phenomenological analysis of being criminally victimized. In A. Giorgi, R. Knowles, & D. L. Smith, (Eds.), *Duquesne studies in phenomenological psychology: Volume III* (pp. 135–158). Pittsburgh, PA: Duquesne University Press.

Fowler, J. W. (1981). *Stages of faith: The psychology of human development and the quest for meaning.* San Francisco: Harper & Row.

Fox, M. (1981). *Whee! We, wee all the way home: A guide to sensual prophetic spirituality.* Santa Fe, NM: Bear & Co.

Geller, L. (1983). Reactions to a woman rabbi. In S. Heschel (Ed.), *On being a Jewish feminist* (pp. 210–213). New York: Schocken.

Giorgi, A. (1970). *Psychology as a human science.* New York: Harper & Row.

Giorgi, A. (1985). (Ed.), *Phenomenology and psychological research.* Pittsburgh, PA: Duquesne University Press.

Goldenberg, N. (1993). *Resurrecting the body: Feminism, religion, and psychoanalysis.* New York: Crossroad.

Halling, S., & Leifer, M. (1991). The theory and practice of dialogal research. *Journal of Phenomenological Psychology, 22*(1), 1–15.

Hoagland, S. L. (1988). *Lesbian ethics: Toward a new value.* Palo Alto, CA: Institute of Lesbian Studies.

Jackson, L. E., & Coursey, R. D. (1988). The relationship of God control and internal locus of control to intrinsic religious motivation, coping and purpose in life. *Journal for the Scientific Study of Religion, 27*(3), 399–410.

Jones, M. J. (1987). *The color of God: The concept of God in Afro-American thought.* Macon, GA: Mercer University Press.

Jung, C. G. (1969a). A psychological approach to the dogma of the Trinity. In H. Read, M. Fordham, G. Adler, & W. McGuire (Eds.), *The collected works of C. G. Jung: Volume 11: Psychology and Religion* (pp. 107–200). Princeton, NJ: Princeton University Press.

Jung, C .G. (1969b). Answer to Job. In H. Read, M. Fordham, G. Adler, & W. McGuire (Eds.), *The collected works of C. G. Jung: Volume 11: Psychology and religion* (pp. 355–470). Princeton, NJ: Princeton University Press.

Jung, C. G. (1980). Jung and religious belief. In H. Read, M. Fordham, G. Adler, & W. McGuire (Eds.), *The collected works of C. G. Jung: Volume 18: The symbolic life: Miscellaneous writings* (pp. 702–744). Princeton, NJ: Princeton University Press.

Kohlberg, L. (1968). The child as moral philosopher. *Psychology Today,* September, 25–30.

Kohlberg, L. (1969). Stage and sequence: The cognitive developmental approach to socialization. In D. Goslin (Ed.), *Handbook of socialization theory and research* (pp. 347–480). Chicago: Rand McNally.

Kohlberg, L. (1974). Education, moral development and faith. *Journal of Moral Education, 4*(1), 5–16.

Kunz, G., Clingaman, D., Kortsep, K, Kugler, B., & Park, M. S. (1987). *A dialogal phenomenological study of humility.* Paper presented at the Sixth International Human Science Research Conference, University of Ottawa, Ontario, Canada.

Lerner, G. (1986). *The creation of patriarchy.* Oxford, England: Oxford University Press.

Nelsen, H. M., Cheek, N. H., Jr., & Au, P. (1985). Gender differences in images of God. *Journal for the Scientific Study of Religion 24*(4), 396–402.

Pagels, E. H. (1976). What became of God and mother? Conflicting images of God in early Christianity. *Signs 2,* 293–303.

Pargament, K. I., Kennell, J., Hathaway, W., Grevengoed, N., Newman, J., & Jones, W. (1988). Religion and the problem-solving process: Three styles of coping. *Journal for the Scientific Study of Religion, 27,* 90–104.

Piaget, J. (1962). *Play, dreams and imitation in childhood.* New York: W. W. Norton.

Piaget, J. (1967). *Six psychological studies.* New York: Random House, Vintage.

Ramge, S. V. (1963). *An introduction of the writings of Saint Teresa.* Chicago: Henry Regnery.

Randour, M. L. (1987). *Women's psyche, women's spirit: The reality of relationships.* New York: Columbia University Press.

Rizutto, A. M. (1979). *The birth of the living God: A psychoanalytic study.* Chicago: University of Chicago Press.

Ronan, M., Taussig, H., & Cady, S. (1986). *Sophia: The future of feminist spirituality.* San Francisco: Harper & Row.

Roof, W. C., & Roof, J. L. (1984). Review of the polls: Images of God among Americans. *Journal for the Scientific Study of Religion. 23*(2), 201–205.

Sandler, J. (1960). The background of safety. *International Journal of Psycho-Analysis 41,* 352–256.

Schafer, R (1968). *Aspects of internalization.* New York: International University Press.

Spretnak, C. (Ed.), (1982). *The politics of women's spirituality.* Garden City, NY: Anchor Press/Doubleday.

St. John of the Cross. (1946). *The complete works of Saint John of the Cross, Doctor of the Church* (E. A. Allison Peers, Ed.), Westminster, England: Newman Bookshop.

Stone, M. (1976). *When God was a woman.* New York: Harcourt Brace Jovanovich.

Underhill, E. (1961). *Mysticism: A study in the nature and development of man's spiritual consciousness.* New York: Dutton.

Valle, R. S. (1989). The emergence of transpersonal psychology. In R. S. Valle & S. Halling (Eds.), *Existential–phenomenological perspectives in psychology: Exploring the breadth of human experience* (pp. 257–268). New York: Plenum Press.

Valle, R. (1995). Towards a transpersonal–phenomenological psychology: On transcendent awareness, passion, and peace of mind. *Journal of East-West Psychology, 1*(1), 9–15.

van Kaam, A. (1975). *In search of spiritual identity.* Denville, NJ: Dimension Books.

Vergote, A., & Tamayo, A. (1980). *The parental files and the representation of God: A psychological and cross-cultural study.* The Hague: Mouton.

von Eckartsberg, R. (1971). On experiential methodology. In A. Giorgi, W. Fischer, & R. von Eckartsberg (Eds.), *Duquesne studies in phenomenological psychology: Volume I* (pp. 66–79). Pittsburgh, PA: Duquesne University Press.

von Eckartsberg, R. (1986). *Life-world experience: Existential phenomenological research approaches in psychology.* Washington, DC: Center for Advanced Research in Phenomenology; University Press of America.

Walker, A. (1982). *The color purple.* New York: Washington Square Press.

Winnicott, D. W. (1965). *The maturational processes and the facilitating environment.* New York: International Universities Press.

Winnicott, D. W. (1971). *Playing and reality.* New York: Basic Books.

Winnicott, D. W. (1975). *Through pediatrics to psycho-analysis.* New York: Basic Books.

III

Transpersonal Dimensions

The final eight chapters, in either their original intent or their findings, address transpersonal/transcendent dimensions of experience. Chapters 13–19 each describe a phenomenological investigation of an experience with recognizable transpersonal elements, the purpose of the research being, in each case, to identify the constituents that comprise the prereflective structure of the particular experience being studied as it manifests in its reflective, conceptual form. These studies, as such, represent the beginning of an emerging transpersonal–phenomenological psychology.

In this way, phenomenological psychology and research offer us a common language and approach to both speak about and more deeply understand sacred experience. This progressive, shared understanding promises to reduce the negative judgments and feelings that often arise in the comparison of differing religious perspectives by enhancing our ability to discern and articulate the common elements in sacred experience as it is described in various spiritual traditions. In addition, these findings suggest areas for further research, for example, identifying those factors that distinguish psychotic from spiritual experience, or discerning the different levels or types of prayerful or meditative experience.

Chapter 12, by Ron Valle, serves as an introduction to this section on transpersonal dimensions. Situating transpersonal psychology within the evolution of the major approaches in psychology, Valle presents the essence of the existential–phenomenological perspective, discusses the distinction between the existential and transpersonal world-views, and then describes the relationship between transpersonal and transcendent awareness. The chapter concludes with his thoughts on the nature of transpersonal–phenomenological psychology.

Chapter 13, by D. Hanson and Jon Klimo, presents an original research study on the experience of a particular type of synchronicity. Although the nature of synchronicity has been addressed historically by a number of psychologists, most notably Carl Jung, this study represents a first in its intent to reveal the essence or structure of the experience in a disciplined empirical phenomenological research setting.

Chapter 14, by Ourania Marcandonatou (Elite), offers us a look through the eyes of a phenomenological psychologist into a realm of experience previously reserved for mystics, theologians, and scholars of comparative religion: the experience of being silent. Consistent with the view of most of the world's great spiritual traditions, Elite finds the heart to be the "innermost silent center of our being" and that being silent facilitates the direct experience of the "sacred union of opposites."

Chapter 15, by Craig Matsu-Pissot, addresses the nature of unconditional love, an issue that has largely been ignored in Western psychology. He reports that, above all else, the experience of being unconditionally loved by a spiritual teacher is profound, intimate, and deeply moving for those who accepted his invitation to describe their experience. The relationship of love and sacred experience is discussed in different ways throughout the chapter.

Chapter 16, by Patricia A. Qualls, chronicles her remarkable journey to the orphanages in Romania after the recent fall of Communism in that country. Her study focuses on the volunteers who served with her and their experience of being with the suffering of the orphaned children they had cared for while there. As Qualls reflects, the experience of being with suffering has not—prior to this study—been investigated in any formal way, perhaps because of its intense and emotionally painful nature.

Chapter 17, by Thomas B. West, examines an experience that touches us at our deepest levels—being with a dying person. The profound mystery that dying represents and the intense feelings that often arise when facing this issue are acknowledged and addressed in West's presentation.

Chapter 18, by Paul Gowack and Valerie A. Valle, reports research findings on the experience of feeling grace in voluntary service to the terminally ill. This research naturally follows West's study on the dying process reported in the preceding chapter, and offers an empirical phenomenological approach to an experience that has been the focus of discussion almost exclusively in spiritual and religious circles prior to this particular research project.

Chapter 19, by Timothy West, also addresses an experience previously considered only by those with an interest in mystical and/or spiritual matters—encountering a divine presence. In addition, by exploring this experience in the context of near-death experience, West's study also represents the first empirical phenomenological investigation of any aspect of near-death experience.

12

Transpersonal Awareness

Implications for Phenomenological Research

Ron Valle

When this life ends, the mystery of love begins.
—Swami Rama

Phenomenological psychology invites us not just to an awareness of another perspective with a previously unrecognized body of knowledge, but to a radically different way of being-in-the-world. In addition, this different way of being leads naturally to a different mode or practice of inquiry—the methods of phenomenological research. This chapter will compare phenomenological psychology to the more mainstream behavioral and psychoanalytic approaches (Valle, 1989), present the essence of the existential–phenomenological perspective (Valle, King, & Halling, 1989), discuss the distinctions between the existential and transpersonal world-views, and then describe the nature of an emerging transpersonal–phenomenological psychology (Valle, 1995).

PHILOSOPHY AND APPROACHES IN PSYCHOLOGY

Existentialism as the philosophy of being became intimately paired with phenomenology as the philosophy of experience, since it is our experience alone that serves as a means

Ron Valle • Awakening: A Center for Exploring Living and Dying, Walnut Creek, California 94598.

Phenomenological Inquiry in Psychology: Existential and Transpersonal Dimensions, edited by Ron Valle. Plenum Press, New York, 1998.

or way to inquire about the nature of existence—what it means *to be.* Existential-phenomenology as a specific branch or system of philosophy was therefore the natural result, with what we have come to know as phenomenological methods being the manifest, practical form of this inquiry. Existential-phenomenology, when applied to experiences of psychological interest, became existential–phenomenological psychology and has taken its place within the general context of humanistic or "Third Force" psychology, since it is humanistic psychology that offers an openness to human experience as it presents itself in awareness.

From a historical perspective, the humanistic approach has been both a reaction to and progression of the world-views that constitute more mainstream psychology, namely, behavioral–experimental and psychoanalytic psychology. It is in this way that the philosophical bases that underlie both existential–phenomenological and transpersonal ("Fourth Force") psychology have taken root and grown in this field.

In classic behaviorism, the human individual is regarded as a passive entity whose experience cannot be accurately verified or measured by natural scientific method. This entity, seen as implicitly separate from its surrounding environment, simply responds/reacts to stimuli that impinge upon it from the external physical and social world. Since only that which can be observed with the senses, quantified, and whose qualities and dimensions can be agreed to by more than one observer is recognized as acceptable evidence, human behavior (including verbal behavior) became the focus of psychology.

In a partial response to this situation, the radical behaviorism of Skinner (e.g., Skinner, 1974) claims to have resolved this classic behavior–experience split by regarding thoughts and emotions as subject to the same laws that govern operating conditioning and the roles that stimuli, responses, and reinforcement schedules play within this paradigm. For the purposes of natural scientific inquiry, thoughts and feelings are, simply, behaviors.

In the psychoanalytic world, an important difference with behavioral psychology stands out. Experience is recognized not only as an important part of being human, but also as essential in understanding the adult personality. It is within this context that both Freud's personal unconscious and Jung's collective unconscious take their place. The human being is thereby more whole, yet is still treated as a basically passive entity that responds to "stimuli" from within (e.g., childhood experiences, current emotions, unconscious motives), rather than the pushes and pulls from without. Whether the analyst speaks of one's unresolved oral stage issues or the subtle effects of the shadow archetype, the implicit separation of person and world remains unexamined, as does the underlying causal interpretation of all behavior and experience. Both behavioral and analytic psychology are grounded in an uncritically accepted linear temporal perspective that seeks to explain human nature via the identification of prior causes and subsequent effects.

EXISTENTIAL–PHENOMENOLOGICAL PSYCHOLOGY

It is only when we reach the existential–phenomenological approach in psychology that the implicitly accepted causal way of being is seen as only one of many ways human beings can experience themselves and the world. More specifically, our being presents itself to awareness as a being-in-the-world, so that the human individual and his or her surrounding environment are regarded as implicitly interdependent, as inextricably intertwined. The person and world are said to *co-constitute* one another. One has no meaning when regarded independently of the other. Although the world is still regarded as essentially different in kind from the person, the human being, with his or her full experiential depth, is seen as an active agent who makes choices within a given external situation (i.e., human freedom always presents itself as a situated freedom). A number of other concepts come from existential–phenomenological psychology, including the prereflective, lived structure, the life-world, and intentionality (see Valle et al., 1989). All of these concepts represent aspects or facets of the deeper dimensions of human being and human capacity.

The prereflective level of awareness is central to understanding the nature of phenomenological research methodology. Reflective, conceptual experience is regarded as literally a "reflection" of a preconceptual, and therefore prelanguaged, foundational, bodily knowing that exists "as lived" before or prior to any cognitive manifestation of this purely felt-sense. Consider, for example, the way a sonata exists or lives in the hands of a performing concert pianist. In fact, if he or she begins to think about which note to play next, the style and power of the performance are very likely to suffer noticeably.

This prereflective knowing is present as the ground of any meaningful (meaning-full) human experience, and exists in this way, not as a random, chaotic inner stream of subtle senses or impressions, but as a prereflective structure. This embodied structure or essence exists as an aspect or dimension of each individual's *Lebenswelt,* or life-world, and emerges at the level of reflective awareness *as* meaning. Meaning, then, is regarded by the phenomenological psychologist as the manifestation in conscious, reflective awareness of the underlying prereflective structure of the particular experience being addressed. In this sense, the purpose of any empirical phenomenological research project is to articulate the underlying lived-structure of any meaningful experience on the level of conceptual awareness. In this way, understanding for its own sake is the purpose of phenomenological research. The results of such an investigation usually take the form of basic themes or constituents that collectively represent the structure of the experience for that study. They are the notes, if you will, that comprise the melody of the experience being investigated.

Possible topics for a phenomenological study therefore include any meaningful human experience that can be articulated in our everyday language such that a reasonable number of individuals would recognize and acknowledge the experience being described (e.g., "being anxious," "really feeling understood," "learning"). These many different experiences comprise, in a very real sense, the fabric of our existence as experienced. In this way, phenomenological psychology with its attendant research methods has been, to date, an existential–phenomenological psychology. From this perspective, reflective and prereflective awareness are essential elements or dimensions of human being as a being-in-the-world. They co-constitute one another. One cannot be fully understood without reference to the other. They are truly two sides of the same coin.

TRANSPERSONAL/TRANSCENDENT AWARENESS

There are experiences or certain kinds of awareness, however, that do not seem to be captured or illuminated by phenomenological reflections on descriptions of our conceptually recognized experiences and/or our prereflective felt-sense of things. Often referred to as transpersonal, transcendent, sacred, or spiritual experience, these types of awareness are not really "experience" in the way we normally use the word, nor are they the same as our prereflective sensibilities. The existential–phenomenological notion of intentionality is helpful in understanding this distinction.

Please note that the words "transpersonal," "transcendent," "sacred," and "spiritual" themselves connote subtle distinctions. For example, "transpersonal" currently refers to any experience that is trans-egoic, including both the archetypal realities of Jung's collective unconscious and radical transcendent awareness. While notions such as the collective unconscious refer to states of mind that are deeper than or beyond our normal ego consciousness, "transcendent" refers to a completely sovereign or soul awareness without the slightest inclination to define itself in terms of anything outside itself including contents of the mind, either conscious or unconscious, personal or collective (i.e., awareness that is not only trans-egoic, but *trans-mind*). This distinction between transpersonal and transcendent awareness may, in fact, lead to the emergence of a "Fifth Force" or more purely spiritual psychology.

Returning to existential–phenomenological psychology, intentionality refers to the very nature or essence of consciousness as it presents itself. Consciousness is said to be intentional, meaning that consciousness always has an object, whether that intended object be a physical object, another person, or an idea or feeling. Consciousness is always a "consciousness of" something that is not consciousness itself. This particular way of defining/describing intentionality directly implies the deep, implicit interrelatedness between the perceiver and that which is perceived that characterizes consciousness in this approach. It is this very inseparability that enables us, through disciplined reflection, to illumine the meaning that was previously implicit and unlanguaged for us in the situation as it was lived.

Transpersonal/transcendent awareness, on the other hand, seems somehow "prior to" this reflective–prereflective realm, presenting itself as more of a space or ground from which our more common experience and felt-sense emerge. This space or context does present itself in awareness, however, and is thereby known to the one who is experiencing. Implicit in this awareness, moreover, is the direct and undeniable realization that this foundational space is not of the phenomenal realm of perceiver and the perceived. Rather, it is a noumenal, unitive space within/from which intentional consciousness and phenomenal experience both manifest. Personally, I have come to recognize the following six qualities or characteristics of transpersonal/transcendent awareness (usually during the practice of meditation) (from "A Beginning Phenomenology of Transpersonal Experience"; Valle, 1989, pp. 258–259):

1. There is a deep stillness and peace that I sense as both existing as itself and, at the same time, as "behind" all thoughts, emotions, or felt-senses (bodily or otherwise) that might arise or crystallize in or from this stillness. I experience this as an *isness* or *amness* rather than a state of *whatness* or "I am this or that." This stillness is, by its very nature, neither active nor bodily and is, in this way, prior to both the prereflective and reflective levels of awareness.
2. There is an all-pervading aura or feeling of love for and contentment with all that exists, a feeling that exists simultaneously in my mind and heart. Although rarely focused as a specific desire for anyone or anything, it is nevertheless experienced as an intense, inner energy or inspired "pressure" that yearns, even "cries," for a creative and passionate expression. I sense an open embracing of everyone and everything just as they are, which literally melts into a deep peace when I find myself able to simply "let it all be." Peace of mind is, here, a heart-felt peace.
3. Existing as or with the stillness and love is a greatly diminished, and on occasion absent, sense of "I." The more common sense of "I am thinking or feeling this or that" becomes a fully present "I am" or simply, when in its more intense form, an "Amness" (pure Being in the Heideggerian sense). The sense of a "perceiver" and "that which is perceived" has dissolved, for there is no longer any "one" to perceive as we normally experience this identity and relationship.
4. My normal sense of space seems transformed. There is no sense of "being there," of being extended in and occupying space, but, similar to the previously mentioned, simply Being. Also, there is a loss of awareness of my body sense as a thing or spatial container. This loss ranges from

an experience of distance from sensory input to a radical forgetfulness of the body's very existence. It is here that my everyday, limited sense of body-space touches a sense of the infinite.

5. Time is also quite different from my everyday sense of linear passing time. Seemingly implicit in the sense of stillness described here is also a sense of time "hovering" or standing still, of being forgotten (i.e., no longer a quality of mind) much as the body is forgotten. No thoughts dwelling on the past, no thoughts moving into the future, hours of linear time are experienced as a moment, as the eternal Now.
6. Bursts or flashes of insight are often part of this awareness, insights that have no perceived or known antecedents but that emerge as complete or "full-blown." These insights or intuitive "seeings" have some of the qualities of more common experience (e.g., although "lighter," they have a felt weightiness or subtle "content"), but they initially have an "other-than-me" quality about them, as though the thoughts and words that emerge from the insights are being done to or, even, through me—a sense that my mind and its contents are vehicles for the manifestation as experience of something greater and/or more powerful than myself. In its most intense or purest form, the "other-than-me" quality dissolves as the "me" expands to a broader, more inclusive sense of self that holds within it all that was previously felt as "other-than-me."

As I came to recognize these qualities or dimensions over the years, I found myself recontextualizing the existential–phenomenological concept of intentionality by acknowledging a field of awareness that appears to be inclusive of the intentional nature of mind but, at the same time, not of it. It therefore seemed necessary to posit a "transintentionality" that addresses this consciousness *without* an object (Merrell-Wolff, 1973). I soon came to realize, as my colleague Steen Halling (personal communication) rightfully pointed out, that consciousness without an object is also consciousness without a subject, and that transintentional awareness represents a way of being in which the separateness of a perceiver and that which is perceived has dissolved, a reality not of (or in some way beyond) time, space, and causation as we normally know them.

This, for me, is the bridge between existential/humanistic and transpersonal/transcendent approaches in psychology, for it is here that one is called to recognize the radical distinction between the reflective/prereflective realm and pure consciousness, between rational/emotive processes and transcendent/spiritual awareness, between intentional knowing of the finite and being the infinite. It is, therefore, mind, not consciousness per se, that is characterized by intentionality, and it is our recognition of the transintentional nature of Being that calls us to investigate those experiences that clearly reflect or present these transpersonal/transcendent dimensions in the explicit context of phenomenological research methods.

FURTHER REFLECTIONS AND RECENT RESEARCH ON TRANSPERSONAL EXPERIENCE

What follows are some personal reflections on these dimensions (from Valle, 1995). My purpose and hope in offering these reflections is to deepen our understanding of transpersonal and transcendent awareness through the application of phenomenological research methodology and to facilitate the emergence of a new approach: transpersonal–phenomenological psychology.

This presentation and discussion are based on the following thoughts regarding the meaning of "transpersonal" in this context. Based on the themes that Huxley (1970) claimed to comprise the "perennial philosophy," I presented five premises that characterize any philosophy or psychology as transpersonal (Valle, 1989, p. 261):

1. That a transcendent, transconceptual reality or Unity binds together (i.e., is immanent in) all apparently separate phenomena, whether these phenomena be physical, cognitive, emotional, intuitive, or spiritual.
2. That the individualized or ego-self is not the ground of human awareness but, rather, only one relative reflection–manifestation of a greater trans–personal (as "beyond the personal") Self or One (i.e., pure consciousness without subject or object).
3. That each individual can directly experience this transpersonal reality that is related to the spiritual dimensions of human life.
4. That this experience represents a qualitative shift in one's mode of experiencing and involves the expansion of one's self-identity beyond ordinary conceptual thinking and ego–self awareness (i.e., mind is not consciousness).
5. This experience is self-validating.

It has been written and taught for millennia in the spiritual circles of many different cultures that sacred experience presents itself directly in one's awareness (i.e., without any mediating sensory or reflective processes) and, as such, is self-validating. The direct personal experience of God is therefore the "end" of all spiritual philosophy and practice.

Transcendent/sacred/Divine experience has been recognized and often discussed, both directly and metaphorically, in terms of either intense passion or the absolute stillness of mind. In our day-to-day experience, a harmonious union of passion and stillness or peace of mind is rarely experienced. In fact, passion and stillness are regarded as somehow antagonistic to each other. For example, when one is passionately involved with some project or person, the mind is quite active and intensely involved. On the other hand, the calm, serene, and profoundly peaceful quality of mind that often accompanies deep meditation is fully disengaged from and thereby disinterested in things and events of the world.

I suggest that what presents itself as quite paradoxical on one level offers us a very real way to approach the direct personal experience of the transcendent—that is, to first recognize and then deepen any experience in which passion and peace of mind are simultaneously fully present in one's awareness. If, in fact, Divine presence manifests in human awareness in these two ways, and sacred experience is what one truly seeks, it becomes important to approach and understand those experiences wherever these two dimensions exist in an integrated and harmonious way. In this way, one comes to understand the underlying essence that they share, rather than simply being satisfied with the seeming opposites they first appear to be.

The relationship between passion and peacefulness is addressed in many of the world's scriptures and other spiritual writings. These two threads, for example, run through the *Psalms* (May & Metzger, 1977) of the Judeo-Christian tradition. At one point, we read: "Be still and know that I am God" (Psalm 46:10), and "For God alone my soul waits in silence" (Psalm 62:5), and at another point: "For zeal for thy house has consumed me" (Psalm 69:9), and "My soul is consumed with longing for thy ordinances" (Psalm 119:20). Stillness, silence, zeal, and longing all seem to play an essential part in this process.

In his teachings on attaining the direct experience of God through the principles and practices of yoga, Paramahansa Yogananda (1956) affirms: "I am calmly active. I am actively calm. I am a Prince of Peace sitting on the throne of poise, directing the kingdom of activity." And, more recently, Treya Wilber (Wilber, 1991, pp. 338–339) offers an eloquent exposition of this integration:

> I was thinking about the Carmelites' emphasis on passion and the Buddhists' parallel emphasis on equanimity. It suddenly occurred to me that our normal understanding of what passion means is loaded with the idea of clinging, of wanting something or someone, of fearing losing

> them, of possessiveness. But what if you had passion without all that stuff, passion without attachment, passion clean and pure? What would that be like, what would that mean? I thought of those moments in meditation when I've felt my heart open, a painfully wonderful sensation, a passionate feeling but without clinging to any content or person or thing. And the two words suddenly coupled in my mind and made a whole. Passionate equanimity—to be fully passionate about all aspects of life, about one's relationship with spirit, to care to the depth of one's being but with no trace of clinging or holding, that's what the phrase has come to mean to me. It feels full, rounded, complete, and challenging.

It is here that existential–phenomenological psychology with its attendant descriptive research methodologies comes into play. For if, indeed, we each identify with the contents of our reflective awareness and speak to and/or share with one another from this perspective in order to better understand the depths and richness of our meaningful experience, then phenomenological philosophy and method offer us the perfect, perhaps only, mirror to approach transcendent experience.

Experiences that present themselves as passionate, peaceful, or an integrated awareness of these two become the focus for exploring in a direct, empirical, and human scientific way the nature of transcendent experience as we live it. Here is the "flesh" and promise of a transpersonal–phenomenological psychology.

At this point in time, I am pleased to note that a more formal emergence of transpersonal–phenomenological psychology has already begun. The following seven chapters in this volume report recent research studies that employed an empirical phenomenological approach to investigate experiences with transpersonal/transcendent qualities or dimensions. Existential–phenomenological and transpersonal psychology have remained, to date, relatively separate in both their theoretical and research endeavors. It is my hope that the perspective offered in this chapter and the research described in the chapters that follow will serve to deepen our understanding of human experience and contribute to the systematic inclusion of transpersonal issues within the philosophical and research projects of phenomenological psychology in particular, and the broader field of psychology in general.

REFERENCES

Huxley, A. (1970). *The perennial philosophy.* New York: Harper & Row.

May, H., & Metzger, B. (Eds.). (1977). *The new Oxford annotated Bible with the Apocrypha: Revised standard version.* New York: Oxford University Press.

Merrell-Wolff, F. (1973). *The philosophy of consciousness without an object.* New York: Julian.

Skinner, B. F. (1974). *About behaviorism.* New York: Vintage.

Valle, R. (1989). The emergence of transpersonal psychology. In R. Valle & S. Halling (Eds.), *Existential–phenomenological perspectives in psychology: Exploring the breadth of human experience.* (pp. 257–268). New York: Plenum Press.

Valle, R. (1995). Towards a transpersonal–phenomenological psychology: On transcendent awareness, passion, and peace of mind. *Journal of East–West Psychology, 1*(1), 3–15.

Valle, R., King, M., & Halling, S. (1989). An introduction to existential–phenomenological thought in psychology. In R. Valle & S. Halling (Eds.), *Existential–phenomenological perspectives in psychology: Exploring the breadth of human experience* (pp. 3–16). New York: Plenum Press.

Wilber, K. (1991). *Grace and grit: Spirituality and healing in the life and death of Treya Killam Wilber.* Boston: Shambhala.

Yogananda, P. (1956). Affirmation. *In Self-realization fellowship lesson S-1 P-9.* Los Angeles: Self-Realization Fellowship.

13

Toward a Phenomenology of Synchronicity

D. Hanson and Jon Klimo

INTRODUCTION

The purpose of this chapter is to share the results of a research study of the experience of a particular type of synchronicity. Carl Jung (1973) defined synchronicity as coincidence that is acausal, wherein certain events do not seem to be connected by normal causal means, and that is also particularly meaningful to the experiencer. Specifically, this is a study of a kind of synchronicity that appears to be goal-directed, although not under the conscious control of the experiencer. Consider the following example:

The young man sat before me with a soft peaceful expression. I asked him how it came to be that he was a student at Steiner College studying to become a Waldorf schoolteacher. His eyes took on a glow as he described falling in love with Steiner philosophy, watching his young children bloom and thrive in a Waldorf school, and his admiration of one of the school's master teachers. Albeit innocent of Joseph Campbell's suggestion that we "follow our bliss," the young man had identified his "bliss." But what to do? His bachelor's degree was in an unrelated field; his full-time career as a county planner was well under way. Head of household for a single-income family of four, he had initially seen a career change as out of the question. After much soul-searching, often shared until the wee hours of the morning with his wife, the young man fearlessly gave 30 days' notice to job and career. He describes the process of this move with these words:

> During the past two years, I have struggled with making the decision to leave this type of work [county planner] and pursue an avenue that I felt was more spiritually correct for me. It was not

D. Hanson and Jon Klimo • Rosebridge Graduate School of Integrative Psychology, Concord, California 94518.

Phenomenological Inquiry in Psychology: Existential and Transpersonal Dimensions, edited by Ron Valle. Plenum Press, New York, 1998.

> until I was secure in this decision that I could surrender to all the chaos surrounding the decision. . . . The experience would not have occurred without this surrendering. It allowed me to see my life clearly, and allow the Spirit for us to have an effect.

He submitted his application to Steiner College and relegated the issue of part-time employment to a future hurdle that he would jump when he came to it. Unexpected and unforeseen doors began to open. Forces seemingly reached out to carry him toward his goal. In his words, "It became obvious to me that this was my destiny."

Of particular significance was the following occurrence: When the young man submitted his notice, his supervisor immediately sought to hire a replacement. After weeks of disappointing interviews, he decided to keep the young man on as planner, part-time, a move unprecedented by the county for this position. For this young man, "The event itself of the renewed job opportunity is not exciting. The excitement occurred in the original decision to leave."

What happened to this young man? On one level, with the identification of his bliss and the dissolution of his fears, he took action. He then perceived the world to respond in unexpected ways to support him in his goal. In his words, "The balance of the tale is destiny. It occurred when I began to trust and love myself." Were the doors that opened luck or random chance? Was his experience nothing more than a series of coincidences? Or could he have experienced something more?

This example ends with the kind of questions that led us to using a formal phenomenological methodology in order to investigate the experience of "being carried along by a series or flow of unforeseen circumstances or events culminating in a right and desired outcome." Since there had never been a survey-type study of such an experience, we did not know what percentage of the population may have had such an experience. But it seems to us that many individuals at some time or other have had an experience of something like this at some point in their lives.

As described, this experience would seem, on the face of it, to have a number of possible characteristics or associations. Human experience on occasion appears to be embedded in a context that is larger than or different from ordinary day-to-day experience and the reigning scientific paradigm of how things are supposed to be. From another perspective, a transpersonal or transcendent quality is implied with a deeper dimension of meaning and purpose, wherein something is taking place outside normal ego consciousness and decision making, and outside a world running just on the principles of traditional classic physics.

The chief identifying characteristic of such an experience, however, may well be what Jung (1973) termed synchronicity. Over the years, we have come to use this term to indicate an acausal—or not normally causally connected—and personally meaningful coincidence or relationship among events. Today, we often hear people talk in terms of feeling as though they are "riding a wave" of synchronicities. These synchronicities seem to display a fabric of how events, directions, and destinies may be woven together on some deeper level. What might be the nature of this deeper loomwork? What or who is the weaver? The experience chosen for this study serves as a kind of Rosetta Stone for a group of interrelated conceptual and experiential domains that are absolutely central to the theme of human existence, such as intention and free will, purpose, direction in life, and destiny or nonordinary guidance in human affairs.

RELATED THEMES

Before describing our study of the experience being considered here, we first address a number of relevant interrelated subjects. These areas include: transpersonal psychology, the study of spiritual/mystical experiences, meaning, synchronicity, teleology, peak experience, and a holistic worldview pointed to by both traditional Eastern philosophy and post-classic physics. It is our hope that by briefly considering these themes, we can set the scene for better appreciating some of the facets involved in the experience we studied. We did not know initially whether the experience of being carried along by a series or flow of unforeseen circumstances or events culminating in a right and desired outcome would prove to be more difficult to research than other kinds of experience, because of its subjective, relatively unverifiable, and idiosyncratic nature, or whether human experiences of whatever kind equally challenge the researcher to arrive at its essential characteristics. However, there appears to be something about the subject of this study that spreads into a number of interrelated realms in ways that most other experiences do not. The attempt to investigate the experience in question, for example, parallels attempts to understand paranormal and mystical/spiritual types of experiences. Pioneer works in the mystical/spiritual field of inquiry include C. D. Broad's (1953) *Religion, Philosophy and Psychical Research,* Evelyn Underhill's (1955) *Mysticism: A Study in the Nature and Development of Man's Spiritual Consciousness,* W. T. Stace's (1960) *Mysticism and Philosophy,* William James's (1961) *The Varieties of Religious Experience,* Richard Bucke's (1969) *Cosmic Consciousness,* and Aldous Huxley's (1970) *The Perennial Philosophy*—to name just some of the modern classics in this field.

With regard to mystical experience, the philosopher W. T. Stace (1960) presents what he calls the "argument for unanimity," or "the argument from analogy," which contends "that there is an analogous agreement among mystics everywhere about what they experience, and that this supports belief in the objectivity of the experience" (p. 16). This argument, however, requires the assumption "that the witnesses are telling the truth . . . [and] that in their reports of their experiences they have not unintentionally misdescribed the nature of their experiences" (p. 33). Since we each proceed alone through our respective mystical or other nonordinary experiences, proving the objective reality or potentially public nature of such experiences is beyond a natural scientific grasp. What we can do, however, is seek the commonalties and regularities within the self-reports of such nonordinary experiences. Indeed, this is what the phenomenological approach attempts to do with any human experience it chooses to investigate.

Including the study of nonordinary and transcendent experience within its scope, transpersonal psychology, the so-called "fourth force" in psychology, addresses both ego and the related trans-egoic concepts of spirit in a framework larger than simple physical reality. Transpersonal theory was pioneered by Jung (1976), who asked us to pay attention to spiritual symbols, seen as embodiments of the numinous archetypes that, rooted in the unconscious, guide us on our odyssey of self-realization. Examples of such archetypal symbols are the mandala; the shadow; the divine child, and the initiation mysteries; the anima, animus, and the sacred marriage; the hero; the wise man; and the great mother. These spiritual symbols point to the future and highest stages of human development; they are signposts along the way to full spiritual awareness.

Another related theme is that of finding meaning in our experience. It appears that those who had the experience we studied were moved to make meaning for themselves out of their experience in particularly distinctive ways. Humanistic psychology and existential psychology address such value-laden domains, which include meaning-making, choice, and intention. Humankind's awareness of itself and the meaning of its existence in the context of the larger world is central. For Bugental (1981), a leading existential psychologist, it is the supreme achievement of our choice-making possibilities to create meaning. Vicktor Frankl (1963, p. 154), another existential pioneer, wrote that "the striving to find a meaning in one's life is the primary motivational force in man."

One person's experience of a series of events may be that the events are simply random chance or no more meaningful than their pedestrian qualities suggest. On the other hand, another person might experience the same series of events and find them deeply and personally interrelated and meaningful. It would seem that meaning is truly in the eye of the beholder.

In attempting to come to terms with the enigma of meaning, the contemporary psychologist Csikszentmihalyi (1990) provides three perspectives on what he calls *applied meaning*. First, meaning points toward the end, purpose, or significance of something, as in "the meaning of life." This definition reflects the assumption that events are linked to each other in terms of an ultimate goal. It assumes a temporal order, a causal connection between events. In this perspective, meaning assumes that phenomena are not random but fall into recognizable patterns directed by a final purpose. Second, meaning refers to a person's intentions. For example, "He means well." This implies that people reveal their purposes in their actions. Their goals are expressed in predictable and consistent ways. Third, meaning is used to order information. For example, "Aquatic entomology means the study of insects that live in water" or "Pain in arthritic joints means it's going to rain." This use of meaning indicates the identity of different words, or the relationship between events, and thereby helps to clarify and establish order among unrelated or conflicting pieces of information (Csikszentmihalyi, 1990).

Teleology and synchronicity are both concepts that address the nature of meaning in our lives. Let us look at each in turn.

Teleology attributes a character to nature or natural processes of being directed toward an end or shaped by a purpose. Basing his view on the concept of teleology, evolutionary thinker and Jesuit scientist Teilhard de Chardin regarded the human individual as an open-ended creature who may evolve indefinitely. He perceived a teleological pull, in the evolutionary process, toward increased awareness, complexity, and freedom. In this view, evolution is characterized both by the organism's freedom to choose and by its inner sense of right direction. In short, he saw evolution moving in a preferred direction (de Chardin, 1961).

In the past, the natural sciences have tended to resist teleological propositions, insisting that they are value-laden and therefore inappropriate for study. Within the field of psychology, Jung addressed teleology through his theory of individuation, which portrays the psyche as essentially goal- or aim-directed and therefore requiring an understanding in teleological terms (i.e., in the language of goals and purposes) (in Clarke, 1992). Further, Jung (1962, p. 687) tells us: "The nature of the human mind compels us to take the finalistic [i.e., the teleological] view. It cannot be disputed that, psychologically speaking, we are living and working day by day according to the principle of directed aim or purpose as well as that of causality."

Jung's thoughts (Clarke, 1992) correspond with those of de Chardin in applying the concept to life. He wrote: "Life is directed towards a goal. . . . Life is teleology par excellence; it is the intrinsic striving toward a goal, and the living organism is a system of directed aims which seek to fulfill themselves" (p. 170).

He prefaced his document on synchronicity with the desire to "open up a very obscure field which is philosophically of greatest importance," and he defined synchronicity as "the timely coincidence of two or several events which cannot be causally related to each other, but express an identical or similar meaning" (Jung, 1976).

Jung described synchronicity in terms of four categories: (1) "coincidental" experiences between one's mental content (thought or feeling) and an outer event; (2) experiences wherein one has a dream or vision that coincides with an event that is taking place at a distance; (3) instances in which one has an image, dream, vision, or premonition about something that will happen in the future and it does then occur; and (4) a series of synchronistic connections that culminate in meaning that unites them in terms of their inherent connection to a significant meeting, process, or goal (Bolen, 1979). An example of this last type would be the first meeting of two people who ultimately commit to an intimate relationship.

Jung came to the conclusion that the best way to describe the development of personality is in terms of an unfoldment from within. It is interesting to speculate on how such individual unfoldment and individuation might be acausally connected with this unfolding from within and the individuation of others as well. Such a connection would appear to comprise complex syntheses of synchronicities that take the form of teleologically guided sequences of events involving more than one person.

Progoff (1973, p. 64), a leading Jungian, said: "The coming together by apparent chance of factors that are not causally linked, but that nevertheless show themselves to be meaningfully related, is at the very heart of the process by which the purpose of the individual's life unfolds and becomes his 'fate.'" Jung (1976, pp. 431–432) wrote: "Just as in a living body the different parts work in harmony and are meaningfully adjusted to one another, so events in the world stand in a meaningful relationship which cannot be derived from any immanent causality. The reason for this is that in either case the behavior of the parts depends on a central control which is superordinate to them" What, then, is this superordinate, guiding, connection-making force that appears to be operating in our lives?

Although we may not be able to understand the source or mechanisms responsible for such experiences in our lives, we can deepen our understanding by considering Maslow's (1962) term *peak experience,* which addresses deeply meaningful experiences that transcend the ordinary. Peak experiences characterize the lives of individuals whom Maslow described as "self-actualizing," those who had achieved a high level of maturation, health, and self-fulfillment. In order to attain such a high level, Maslow maintained, an individual must undergo such a peak experience as a necessary prerequisite to establishing a more integrated and fully functional identity.

Maslow (1962) described peak experiences as having the following five general characteristics: (1) Perception in the peak moment tends to be nonclassificatory, in that we transcend our normal perception of the world, which rests essentially on generalization and on a division of the world into classes of various sorts. Perception can be relatively ego-transcending and self-forgetful. One may even speak of identification of the perceiver and the perceived, a fusion of what were two into a new and larger whole, a superordinate unit. We may see the world for a moment as in its own being. (2) There can be a characteristic disorientation of time and space. During these moments, the person is subjectively outside time and space. The peak experience can manifest, for example, in the creative fervor of the poet or artist who becomes oblivious to his or her surroundings and to the passage of time. (3) There can be a complete, though momentary, loss of fear, anxiety, inhibition, defense, and control, a surrender of renunciation, delay, and restraint. The fear of disintegration and dissolution, the fear of being overwhelmed by the instincts, the fear of death and of insanity, the fear of giving in to unbridled pleasure and emotion, all tend to disappear or go into abeyance during the peak experience. This, too, implies a greater openness of perception, since fear tends to distort normal perception. (4) By definition, the peak experience is only good and desirable. It is intrinsically valid, perfect, complete, and inevitable. It is just as good as it should be. (5) It involves a sudden surge of meaning. An individual does not choose to have it. One is surprised by it. In the cognition of the peak experience, the will does not interfere. We cannot command the peak experience. It happens to us (Wilson, 1972).

In light of these related issues, the experience investigated in this study can be seen to be of particular interest to those who also are involved with the study of a variety of nonordinary and transcendent human experiences.

THE RESEARCH: OVERVIEW, SUBJECTS, AND DATA GATHERING

We now turn to a description of the research study itself. First we share the overall design of the research project with references to the phenomenological research tradition within which it is situated. Next, we describe the subjects and the data-gathering procedures. We then give a detailed presentation of the data-analysis process. We then share with the reader how a sample subject's protocol was taken through the complete data-analysis procedure, with the hope, in doing so, of providing insight into the intricacies of this phenomenological process.

The study asked the following three questions (Giorgi, 1975): (1) Do people have the experience of being carried along by a series or flow of unforeseen circumstances or events culminating in a right and desired outcome? (2) If they do, what is the structure (nature, essence, or form) of the experience? (3) What is the style or "how" of the experience?

Heidegger (1962) states that in seeking to understand a phenomenon, "the phenomenologist thinks meditatively about its meaning." The phenomenological researcher is concerned with the meaning of the experience proposed for study, with the way things are experienced by the experiencer, and with how events are integrated into a dynamic, meaningful experience. As Giorgi (1975, p. 72) writes: "Phenomenology is the study of the structure, and the variations of structure, of the consciousness to which any thing, event, or person appears." Phenomenological method is used to elucidate both that which appears and the manner in which it appears, as well as the overall structure that relates what appears with its mode or manner of appearance.

The following five tenets of the existential–phenomenological approach were central to choosing and carrying out this study: (1) In the existential–phenomenological view, certain kinds of thoughts, feelings, and perceptions of the world express underlying, holistic ways of knowing (Bohm & Welwood, 1980; Valle & Halling, 1989). (2) The existential–phenomenological psychologist speaks of the total, indissoluble unity or interrelationship of the individual and his or her world. Further, the person is viewed as having no existence apart from the world and the world as having no existence apart from the person. Each individual and his or her world are said to co-constitute one another (Valle & Halling, 1989). (3) It is through the world that the very meaning of the person's existence emerges for both himself or herself and others (Valle & Halling, 1989). (4) People and the world are always in dialogue with each other (Valle & Halling, 1989). (5) A major focus of attention is with the world as given in direct and immediate experience (Husserl, 1970). All experience is regarded as having its basis or foundation in the *Lebenswelt,* or life-world, the world as lived by the person and not the hypothetical external entity separate from or independent of him or her. The prereflective structure of any identifiable experience, including the experience being studied here, is of this realm.

Turning now to the research subjects, or reporters, the following criteria of selection were used. Subjects had to: (1) be over 18 years of age; (2) have had the experience being studied; (3) have the ability to describe the experience; and (4) be willing to share their experience by participating in the study. For the study, 24 subjects were chosen who satisfied these criteria.

The data collection was conducted in two parts. Part I was designed to obtain a written report of the experience as recalled by the subjects. These written descriptions were elicited by a questionnaire. Part II of the research consisted of a walk-through interview designed to clarify the written descriptions and provide a richer description of the experience. Descriptions derived from Parts I and II were then brought together as the final protocol.

The initial questionnaire asked each subject to:

> 1. Select a time and space where you can be alone, uninterrupted, and relaxed.
> 2. Recall an episode when you experienced being carried along by a series or flow of unforeseen circumstances or events culminating in a right and desired outcome. Please describe the experience so that someone reading or hearing the report would know just what the experience was like for you. Keep your focus on the experience, not just the situation itself. Do not stop until you feel you have completely described the experience.

In addition, information regarding age, sex, occupation, and highest education completed was also collected.

The following instructions were given in person to each subject in the walk-through interview phase of the study:

> We're going to walk through the experience together just the way you wrote it. I'll read a part of it, then stop, and I would like for you to describe for me anything that comes to your mind about the part of the experience I have just read. You may provide further details about it, describe any thoughts or feelings you have, relate any images or metaphors that may come to your mind; in short, just let yourself go and associate to what I've just read to you in whatever way you choose until you feel ready to stop. We'll then move on to the next part.

These interview reports were recorded by audio tape recorder and then transcribed verbatim for analysis. The actual analysis was based very closely on Colaizzi's (1978) method. A final step used by van Kaam (1959) wherein he calculated the percentages of the occurrence of constituents across protocols was added. Figure 1 depicts the various stages in the data analysis. These steps in the analysis were as follows:

1. The written reports and transcribed walk-through interviews were first combined. These original protocols were read several times to acquire a feeling sense of each individual's experience in its totality as well as to obtain a beginning sense of the overall pattern.

2. The protocols were then condensed and summarized. During this process, repetitive statements were eliminated. Statements that directly related to the reporter's experience

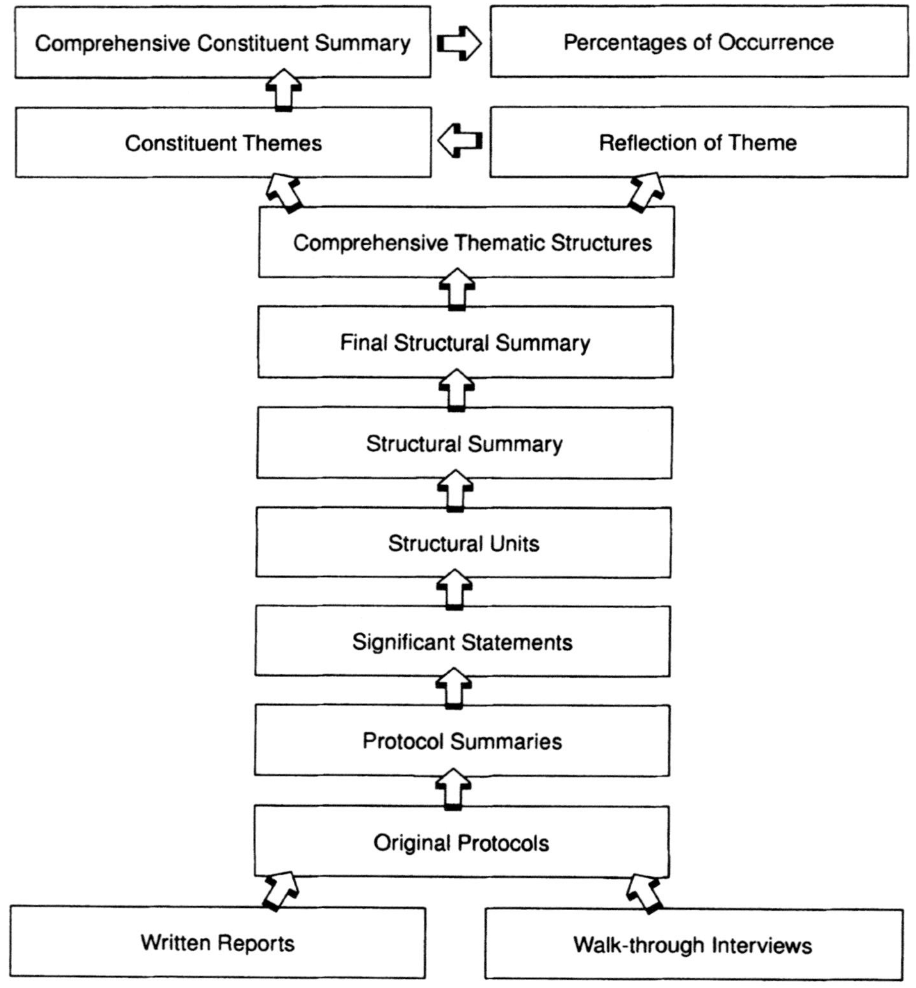

Figure 1. Flow chart of the steps in the analysis.

were extracted and combined into a flowing account of that experience. This process was repeated for several rounds, culminating in a Protocol Summary.

3. Extracted from each Protocol Summary were significant words, phrases, and sentences that directly pertained to the experience. These Significant Statements provided a further reduction of the original.

4. The extracted Significant Statements were clustered to form Structural Units. These units expressed natural components of the experience for each protocol.

5. The Structural Units were then synthesized and woven together to form a Structural Summary for each protocol. This Structural Summary contains a description that delineates

the unfolding of the experience in terms of essential elements and stages (Ricoeur, 1970). This step requires the process of an intuitive eidetic grasping of the whole of the experience, and was repeated through several rounds. Each resulting formulation was then compared to the Structural Units to check for consistency. With each repetition, the formulation became more refined until the Structural Units were clearly harmonious (i.e., integrated) in the Structural Summary (Polkinghorne, 1989).

6. In a short follow-up interview, a step formulated by Colaizzi (1978), each reporter was recontacted and asked by the researcher, "How do my descriptive results compare with your experience?" and "Have any aspects of your experience been omitted?" Any relevant new data that emerged from these follow-up interviews were synthesized into the corresponding description or Final Structural Summary (Polkinghorne, 1989).

7. Extracted from the Final Structural Summary were words, phrases, and sentences that together represented Comprehensive Thematic Structures for each protocol.

8. Each cluster of Comprehensive Thematic Structures was then interpreted by the researcher in terms of its psychological meaning to be a theme. It was given a tentative label and listed. Moving back and forth from each Comprehensive Thematic Structure to the developing list of themes, the researcher assigned each such cohesive whole to its descriptive appropriate theme, or added a new theme as needed, until all cohesive wholes from all protocols were reflected in a preliminary list of themes with subcategories as needed. In this process, each of the descriptive themes is reflected in the Constituent Themes that result.

In Colaizzi's view, reflection upon the basic descriptions (Comprehensive Thematic Structures) yields fundamental structures or the fundamental reality of the phenomenon (themes). In this stage, there is a leap from what the reporters say in their interviews to what the researcher interprets what they say to mean. Colaizzi (1978, p. 59) states: "This is a precarious leap because, while moving beyond the protocol statements, the meanings he [the researcher] arrives at and formulates should never sever all connection with the original protocols; his formulations must discover and illuminate those meanings hidden in the various contexts and horizons of the investigated phenomenon which are announced in the original protocols."

The preliminary list of themes was then scrutinized by the researcher for duplication in meaning and for subcategory overlap until a final list of Constituent Themes emerged. These Constituent Themes represent, across protocols, the essence of the Comprehensive Thematic Structures.

9. The Constituent Themes were then woven together, culminating in a flowing Comprehensive Constituent Summary.

10. The Percentages of Occurrence of themes across protocols were then calculated (van Kaam, 1959).

SAMPLE PROCEDURE

Following is a detailed example of the data analysis process for one subject's data. Please note that the contents of steps 2 and 6 are omitted, since the results of each are virtually identical to those of the prior step in each case. This example is offered because this step-by-step process is not usually shared by other phenomenological researchers, who usually jump from the generic description of the overall data-analysis process to the penultimate step of listing the final Constituent Themes. This tends to leave the reader with a vague sense regarding the highly rigorous identifying, winnowing, clustering, condensing, paraphrasing, and interpreting process that takes one from the original verbatim protocols to the final Constituent Themes of the experience or phenomenon in question.

Step 1: Original Protocol

When I was 19 I was living on a farm with some friends, They weren't real old friends. I had been living with them for about two months. We were renting the house and property up in Northern Oregon. I had a very strong desire to live communally with people seeking spiritual values and when I had come upon this farm it was after a lot of searching. I had left Ohio right after high school. I did a lot of hitch hiking around and looking for that and never really finding it. There was so much of it going on at that time, but I just never felt comfortable in the different places I ended up. I thought, 'this is it' when I first moved in. I did because what I thought about on the exterior. There were about 12 people there on a cozy farm in Northern Oregon, which is where I always wanted to end up. Plus I was tired of being rootless. I was getting tired of being shiftless, which I had been for two years at that point. But after a few months I was beginning to feel perhaps this was not it. Basically because there wasn't a real emphasis on spiritual . . . there wasn't any cohesive factor to the group, it just wasn't really what I was looking for. There wasn't the spiritual focus I had been hoping for, but I didn't want to admit it to myself because I really wanted to be settled.

One day I woke early and decided to go on a day hike by myself. Just an impulse. Hadn't really planned it, There were miles and miles of logging roads through the forest that surrounded the area. It was real pretty. It was a wonderful hike. It was. That day was . . . I remember that day very vividly. As I was going for the walk I was thinking that I needed to make a decision on what was happening there and that I needed my own space to think about it and to figure out what I was going to do with it. I remember coming to the conclusion that I had to leave the farm. It was real evident. All of a sudden in my mind I realized that I had to do that. But I felt very good about it and decided to cast my fate to the wind. I cast my fate to the wind a lot because I really believe in the flow and I wanted to become a part of it. But I never really ended up where I wanted to go and I was really ready to land somewhere. On that day I said here we go again, I'm just going to have to do it again because this just isn't right. And it felt OK for me to do that.

When I returned I found everyone had been packing up to leave throughout the day. It was a very big surprise. Earlier that morning the landlord came out and demanded everyone leave by nightfall. Seems he got a notion about not having a bunch of hippies on his land. Never bothered him before. He knew that we were all called hippies. For some reason he decided he didn't want us living there and he just came and kicked everyone out. No motive.

My good friend Marcia was waiting for me to return and had packed up my things which filled my back pack. She was very upset and afraid at not having a place to go. She and I had traveled for about a year. We met at Grand Canyon. We were both kind of looking for the same thing. She was really ready to stay. She liked the farm and the people. She was really happy with it. It was about 7:30 when we found ourselves hitchhiking into Portland thinking we would spend the night with some friends we knew. We had met people in Portland before we stayed at the farm. I remember feeling very peaceful inside. I was. I remember we were hitch hiking on this mountain road in Oregon and I remember thinking I probably should be afraid, or on my guard and we were far from the city. When I got back to the farm I saw everything just as a confirmation to what I knew I needed to do. It was kind of like all this other time I had been traveling I always had a destination, but this time I really knew that I was really casting my fate to the wind. I had absolutely no idea where we were going to go. I really felt like I finally was able to move with it. It was an exhilarating feeling, very powerful. I really knew that everything was happening the way it should be. I said, 'Marcia, don't worry, the Father will take care of us.' I had complete faith that everything would be all right. It was very much like being caught up in a wave. It was like not being in control of what was happening, but it was all right. It wasn't being victimized. It was OK to feel like that. I knew if I surrendered to it I could ride on the power of it. And it was a powerful exhilarating feeling.

When we arrived at the house where our friends lived one of them, Peter, had come back from California where he had found a spiritual community. He was planning to leave the very next morning to go down to live there. He had bought a station wagon, so we could fit in there too. He was up there going to school. He found this place and sold everything so that we could fit in there too and was going to leave the next morning. I knew that's why we ended up at his doorstep that night.

> I went down to the community and spent the next 10 years there. Ten pretty amazing years. These years were very formative to my own spirituality, to undying friendships and also to my marriage. It's where I met my husband, I really know that I was meant to be in that community. That was the experience that got me there. That experience was waiting for me and I was waiting for it.

Step 2: Protocol Summary

Here, the original protocol is condensed and summarized, eliminating repetition. All statements directly relating to the experience are extracted and combined into a flowing account. Due to the almost identical wording of this protocol summary compared to the original protocol, this particular step is not illustrated here.

Step 3: Significant Statements

Here, significant words, phrases, and sentences that directly pertain to the experience are extracted and numbered. There were 57 Significant Statements derived from this particular protocol summary. As illustrations, numbers 1–10 and 17–26 of these 57 Significant Statements follow (please refer back to the original protocol to see the relationship between the original material and the identified Significant Statements):

1. When I was 19 I was living on a farm with some friends.
2. I had a very strong desire to live communally with people seeking spiritual values.
3. When I had come upon this farm it was after a lot of searching.
4. There was so much of it going on at that time, but I just never felt comfortable in the different places I ended up.
5. I thought 'this is it' when I first moved in.
6. I was tired of being rootless.
7. Which I had been for two years at that point.
8. But after a few months I was beginning to feel perhaps this was not it.
9. There wasn't the spiritual focus I had been hoping for, but I didn't want to admit it to myself because I really wanted to be settled.
10. One day I woke early and decided to go on a day hike by myself.
17. I really believe in the flow and I wanted to become a part of it.
18. But I never really ended up where I wanted to go and I was really ready to land somewhere.
19. On that day I said here we go again, I'm just going to have to do it again because this just isn't right.
20. It felt OK for me to do that.
21. When I returned I found everyone had been packing up to leave throughout the day.
22. It was a very big surprise.
23. Earlier that morning the landlord came out and demanded everyone leave by nightfall.
24. Seems he got a notion about not having a bunch of hippies on his land. Never bothered him before.
25. No motive.
26. My good friend Marcia was waiting for me to return and had packed up my things.

Step 4: Structural Units

In this step, Significant Statements are clustered to form Structural Units. These are units that seem to express the natural structure of the experience. Again, consider just a sampling of how the Significant Statements became clustered into Structural Units. Significant Statements 3, 4, 6, 7, and 8 become one larger Structural Unit:

> When I had come upon this farm it was after a lot of searching. There was so much of it going on at that time, but I just never felt comfortable in the different places I ended up. I was tired of being rootless, which I had been for two years at that point. But after a few months I was beginning to feel perhaps this was not it.

Significant Statements 5, 9, and 10 became clustered into another Structural Unit:

> I thought 'this is it' when I first moved in. There wasn't the spiritual focus I had been hoping for, but I didn't want to admit it to myself because I really wanted to be settled. One day I woke early and decided to go on a day hike by myself.

Significant Statements 17 and 18 clustered as:

> I really believe in the flow and I wanted to become a part of it. But I never ended up where I wanted to go and I was really ready to land somewhere.

Significant Statements 19, 20, and 21 became:

> One day I said here we go again, I'm just going to have to do it again because this just isn't right. It felt OK for me to do that. When I returned I found everyone had been packing up to leave throughout the day.

Significant Statements 22 through 25 became:

> It was a very big surprise. Earlier that morning the landlord came out and demanded everyone leave by nightfall. Seems he got a notion about not having a bunch of hippies on his land. Never bothered him before. No motive.

Step 5: Structural Summary

Recall that the Structural Summary contains a description that delineates the unfolding of the experience in terms of essential elements and stages, and requires a process of an intuitive, eidetic grasping of the whole of the experience:

> When I was 19 I was living on a farm with some friends.
>
> I had a very strong desire to live communally with people seeking spiritual values.
>
> When I had come upon this farm it was after a lot of searching. There was so much of it going on at that time, but I just never felt comfortable in the different places I ended up. I was tired of being rootless, which I had been for two years at that point. But after a few months I was beginning to feel perhaps this was not it.
>
> I thought 'this is it' when I first moved in. There wasn't the spiritual focus I had been hoping for, but I didn't want to admit it to myself, because I really wanted to be settled. One day I woke early and decided to go on a day hike by myself.
>
> Just an impulse. Hadn't really planned it. I remember that day very vividly. I remember coming to the conclusion that I had to leave the farm.
>
> As I was going for the walk I was thinking that I needed to make a decision on what was happening there and that I needed my own space to think about it and to figure out what I was going to do with it.
>
> All of a sudden in my mind I realized that I had to do that. But I felt very good about it and decided to cast my fate to the wind.

I really believe in the flow and I wanted to become part of it. But I never really ended up where I wanted to go and I was really ready to land somewhere.

On that day I said here we go again, I'm just going to have to do it again because this just isn't right. And it felt OK for me to do that. When I returned I found everyone had been packing up to leave throughout the day.

It was a very big surprise. Earlier that morning the landlord came out and demanded everyone leave by nightfall. Seems he got a notion about not having a bunch of hippies on his land. Never bothered him before. No motive.

My good friend Marcia was waiting for me to return and had packed up my things.

It was about 7:30 when we found ourselves hitch hiking into Portland thinking we would spend the night with some friends we knew.

I remember feeling very peaceful inside.

We were hitch hiking on this mountain road and I remember thinking I probably should be afraid, or on my guard.

We were far from the city.

When I got back to the farm I saw everything just as a confirmation to what I knew I needed to do.

All this other time I had been traveling I always had a destination, but this time I really knew that I was really casting my fate to the wind. I had absolutely no idea where we were going to go.

I really felt like I finally was able to move with it. It was an exhilarating feeling, very powerful. I really knew that everything was happening the way it should.

I said 'Marcia, don't worry, the Father will take care of us.' She was very upset and afraid at not having a place to go. She and I had traveled for about a year. We were both kind of looking for the same thing. I had complete faith that everything would be all right.

It was very much like being caught up in a wave. It was like not being in control of what was happening, but that was all right. I knew if I surrendered to it I could ride on the power of it. And it was a powerful exhilarating feeling.

When we arrived at the house where our friends lived one of them, Peter, had come back from California where he had found a spiritual community. He was planning to leave the very next morning to go back down to live there. He had bought a station wagon, so we could fit in there, too.

I knew that's why we ended up at his doorstep that night.

I went down to the community and spent the next 10 years there.

Ten pretty amazing years. These years were very formative to my own spirituality, to undying friendships and also to my marriage. It's where I met my husband.

I really know that I was meant to be in that community.

That was the experience that got me there.

That experience was waiting for me and I was waiting for it.

Step 6: Final Structural Summary

The Structural Summary contains a description that delineates the unfolding of the experience in terms of essential elements and stages. This description is done in light of an intuitive grasp of the whole experience on the part of the researcher. One continues to review the Structural Units that support this summary. Then each subject is recontacted and asked, "How do my descriptive results [now the Structural Summary] compare with your experience? Have any aspects of your experience been omitted?" Any new data that emerge are then synthesized with the existing Structural Summaries. The Structural Summary reads quite similarly to the original Protocol Summary, except that now the material is grouped not in terms of the original six prose paragraphs or in terms of the later 57 separated-out Significant Statements, but rather in terms of the Structural Units. In the sample protocol being used here, the Final Structural Summary and the prior Structural Summary are identical.

Step 7: Comprehensive Thematic Structures

A total of 16 Comprehensive Thematic Structures were derived from the Final Structural Summary. They are listed under step 8 as they are grouped in terms of the original 57 Significant Statements and according to the final 16 Constituent Themes that they led to and helped comprise.

Step 8: Constituent Themes

In order to give the reader a more direct sense of these results, the presentation of each of the 16 Comprehensive Thematic Structures begins with the final Constituent Theme (in **boldface**) for which they are responsible.

Following each set of grouped Comprehensive Thematic Structures is the researchers' paraphrasing of them as Reflection of a Theme, if applicable, identified by *italics* to distinguish it from the subject's own directly quoted material in the Comprehensive Thematic Structures and their numbered Significant Statements.

Notice that some Comprehensive Thematic Structures are repeated, contributing to more than one final Constituent Theme. Also, keep in mind that the 16 final Constituent Themes were derived from all 24 subjects, not from just the single sample protocol being offered here. Please note that each Comprehensive Thematic Structure led to a particular Constituent Theme as indicated.

Comprehensive Thematic Structure (basis of **Constituent Theme 1: Point of Embarkation**):

1. When I was 19 I was living on a farm with some friends.
2. I had a very strong desire to live communally with people seeking spiritual values.

Comprehensive Thematic Structure (basis of **Constituent Theme 2: Identification of the Desire, Need, or Goal**):

2. I had a very strong desire to live communally with people seeking spiritual values.

Comprehensive Thematic Structure (basis of **Constituent Theme 3: Gather Information**):

3. When I had come upon this farm it was after a lot of searching.
4. There was so much of it going on at that time, but I just never felt comfortable in the different places I ended up.

   Reflection of Theme (re: statements 3, 4): In her awareness of and searching for spiritual communities, she is gathering information regarding the achievement of her goal.

Comprehensive Thematic Structure (basis of **Constituent Theme 4: Action Taken Toward Desire, Need, or Goal**):

3. When I had come upon this farm it was after a lot of searching.

   Reflection of Theme: Her searching for a spiritual community entailed conscious direction on her part toward the desired goal.

51. I went down to the community and spent the next 10 years there.

   Reflection of Theme: Her departure for the spiritual community discovered by her friend was a conscious action taken toward her goal.

11. Just an impulse. Hadn't really planned it.
13. As I was going for the walk I was thinking that I needed to make a decision on what was happening there and that I needed my own space to think about it and to figure out what I was going to do with it.
14. I remember coming to the conclusion that I had to leave the farm.
    Reflection of Theme (re: statements 11, 13, 14): Following her impulse to take a day hike was an unconscious action taken that resulted in the privacy to process her situation and conclude she must leave the farm.

Comprehensive Thematic Structure (basis of **Constituent Theme 5: Change Occurs**):
14. I remember coming to the conclusion that I had to leave the farm.
15. All of a sudden in my mind I realized that I had to do that.
16. But I felt very good about it and decided to cast my fate to the wind.
    Reflection of Theme (re: statements 14–16): The change occurs as she moves to being at ease with leaving without a destination.

Comprehensive Thematic Structure (basis of **Constituent Theme 6: Resistance to Possibilities or Changes**):
9. There wasn't the spiritual focus I had been hoping for, but I didn't want to admit it to myself because I really wanted to be settled.
    Reflection of Theme: She is resistant to acknowledging the lack of a spiritual focus at the farm out of her desire to be settled.

Comprehensive Thematic Structure (basis of **Constituent Theme 7: Surrender of Control**):
14. I remember coming to the conclusion that I had to leave the farm.
15. All of a sudden in my mind I realized that I had to do that.
16. But I felt very good about it and decided to cast my fate to the wind.
    Reflection of Theme (re: statements 14–16): She surrenders control of the form her desire for a spiritual community will take and decides to leave the farm without a known destination.
20. It felt OK for me to do that.
    Reflection of Theme: She experienced a change from her resistance to moving on to a feeling of it being okay to do that.

Comprehensive Thematic Structure (basis of **Constituent Theme 8: Time Lapse**):
There is no Comprehensive Thematic Structure in this sample protocol that relates to Constituent Theme 8.

Comprehensive Thematic Structure (basis of **Constituent Theme 9: Promptings**):
10. One day I woke early and decided to go on a day hike by myself.
11. Just an impulse. Hadn't really planned it.
    Reflection of Theme (re: statements 10, 11): She experienced an impulse that led to a decision to go on a day hike alone.

Comprehensive Thematic Structure (basis of **Constituent Theme 10: Synchronistic Occurrences**):
10. One day I woke early and decided to go on a day hike by myself.
11. Just an impulse. Hadn't really planned it.
12. I remember that day vividly.

13. As I was going for the walk I was thinking that I needed to make a decision on what was happening there and that I needed my own space to think about it and to figure out what I was going to do with it.
14. I remember coming to the conclusion that I had to leave the farm.
15. All of a sudden in my mind I realized that I had to do that.
16. But I felt very good about it and decided to cast my fate to the wind.
17. I really believe in the flow and I wanted to become a part of it.
21. When I returned I found everyone had been packing up to leave throughout the day.
22. It was a very big surprise.
23. Earlier that morning the landlord came out and demanded everyone leave by nightfall.
24. Seems he got a notion about not having a bunch of hippies on his land. Never bothered him before.

    Reflection of Theme (re: statements 10–17, 21–24): The meaningful coincidence occurred between her inner decision to leave the farm, while unbeknownst to her, the landlord had evicted the group without warning or apparent motive.
30. It was about 7:30 [P.M.] when we found ourselves hitchhiking into Portland thinking we would spend the night with some friends we knew.
37. I had absolutely no idea where we were going to go.
47. When we arrived at the house where our friends lived one of them, Peter, had come back from California where he had found a spiritual community.
48. He was planning to leave the very next morning to go back down to live there.
49. He had bought an old station wagon, so there was room for us to go too.
50. I knew that's why we ended up at his doorstep that night.

    Reflection of Theme (re: statements 30–37, 47–50). The meaningful coincidence occurred between her inner surrender of control regarding a known destination and, without foreknowledge, finding herself at the residence of a friend who was leaving the following morning for a spiritual community and there was room for her to accompany him.
51. I went down to the community and spent the next 10 years there.
53. These years were very formative to my own spirituality, to undying friendships and also to my marriage.
54. It's where I met my husband.
55. I really know that I was meant to be in that community.
56. That was the experience that got me there.
57. That experience was waiting for me and I was waiting for it.

    Reflection of Theme (re: statements 10–17, 21–24, 30, 37, 47–51, 53–57): Two synchronistic occurrences combine in a meaningful way to culminate in her arrival to a right and desired spiritual community. She also met her future husband while there.

Comprehensive Thematic Structure (basis of **Constituent Theme 11: Risk**):

19. On that day I said, here we go again, I'm just going to have to do it again because this just isn't right.
35. All this other time I had been traveling I always had a destination.

36. This time I really knew that I was really casting my fate to the wind.
    Reflection of Theme (re: statements 19, 35, 36): The risk is contained in her decision to leave without a known destination. This is new behavior for her and therefore carries the risk of unknown consequences.

Comprehensive Thematic Structure (basis of **Constituent Theme 12: Peak Experience**):

34. When I got back to the farm I saw everything just as a confirmation to what I knew I needed to do.
31. I remember feeling very peaceful inside.
32. We were hitch hiking on this mountain road and I remember thinking I probably should be afraid, or on my guard.
33. We were far from the city.
36. This time I really knew that I was really casting my fate to the wind.
37. I had absolutely no idea where we were going to go.
38. I really felt like I finally was able to move with it.
39. It was an exhilarating feeling, very powerful.
40. I really knew that everything was happening the way it should be.
41. I said, 'Marcia, don't worry, the Father will take care of us.'
42. I had complete faith that everything would be all right.
43. It was very much like being caught up in a wave.
44. It was like not being in control of what was happening, but it was all right.
45. I knew if I surrendered to it I could ride on the power of it.
46. And it was a powerful exhilarating feeling.
    Reflection of Theme (re: statements 34, 31–33, 36–46): While experiencing heightened awareness of her connection to a non-ego-based power, she has thoughts of her vulnerability, but is simultaneously without fear. She experiences being caught up in and moving with a non-ego-based power that requires a surrender of her control. It was non-threatening for her to surrender control of what was happening. The experience evokes in her complete faith in the rightness of the process and that she will be taken care of by a transcendent element.

Comprehensive Thematic Structure (basis of **Constituent Theme 13: Assignment of Meaning**):

4. There was so much of it going on at that time, but I just never felt comfortable in the different places I ended up.
   Reflection of Theme: Although she was aware of the existence of many spiritual communities, at this point her search to find a comfortable one had been fruitless.
5. I thought, 'this is it' when I first moved in.
   Reflection of Theme: She thought that she had found a community that fit her needs when she first moved in at the farm.
6. I was tired of being rootless.
8. But after a few months I was beginning to feel perhaps this was not it.
   Reflection of Theme (re: statement 8): After a few months, she began to feel that perhaps the community did not fill her needs.
34. When I got back to the farm I saw everything just as a confirmation to what I knew I needed to do.

Reflection of Theme: Upon returning from her hike to find that she and her friends had been evicted, she saw everything happening as a confirmation of her decision to leave the farm.

55. I really know that I was meant to be in that community.
56. That was the experience that got me there.
57. That experience was waiting for me and I was waiting for it.
    Reflection of Theme (re: statements 55–57): The experience in its totality was waiting for her as she waited for it to bring her to the spiritual community that was to offer the culmination of her right and desired outcome.
53. These years were very formative to my own spirituality, to undying friendships and also to my marriage.
54. It's where I met my husband.
    Reflection of Theme (re: statements 53, 54): The years that she spent in the spiritual community were very formative to her spirituality, to undying friendships, and also to her marriage as she met her husband there.
32. We were hitch hiking on this mountain road and I remember thinking I probably should be afraid, or on my guard.
33. We were far from the city.
37. I had absolutely no idea where we were going to go.
41. I said, 'Marcia, don't worry, the Father will take care of us.'
42. I had complete faith that everything would be all right.
    Reflection of Theme (re: statements 32, 33, 37, 41, 42): She expresses to her friend her faith and trust in a transcendent element to take care of them regarding their safety and their ultimate destination.
40. I really knew that everything was happening the way it should be.
    Reflection of Theme: She really knew that the events and circumstances that had transpired to take them from the farm were part of a predetermined order.
55. I really know that I was meant to be in that community.
    Reflection of Theme: She really knows that her presence in the spiritual community was part of a predetermined order.

Comprehensive Theme Structure (basis of **Constituent Theme 14: Faith or Trust**):

41. I said, 'Marcia, don't worry, the Father will take care of us.'
42. I had complete faith that everything would be all right.
    Reflection of Theme (re: statements 41, 42); The meaning that emanates is that she can trust the process and outcome of her experience. The meaning that emanates is trust and faith that a transcendent element will provide both safety and right outcome.

Comprehensive Thematic Structure (basis of **Constituent Theme 15: Circumstances or Events Leading to a Right and Desired Outcome**):

1. When I was 19 I was living on a farm with some friends.
2. I had a very strong desire to live communally with people seeking spiritual values.
3. When I had come upon this farm it was after a lot of searching.
   Reflection of Theme (re: statements 1–3): Considerable searching for her desire to live communally with people seeking spiritual growth led to the farm where she currently lived with friends.

10. One day I woke early and decided to go on a day hike by myself.
11. Just an impulse. Hadn't really planned it.
13. As I was going for the walk I was thinking that I needed to make a decision on what was happening there and that I needed my own space to think about it and to figure out what I was going to do with it.
14. I remember coming to the conclusion that I had to leave the farm.
19. On that day I said here we go again, I'm just going to have to do this again because this just isn't right.
20. It felt OK for me to do that.

Reflection of Theme (re: statements 10, 11, 13, 14, 19, 20): Her impulse decision to go on a day hike gave her the opportunity to assess the situation as inadequate for her needs, come to the conclusion to leave the farm, and be open to another possibility.

21. When I returned I found everyone had been packing up to leave throughout the day.
23. Earlier that morning the landlord came out and demanded everyone leave by nightfall.
25. No motive.

Reflection of Theme (re: statements 21, 23, 25): On returning from her hike, she found that the landlord had synchronistically, for no apparent reason, evicted her and her friends, demanding that they leave by nightfall.

30. It was about 7:30 when we found ourselves hitch hiking into Portland thinking we would spend the night with some friends we knew.

Reflection of Theme: At nightfall, the circumstances of the eviction from the farm called for her and her friends to hike into town in hopes of spending the night with friends there.

31. I remember feeling very peaceful inside.
32. We were hitch hiking on this mountain road and I remember thinking I probably should be afraid, or on my guard.
33. We were far from the city.
34. When I got back to the farm I saw everything just as a confirmation to what I knew I needed to do.
35. All this other time I had been traveling I always had a destination.
36. This time I really knew that I really was casting my fate to the wind.
37. I had absolutely no idea where we were going to go.
38. I really felt like I finally was able to move with it.
39. It was an exhilarating feeling, very powerful.
40. I really knew that everything was happening the way it should be.
41. I said, 'Marcia, don't worry, the Father will take care of us.'
42. I had complete faith that everything would be all right.
43. It was very much like being caught up in a wave.
44. It was like not being in control of what was happening, but it was all right.
45. I knew if I surrendered to it I could ride on the power of it.
46. And it was a powerful exhilarating feeling.

Reflection of Theme (re: statements 31–46): The peak experience that had ensued on her hike held at bay her fears for safety en route, as well as her lack of control regarding an ultimate destination.

47. When we arrived at the house where our friends lived one of them, Peter, had come back from California where he had found a spiritual community.
48. He was planning to leave the very next morning to go back down to live there.
49. He had bought a station wagon, so we could fit in there too.
Reflection of Theme (re: statements 47–49): On arrival, she found that one of her friends had synchronistically just returned from a trip wherein he had discovered a spiritual community, was returning there, had bought a station wagon with extra room, and would take them with him.
50. I knew that's why we ended up at his doorstep that night.
51. I went down to the community and spent the next 10 years there.
Reflection of Theme (re: statements 50, 51): She gave the offer significant meaning, and took the opportunity to travel to a spiritual community, where she remained for 10 years.

Comprehensive Thematic Structure (basis of **Constituent Theme 16: Speculation on the Future**):
There is no Comprehensive Thematic Structure in this sample protocol that relates to Constituent Theme 16. This theme was contributed to by enough of the other 23 subjects for it to be one of the 16 Final Constituent Themes.

Step 9 and 10: Comprehensive Constituent Summary with Percentages of Occurrence

The 16 Constituent Themes that were generated, most of which contained a number of subcategories, are listed here in the order that makes the most sense, given the unfolding of most of the subjects' experiences. Percentages of occurrence of each theme across all subjects are included. In addition, examples of Significant Statements, drawn from various subjects, are given, one such statement for one subcategory for each theme. These examples will provide the reader with a sense of how the final Comprehensive Constituent Themes have their roots in, and were derived from, the primary data.

1. Point of Embarkation (100%). Represents the beginning point of the experience and is found within the first Significant Statements. Significant is the wide range of subcategories: (a) homeostasis; (b) trauma; (c) stress; (d) premonition; (e) change; (f) identification of desire, need, or goal; (g) figure–ground shift. An example of (b): "I participated in a Healing the Child Within seminar."
2. Identification of the Desire, Need, or Goal (100%). Subcategories: (a) preexperience; (b) following figure–ground shift, major life change, or trauma; (c) through focused attention; (d) post-experience. Example of (a): "I had a very strong desire to live communally with people seeking spiritual values."
3. Gather Information (includes intuition and precognition) (100%). Subcategories: (a) regarding possible routes to desired goal; (b) regarding location of self in lifeworld. Example of (b): "I began to look seriously at my work situation, with which I had been dissatisfied for years."
4. Action Taken Toward Desire, Need, or Goal (100%). Subcategories: (a) conscious striving; (b) despite resistance; (c) without recognition of reason/logic, unconscious. Example of (a): "I practiced my breathing with great diligence for months in preparation for my daughter's birth."

5. Change Occurs (100%). Subcategories: (a) self-induced; (b) other than self-induced. Example of (a): "Things went back to normal."
6. Resistance to Possibilities or Changes (70%). Subcategories: (a) within self; (b) by others; (c) by circumstances or events. Example of (a): "I was going off and didn't want to be tied down."
7. Surrender of Control (45%). Subcategories: (a) regarding what form desire or goal will take; (b) regarding change; (c) to other than ego-bound power. Example of (a): "I consciously decided to surrender to the pain."
8. Time Lapse. (22%). Example: "After five months I had experienced no change."
9. Promptings (61%). Subcategories: (a) by an inner urge; (b) by feeling compelled; (c) by premonitions; (d) by intuitive knowing; (e) by attraction. Example of (b): "It was very uncomfortable, but I was still being drawn by his absolute uncompromising clarity that he wanted to have a relationship with me."
10. Synchronistic Occurrences (56%). Subcategories: (a) meaningful coincidence between mental content and an outer event; (b) an image (dream, vision, or premonition) about the future that does occur; (c) a series of synchronistic events culminating in a right and desired outcome. Example of (a) that reflects five Significant Statements comprising one Structural Unit and one Comprehensive Thematic Structure: "Maria said during the June Life's Work course, 'Let's write a book on life's work together.' This was unforeseen. About this same time a manuscript was returned to me. And I saw that my thesis [for the article], that Alice's journey to Wonderland was a journey to the accordance of individual self/identity and universal self/identity, could be used as a sort of framework, or guiding inspiration, for the book on discovering one's life-work or purpose. It was as though both were the same."
11. Risk (action taken against unknown odds) (96%). Example: "Against my better judgment, I began to do what was suggested to me, based on my 'willingness to be willing.'"
12. Peak Experience, Spiritual Experience, or Transpersonal State of Consciousness (46%). Subcategories: (a) relative; (b) absolute. An example of (b) is represented by the following four Significant Statements: "It was an exhilarating feeling, very powerful. I really knew that everything was happening the way it should be. I said, 'Marcia, don't worry, the Father will take care of us.' I had complete faith that everything would be all right."
13. Assignment of Meaning (100%). Subcategories: (a) regarding process of experience; (b) regarding self; (c) regarding others; (d) regarding gains; (e) regarding life; (f) regarding human experience; (g) regarding the transcendent (includes faith); (h) regarding mystery; (i) regarding destiny or fate. Example of (b): "I was tired of being rootless."
14. Faith or Trust (43%). Indicates faith or trust in the final outcome, or in a non-ego-based element of the experience. Subcategories: (a) in outcome; (b) in non-ego-based element. Example of (a): "I had complete faith that everything would be all right."
15. Circumstances or Events Leading to a Right and Desired Outcome (100%). These represent stepping-stones of the experience from the Point of Embarkation to its culmination in a Right and Desired Outcome. Significant Statements reflecting

this theme were specific to the flow of the experience of each reporter. Subcategories: (a) with accompanying awareness; (b) with retrospective awareness. An example of (b): "I was giving birth at home and went into my back yard."

16. Speculation on the Future (57%). Subcategories: (a) regarding self; (b) regarding self and other(s). Example of (a): "I have no doubt that I'm going to succeed in everything I do."

INTEGRATION, ANALYSIS, AND SPECULATION

After careful reflection on the 16 Constituent Themes, three primary categories were formulated in order to understand these findings on a more integrated level:

1. Meaning, which includes choice and intention, and the concept of subjects "co-creating" their own lives.
2. "Synchronistic teleology," which includes a dynamic pattern of directional, purposeful movement with synchronistic aspects, and with less or no presence or efficacy of conscious intention. Here, synchronicity appears to be one of the key ways in which the teleological flow manifests itself in our experience. Synchronicity is the acausal connector, matchmaker, "cosmic glue," or transcendent intelligence that brings people and events together in nonordinary ways [Maslow's (1962) concept of peak experience appears to characterize this aspect of the experience for the subjects].
3. What might he termed the New Metaphysics, or the emerging paradigm, that might help us better understand the kind of synchronistic experience focused on in this study.

Meaning

As mentioned above, all of the subjects, to varying degrees and in varying ways, were moved to make meaning for themselves out of the experiences they described. That is, the meaning-making aspect was an integral part of "the experience of being carried along by a series or flow of unforeseen circumstances or events culminating in a right and desired outcome." In the aforecited words of leading existential thinker Frankl (1963, p. 154): "The striving to find a meaning in one's life is the primary motivational force in man." Recall, also, that Jung originally defined synchronicity as being "meaningful coincidence," or coincidental or "acausal" connections that are deemed personally meaningful by those who are experiencing them.

Consider, then, this subject of personal meaning making in light of the Constituent Themes (CTs): CT 2—Identification of the Desire, Need, or Goal, CT 3—Gather Information, CT 16—Speculation on the Future, and CT 14—Faith or Trust are all associated with how subjects assigned meaning to their experience, as opposed to what meanings they might have assigned. With regard to the latter (i.e., what meanings), CT 1—Point of Embarkation, CT 5—Change Occurs, CT 10—Synchronistic Occurrences, CT 12—Peak or Spiritual Experiences, and CT 15—Circumstances or Events Leading to a Right and Desired Outcome, are all, to varying degrees, categories of meaning that subjects assigned to aspects of their experience.

CT 3—Gather Information and CT 13—Assignment of Meaning are found across all protocols and represent the circulatory system of the experience. These themes occur repeatedly and seemingly at random throughout the experience. In combination, they portray a process wherein the reporter takes in information regarding both location of self in his or her life-world and of possible routes to a desire, need, or goal that he or she is aware of, and then applies meaning to the information.

Consider CT 13—Assignment of Meaning: Csikszentmihalyi's (1990) breakdown of the common usage of applied meaning was evidenced in the protocols. Reporters assigned meaning that pointed toward an end, purpose, or significance of their experiences, as in "the meaning of life." This use of meaning reflected their assumption that events are linked to each other in terms of an ultimate goal. That this aspect of meaning implies teleology as well shows how deeply interrelated these two primary categories are and demonstrates how difficult it is to separate them out in certain respects. Also within the domain of meaning making is the complex issue of the extent to which we have free will, choice, and intention that can affect the nature of our lived world and the extent to which it is, so to speak, out of our hands.

Variations of CT 2—Identification of the Desire, Need, or Goal describe a holistic, nonordinary model of the process by which individuals may be seen to cocreate their lives. Recall that 14 of the 16 final Constituent Themes were comprised of two or more subcategories. For example, under CT 2, the subcategory of preexperience identification describes the process by which participants have already arrived at a chosen focus of attention and then proceed to go about their role in achieving it. In other words, the reporters gather information regarding what their positions are in relation to the goal, and regarding possible routes to the goal (CT 3), apply meaning to that information (CT 13), and take action (CT 4), and change occurs (CT 5), moving them one step closer to the desired outcome (CT 15). That process is then repeated for each successive step in arriving at the right and desired outcome. At the same time, within that process, they will be repeating miniversions of the same process with each decision made. As the individuals move along, they may or may not be aware of other factors participating in their progress beyond their own efforts. This realization often comes after their arrival at the goal. Along with this realization may come recognition that within the process of achieving the preidentified goal, they have also achieved a right and desired outcome held within it that was not necessarily anticipated.

Continuing with this process of cocreating one's life, identification of one's desire, need, or goal is sudden and charged with energy and meaning. The picture changes with this charge of energy. In CT 2, the individual may have a strong attachment to the form that the desired outcome will take, may resist possibilities or changes that do not fit that preconceived form (CT 6a), and may need to surrender control of the form (CT 7a) in order to reach a desired outcome. Often, in protocols in which this scenario presented itself, synchronistic occurrences (CT 10) and/or spiritual experiences (CT 12) followed surrender and a perceived standstill in progress (CT 8).

In subcategory (d) of CT 2—Identification of the Desire, Need, or Goal comes with a "postexperience recognition." The individual seems to move from stepping-stone to stepping-stone (CT 15) without consciously choosing to do so with any specific cumulative outcome in mind (CT 4c) change occurs (CT 5), one may or may not identify an intermediate goal along the way (CT 2), one experiences synchronistic occurrences (CT 10) as well as spiritual experiences (CT 12), and arrives at the time of the report recognizing a right and desired outcome in culmination of the experience. This particular facet of the experience, as described by subcategory CT 2d, often covers many years. In such cases, meaning assigned is primarily regarding personal growth.

Subcategories CT 2a and 2b reflect the experiences of those who are more conscious of both their personal power to create their lives and what the desired outcome is. Many spend a lot of energy striving to achieve that outcome on their own, only to recognize along the way that it takes more than their individual efforts, that some kind of trans-egoic or transcendental processes are needed in order to reach their goal. Surrender is an important issue as these subjects open to possibilities beyond the form of the desire that they are attached to. In addition, when this surrender of control is charged with risk, it often results in a peak experience. These individuals perceive themselves as getting to where they needed to go, but were often not aware of their need until it was fulfilled.

Synchronistic Teleology

Recall that teleology attributes a character or quality to nature and/or natural processes of being directed toward an end or shaped by an inherent purpose. De Chardin (1961) perceived a teleological "pull," in the evolutionary process, toward increased awareness, complexity, and freedom. In this view, evolution is characterized both by the organism's freedom to choose and by its inner sense of "right" direction. Jung's process of individuation (Clarke, 1992), as part of its definition, portrays the psyche as essentially goal- or aim-directed, and therefore requires understanding in teleological terms.

We see this category, which we are calling synchronistic teleology, as expanding to include an entire kind of Bohmian holomovement or Eastern-type holism (described below), a dynamic pattern of goal-oriented, purposeful, nonordinary action that appears to contain a number of stages or steps that are related by the acausal connectionism of synchronicity. The following CTs reflect this interpretation:

Subcategories CT 2a and 2b reflect the reporters' self-assertion in their choices of desired outcomes in their lives. Increased awareness is seen in their Assignment of Meaning (CT 13) to the pattern of events that unfold following that self-assertion. A correlation is also seen in subcategory CT 2d, where a teleological pull is evidenced by the self-defined "personal growth" of the reporters.

The inner sense of "right direction" is visible in Assignment of Meaning (CT 13) in the applying of meaning to the Gathering of Information (CT 3) in terms of the "rightness" of the process of the experience and one's role in it. CT 9—Promptings can be seen as contributing to an inner sense of "right" direction in terms of resulting Action Taken (CT 4) occurring across the stepping-stones of CT 15 in reaching the desired outcome.

CTs 4 and 5 represent the mechanics of the teleological flow experience and appear to indicate the movement of the experience, as well as the way in which movement occurs. These two CTs appear to comprise a set that describes one mode of movement inherent to the experience by revealing movement through a cause-and-effect perspective, or what could be referred to as "ordinary movement." In other words, the reporters repeatedly experienced change following action taken toward a desire, need, or goal culminating in movement. Teleological flow, however, is not necessarily comprised of "ordinary movement."

CTs 9 and 10 also comprise a set that can be seen as describing movement, and, in this way, revealing movement of an acausal nature, or what could be referred to as "nonordinary movement." In other words, the reporters experienced a synchronistic occurrence during the experience following one or more promptings to act that culminated in movement (as revealed by CT 15).

CTs 6 and 7 comprise a set that indicates the rhythm or wave action of the experience. In CT 6, the reporters resist possibilities or change (contraction); in CT 7, they surrender control (expansion), which involves risk (CT 11).

The "New Metaphysics": Bohmian Holomovement and Eastern Holism

This investigation of the experience of being carried along by a series or flow of unforeseen circumstances or events culminating in a right and desired outcome produced findings that are consistent with what both age-old Eastern philosophy and contemporary postclassic physics tell us about the way the world may indeed be operating. According to the theoretical physicist David Bohm (1980), the phenomenal world that we observe in our ordinary states of consciousness represents only one partial aspect of reality—what he calls the *explicate* or *unfolded* order. Its generative matrix—the *implicate* or *enfolded* order—exists on another, subsuming level of reality and cannot be directly observed. In Bohm's theory, the term "implicate" refers to an order of undivided wholeness wherein many elements are holistically compressed or enfolded together. They are implicate; that is, they have not yet become apparent, defined, or explicit as separate elements in reflective awareness.

In CT 4 (Action Taken Toward Desire, Need, or Goal), CT 5 (Change Occurs), CT 7 (Surrender of Control), and CT 15 (Circumstances or Events leading to a Right and Desired Outcome), there may be a relationship between the experience of being carried along by a series or flow of unforeseen circumstances or events culminating in a right and desired outcome, on one hand, and David Bohm's concept of what he calls the "holomovement," on the other. Bohm embeds this holomovement in what he calls a *super-implicate* order, a higher order that serves to organize the implicate order, which in turn is responsible for the organization of all relatively stable forms with complex structures on the explicate level of time, space, matter, and energy. In many ways, the experience in question may be the epitome of how we as human beings become aware of this holomovement as it unfolds in our conscious experience.

Bohm's view of such an unfolding holomovement is related to a number of other perspectives, including Bergson's (1975) writing on duration and Whitehead's (1925) philosophy of process. However, it is Eastern philosophical perspectives on the nature and workings of the universe that bear the closest resemblance to Bohm's holomovement, the worldview of contemporary physics, and our earlier discussions of meaning and synchronicity.

Aspects of the various Eastern worldviews, as characterized by the Vedic literature, for example, are intrinsically dynamic, and contain time and change as essential features. The cosmos is seen as one inseparable reality—forever in motion, alive, organic, spiritual and material at the same time. Motion and change are seen as essential properties of things. According to the Indian philosopher Coomaraswamy (1977):

> We are deceived if we allow ourselves to believe that there is ever a pause in the flow of being, a resting place where positive existence is attained for even the briefest duration of time. It is only by shutting our eyes to the succession of events that we come to speak of things rather than processes.

Quantum physics seems to be in agreement with this view. In the words of Werner Heisenberg (1962), a pioneer of subatomic physics:

> This agreement (in concepts) between modern physics and the deeper understandings of the mystics demonstrate that in the real world surrounding us, it is not the geometric forms, but the dynamic laws concerning movement, or coming into being and passing away, which are permanent. . . . The world appears as a complicated tissue of events, in which connections of different kinds alternate or overlap or combine and thereby determine the texture of the whole.

This is echoed by Tibetan Buddhist Lama Govinda (1973, p. 93) who writes:

> The external world and his inner world are for [the Buddhist] only two sides of the same fabric, in which the threads of all forces and of all events, of all forms of consciousness and their objects, are woven into an inseparable net of endless, mutually conditioned relations.

CONCLUSION

This study has attempted to elucidate the meaning of one kind of experience and reveal its essence. This experience—being carried along by a series or flow of unforeseen circumstances or events culminating in a right and desired outcome—is idiosyncratic, complex, and difficult to objectify and analyze. It appears to be nonordinary or paranormal in nature and may possess transcendent or spiritual characteristics, satisfying many of Maslow's (1962) criteria for both peak experience and self-actualization. It has transpersonal implications for understanding the true nature of human experience and the lived world in a nonpathological, even optimum-performance or normal-reality-transcending, sense. Studying the experience has led us to a new concept "synchronistic teleological flow," suggesting a new field of study we might call "teleological synchronicity." It is a real test of any phenomenological investigative process to attempt to study such an experience and reflect on the varieties of meanings revealed. We stress the word *attempt* here, because even the kind of explorations and speculations offered above are just a beginning in our attempts to understand the far-reaching implications of our findings.

We welcome you to join us in reflecting further on this process. It is noteworthy that fully half of the first hundred people approached to take part in this study felt that they had had such an experience, leading us to reasonably suspect that this kind of experience is more widespread than some might think. This, in turn, suggests that the universe, as we human beings are capable of experiencing it, does have some transcendental, purposeful qualities that are not only real, but identifiable. We may be face-to-face here with empirical, experiential evidence, not just for the ego-based, material world as we normally perceive it, but for a world that is basically holistic, even spiritual, in nature.

REFERENCES

Bergson, H. (1975). *Creative evolution.* Westport, CT: Greenwood.

Bohm, D. (1980). *Wholeness and the implicate order.* London: Routledge & Kegan Paul.

Bohm, D., & Welwood, J. (1980). Issues in physics, psychology, and metaphysics: A conversation. *Journal of Transpersonal Psychology, 12*(1), 25–36.

Bolen, J. S. (1979). *The tao of psychology: Synchronicity and the self.* New York: Harper & Row.

Broad, C. D. (1953). *Religion, philosophy, and physical research.* New York: Harcourt, Brace.

Bucke, R. M. (1969). *Cosmic consciousness.* New York: E. P. Dutton.

Bugental, J. (1981). *The search for authority.* New York: Irvington Publishers.
Clarke, J. (1992). *In search of Jung.* New York: Routledge, Chapman, and Hall.
Colaizzi, P. F. (1978). Psychological research as the phenomenologist views it. In R. S. Valle & M. King (Eds.), *Existential–phenomenological alternatives in psychology* (pp. 48–71). New York: Oxford University Press.
Coomaraswamy, A. K. (1977). *Coomaraswamy. 2: Selected papers.* R. Lipsey (Ed.), Bollingen Series. Princeton, NJ: Princeton University Press.
Csikszentmihalyi, M. (1990). *Flow: The psychology of optimal experience.* New York: Harper & Row.
de Chardin, T. (1961). *Phenomenon of man.* New York: Harper Torchbooks.
Frankl, V. (1963). *Man's search for meaning.* New York: Simon & Schuster.
Giorgi, A. (1975). Convergence and divergence of qualitative and quantitative methods in psychology. In A. Giorgi, C. Fischer, & E. Murray (Eds.), *Duquesne studies in phenomenological psychology: Volume II* (pp. 72–79). Pittsburgh, PA: Duquesne University Press.
Govinda, L. (1973). *Foundations of Tibetan mysticism.* New York: Samuel Weiser.
Heidegger, M. (1962). *Being and time.* New York: Harper & Row.
Heisenberg, W. (1962). *Physics and philosophy.* New York: Harper & Row.
Husserl, E. (1970). *The idea of phenomenology.* The Hague: Martinus Nijhoff.
Huxley, A. (1970). *The perennial philosophy.* New York: Harper & Row.
James, W. (1961). *The varieties of religious experience.* New York: Collier.
Jung, C. G. (1962). *The structure and dynamics of the psyche: Collected works: Volume 8.* Princeton, NJ: Princeton University Press.
Jung, C. G. (1973). *Synchronicity: An acausal connecting principle.* Princeton, NJ: Princeton University Press.
Jung, C. G. (1976). *The portable Jung.* J. Campbell (Ed.). New York: Penguin Books.
Maslow, A. (1962). *Toward a psychology of being.* New York: Van Nostrand Reinhold.
Polkinghorne, D. E. (1989). Phenomenological research methods. In R. S. Valle & S. Halling (Eds.), *Existential–phenomenological perspectives in psychology: Exploring the breadth of human experience* (pp. 41–60). New York: Plenum Press.
Progoff, I. (1973). *Jung, synchronicity, and human destiny.* New York: Dell Publishers.
Riceour, P. (1970). *Freud and philosophy: An essay on interpretation.* New Haven, CT: Yale University Press.
Stace. W. T. (1960). *Mysticism and philosophy.* London: Macmillan.
Underhill, E. (1955). *Mysticism: A study in the nature and development of man's spiritual consciousness.* New York: Meridian Books.
Valle R. S., & Halling, S. (Eds.). (1989). *Existential–phenomenological perspectives in psychology: Exploring the breadth of human experience.* New York: Plenum Press.
van Kaam, A. (1959). Phenomenal analysis: Exemplified by a study of the experience of "really feeling understood." *Journal of Individual Psychology, 15*(1), 66–72.
Whitehead, A. N. (1925). *Science and the modern world.* New York: Macmillan.
Wilson, C. (1972). *New pathways in psychology: Maslow and the post-Freudian revolution.* New York: New American Library.

14

The Experience of Being Silent

Ourania Marcandonatou (Elite)

> *I am a Voice speaking softly.*
> *I exist from the first.*
> *I dwell within the Silence . . .*
> *And it is the hidden Voice that dwells within me,*
> *Within the incomprehensible, immeasurable Thought,*
> *Within the immeasurable Silence.*
>
> —Trimorphic Protennoia

SILENCE IN PSYCHOLOGY, PHILOSOPHY, AND SPIRITUAL TRADITIONS

Silence has captivated the human psyche from time immemorial. In the fables of the Golden Age, it is said that primordial humankind understood the language of all animals, trees, rocks, and natural elements. This first language, it is believed, came from the fullness of silence and communion with the Divine, in which the Divine and the individual existed as one. The biblical "fall" refers to the awareness of the separation between the individual and the Divine. It was precisely because of this standing out of individuals from the primordial wholeness and oneness of all things in the Divine that this union was broken and the dual nature of reality emerged. In turn, the self or ego was born and the incessant struggle for reconnection or union with the Divine was initiated.

The ego has played a pivotal part in understanding the workings of one's inner life. In psychotherapy, the search for truth takes place in a supportive environment where being silent and listening is widely used by both the therapist and the client. In that environment, being silent helps bring into the light aspects of the individual's true self. Consequently, being silent in psychotherapy often leads one to understanding and accepting oneself.

Ourania Marcandonatou (Elite) • School of Liberal Arts, John F. Kennedy University, Orinda, California 94563.

Phenomenological Inquiry in Psychology: Existential and Transpersonal Dimensions, edited by Ron Valle. Plenum Press, New York, 1998.

Transpersonal psychology, a more recent development in the field of psychology, views the individual as a whole: body, mind, and spirit. It goes beyond the limits of the ego-self toward the transcendent/spiritual Self. Anthony Sutich (1969), a cofounder of this new psychology, describes transpersonal psychology in terms mostly concerned with individual and species-wide meta-needs, values, transcendental phenomena, transcendence of the self, mystical and peak experiences, ultimate meaning, cosmic awareness, unitive consciousness, oneness, spirit, ecstasy, bliss, playfulness, maximal sensory awareness, and related concepts, experiences, and activities.

In particular, Abraham Maslow studies different aspects of the human being (e.g., creativity, values, need hierarchy, deficiency versus being motivation). He is best known for his work on self-actualization. Maslow (1971) defined self-actualization as "experiencing fully, vividly, selflessly, with full concentration and total absorption" (p. 45), while he termed peak experiences as "the best moments of the human being . . . the happiest moments of life . . . experiences of ecstasy, rapture, bliss . . . the greatest joy" (p. 105). He went on to say that perception in the peak experiences can be relatively ego-transcending, self-forgetful, egoless, detached, desireless, and unselfish.

Inherent in true silence is a state of consciousness that includes the aforementioned transpersonal aspects of Self. This state is attained through the control of one's own body, mind, heart, and speech. Silence, as an aid to worship and/or a method of understanding the working of one's mind, has been practiced among larger or smaller groups of people throughout history and in almost all parts of the world. In the world of philosophy, Pythagoras and Socrates placed great importance on silence as a means of initiation and as a way of acquiring inner knowledge (Hastings, 1924).

In more recent times, Max Picard (1948), a renowned Swiss philosopher, examined and studied in detail the positive phenomenon of silence. He believes that silence is profound because it simply *is* and thereby belongs to the basic structure of being human. For him, silence is the origin of all speech. In his classic and poetic book, *The World of Silence,* Picard (1948) wrote that there is no beginning or end to silence because it contains everything within itself. He also believes that silence points to the state of the Divine and that this is why one finds most practitioners of silence in monastic settings.

In the Eastern Orthodox Christian tradition, there are the *hesychasts,* individuals who have attained inner stillness, tranquility, repose, concentration, and/or silence (Colliander, 1982). In the Western Christian tradition, there are the Carmelites and the Trappists. In the Buddhist tradition, one finds monastics and lay people alike practicing meditation for extended periods of time. In the Native American tradition, one finds shamans as well as lay people undertaking the sacred experience of vision questing in remote mountains and deserts. The vision quest is a rite of passage in which for four days and nights questers remain in a solitary spot, fasting and being inwardly centered and silent in order to have a vision or a sign that will trigger insight (Foster & Little, 1987).

The most important aspect of silence is that of *listening.* It is curious to note that the word *silent* is an anagram of the word *listen,* so that one cannot be present without the other. They both bring one to the absolute centeredness of one's being. Brother David Steindl-Rast is a Benedictine monk and a pioneer in the dialogue that is taking place between the East and the West. Steindl-Rast (1984) talks about *silence as being an attitude of listening*—a gift that each one of us is invited to give the other. This is a special kind of listening; it is a listening with one's heart, one's whole being. By listening in this way, one opens to the mystery and meaning of personal existence.

One finds that being silent for extended periods of time is deeply and intimately connected to the *heart.* Most spiritual traditions point the way to the heart, where knowledge, meaning, and ultimate transcendence emerge. It is considered to be the ground for the Divine as well as the ground of one's being. Sogyal Rinpoche (1992), in his book *The Tibetan Book of Living and Dying,* writes that the mind cannot realize absolute truth. It is only through the heart and beyond the ordinary mind that one can attain absolute truth. In the Buddhist tradition, compassion/loving-kindness and wisdom are the cornerstones of an enlightened being. Compassion is manifested in the Bodhisattva vow, which one takes in order to gain enlightenment not just for oneself, but for the sake of all sentient beings (Goleman, 1988).

In the same vein, the Eastern Orthodox Christians talk about the *nous,* which is the spiritual mind—where the mind gets baptized within the heart in order to manifest the innermost aspect of the heart. Communion with the *nous* leads one to enlightenment (*photismos*) or deification (*theosis*) (Palmer, Sherrard, & Kallistos, 1984).

Sri Ramana Maharshi, a renowned yogi in the Hindu tradition, believed that the Self resides within the heart. Symbolically, the "heart" is the same as God, Self, existence, the Great Mystery, bliss, the eternal, and the seat of consciousness. It is not only the very core of one's being but also its spiritual center. It is the consciousness of consciousness, the thinking of thinking, and the feeling of feeling (Ramana Maharshi, 1978).

In much the same way, being silent during a vision quest creates a certain monotony for the ordinary mind, which may lead to boredom and then sleep. This allows the real Self, the Self directed by the heart, to emerge and speak. At this point, reality becomes pure and the vision becomes manifest (Brown, 1988). The Native American tradition is considered to be an earth tradition. It is interesting to note that the anagram of the word *earth* is *heart.*

These particular spiritual traditions became venues for my own search for the ground of being or the Divine. My interest and passion for silence grew from my early experiences of growing up in East Africa, where silence and nature were an integral part of my existence. I remember reveling in silence and playing in the abundance of the natural beauty that surrounded me. At a very young age, I began paying attention to minute occurrences of my psyche's constant fluctuations among stillness, peace, and loss. I learned that nature was my container and healer and that I could retreat into the quiet natural beauty of it all and be safe and peaceful. I found that by being silent I was connected and in communion not only with the Divine but also with the centermost part of my being. At an early age, I experienced that they were both one and the same.

In my adult years, I have continued exploring this relationship with my inner self through vision quests, spending time in solitude in various monastic settings and retreat centers, as well as observing silence within my own home. Since my time in silence has been so personally profound, I wanted to inquire further into this phenomenon to see whether others did, indeed, have similar experiences.

RESEARCH TOPIC AND METHOD

The actual research topic was "on the experience of being voluntarily silent for a period of four or more days." I chose the existential–phenomenological methodology because this type of approach seeks to understand a phenomenon in its pure essence, prior to any reflective interpretation, scientific or other. It has to do with the world of direct and immediate everyday experience as it is expressed through one's language. Husserl (1962) describes the "life-world" (*Lebenswelt*), wherein the individual and the world are always in a dialogue with each other, where one has no meaning without consideration of the other. It is said that each individual and his or her world co-constitute one another. In this sense, the research subject is seen as a co-researcher because both the researcher and the subject, as a unit, co-constitute the emerging meaning of the experience.

My research was an adaptation of Colaizzi's (1978) method, utilizing a 12-step procedure. I interviewed 12 co-researchers, 9 women and 3 men. They all came from different spiritual backgrounds and had experienced being silent either in a monastery, a retreat center, a home, or in nature. They were all instructed to write a description of their experience in the following way: (1) to recall a specific time when they were silent for a period of four or more days; (2) to describe how they felt during that period; (3) to try to describe their feelings just as they were, so that someone reading or hearing the report would know exactly what that experience was like for them; (4) to keep their focus on the experience itself, and not just the situation; and (5) to not stop until they felt that they had described their feelings completely, taking as long as they needed to complete their description.

After reading their written descriptions, or protocols, to acquire a beginning understanding of their experience, I proceeded with the "walk-through" interview. This face-to-face interview was to make explicit and examine more deeply the descriptions that the co-researchers had written. The intention of the interview was to "walk them through" their written description by reading it back to them in a slow and relaxed manner. There was a pause at the end of each statement, so that the co-researchers had a chance to dwell and reflect on their experience and describe it in greater detail if they wished. In this way, I feel that the descriptive material was pure in the sense that it was unprompted by any questions that might have biased the co-researchers' descriptions. The taped interviews were then transcribed.

I proceeded with the analysis in the following manner: All 12 protocols were read carefully for content as well as to acquire a feeling sense for the material presented. Phrases or sentences that directly pertained to the investigated phenomenon were underlined and extracted from each protocol. These were collected and logged as *extracted significant statements.* The ones with the same meaning were then integrated and translated into a clearer *meaning statement,* in which I translated the co-researcher's words in a way that remained true to the underlying essence of the experience itself without severing any connection with the original protocol. Colaizzi (1978) describes this procedure as a "precarious leap" that the phenomenologist makes in order to bring an interpretive–psychological meaning to the extracted statements.

For each protocol, these meaning statements were first listed and then sorted as *theme clusters.* The theme clusters from all 12 protocols were then combined to form 39 *constituent themes.* From the 39 constituent themes, the ones that were found to contain the same meaning were then combined to form 17 *comprehensive constituent themes.* Upon further reduction, nine *final comprehensive constituent themes* emerged, reflecting the pre-reflective structure of the experience itself. A list of the final themes was given to each co-researcher for *feedback* and *verification.*

RESULTS

The essential structural definition or fundamental essence of the research is portrayed in the following nine final comprehensive constituent themes:

1. Experiencing the essence of one's being.
2. Experiencing one's inner life with a heightened sense of awareness.
3. Experiencing more acutely through the senses.
4. Experiencing auditory, visual, perceptual, and/or other sensory alterations.
5. Feeling connected and/or unified with various aspects of existence.
6. Feeling intensely a wide range of feelings and emotions.
7. Feeling rejuvenated.
8. Perceiving a change in the ontological meaning and/or significance of ideas and the nature of personal reality.
9. Perceiving the experience as ineffable.

The 12 co-researchers *experienced, felt,* and *perceived* themselves with intensity and a deep sense of awareness. They also experienced, felt, and perceived the multifaceted aspects of their reality in a new way. They all reported that they were transformed on levels never before realized. Interestingly enough, these nine themes parallel Valle's own personal reflections on transcendent awareness presented in Chapter 12 in this volume. Valle suggests that there are six qualities or characteristics of transpersonal awareness: (1) deep stillness and peace; (2) an all-pervasive aura or feeling of love; (3) a sense of egolessness; (4) a transformed sense of space; (5) a sense of timelessness; and (6) bursts or flashes of insight. These characteristics seem quite harmonious with six of the nine final comprehensive constituent themes of this research (numbers 1, 2, 4, 5, 6, and 8).

These nine themes, although relatively distinct in their categories of experiencing, feeling, and perceiving, nonetheless represent the essence of the experience of being silent for these co-researchers. Consistent with the phenomenological perspective, this essence is reflected in the holistic fashion in which these nine comprehensive constituent themes interrelate/interweave.

Final Comprehensive Constituent Themes

1. All 12 co-researchers *experienced the essence of their being* in an extraordinarily conscious and profound way. They reported rediscovering who they really were, where the essence of love was experienced deeply within themselves. For some, the experience was found to be equivalent to that of being "home" where they felt not only self-embraced, but also part of the Universe. During their time in silence, different qualities or states emerged, such as freedom, voidness, egolessness, nonjudgment, inner power, creativity, purity, clarity, acceptance, control, and integration. The physical body was experienced on several different levels. It was predominantly reported as being light—light exuding from within the body as well as surrounding it. It was also experienced as energy that was felt in a most intense way.

One co-researcher's own words clearly describe a few of these experiences. For example:

> Having much less of a sense of effort or striving in any way—just more of a sense of being; of being able to be by myself and exist and not have to do anything, certainly not having to talk. . . . I'm more relaxed and more attentive to what people are saying. . . . I was more into that internal space in general. . . . [It felt] like an increased awareness of everything around me—interactions, my own energy, how I am, how I operate, how I think. . . . Everything became more clear to me and I became more conscious of it. . . . My body felt a lot more full of light.

In a sense, they began the process of relying on and paying attention to their own inner guidance rather than to external feedback alone.

2. They all *experienced their inner life with a heightened sense of awareness.* They reported facing and observing themselves with a clear insight into the workings of their mind. They related experiencing both a decrease and an increase in thought processes, which in many instances were intense and sometimes even unbearable. This process was accompanied by a variety of powerful emotional states, such as frustration and anger.

Eight co-researchers reported having remembrances of their childhood that were accompanied by strong feelings and emotions, positive and negative. Some of them had satisfying and warm nonverbal communication with other individuals whom they encountered during their silent period. They also felt nourished and healed.

Co-researchers stated that being silent brought about a strong sense of clarity for them. They discovered that noise and sound were part of silence. For example:

> [Being silent] is a process. . . . Communication begins then with the self. . . . In silence [there is] a lot of revealing of core beliefs. . . . I found that the inner life becomes very lucid and thoughts become very clear, and images become very clear, and imagination, and memory. . . . There is the energy and the time to review the structure of personality and makeup. . . . There is a buildup of energy when not speaking. Energy being put back into the inner life, becoming clearer. . . . Noise is a part of the world, it is not separate from silence.

They reported becoming involved with both the internal and external world in a deeper and more intense way, appreciating things for what they truly were, rather than for what they appeared to be. Ultimately, they understood the way they acted or reacted in the world.

Being silent was also experienced as a process in which the present was seen as playing a vital role in one's life. By focusing attention on the present moment, seven of the co-researchers discovered a plethora of inner messages that were otherwise ignored during their everyday lives. By remaining silent, they experienced energy as being conserved and recycled in their inner life, thereby heightening their sense of awareness. One response was:

> My whole childhood starts to come in [when I'm silent]. . . . There was this incredibly predominant sense of a tremendous amount of energy saved, on a moment-to-moment basis, from not having to talk and not having to figure things out, not having to say the words. . . . I could be more directly in the experience and less in the words, the mental, whatever distractions were in the way. Even opinions and thoughts were much less there. I could just be observing. And I could observe my own feelings and my feelings interacting with people with a heightened awareness—with much more awareness than when I'm speaking. . . . All the pressures and pulls and pushes on how I am in the world became very clear to me and I became more conscious of it.

As a whole, their silence opened up new insights by which they could better understand themselves, others, and the world around them.

3. All 12 co-researchers *experienced more acutely through the senses.* Not only did their senses become more acute, vibrant, and heightened, but they also experienced themselves and their environment in a more intense way. Colors, sounds, textures, and motion became extremely vivid, rich, and brilliant. Hearing, seeing, touching, smelling, and tasting became sharpened. For example:

> Colors were rich and clear, music was exquisite, my cats felt so soft . . . very sensual. . . . I felt good in my body, enjoyed feeling the air, smells and sun of spring days. . . . My perceptions were much sharper than usual.

For some co-researchers, their physical senses and instincts became so acute that they developed an alertness that went beyond ordinary perception. They reported feeling and sensing things around and within the areas of the heart and the solar plexus. Their bodies felt stronger and more attuned to the natural environment, not unlike the behavior of animals in the wild.

Interestingly enough, a few considered their mind as being part of their sense experience. They reported that they experienced mental activities in an intense way, whereby thoughts took on a profound meaning. One co-researcher in particular felt a pull on the brain or the body when she needed to verbally communicate with others. In general, all of them reported that their physical senses provided a new and rich avenue for knowing themselves and their environment in a nonordinary way. One response was:

> My senses were heightened to an awareness level that offered me a world full of sensory details I had never experienced before. . . . My fingertips became a conduit for the current which coursed its way to the sea and through my veins. . . . It [water] felt like liquid gold going into my system.

4. Of the 12 co-researchers, 11 *experienced auditory, visual, perceptual, and/or other sensory alterations.* These experiences fall into the general category of altered states of consciousness in which there was a quantitative, as well as a qualitative, shift in the co-researchers' pattern of mental functioning. These states of consciousness are not a common occurrence in everyday living.

The co-researchers reported experiencing an expansion of personal boundaries, with no distinction between inner and outer reality; they became fused into a single and inseparable reality (e.g., the dream state and the waking state were experienced as one). For others, the state of expanded boundaries took the form of timelessness, where past, present, and future were one and the same.

Six co-researchers reported experiencing an awareness of the cells and molecular structure of their body to the extent that they became one with the atmosphere and the environment around them. For example:

> It's as if there becomes an opening between my cells or my molecules that let my physical being intermingle with the atmosphere, so that the wind passes through me instead of bouncing off me.

Co-researchers also described seeing light surrounding them. They reported seeing energy patterns in nature revealing the inner life of plants and trees. They also reported that they had clairvoyant experiences of events that actually occurred at a later time. There were instances of visual and auditory hallucinations. One co-researcher in particular reported seeing a man's head rise from the river, while others had visions of guides. Six of the 12 co-researchers heard music and/or voices during their silent retreat. Independently, two of them heard two different types of music reverberating from mountains around them. Another reported hearing harmonic sounds emanating from an audiotaped lecture. One description was:

> I began to have experiences—unassisted by any drugs—of some sort of supernormal auditory phenomena. In the evening from one range of hills I began to consistently hear polyphonic "chanting." I would also occasionally, at any point during the daytime, become aware of a single bass note. . . . One time I very clearly heard what seemed to be an AM radio station, although there were no radios within at least a couple of miles.

Finally, one co-researcher had an unexpected out-of-body experience, during which she lay suspended and cradled among the stars. The instant that she became fearful, she found herself back into her body. It was an experience that she claimed will forever be imprinted in her memory.

5. *Feeling connected and/or unified with various aspects of existence* is another state that is considered to be an altered state of consciousness. All 12 co-researchers felt "at-onement" and unified with everything. Connectedness was the key to the feeling of oneness that they characterized as a mystical experience They described feeling connected with either animals, nature, people, the Universe, a Greater Force, Spirit, and/or God. They felt being an integral part of the universe and experiencing a sense of belonging where the ego was felt to be dissolved. Some of them experienced being either in God or part of God. For example:

> The only way, I think, that I can put [it] into words is that I feel that I am part of God's creation and that everything I am, and everything that I experience, and everything that surrounds me, and everything that I can imagine is God. We are all of the same—all connected.

Another response was:

> That was probably the beginning when I began to realize that we are one . . . I was tapping into an innate intelligence, plugged into a source beyond my comprehension. . . . I felt that it [my physical being] blended into everything.

6. The silent experience, in general, allowed the co-researchers to *feel intensely a wide range of feelings and emotions.* This particular theme was found to be an integral part of each of the other eight final comprehensive constituent themes. Feelings and emotions were manifested throughout their experience of silence. Although the negative feelings and emotions covered a broader range than the positive ones, more co-researchers felt positive during their silence than felt negative. The positive feelings reported, in order of most to least frequent occurrence, were: love, happiness, peace, calmness, joy, gratefulness, serenity, "wonderfulness," elation, grace, thankfulness, bliss, security, euphoria, contentment, groundedness, and feeling blessed. For example:

> Part of the experience was that after a few days I started to feel extremely happy—happier than usual. I'm usually a pretty happy person but this state of not speaking made me feel very loving and very happy and quite contented with life and everything that was going on around me.

The negative feelings and emotions reported, in order of most to least frequent occurrence, were: fear, anger, loneliness, emotional pain, worry, frustration, anxiety, despair, depression, vulnerability, physical pain, "churning" of emotions, restlessness, sadness, timidity, paranoia, loss, startle, nervousness, rejection, aggravation, stupor, embarrassment, suffering, feeling inconsequential, inadequate, inferior, and wrestling with the "demons" within. Five co-researchers who found that the silent experience was difficult were meditators who did their silent retreat either in a monastery, in a retreat center, or at home. Following is one co-researcher's response:

> In these four days of silence I felt a myriad of feelings. . . . I also remember struggling with challenging mind states of inferiority, of despair, of sadness, of longing for something that I just can't seem to find. . . . I remember feeling the grace of all the wisdom traditions that have come before me and that will come after me. I remember feeling a speck of dust in a universe of infinite breadth, unseen to normal seeing. I remember feeling blessed and showered with this grace.

7. The silent experience also brought a distinct *feeling of rejuvenation* that was reported by 11 co-researchers. Some of them related seeing the world "in a new light." Others reported feeling a sense of rediscovery, while others felt as if they were being born again. The experience was expressed in terms of strong positive feelings from which they felt a new surge of

energy and a positive outlook toward life. This theme was inferred from a variety of different descriptions rather than emerging directly from specific statements expressing "rejuvenation." Two co-researchers described the feeling of rejuvenation in the following manner:

> I wrote a note to a friend saying, "I just gave birth to myself." There was very much a sense of rebirth that came out of the birthing, that came out of the struggle of those days [of silence].

> You often hear that saying, "This is the first day of the rest of your life." It's kind of like that. . . . Everything is as if you are seeing it for the first time.

8. The co-researchers not only saw the world or themselves in a new way or in a new light, but also *perceived a change in the ontological meaning and/or significance of ideas and the nature of personal reality.* In other words, 11 co-researchers perceived a radical shift in consciousness, acquiring meaningful and clear insights into their personal reality. This shift was perceived to be closely akin to the spiritual realm. One co-researcher had insights into the essence of God as "arid" and "desert-like, " while another reported perceiving God's essence to be her own. One described the experience as follows:

> The changes that silence brought were clearly of the spirit. I felt peace and harmony in the latter part of the silent practice, a strong sense of being right where I should be, of being on my path.

Another response was:

> During this time of solitude and silence, my spiritual life surrounded me. It wasn't a separate thing I had to reach for through ritual or prayer. It was simply my entire life. They were one and the same. . . . The silence helps to counteract the feeling that everything is random or everything is just chaotic. . . . [That] period of silence and solitude was a turning point almost ten years ago.

Some reported perceiving the state of silence as being a "space" or a place to which they could have direct access to themselves and the Divine. In that place, they all felt safe, harmonious, and whole. They reported perceiving that a personal transformation had taken place and that they became more peaceful and "open." One co-researcher described the perception as follows:

> It was just [that] the consciousness of how my attention is usually split between listening and preparing my next sentence was very obvious. Normally, this is going on all the time, but I'm unconscious of it. And suddenly there it was very clear. . . . [It was] a real shift in consciousness of the burden that [talking] puts on me . . . completely unconsciously all the time. And it gave me a different sense of myself in the world . . . that the universe and people provide what I need without having to make a tremendous effort to get that. . . . And if I'm not in control things seem to work out just fine without me doing that. And I think that my life changed after that.

In addition, three co-researchers reported perceiving the dual nature of reality. Some perceived talking or noise as highly distractive. They observed that words were "clothes" that cover one's ability to feel. Nonviolent speech—use of language devoid of violent and negative connotations—was considered by one co-researcher as being a desirable alternative source of communication.

One co-researcher reported perceiving a shift in the direction of her life. She subsequently decided to make the arts her vocation rather than to pursue a career in the sciences.

9. All 12 co-researchers perceived their silent experience to be *ineffable.* They were unable to find appropriate words to adequately describe this particular experience. In many instances, they were unable to complete sentences when trying to describe how they felt about it or how they perceived it. They all observed that they were unable to articulate their

silent experience. In its essence, it is perceived to be wordless. The following responses describe this aspect:

> On speaking of silence, I am struck with the difficulty at hand. How to speak of what feels unspeakable? . . . . How to speak of a world without words? How to put words to the sweetness of a moment in time when one feels the beauty and depth of being alive and aware of what is delicate and precious, of what is life? How to convey the subtle nuances of perception in movement, a footstep perhaps, or of thought, or of seeing into one's life, or the feeling of the presence of a tree? And how to do this so that you know this experience to exist for yourself also, in our world of talking and of noise?
>
> It's so hard to explain. . . . There was a connectedness that was accomplished, a communication that took place that wasn't verbal at all.

In short, I feel that the essence of ineffability is at the very core of the experience of silence. The word itself denotes or points to the fact that this experience goes beyond words or any descriptive efforts that might be undertaken by any individual. From these results, I now see the essential structure of this experience to be a mere reflection of what the experience of silence truly is—a mystery. The spiritual writings of the ages repeatedly describe its essence as masked by paradox and riddles. Ironically, only silence itself can best describe the silent phenomenon. It can be described as the sacred silence—an ineffable experience indeed.

TRANSPERSONAL ASPECTS

The research findings invite us to look more closely at concepts such as a higher self, personal transformation, union, transcendence, mystical states, altered states, and love, to name a few. They all represent transpersonal dimensions of existence wherein the co-researchers moved from their personal ego-self perspective toward aspects of experience that can appropriately be named *trans-egoic* (i.e., transpersonal).

Many of the co-researchers' experiences included "intense realness, unusual sensations, unity, ineffability and trans-sensate states" (Deikman, 1990, p. 47). According to Deikman, these experiences are considered to be mystical and transpersonal in nature. In this context, the silent phenomenon can be characterized as being a transpersonal, transcendent, and mystical experience.

Valle (1989, p. 261) offers five characteristics of the transpersonal realm: (1) there is a transcendent unity of all aspects of existence; (2) the ego is not the ground of human awareness; (3) each one of us can directly experience this realm; (4) there is an expansion of one's self-identity beyond the ordinary mind and its conceptual thinking; and (5) this experience is self-validating.

All 12 co-researchers not only experienced all of these characteristics, but also came out of their silent experience transformed, considering it to be a peak experience. Being silent involves the whole human being, a state in which one experiences, feels, and perceives oneself, others, and the cosmos in a nonordinary and/or nondualistic way. Insight and awareness increase to the point where new meanings are recognized along with a deeper understanding of how one acts in the world.

On the basis of the results, it can be said that the silent experience not only is transpersonal, transcendent, and mystical in nature, but also has the potential of leading one to a place of self-discovery and transformation where self-understanding and the essence of one's being become manifest. The co-researchers experienced the essence of their being as light, loving, pure, free, egoless, and integrated. They experienced a profound union and connection with the whole of creation. They all had a definite sense of a place, a "coming home" of sorts, a sacred space where they felt safe and trusting. They also became highly attentive to the world around them with a strong, vibrant, and acute sensorial awareness. At the same time, they experienced intense feelings and emotions to a point where they felt a sense of rejuvenation, seeing their lives in a different or new light.

What became explicit from their protocols was that the experience of love was found to be interwoven throughout all of the nine themes. Being silent seems to be a state in which one's heart expands to such dimensions that one feels interconnected, part of the whole—in which there is no longer any alienation or disconnectedness between oneself, others, and/or the Universe. Through silence, the ego transcends itself to come into communion with a transpersonal/transcendent reality, thus healing the biblical "fall" and suggesting a reuniting of the individual with the Divine.

The results further suggest that by being silent, one is brought directly in contact with the *here and now*. When one is in the here and now, the ever-present, one finds oneself nowhere (now+here) because all time (past, present, and future) is contained in the now. This is the point of timelessness, the still point, the void. This is the point at which fullness and emptiness become one. It is the sacred silence. In other words, the silent experience has the potential of igniting each individual's inner sense or spark of the timeless dimension of the Divine, which is waiting to be rediscovered, recognized, awakened, and ultimately illumined from within.

On a more personal level, engaging this type of phenomenological research brought about a shift in my own consciousness. By immersing myself in the protocols, I became more acutely aware of my own deep and internal experiences with silence. I discovered that one of the three elements in phenomenology, namely, that of the essence of the phenomenon, is none other than the transpersonal–spiritual dimension of reality. I believe, like Valle (1989), that existential–phenomenologists fall short of treating consciousness in its purest form—the formless, transcendent–spiritual aspect of reality, which is none other than our true Self, that which remains constant and unchanged within as we journey through our lives.

Ultimately, I began to realize that what was happening was that my heart was beginning to expand in such a way that it embraced my own personal self. I have observed that, as a whole, we tend to put ourselves somewhere outside the center of our own hearts. We are, rather, outer-directed, being more loving, compassionate, or kind toward others than we are to ourselves. Rarely do we tend to ourselves in this way. By including ourselves in this journey of loving and ultimate healing, we acquire the essence of unconditional love, which I believe is the cornerstone of the transpersonal realm.

I feel that we as research psychologists tend to be intimidated from speaking or writing about love. We speak of many different states of being (e.g., transcendent, transpersonal, mystical, peaceful), but rarely about love. Love is the ground, or better the container, for all those states. Without a sense of a loving heart, these states become empty vessels that do not correspond to or reflect the total/whole human experience. The co-researchers themselves experienced love in a most profound way. Several of their responses were:

> Love appears to be in all things, embedded deeply within every atom, every living cell of all living things, including myself, my cells . . . within all solid things, even rocks, and asphalt, and porcelain.

> . . . When the sense of an important self, of a demanding one, when that settles down and becomes quiet, there is a sense of grace. And in that sense of grace there is love.
>
> During some of these times, I felt that something was still being communicated on a very deep level, something beautiful and profound—a spiritual love.
>
> My experience of being silent is like a mother's love—an unqualified acceptance.

As I pointed out in the beginning of this chapter, most of the world's spiritual traditions consider the heart to be the seat of consciousness/Divine/Self/lover–beloved. The Divine is believed to be found neither in heaven nor on earth but in the heart. The heart is the innermost silent center of our being, where the sacred union of opposites (e.g., earth/heaven, life/death, body/spirit, feminine/masculine) takes place, thereby healing the proverbial split/separation/biblical "fall" that begins for each of us at our birth. Being silent facilitates the direct experience of this sacred union.

I feel that as researchers we are presented with the challenge of beginning to integrate our whole being into what we are doing and who we are vis-à-vis our work. I contend that this new and emerging area of transpersonal–phenomenological psychology can transform our educational process into a more meaningful, vibrant, and creative endeavor.

In phenomenological research, we are looking for the essence of any meaningful experience. By conducting research with a transpersonal eye, we not only find meaning and the essence of the subject of our investigation, but also emerge from it transformed. That is precisely what I wish to elucidate here, that transpersonal–phenomenological research in its very essence/core has the seed of transformation, for the researcher and the co-researcher alike.

REFERENCES

Brown, T. (1988). *The vision: The dramatic true story of one man's profound relationship with nature.* New York: Berkeley Publishing.

Colaizzi, P. F. (1978). Psychological research as the phenomenologist views it. In R. S. Valle & M. King (Eds.), *Existential–phenomenological alternatives for psychology* (pp. 48–71). New York: Oxford University Press.

Colliander, T. (1982). *Way of the ascetics: The ancient tradition of discipline and inner growth.* Crestwood, NY: St. Vladimir's Seminary Press.

Deikman, A. (1900). Deautomatization and the mystic experience. In C. Tart (Ed.), *Altered states of consciousness* (pp. 34–57). San Francisco: Harper & Row.

Foster, S., & Little, M. (1987). *The book of the vision quest: Personal transformation in the wilderness.* New York: Prentice Hall.

Goleman, D. (1988). *The meditative mind: The varieties of meditative experience.* Los Angeles: Jeremy P. Tarcher.

Hastings, J. (Ed.) (1924). *Encyclopaedia of religion and ethics: Volume XI.* New York: Scribner's.

Husserl, E. (1962). *Ideas: General introduction to pure phenomenology.* New York: Collier.

Maslow, A. (1971). *Motivation and personality.* New York: Harper & Row.

Palmer, G., Sherrard, P., & Kallistos, W. (Eds.) (1984). *The Philokalia: The complete text compiled by St. Nikodimos of the Holy Mountain and St. Makarios of Corinth: Volumes 1–3.* London: Faber & Faber.

Picard, M. (1948). *The world of silence.* Washington, DC: Regnery Gateway.

Ramana Maharshi (1978). *The teachings of Bhagavan Sri Ramana Maharshi in his own words* (A. Osborne, Ed.). New York: Samuel Weiser.

Rinpoche, S. (1992). *The Tibetan book of living and dying.* San Francisco: Harper.

Steindl-Rast, D. (1984). *Gratefulness, the heart of prayer: An approach to life in fullness.* New York: Paulist Press.

Sutich, A. (1969). Some considerations regarding transpersonal psychology. *Journal of Transpersonal Psychology.* Spring, 11–20.

Valle, R. S. (1989). The emergence of transpersonal psychology. In R. S. Valle & S. Halling (Eds.), *Existential–phenomenological perspectives in psychology: Exploring the breadth of human experience* (pp. 257–268). New York: Plenum Press.

15

On the Experience of Being Unconditionally Loved

Craig Matsu-Pissot

INTRODUCTION

Unconditional love. Rumi, the great Sufi poet-mystic, wrote of this love: "I've heard it said, there's a window that opens from one mind to another, / but if there's no wall, there's no need for fitting the window, or the latch" (Barks & Bly, 1981). Such beautiful words. Much of Rumi's work directly illustrates the loving relationship that he shared with his spiritual teacher, Shams of Tabriz. Indeed, spiritual traditions from all around the world proclaim unconditional love to be among the highest forms of human expression. These traditions have gone on to describe this expression in the form of the affection a spiritual teacher holds for students or disciples. Buddhist (Rinpoche, 1992; Wangyal, 1978) and Hindu (Muktananda, 1980; Venkatesananda, 1989) sources, the words of the Old and New Testament, tales from indigenous or shamanic cultures (Castaneda, 1968, 1972) all show us that the love given us by spiritual guides is of utmost importance. Experiencing a spiritual teacher's unconditional love can teach us how to love and accept ourselves. This self-acceptance and self-love can then find expression in love for our families, friends, colleagues, and even our enemies. These expressions of love help us to live a rich and meaningful life.

It was a combination of traditional spiritual teachings and personal experience that led me to conduct a phenomenological investigation into the experience of receiving unconditional love from a spiritual teacher. And it is this investigation that will be the focus of this chapter. First we will look at some of the existing relevant research and writings. This review will be followed by a brief description of my own experience of receiving unconditional love

Craig Matsu-Pissot • 15303 Northeast 166th Lane, Woodinville, Washington 98072.

Phenomenological Inquiry in Psychology: Existential and Transpersonal Dimensions, edited by Ron Valle. Plenum Press, New York, 1998.

from a spiritual teacher. The final section will concentrate upon the research project itself and will include a brief description of the methodology and a discussion of the results.

PAST RESEARCH

With such widespread recommendation from spiritual traditions, one might wonder why there is so little psychological research in the topic of unconditional love, especially regarding such love from a spiritual teacher. Considering the implications such research would have regarding relationships on every level, this seems to be something of an oversight. Though a small handful of studies address the topic in peripheral ways, only a single study directly examined unconditional love of any kind. Hawka (1985) conducted a heuristic study on the experience of feeling unconditionally loved. Not all the sources of love were identified, but those that were included a seminar leader, a group of colleagues, a coworker, and a "personal, private knowing of being loved unconditionally by God" (p. 76).

Hawka reported four major findings. Fundamental to unconditional love was a profound acceptance that included recognition, acknowledgment, and respect free of any demands or obligations. This acceptance relaxed barriers within the beloved to inner resources, sparked new awarenesses, and facilitated a more creative approach to life, all of which were part of a greater sense of freedom and openness. The experience was also described as revitalizing or empowering. These three factors led to the final primary experience—a movement from a negative to a positive self-image.

A second study relevant to the experience of unconditional love from a spiritual teacher was directed by Glick (1983). Glick, also employing a phenomenological method, explored the change experienced by disciples due to their relationship with a guru. Among the qualities he cited as being important in the makeup of a guru was unconditional love. Ten of the subjects explained that their guru's love was instrumental in helping them to grow as human beings. Glick went on to discuss his finding that the guru's unconditional love was the foremost component in facilitating positive change in his students.

The loving teacher–student relationship was investigated by Monnich (1985). Among her conclusions was that such a relationship offered support and encouragement to the student. This positive environment helped the student to reevaluate and overcome perceived personal liabilities. The student was instilled with a greater courage and expansion. Again, all of these factors combined to form a greater self-acceptance and self-confidence in the student.

Solimar (1987) has also contributed to our understanding of unconditional love with her phenomenological investigation into the experience of self-love. She first concluded that within the context of self-love, the self can be experienced in three ways: in the sensations, energy, and aliveness of the body; as mind, in ways that express self-confidence, self-esteem, purposeful interaction with life, and the ability to initiate life changes; and as spirit or essence, characterized by unity with a larger reality.

Her second finding was that self-love is important to a healthy, joyful life in which mental imbalances such as depression and neurosis give way to such positive mental states as joy, peace, and well-being.

Solimar's third conclusion was that self-love is a necessary factor in discovering and realizing one's true nature, inner potential, and creative expression through initiating a relationship with a deeper self.

The fourth finding was that self-love is closely related to the ability to love others. Participants felt a reduced inclination to be defensive or judgmental. They experienced more positive relationships with friends, family, and humanity. And they "experienced self-love as unconditional love for self and for others" (p. 122).

A handful of other studies also approach the question of unconditional love, albeit more indirectly. Motto and Stein (1973) found that receiving unconditional love is critical in allowing the expression of hostility and other emotionally charged feelings. The subjects found themselves in a safe and accepting environment. This accepting atmosphere allowed them to be more accepting of themselves, thereby becoming more willing to communicate in a more completely authentic way.

In researching Maslow's theories, Dietch (1978) found correlations between the essence of a B-love relationship and self-actualization (see also Bracket, 1975). Weinmann and Newcombe (1990) reported that the amount of parental love parallels a significant linear increase in an adolescent's sense of identity. These adolescents were also more likely to perceive and return their parents' love than were their peers who had not developed as tangible a sense of self.

Writers from existential psychology have also written about the nature of unconditional love. For example, although he did not allude to unconditional love directly, Martin Buber wrote extensively on the concept of genuine dialogue in his books *The Knowledge of Man* (1965) and *I and Thou* (1970). This form of conversation, which does not necessarily involve the exchange of words, is based upon a mutual regard. The participants in the dialogue are open to each other in a manner completely free of analysis or interpretation, demand or objectification, greed or anticipation. To perceive others in this way is basic to unconditional love. Each person, with genuine expression and perception, becomes aware of the other person in his or her wholeness and authenticity. It is a relationship of shared beingness, a dialogue of direct contact.

In his writings, Harper (1966) explored the similarities and differences in the love between God and the human being, on one hand, and the love that one human being holds for another or some other aspect of the material world. Harper relied upon a number of historical and literary figures as examples of these forms of love. Among the examples are St. John of the Cross, St. Teresa of Avila, Heloise and Abelard, and Cathy and Heathcliff of *Wuthering Heights*.

Two of Harper's conclusions are of particular interest here. The first point is that being touched by love, whether mystical or existential, can bring about intense confusion, disintegration, and emotional suffering. St. John of the Cross (Harper, 1966) described this type of response as the dark night of the soul, and wrote of this dark night as "the night of absence, loss, emptiness which succeeds the peak of ecstasy in the encounter with a living vision" (p. 118).

This absence, though painful, is part of a process that leads to a purified heart and mind. One experiences a change of values in which the trivial and unsubstantial can be more clearly distinguished from what has true value in this life. The heat from the fire of longing burns away the dross and the base, leaving only what is golden.

The second point Harper raises is that "the root of all evils is the frustration of love" (p. 169). In this light, he advises us not to repress existential love for the sake of mystical love. Such a repression leads only to a contracted state of being. These forms of love exist on a continuum or, better stated, mystical love is found within existential love.

John Welwood (1990) is another writer who integrates existential psychology with spiritual thought. It is his contention that unconditional love is always with us, but lies hidden under layers of conflicting thoughts and emotions. He writes that we reject within ourselves aspects of our experience that cause us pain and therefore live only partial lives. But there is a part of us that is aware of this partial living and seeks to find or rediscover wholeness. This faculty of awareness is our inherent unconditional nature.

To heal our segmented or compartmentalized selves, we must be open to what is, we must encounter it fully without relying on concepts or conditions that lead to evaluation or judgment. By opening unconditionally to all aspects of our lives, including those we find painful, we access that intrinsic unconditional nature. When we do so, unconditional love springs forth naturally to be offered to ourselves and to those around us.

PERSONAL REFLECTIONS

My own experience of unconditional love was very similar to findings in the collection of studies cited above. During my first quarter of graduate school, and continuing to the present, I have been fortunate to have formed relationships with two teachers, Dr. Rina Sircar, a Buddhist nun, and Dr. Angeles Arrien, an anthropologist who has the role of elder in her Basque community. More than eight years have passed since those initial contacts, but in the course of those years, much has unfolded due to their presence in my life. They certainly imparted to me many scholastic lessons within the formal context of my schooling. But the deepest impact came from their love and acceptance of me as the person that I am. And it was, and continues to be, an impact that favorably touches every aspect of my relationship with myself and the world in which I live. It was this gentle, subtle, and pervasive regard they always gave to me that inspired me to pursue an exploration into the experience of unconditional love in others.

In conducting an existential–phenomenological study of the experience of receiving unconditional love from a spiritual teacher—a study that, together with its results, is the main focus of the remainder of this chapter—I found my personal experiences to be both beneficial and challenging. Among the benefits were a familiarity with the subject and the inspiration to investigate it. The challenge was to not let this familiarity and inspiration cloud my reflection and interpretation of the meaning contained in the descriptions of the co-participants.[1]

RESEARCH METHODOLOGY

I decided to use an existential–phenomenological approach for this project's research methodology. For the purposes of the project, unconditional love from a spiritual teacher was defined as "the experience of being loved and accepted by someone one recognizes as a spiritual teacher without the possibility that that love and acceptance will be withdrawn due to something one might or might not say or do." I enlisted 12 co-participants to take part in the study. Seeking representation from different spiritual traditions, I found two co-participants

[1]The term *co-participant* rather than the usual term *subject* is used in recognition that those who contribute to the research project are involved in a process of co-creating the meaning revealed through the researcher's analysis and focused intent.

with relationships with Hindu, Christian, or Native American teachers. Among the remaining teachers were a member of the Jewish faith; a psychologist who was nondenominational but who, through his love, exemplified for his student the teachings of past Jewish masters; a minister of a congregation of Jews who believe in Jesus; a Sufi; a Buddhist; and a person who was identified by his student as being Sufi/other.

Ten of the co-participants wrote descriptions of their experience and two tape-recorded their initial descriptions; the latter were then transcribed. These initial protocols were then followed by a walk-through interview. In the walk-through interview, the initial protocol was read back to the co-participant with selected pauses in the reading that offered the co-participant opportunities to provide any new insights or further description. A transcript of the walk-through interview was added to the initial protocol to form the final protocol.

Analysis of the 12 final protocols, based upon procedures outlined by Colaizzi (1978), involved seven steps:

Step 1. The first step was to carefully read each protocol individually in order to gain as great an overall understanding of the experience as possible. At this point, and throughout the analysis, the practice of bracketing became very important. Bracketing is a procedure by which the researcher suspends personal biases or preconceptions by becoming conscious of them and identifying them. By holding these biases in awareness, I was able to look at the experiences of the co-participants with a clearer eye, thereby allowing their descriptions to present themselves in a way less colored by my own biases and/or agendas.

Step 2. After reading all the protocols, I began the work of extracting the significant statements in each individual's protocol. Extracted significant statements are those phrases, sentences, or groups of sentences that pertained directly to the experience of receiving unconditional love from a spiritual teacher, and not to the situation itself.

Step 3. The significant statements were then formulated into "constituent themes." This process of explication transforms the statements of the co-participants in such a way that their meaning or insight is retained and clearly expressed. Explication therefore involves a creative interaction between the researcher's understanding and the words of the co-participants, whereby the researcher looks deeply within to find and clearly state the meaning and significance of the experience. Constituent themes were thereby formed for each co-participant.

Step 4. The constituent themes were then arranged into "comprehensive constituent themes." Comprehensive constituent themes are the result of a synthesis and integration process in which like constituent themes are grouped together across all protocols. In this step, the fundamental aspects of the experience being researched are brought to light.

Step 5. The comprehensive constituent themes were then compared with the original descriptions of the co-participants to check for any possible inconsistencies. The purpose of this step is to ensure that the comprehensive constituent themes do in fact reflect the experience of the co-participants.

Step 6. A "fundamental structural definition" was then developed by combining the comprehensive constituent themes into a final defining paragraph. This definition provides a concise description of the essential aspects of the experience.

Step 7. Each of the co-participants was then contacted to discuss the results in order to insure that the findings were consistent with their experiences. In this light, the co-participants were asked for any feedback that they might wish to offer. This final step also provided an opportunity to thank the co-participants for their involvement in the project.

COMPREHENSIVE CONSTITUENT THEMES

The comprehensive constituent themes that emerged from the analysis were:

1. The teacher as having nurturing qualities.
2. Encouragement to find/manifest/express one's authentic self-nature.
3. The experience as profound.
4. Mutuality.
5. Transpersonal/metaphysical properties.
6. The experience as psychologically therapeutic.
7. A heightened awareness of one's authentic self-nature.
8. A heightened awareness of unconditional love within one's own being.
9. A shift in perception of self in relationship to other/world.
10. A heightened sense of understanding/commitment to the spiritual/sacred.
11. The experience as continually influencing/inspiring/unfolding.

The fundamental structural definition that emerged from the combination of these comprehensive constituent themes is as follows:

The experience of receiving love and acceptance from someone whom you recognize as a spiritual teacher without any possibility that that love and acceptance will be withdrawn because of anything you might or might not say or do is a profound and continually unfolding, influential, and inspiring experience. Primary aspects of the experience are transpersonal or mystical qualities, nurturance, psychological healing, and encouragement in finding, manifesting, and expressing one's authentic self-nature. The experience fosters a heightened sense of understanding or commitment to the sacred or spiritual, greater awareness of unconditional love within one's own being, greater awareness of one's authentic self-nature, and a shift in perception of self in relationship to other and the world.

These themes were found to express three different aspects of the experience. The first two themes addressed characteristics of the teacher. The characteristics of the experience itself are found in themes 3 through 5. The remaining themes pertain to what the experience evoked within the student.

Themes 1 and 2

What, in addition to unconditional love, did the teacher offer to the student? From the experiences of the co-participants, we find that these teachers carried in their pockets other treasures as well. Among these treasures were other nurturing qualities of the teacher and the encouragement the teacher provided to the students to find, manifest, and express their authentic self-nature.

All 12 co-participants experienced one or more of the following nurturing qualities from their teachers: being appreciated, protected, approved of, compassionately comforted, cared for, forgiven, and disciplined. Any vibrant memory we may have of any one of these qualities can help one to appreciate the value of that particular characteristic of unconditional love. The following interaction between a co-participant and her uncle provides a poignant example. It took place as she was mourning her father's recent death.

> I threw myself into his arms, and he held me very tightly, and his first words were in Yiddish, "Sha, sha, mien kind," which are the comforting words which mean, "Quiet, quiet, my child." Then, in English, he said, "Now that your father is dead, I am your father."

This quote is particularly appropriate because it demonstrates the parental nature of the nurturing expressed by the teachers toward their students. This quiet tenderness affords nourishment, deep caring, and a sense of safety, qualities that provide a vital environment that allows other features of the experience to take root.

But just as a garden must receive water and sun and nutrition, so must the dead twigs and weeds be removed. This being the case, discipline was also a part of this nurturing, as attested to by six of the co-participants. In all these cases, however, there was no sense of stifling or control, but rather the fostering of personal growth and valuable insights. These experiences were considered to be great gifts given them by their teachers, though it is true that the students might not have felt this to be the case at the time of the incident.

It is important to remember that the discipline that the co-participants described was recognized to be given within the context of unconditional love. The discipline was experienced as being in and a part of unconditional love. Perhaps it is the memory of this unwavering affection, this clear blue sky, that allows the temporary rainstorm of sometimes painful correction to be fully appreciated. There is a knowing that the moisture is needed and therefore allowed to penetrate deep to the roots, allowing for the rich and productive insights that the students later harvested. When problematic behaviors are addressed so genuinely and with such respect for the student, not only outer change but also inner transformation result.

It does not take too great a leap of the imagination to see that the teachers' unconditional love combined with their nurturing qualities would form the basis for an encouragement of the student to find, manifest, and express a more authentic self-nature. For some, this encouragement took the form of empowerment, for others of confirmation. Where one co-participant felt supported, another received respect or acknowledgment. Whatever the case, all 12 described the experience of being encouraged to somehow come into a closer communion with who they really were. Quotes from their descriptions offer a flavor of what the experience was like:

> [My teacher] created an atmosphere for me to become who I shall become. . . . It is clear that I will never walk in my mentor's footsteps, but my own.
>
> Pleasing him is only doing that which will benefit my soul.
>
> The experience of a love that acknowledges the expression of the truth of myself.
>
> His love provided the basis of strength and his guidance provided ways of developing strength without reliance on him.
>
> This feeling of love was without attachment.

Interesting points come forward in these quotes. First, the student is encouraged to live more genuinely in the everyday world. This statement speaks to an incitement to carve one's own niche, follow one's own heart, march to the beat of one's own drummer. It is true that these phrases seem terribly clichéd, but clichés evolve out of experiences that are universal in nature. Who among us has not felt that the wishes others have for our lives run contrary to the intuitions of our own hearts? How many of us have not felt the thrill of having someone significant in our lives support these intuitions?

The teacher's encouragement also points toward an inner movement, an intimate authenticity from a spiritual perspective. This movement shows us how limited is the ego as a form of identity, shows us that we are much more than we ever imagined, that our potentialities do not stop at the door of material existence. It seems that this inner knowledge is of prime necessity if one is to realize one's true meaning in the world, that clearer understanding of ego, inner knowledge, and realizing one's true meaning are but a single enterprise. It is also quite understandable, then, that a spiritual teacher would encourage a student to pursue a course that leads to a warmer embrace of or identification with a transpersonal or spiritual dimension.

The key to the power of this encouragement lies in the engaged detachment of the teacher. The term "engaged detachment" may seem an oxymoron, but it is the teacher's primary investment that the student have an open and expansive pasture in which to run, explore, and test his or her strength, and that this exploration lead to the student's experience of a positive development. The form or direction of this positive development is not for the teacher to determine; the student must make this decision. How else can one come to know one's genuine self than through a significantly unbridled freedom to find and form that genuine self? As has already been discussed, guidance is also a part of the teacher's role, but that guidance is given in an open way, free of the contraction of selfish stipulations.

Themes 3–5

Turning to characteristics of the experience itself, three fundamental features emerged from the descriptions. Receiving unconditional love from a spiritual teacher was described as being profound, was felt to evoke a sense of mutuality, and carried with it transpersonal and/or metaphysical qualities.

All of the co-participants expressed the profundity of the experience. Many described receiving unconditional love as extraordinary and as being a great gift or blessing. Again, brief reference to the protocols themselves is probably the best way of portraying the experience:

> What had I done to deserve such bounty?
>
> There no greater gift.
>
> I don't know how anybody could compare it with anything because it's not like anything I've ever known.
>
> It's impossible for me to convey its importance.
>
> It's a mystery, it's just a mystery.
>
> I can only attribute my change of attitude to the Grace of Allah.
>
> The creator has given me a very precious gift.

Obviously, these people have all been touched in a very deep way. The nature of these gifts and what makes the experience extraordinary will become increasingly apparent as we look at the remaining comprehensive constituent themes, the next of which concerns a sense of mutuality.

In this sense, the student responds in some way to the positive attention of the teacher. Something is touched and awakened in the student. A spark of affection for the teacher is kindled. A relationship is formed. Even if a stranger smiles at us guilelessly and congenially, it is only natural, even if we are in the foulest mood, to feel a glimmer of warmth in our hearts.

This emerging relationship might best be described as a psychological and emotional opening toward the teacher. Some co-participants described it in just that way, an opening. Additional terms include feeling connected to the teacher, and sharing love, caring, or acceptance with the teacher. The unconditional nature of the love has a truly penetrative quality. Since it is a love that does not originate from an egoic place, with egoic motivations and concerns, it reaches a place within the student that transcends the ego. Attachment to limited and therefore limiting identities begins to loosen; a more open, assured, and fluid personality is cultivated. A process begins in which receiving such love and acceptance helps us to more fully love and accept ourselves. The more we love ourselves, the more we can love others.

As will be seen, this process, initially realized or brought to life in this sense of mutuality, is truly a pivotal aspect of the experience of receiving unconditional love from a spiritual teacher. This spiraling reciprocation of love, caring, and acceptance is, in a way, a training ground, a prototype of things to come; that is, this expansion of positive and affirming sentiment is ultimately directed toward self, others, and the world.

The transpersonal dimensions of the experience become more explicit in the next comprehensive constituent theme. As defined by Schneier (1989, p. 322): "Transpersonal means 'over, beyond, or through' the personal and refers to all experiences in which we transcend the ordinary boundaries and limitations of our personalities." Tillich (1967, p. 20) stated that the word *metaphysical* has lost its original meaning of underlying structures or universal laws and has now taken on the meaning of "a world behind the world . .·. the connotation of pointing to a duplication of this world by a transcendent realm of beings." These two definitions provide a foundation for the following discussion.

The range of experience that makes up this particular theme is far-reaching. Among the experiences recounted by the co-participants are oneness, union, expansion, harmony, perfection, infinity, universality, timelessness, and awakening of the soul. Additional experiences included entering another level of awareness, recognizing the unfolding of destiny, being seen in a transcendental sense within the context of a spatial continuum, knowing without the faculty of the mind, being in the presence of God and looking into the eyes of Christ, being in communion with Jesus through the intercession of the Holy Spirit, encountering past lives, recognizing deceased teachers who have been reborn, and continuing to be taught by teachers who have passed away.

The spiritual impact of being in relationship with a spiritual teacher seems to lie at the foundation of these experiences. It also seems that the traditional context of the teacher's path and message colors the experience of the student. The student of a Christian teacher sees images of Christ, that of a Buddhist teacher grows in understanding karma, and so on. The depth to which the students have been led by the loving articulation of the teacher's message is very evident. As was mentioned above in the discussion of mutuality, a profound opening was experienced. Part of that opening is to realms that are generally unfamiliar, the transpersonal/transcendent/metaphysical realms. A more expanded understanding of the nature of one's personal being comes to the surface here.

Themes 6–11

Let us now look at what was evoked within co-participants during the experience. All of the co-participants felt that receiving unconditional love from their spiritual teacher was psychologically therapeutic. Many wrote that a part of the love they felt from their teacher was feeling psychologically held, authentically seen, understood, and/or supported though periods of distress in their lives. In all of these instances, the teacher in effect steps into the life of the student. The everyday reality that is particular to the student is witnessed by the teacher in an accepting way, free of judgment or evaluation.

The support during difficult periods in their lives that several co-participants felt from their teachers is of obvious significance in the therapeutic setting. This is often why people seek therapy. Who among us has not sought some form of help in enduring and making sense out of the pain of a death in the family, a divorce, unemployment, or intense psychological suffering? Whether that help came from a psychotherapist or not, we can still acknowledge the comfort received in knowing that someone was sharing the burden of our grief, the weight of our pain.

Three additional therapeutic aspects were also brought to light. Many received guidance regarding problematic situations. The supporting presence of the teacher also helped a number of co-participants to face painful personal issues. And several described the experience of intense emotional states, either painful or pleasant.

All three of these aspects can be discussed by turning once again to the poet Rumi (Schimmel, 1992). Rumi likened his relationship with his teacher Shams to that of a potter making a vase. He said that while the outer hand of the potter is molding and kneading and working the clay, inside the vessel the teacher provides the warm hand of support. The loving atmosphere engendered in the relationship with a spiritual teacher allowed the inner life of the student to emerge in a less threatening manner. At the time, the student is perhaps conscious only of the pain involved in the molding and the kneading of the ego and its attachments. But because the student is now in the safe harbor of the teacher's love, fear of the pain has subsided or become manageable. Life has attained a greater meaning; thus, even the painful aspects of that life have meaning as well. The student, bolstered by the teacher's love, is more courageous in facing the storms of life. And so, returning to Rumi's metaphor, eventually a beautiful vessel is formed. This beautiful vessel is largely comprised of a greater psychological integration.

This integration provides a natural segue to the next comprehensive constituent theme, a heightened awareness of one's authentic self-nature. Due to the nurturing encouragement discussed earlier and to the therapeutic nature of the experience, the students of a purely loving teacher gain a greater sense of who they are. They embark upon the path of self-actualization. Accepted, loved, and respected by the spiritual teacher, 11 co-participants described experiences of a heightened awareness of their authentic selves.

One co-participant was particularly expressive in this regard. Following are several quotes from his protocol.

> They keep bringing me back to myself.
>
> They keep giving me opportunities to look at the different layers of what I thought of as myself—my cultural self, my racial self, my, all of those multiplicities of selves that I have.
>
> It was just this peeling and peeling.
>
> A feeling of a quality of love that nourishes the natural unfoldment of my self to my self.

Coming to this newfound self-awareness was sometimes more an ordeal than an epiphany. As discussed earlier, painful realizations and troublesome states of mind often rose to the surface. As we become stronger in who we genuinely are, perhaps we muster the courage to face our shortcomings, our personal foibles. Our restricting and limiting interpretations of being, our confused and distorted ways of relating to ourselves and the world, become more apparent. Also more apparent is the need to cast off these hindering interpretations and behaviors.

This path of self-actualization led 10 of the co-participants quite naturally to a heightened awareness of unconditional love within their own being. Becoming more accepting and loving of themselves, strengthened by a clearer perception of their potentials and actualities and the greater ability to embrace their imperfections, they were increasingly able to accept and love those around them. In fact, many co-participants came to the realization that unconditional love is a quality that is inherent in or organic to our being-in-the-world. Again, I will rely upon the words of the co-participants themselves.

> It is the most natural thing in the world.
>
> Regarding unconditional love, in general . . . it has its natural place in the universe of mankind.
>
> Everything in the universe is longing to go back to this love.
>
> The mind has created a separation from that love.
>
> It is the truth of what we are, the embodiment of love.
>
> It is the birthright of everybody.
>
> It is given . . . with an acknowledgment that this is simply my natural and organic entitlement by being born into this life.

Another contributing factor to the heightened awareness of unconditional love appears to be the mutuality discussed above. The blossoming care and affection that the co-participants felt for their teachers begins to radiate toward a widening circle of individuals. What a wonderful process this is, unconditional love begetting unconditional love begetting unconditional love. Meher Baba, a Sufi saint of this century, is said to have commented:

> Love can never be forced on anyone. It can be awakened in her or him through love itself. Love is essentially self-communicative. Those who do not have it catch it from those who have it. True love is unconquerable and irresistible and it goes on gathering power and spreading itself until eventually it transforms everyone whom it touches.

It is important to mention that four of the co-participants viewed family members as the spiritual teachers from whom they received unconditional love. Two of these co-participants were mother and daughter, so in this particular instance three generations were involved—a mother unconditionally loving her child who, becoming a mother, unconditionally loves her child. The mother described this process as the passing of love "from canoe to canoe." Again, how natural this appears to be.

The mention of this family connection, coupled with the previous observation that unconditional love is inherent in or organic to our individual nature, suggests further realizations. Unconditional love is not just the unreachable domain of gurus, shamans, or clergy. Nor is unconditional love something "out there" that we acquire; it is already part and parcel of our lives, of our very being. We simply need to become aware of it. The judgments and evaluations that are so often a part of human relationships are seen to be ephemeral products of the mind. As those judgments and evaluations are peeled away or held in abeyance, we find that love in its unconditional form has been waiting there all along, an ember glowing ever so faintly, an ember that can be fanned and fueled to greater and greater brightness. We all can open our hearts, we all can love our children, our spouses, and our fellow human beings in this way.

This greater understanding of unconditional love is reflective of a wider dynamic. The co-participants underwent a shift in their perception of self in relationship to others and the world. They felt more capable or responsible, they had a better sense of belonging or of equality, they were more altruistic or reverent, they had more faith or trust or appreciation.

One of the co-participants beautifully portrays this shift. She writes:

> It's not like, maybe what I did would be different, but now I'm able to deal with what happens on a better level, much more appreciative and much less resistance and getting where maybe [I'm] not trying to control things as much. But just acknowledging how special it is just to be aware of

> [her teacher's] presence in my life because that has brought a lot of comfort and a lot of being able to talk myself through whatever's happening. It's given me just a lot more courage and peace of mind. Just accepting instead of trying to change things, to relax sometimes and let things be and relax to the point where it's like, this is okay, and just try to move forward from there.

The transpersonal aspects of the experience also contributed to this change in perception of the external world. Spiritual experiences have often been credited for meaningful changes in people's lives. As we see life in such an expanded manner, even for a short period of time, this expansion triggers a subtle transformation that slowly permeates various aspects of our lives.

Another expression of this transformation, as part of receiving unconditional love from a spiritual teacher, is that the experiencer gains a deeper understanding of or commitment to the spiritual or sacred. This was the case for all 12 co-participants. The following quotes indicate the breadth of this particular aspect of the experience:

> In my mother's presence, I feel . . . the sacredness of life.
>
> A supernatural phenomenon, employing the intercession of the Holy Spirit, who lives in every believer.
>
> The relationship is constantly developing as I open myself to her grace.
>
> Unconditional love was a way of initiating, sustaining, and fostering deep wisdom.

Some co-participants saw life or features of life as more sacred, others grew in appreciation of great spiritual teachers of the past, many found a greater commitment to their teachers and the teacher's principles, and there were those who found a deeper awareness of spiritual virtues such as compassion, humility, understanding, devotion, and gratitude.

Experiencing therapeutic benefits, self-actualizing, finding unconditional love within, shifting toward a new perspective of others and the world, and gaining a heightened comprehension of spiritual or sacred understandings are all interrelated. These experiences share much in both their quality and their texture, and all make up and spring from the journey to self-awareness and inner transformation.

Moreover, for the co-participants, this journey has not ended. The repercussions of receiving unconditional love from a spiritual teacher continue from the time of the experience into the present. The co-participants expressed that the reception of such a positive recognition has continued to unfold in an influential, inspiring way. The experience of all these inner movements evoked within them carries an impact that stirs a rippling momentum.

A vital element that perseveres in the minds of the co-participants is the magic of those moments when they basked in the warmth of their teachers' love. The continuing presence of the teacher is captured in these moments. Once again, I offer their words:

> Even as I write these words, my eyes water.
>
> It touched something in me that I never will forget.
>
> I felt blessed, protected, and this feeling seemed very permanent, as if it would last many lifetimes.
>
> Those dear people who have given unconditional love to me still do so from the other side.
>
> He continues to be with me.
>
> Her blessings are always available.
>
> There is no separation.

It seems somehow natural to conclude this discussion with this theme, for, though this discussion will soon come to an end, the experience lives on within the hearts of the co-participants. To have such a profound experience, if even for a short period of time, is bound to trigger moments of contemplation and unfoldment for a long time to come. While interviewing the co-participants, I was continually struck by their poignant expressions, by the tears in their eyes. Many, if not all, expressed gratitude for being able to take part in the project because taking part brought back to life the intimate contact that was being described. It was obvious that the experience was still very significant to each and every person as he or she spoke—so much so that I could not help but feel deeply moved myself.

Spiritual energy is often compared to the brilliant light of the sun. It shines continually in all directions, bearing with it a subtle yet penetrating warmth. There are perhaps many of us who, due to the world we have constructed, are blind to the sun, to the golden illuminating force that reveals to us our own innate, starlike, spiritual essence. But even when blind, when even purposely closing our eyes, none of us can block out the subtle penetration of the warmth of that sun.

Unconditional love is a most important aspect, if not the primary expression, of the spectrum of spiritual energy. What you have just read regarding the experience of receiving unconditional love from a spiritual teacher is based on the words of people who were open to that experience. These peoples' words describe what it is like to knowingly bask in such light and warmth.

It is precisely because the loving sun shines without judgment that it can so magnificently and completely illuminate our inner and outer worlds. And so, from within this deeply affectionate light, we can see more clearly the nature of our relationships with ourselves, the people in our lives, and the environment in which we live. In this gentle radiance, we can more fully know, pursue, and attain our personal goals, we can become aware of the beauty of our own unique being.

In this sense, the sun of unconditional love is in the world but not of the world. As we experience ourselves in increasingly accepting ways, as our vision of being more completely approaches an expanded view, we begin to experience those around us with an increasingly loving eye. We can see ourselves and others more fully, not just as a collection of characteristics, a mere mixture of emotions and thoughts and feelings expressed in speech and action. We begin to be aware of something greater, something marvelous and ineffable, and we realize that the boundaries that separate us from that ineffable marvel are becoming increasingly permeable. We become involved in a graceful and grace-filled relationship with everything that is encountered.

Unconditional love. Rumi, the great Sufi poet-mystic, wrote of this love: "I've heard it said, there's a window that opens from one mind to another, / but if there's no wall, there's no need for fitting the window, or the latch."

REFERENCES

Barks, C., & Bly, R. (1981). *Night and sleep: Rumi.* Cambridge, MA: Yellow Moon Press.

Bracket, C. (1975). Toward a clarification of the need hierarchy theory: Some extensions of Maslow's conceptualizations. *Interpersonal Development, 6,* 79–90.

Buber, M. (1965). *The knowledge of man: A philosophy of the interhuman.* New York: Harper & Row.

Buber, M. (1970). *I and Thou.* New York: Scribner's.

Castaneda, C. (1968). *The teachings of Don Juan.* Berkeley: University of California Press.

Castaneda, C. (1972). *Journey to Ixtlan: The lessons of Don Juan.* New York: Pocket Books.

Colaizzi, P. F. (1978). Psychological research as the phenomenologist views it. In R. S. Valle & M. King (Eds.), *Existential–phenomenological alternatives for psychology* (pp. 48–71). New York: Oxford University Press.
Dietch, J. (1978). Love, sex roles, and psychological health. *Journal of Personality Assessment, 42*(6), 626–634.
Glick, S. (1983) *An analysis of the change process in the guru–disciple relationship.* Philadelphia: Temple University Press.
Harper, R. (1966) *Human love: Existential and mystical.* Baltimore: Johns Hopkins Press.
Hawka, S. M. (1985). *The experience of feeling unconditionally loved.* Cincinnati, OH: Union Institute.
Monrich, H. (1985). Teaching and learning in love. *Psychological Abstracts, 74*(7), Arlington, VA: American Psychological Association.
Motto, J. A., & Stein, E. V. (1973). A group approach to guilt in depressive and suicidal patients. *Journal of Religion and Health, 12*(4), 378–385.
Muktananda. (1980). *The perfect relationship.* South Fallsburg, NY: SYDA Foundation.
Rinpoche, S. (1992). *The Tibetan book of living and dying.* San Francisco: Harper Collins.
Schimmel, A. (1992). *I am wind, you are fire: The life and work of Rumi.* Boston: Shambala.
Schneier, S. (1989). The imagery in movement method: A process tool bridging psychotherapeutic and transpersonal inquiry. In R. S. Valle & S. Halling (Eds.), *Existential–phenomenological perspectives in psychology: Exploring the breadth of human experience* (pp. 311–327). New York: Plenum Press.
Solimar, V. (1987). *The nature and experience of self-love.* Unpublished doctoral dissertation, California Institute of Integral Studies, San Francisco.
Tillich, P. (1967). *Systematic theology: Three volumes in one.* Chicago: University of Chicago Press.
Venkatesanda. (1989). *The concise srimad bhagavatam.* Albany, NY: State University of New York Press.
Wangyal, G. (1978). *The door of liberation: Essential teachings of the Tibetan Buddhist tradition.* New York: Lotsawa.
Weinmann, L. L., & Newcombe, N. (1990). Relational aspects of identity: Late adolescents' perceptions of their relationships with parents. *Journal of Experimental Child Psychology, 50*(3), 357–369.
Welwood, J. (1990). *Journey of the heart.* New York: Harper Collins.

16

On Being with Suffering

Patricia A. Qualls

> Time to go into the darkness. That place where every fear and disappointment lives. Where everything that frightened and hurt you as a child lies waiting for a chance at revenge. Every self doubt and recrimination simmers on the back burner in hell's kitchen surrounded by pools of tears, sweat, and blood. You can hear cries from children there. Children you can never soothe. There in the impenetrable darkness are the eyes. Two big, dark eyes that look like full round moons silently suspended in the pale night. Bottomless pits that cannot veil their ravenous hunger. I cannot look at them fully. I am embarrassed by the naked need and my inability to ease their pain. I turn to walk away and know that in that moment there is no escape. Each time I look into a mirror I see their reflection in my two big eyes. (Co-researcher Mary)

Each day the television news is filled with film of dying children, slaughtered human beings, starving and war-ravaged humanity. Each day millions of people watch and then do something with that which they have witnessed. They may store it, ignore it, harden to it, self-medicate, or respond to it with action. Suffering is not only on our televisions, it is in our lives.

Some people manage to distance themselves from suffering. Others live with and around suffering every day. Many who have chosen to work in mental and physical health care facilities have a constant exposure to suffering humanity. Family and professional caregivers such as nurses, doctors, psychotherapists, and ministers deal with suffering individuals daily. In my experience, these individuals are striving to alleviate suffering. Yet, with all of this direct personal contact, there remains little understanding of the experience of suffering itself, and there is even less understanding of the experience of "being with" the suffering of another.

As my years of clinical course work in graduate psychology came to an end, I was looking for something to do that would be "meaningful." I was offered an opportunity to work as a volunteer in a Romanian orphanage as a caregiver for children. When this opportunity presented itself, I did what any graduate student worth her salt would do: I went to

Patricia A. Qualls • 26405 Valley View, Carmel, California 93923.

Phenomenological Inquiry in Psychology: Existential and Transpersonal Dimensions, edited by Ron Valle. Plenum Press, New York, 1998.

the library. There, I searched for anything addressing orphans, institutional care, and child rearing. What I found were studies of the ravages of World War II. I armed myself with child development literature, attachment theory, ideas on bonding, self-psychology, and any other insight or perspective that I might add to my arsenal to be a "good helper." I took it upon myself to make copies of this literature to give to my fellow volunteers at an orientation meeting in New York. I worked diligently in a few weeks' time to prepare myself—and, I hoped, those who would work with me—for the job that lay ahead. As a caregiver, I focused on those we would be caring for, their needs, and their lack of basic care, both physically and emotionally.

Missing, however, was a key piece—understanding the physical, emotional, and psychosocial needs of the caregivers themselves in situations of extreme suffering. I had kept myself so busy "doing" what needed to be done for others and trying to ease my own anxiety with academic theory that I had missed, up to this point, a crucial piece of the equation, namely, that the givers of care and the cared for are in a symbiotic relationship in the lived-experience of suffering. It was with this realization that I decided to use my time in Romania as the basis of my doctoral research, a phenomenological investigation of the experience of "being with the suffering of orphaned children one has cared for."

Even with this awareness and design, it was not until I returned to America after eight months in Romania that I even began to understand the impact of living with and responding to the suffering of others. Being with suffering has shown itself to be a rich and complex experience calling for a deeper understanding. It is this deeper understanding that has come from my research that I share in this chapter.

CONTEMPORARY VIEWS OF SUFFERING

Few training programs in the helping professions prepare the student to deal with suffering, to be with it, or to simply allow it to be. Instead, we discuss the alleviation of pain and pain control. The complexities of suffering are rarely honored. As stated by Archer-Copp (1990, p. 35): ". . . no one, including the anthropologist, sociologist, and psychologist, has markedly enriched our awareness of the phenomenon of suffering. The course is almost uncharted. . . ." The moral theologian Franz Bockle (1987) describes society as led astray by the illusion that we might be free of suffering. We therefore try to escape suffering and the embarrassing stigma that often accompanies it. Our progressive society has established an anonymous ban on suffering (Metz, 1995). If there is a denial of suffering, we cannot help those who suffer, nor can we address the needs of the caregivers who are attending to this phenomenon.

Webster's New Collegiate Dictionary (1976, p. 1164) defines "suffer" as "to be forced to endure death, pain, or distress; to sustain loss or damage." It seems that most if not all human beings experience suffering at some time during their lives. It has been said to be an inherent part of the human condition. For some, it is physical suffering and pain; for others, it may be psychic and/or spiritual suffering.

Suffering—the meaning of, the reason for—has been explored by almost all of the world's great religions, and theologians have built grand theories designed to make sense of it. Psychologists with an existential leaning regard suffering as an integral part of how we impart meaning to our existence. Frankl (1959), for example, believed meaning to be in the very fabric of human experience and emphasized the power of the human spirit to overcome suffering. Consistent with this emphasis on the relationship between meaning and suffering, Starck and McGovern (1992, p. 28) point to the possibilities inherent in suffering for the caregiver as well as for the one who is suffering:

> Suffering has the potential to transform the sufferer and/or caregiver and offers an opportunity for the sufferer and/or caregiver to experience discovery and growth and to give and/or receive compassion and love. Thus, those who suffer can find new meaning in life, as can those who care for the sufferers.

One could consider this self-transformative potential or power as the hidden grace in suffering.

Acknowledging the existence of pain in the world, society has attempted to address via, for example, hospitals, foundations, or charitable organizations, the needs of those who suffer. On the front line of personal intimate contact, however, are those who deliver some type of service, care, or protection to the one who is experiencing pain. While most of us understand intellectually why avoidance and denial may accompany the phenomenon of suffering, it is from this more personal contact that a deeper questioning arises. What is the experience of being with suffering? How is it actually lived? How does it come to make itself known to the one who is experiencing it? Does it have texture, color, smell, or flavor? Does it present itself as emotional phases or developmental stages? My research attempted to answer questions such as these, in the hope that it will provide a deeper understanding so that we might become more sensitive to our own pain and to the suffering of others.

AN OVERVIEW OF THE RESEARCH

The purpose of this study was to enter into the inner world of caregivers who knowingly moved toward the suffering of others. More specifically, I investigated the phenomenon of being with suffering in the lives of nine individuals (eight women and one man) who had traveled to Eastern Europe to work as volunteers with the children in a Romanian orphanage. My hope was to allow the descriptive world of these caregivers to enlighten others about their experience and thereby enhance our sensitivities toward the suffering in our society in general, and that of the courageous caregivers who serve in our society's institutions in particular. One way to become more aware of suffering is to communicate directly with and listen to those who are in pain. There are times when those who are suffering have no voice. There are times when the rest of the world has no ears. There are even times when the givers of care have lost their selves.

This study consists of descriptions offered by these nine volunteers; the internal worlds of each participant are brought to light as they describe their own unique way of living with the suffering of these children. These descriptions were then analyzed using a phenomenological research method (described below), the elements so discerned being woven together in a summary form that expresses the essence of the experience of being with suffering in this context.

I personally worked with the co-researchers of this study in an orphanage and hospital in Romania, a small country in Eastern Europe about the size of Oregon, surrounded by Hungary, the former Yugoslavia, Bulgaria, and the former Soviet Union. The long-time

Communist dictator Nicolae Ceausescu, who had resisted the reform movement that was overtaking Eastern Europe, was tried and executed in December 1989. It was shortly after his death that the cameras of the world revealed the atrocities of his reign, including the conditions of the orphanages in Romania. Most of the individuals participating in this study saw this film and began to search for a way to reach the children in these institutions. Also in response, many organizations had begun grassroot efforts to help, one such organization adopting an orphanage in the village of Calarasi. In conjunction with this particular organization, the first group of co-researchers in this study arrived in Romania in April 1991, 16 months after the fall of Communism.

The village of Calarasi was a two-hour drive from Bucharest. I will never forget my first sight of this town. A giant steel factory billowing out black smoke and orange flames welcomed us. Black soot filled the air and covered the town with a film. The local residents were not even aware of it. These smells were the first to assault us. Starving wild dogs roamed the streets. Gypsies and their wagons were a part of the landscape. Every once in a while, one of the few cars would come tearing down a potholed road, scattering people, horses, and dogs.

We were the first Americans the people of this village had ever seen. When we arrived, they followed us in groups, watching our ways, admiring our clothes, some reaching out to touch us. Because of the years of Securitate (secret police) persecutions, they were at first afraid to speak to anyone from another country. It was common practice to wiretap homes of the citizens of this country and to have informants reporting to the Securitate. We had arrived in a country where oppression was the rule, not the exception. Mistrust and the need to survive had taken its toll on the townspeople.

As we began to make friends with some of the local residents, we soon learned that the idea of us volunteering to help in the orphanages was unbelievable. Survival was such a basic premise that the villagers had no time or energy to volunteer anything outside their immediate circle of family and friends. Buying a quart of milk or a loaf of bread might mean waiting in line for an hour or two at each store. Obtaining fuel for the kitchen stove might require a two- or three-day wait out in the freezing weather.

Survival and suffering were apparent for all to experience and to witness. The conditions at the orphanage were such that most of the children had never received basic care, much less the love and tenderness needed for healthy development. The children were left unattended in rooms, lying in their urine and feces, and there was no stimulation provided. Flies and mosquitoes covered their bodies in the summer. Their little bodies were often covered with sores from insect bites. Human touch was rare and most often was handed out only as a harsh, disciplinary action. Many of the children were infected with the HIV or the hepatitis B virus. At times, the children were tied to their beds. The newborns in the hospital section were starving. Drinking water was not given to the children. Hot water was a rare commodity, as was soap, and basic cleanliness was very difficult to achieve. Most of the children were unable to talk. There were many children, four and five years old, who could not walk and had not been introduced to solid foods. Children who could not walk and feed themselves by the age of six were to be sent to homes for the irrecoverable. The conditions in these institutions were known to be even harsher and more horrifying.

These were the conditions that the co-researchers in this study experienced each and every day. The context and details of the study follow.

PERSONAL HISTORY AND CHOOSING A METHOD

The study of suffering has been left primarily to either the religions of the world or the medical sciences, and even in these circles there is minimal research on the phenomenon of being with suffering. Since the purpose of the study was to bring to light the lived experience of individuals being with suffering, I chose the existential–phenomenological approach as the most appropriate perspective to apply in understanding the essential nature of this phenomenon. As Watson (1985, p. 205) says, "In the day-to-day living that brings problems, struggles, pain, and suffering to many people, the existential–phenomenological factors bring personal meaning to the human predicament."

As a person and therapist who cares for what is most essential to being, I chose an existential–phenomenological method since, instead of focusing on statistical relationships between operationalized variables, this approach studies lived-meanings. More specifically, I used a structural-type analysis designed to reveal and articulate the basic elements or constituents of the experience of "being with the suffering of orphaned children one has cared for." This life-world experience was studied and interpreted in this way, revealing a depth, fullness, and richness previously unknown.

Consider this example. There is a difference between our prereflective, lived, felt-sense of "hearing a child cry" and our reflective comprehension of the same. Being a new mother constantly teaches me new ways of thinking, seeing, hearing, and feeling. My sensing now has a broader range of meaning for me. Before becoming a mother, I would hear a child cry and identify it as simply that. Now my ears distinguish such different cries as those of hunger, distress, pain, wet diaper, frustration, and anger. I feel the differences in my body, and I respond accordingly. The sound enters through my ears and gives my body messages to respond immediately, or to take a moment, he will right himself or reach his goal, or "Oh my God!" My reflective responses prior to motherhood were cursory or not as deeply felt or systematically processed. My relationship to the sound of any baby crying was also much different than it is now. The sound of a baby crying now offers a depth and breadth of communication that I never understood before. Meanings and themes regarding my own child are unique and are constantly broadened and enhanced as each new stage in his development is reached. My response to other babies has changed as I have learned from my own child. As I have been brought into the world of a new being and travel from newborn to infant to toddler, I have learned to listen and sense on a level within my being that I never knew existed before. At first glance, "baby crying" elicited such reactions as acknowledgment, searching for its location, and determining whether the child is attended to. Of course, I wonder "why" my baby is crying, but more important, I am listening and asking "what" is my baby saying? What is his experience in this moment? What is he telling me?

"Baby crying" still calls me to hear and attune to tone, meaning in the tone, pitch of the sound, and I still search for and notice what happens, but more is included now in this initial connection. A level of "measuring" is also happening that was not a part of me prior to my "training" in the lived experience of being a mother. In other words, the structure of meaning of this experience has revealed itself in a new way since my own life-world has the added dimension of motherhood. Only now that I have been with my son's crying in various situations on repeated occasions have the meaning and form of his communication become known to me.

On Bracketing

These meanings are now based in my ever-unfolding history; that is, I always have new assumptions. In the phenomenological approach to revealing the prereflective structure of any meaningful experience, self-reflection is central and involves a process known as "bracketing." Bracketing refers to the process of identifying one's preconceived notions (i.e., biases) regarding the experience being investigated in order to minimize their effects on clearly explicating the meaning of the co-researchers' descriptions (Valle, King, & Halling, 1989).

In this study of being with suffering, I was a caregiver who actually had the experience of being with the orphans, an experience that was deeply meaningful for me in many ways. In addition, I assumed that the reality I had perceived was very similar to that perceived by the other caregivers, since we all went to the same village, worked in the same orphanage, and many times were with the same children. We all brought a unique history, personality, energy level, and even purpose to engage this experience. In this context, I would like to share with you a few of my biases regarding this experience as examples of what I held in my awareness (i.e., bracketed) as I read and reread the descriptive protocols of my fellow volunteers. I have many beliefs about suffering both as one who has suffered and as one who has been with those in pain. Having survived a major depression 12 years ago, I found myself forever changed in a positive way. I felt I was given my self, a second chance, a rebirth to experience life more fully. That experience seems distant now, the rawness is gone, the layers between myself and the world that had been stripped away have started to grow back. In many ways, I idealize that time of suffering in my life as extremely rich. Without that experience, I now feel, I would not have developed as deep a level of compassion, of understanding, and of empathy as I now have. It is with this history in mind that I share with you a number of the biases I discovered within: I had believed that (1) all the caregivers would turn to God or some deep spiritual or religious belief system to help them with this experience; (2) humor would reveal itself as important in this experience; (3) being with suffering made one's life rich and meaningful, and this would lead to positive feelings regarding the experience; (4) being with suffering strips away the usual way of presenting oneself to the world; and (5) "doing" is sometimes a way of coping with suffering.

Selection of Co-Researchers

Since this research was about the investigation of individual caregivers' "being with suffering," one of Heidegger's beliefs is an important premise to understand. Von Eckartsberg (1986, p. 12) describes Heidegger's thoughts about being in the world as grounded in care, "in concernful presence and openness to the world and others." Another reason I chose the existential–phenomenological approach is that it would enable me to enter into the lived-experience of the co-researchers in a "caring" way. The co-researchers spent weeks and, in some cases, months living this experience, and then took many hours to write about it, and then share it with me. Our "walk-through" interviews (described below) often involved hours of travel time for many of them. They were not the subjects or objects of my research, but rather co-researchers who kindly agreed to "re-live" their experiences so that we might "re-search" the phenomenon of "being with suffering." This project is the product of reciprocal relationships, a project that could be accomplished only through partnership.

The co-researchers met two requirements: (1) They had had the experience of being with the suffering of orphaned children they had cared for and (2) they had the willingness and ability to communicate this experience (Polkinghorne, 1989).

Eliciting the Protocols

Each co-researcher was given or mailed a sheet of paper with the following request: "Please describe your experience of being with the suffering of the children you cared for in Romania." They were asked to focus on the experience, not just the situation. They were instructed to include enough detail to fully describe their experience, to not stop writing until they had done so, and to take as much time as they needed to complete their description.

After first reading the written protocols, I arranged an in-person "walk-through" interview with each co-researcher. The purpose of this interview was to present each co-researcher with his or her previously written statements, one by one, in order to elicit any more thoughts or feelings he or she might have. The walk-through interview was an opportunity to once again be with the experience and to expand the description of the phenomenon in a deeper way. These interviews were recorded on audio tape and then transcribed. Any words added during the interview plus the original written statements became the final protocol for each co-researcher.

Analyzing the Protocols

The analysis of these data was based on procedures developed by Giorgi (1971, 1975) and Colaizzi (1978). All protocols were read as a way of getting an initial feeling from the descriptions. As I read, any words or statements that "jumped out" were simply noted in the margin. A second reading was approached with a more searching eye, watching for words and phrases that struck me as significant (i.e., meaning units). The protocols were now reviewed with a focus on recognizing themes coming from the meaning units, which allowed for the emergence of a series of themes for each co-researcher's protocol. These themes were then examined across protocols, forming theme clusters. The theme clusters were then referred back to the original protocols for validation. The final themes were then synthesized to form the essential structural definition of the experience of "being with the suffering of orphaned children one has cared for."

Following Becker (1992), I stayed "with the data" each step of the way, dwelling and reflecting on the co-researchers' exact words and meanings. I asked myself these questions: What stood out about this phenomenon? What is the most important aspect? What is the next most important? How do they fit together?

These co-researchers repeatedly demonstrated their courage by participating in the care of these children, and then by recounting their experiences for this study. These nine individuals went to Romania with the hope of alleviating suffering in some way. In what follows, I invite you to participate in their lived experiences and in all of the many ways in which these experiences have had an impact on their lives.

THE FINDINGS

The following presentation describes the co-researchers' experiences by synthesizing their own words. I have explicitly trusted the co-researchers' descriptive narrations to be rich enough to take us deeper into their experiences. This presentation is not meant to offer an exhaustive analysis of the experience of being with suffering. My intention here is to elucidate new meanings and thereby to broaden our understanding of what it is like to be with the suffering these individuals experienced. Although another researcher may uncover different meaning structures, my hope is that from this presentation, one will be able to see how I moved from the co-researchers' words to the final description of the experience.

Reading and rereading the protocols led me to view the data in two general ways: (1) in a temporal, more linear fashion, and (2) as revealing their underlying elements or constituent themes. In terms of time, the experience emerged in four recognizable components or phases: (1) the initial desire to go to Romania, (2) the introductory time of actually being there in Romania, (3) the day-to-day living with the experience, and (4) living with the experience today. While "being with suffering" takes on a particular quality in each of these four phases, seeing the process of the experience unfolding in this way was helpful in the early stages of organizing the meaning units and themes. In many ways, this subdivision of responses in time facilitated my recognizing important aspects of the experience and how they interweave in the fabric of the descriptions.

When I say "interweave," I imagine a beautiful rug or tapestry in which the threads move back and forth across the width of the rug, sometimes touching in just the right place to make an immediate connection. At other times, the threads do not touch, yet are connected by still other threads. Sometimes the exact colors repeat themselves, while at other times a totally different color appears, or something of a lighter shade makes itself known. In their weaving together, they offer us a pattern to perceive. Whatever the variant forms, each thread is part of the whole. Although analogous to this physical example, the process of unveiling the meaning of an experience from written protocols is, of course, another matter. In phenomenological analysis, there is never *one* way to interpret the essence or meaning of any given set of written or transcribed descriptions. Rather, on the basis of his or her own self-reflective sensitivities and interpretive sense, each researcher offers *a* way in the hope of deepening one's understanding of the experience being investigated.

With this tapestry image in mind, I began to envision the experience of being with suffering as a circle with no beginning and no end. I then watched as definitive patterns began to make themselves known along this circular path. In this way, I began to find ways to understand and embrace the data, ways that seemed to keep growing and expanding.

The process gradually revealed aspects of the experience that were not limited by time, dimensions that did not easily fit into a slot or category. The descriptions showed an ebb and flow with the past, present, and future. A variety of external and internal forces alike seemed to shape each instance described, as constituents and themes began to make themselves known. This nonlinear, dynamic, more feminine way of articulating the threads of the tapestry is in line with the more classic approach of protocol analysis. These two ways of looking at the descriptions offered by the co-researchers helped to articulate the following ten themes, collectively representing the essential nature of "being with the suffering of orphaned children one has cared for" in this study:

1. Emotional and physical responses.
2. Continual questioning.
3. Indelible and unforgettable experiences.
4. Making a difference, often with hope.
5. Being transformed.
6. Experiencing an altered reality.
7. Relationships with the children or relating *with* the children.
8. Communicating with God.
9. Developing ways of coping.
10. Reality orientation: coming to terms.

Let us look at each of these themes in turn.

1. Emotional and physical responses. Entering the orphanage was the beginning of an emotional and physical roller coaster. A wide range of bodily sensations were apparent. There was the struggling with nausea and the desire to run. The shock of such suffering was overwhelming, devastating, appalling, frightening, and deeply moving. Even on the first day, feelings of hopelessness and helplessness began to appear. Arriving at the orphanage opened a wide range of unexpected emotions. Feelings of fear and disgust toward the children were a surprise to some of the caregivers. Co-researcher Mary describes her feelings:

> [I was] overwhelmed by the sadness and desperate needs of these children. I hear the muffled crying from the children upstairs.

a. Not prepared. While the participants in this study had believed themselves to be prepared for anything, they were caught off guard by the sickening smell, the sounds, the flies and mosquitoes, and the actual state of the children. There was no way to prepare for this experience. Co-researcher Rachel shares her initial experience:

> I don't know if I can describe to you how I felt walking through those doors. I thought I had prepared myself for what I was about to see, but I guess there is never really any way to mentally prepare yourself for such suffering. . . .

Another example is described by co-researcher Krystal:

> I tried to prepare myself as best as possible for the sights but you can't prepare for the sounds and you can't prepare for the smells.

The caregivers not only were not prepared for what they were witnessing, but also were not prepared for the wide range of emotional and physical responses that were arising within them.

Co-researcher Rachel's description continues:

> I think the only way I can describe how I felt at that very moment was that I wanted to vomit. I felt like I was going to lose everything in my stomach. I couldn't even cry. I just wanted to run. I didn't want to be there; I didn't want to see this; I didn't want to hear the children crying. All of a sudden, I didn't want any part of this. I just wanted to go home.

b. Hopelessness and helplessness. The hopelessness and helplessness came in waves. The co-researchers' deepest sense of themselves was challenged; at times, feelings of inadequacy as caregivers flooded or overwhelmed them. Consider co-researcher Krystal's experience:

> I felt so helpless, because of my fear and I wasn't expecting that. I didn't think that would be a problem and I was just amazed that I was afraid of this little, little, teeny little wispy child.

Co-researcher Mary describes feeling a constant drone of helplessness and a feeling of being ineffectual that at times was paralyzing:

> I was knocked out of the illusion that I was here to help these children and these people. Hell, I couldn't even help myself.

c. Anger, guilt, sadness, and pain. Anger at the Romanian workers, the United States, and God appeared in the beginning. There was guilt for not being able to do enough, for not being able to be the persons that they thought they were or could be. Shame was added to the guilt for having feelings of disgust at and fear of these animal-like children. The feelings of guilt, which were present throughout much of the experience, are even more alive today for a number of the co-researchers. They describe a continuing struggle with guilt about leaving the children in Romania, as well as going on with their own lives. Co-researcher Marcos describes his feelings:

> Guilt often nagged at me when I think of them; it still does. Anger at the orphanage workers and my fellow man in general, particularly the privileged Americans, often filled my heart and still does. Who are we to go on when others suffer?

Co-researcher Lisa states:

> This [not feeling the children's pain is when the part of feeling ashamed comes in.

A lingering sadness for the children seems always present, and the experience remains very painful. Co-researcher Jeanann states:

> My memories to share with you about Calarasi will be only peppered with my sorrow, frustrations, and a whole gamut of emotions I had never experienced before in my life. Because to dig so deep into these extraordinary three months and come in full touch with what I felt, as you have requested, would be too painful.

There is ambivalence about the pain. The caregivers, at times, wanted the pain to leave, while at other times they wanted to hold onto it. Co-researcher Jeanann recounts an incident in the interview and begins to cry as she asks:

> How can it [pain] ever leave? How can I ever . . . in some ways I never want it to. I ache for children that are not loved and [not] treated like little human beings.

2. Continual questioning. Throughout their descriptions, the caregivers asked themselves questions as they attempted to sort out and make sense of their experience. In the initial contact with the suffering, the questions included these: "What had I gotten into?" "Why am I here?" Co-researcher Rachel describes herself as a coward because she couldn't face how horrid the conditions were, and she began to ask herself, "What's wrong with me?"

At some point, their focus shifted from questions about themselves such as "What have I done?" or "What am I doing here?" to others, the world, and the reality of the situation itself: How could the Romanians let this happen? Why do we as human beings let others suffer? Why do others not respond? In this modern era, how could this be happening? Just how could this be?

Over and over again, the questions of how and why came from the caregivers' lips. As co-researcher Marcos asked:

> How did the workers there and the people there live with seeing those kids suffer at their hands?

Co-researcher Jane directly states:

> My soul was screaming "How?" and "Why?"

The seemingly endless questions go on: Did I do enough? Did I make a difference? Did I make it worse or cause harm? Why did I go? What was it I did do? Did we do the right thing? Why do people cause suffering? Who are we to go on while others suffer? Why don't we do something? What's going to happen to them? Will it ever end?

Co-researcher Mary describes the continual struggle:

> As far as improving the lives of the children, what were we really doing here?!. . . . In retrospect our coming seems almost cruel. We held and loved these children, slowly coaxing them into their world and beyond down by the river . . . return them crying and begging not to be left in the four walls of their prison. Would a blind man be willing to see for an hour knowing that the rest of his life would be spent in darkness? Would the hour of sight comfort him or would it become a torture and exaggerate his darkness? Oh God, did I do the right thing? It's so sad! Do you think we made it worse? Is it worse? How are we ever going to change the world?

3. *Indelible and unforgettable experiences.* While the whole experience is certainly one the caregivers will not forget, there are specific moments or incidences that stand out for each one of them. Many such moments happened on the first day. Co-researcher Marcos's description captures the physical responses as well as the unforgettable quality of the experience:

> I remember the pit in my stomach growing deeper when I saw the dingy appearance of the orphanage—there it stood, desert brown in the middle of a bone-chilling winter. Probably, one of the most powerful images for me was stepping into the building, like here it goes, here we go. Now it begins, now what you really came here for starts to happen. So, that really is a stark memory.

Co-researcher Jeanann states:

> That first day, a chill had just come over me as I remember so vividly walking up the steps and into a dingy dark hall. My senses were alive with anticipation.

Many report a distinct memory of a child or an incident that is still with them:

> I will carry her face through eternity.
>
> I will never forget his eyes, the trembling of his body, the gentle thud of his body next to mine.

In the following description by co-researcher Lisa, one can hear the unforgettable nature of the experience, along with the wide range of physical and emotional responses, and her self-questioning:

> The first day I screamed for help—fifty little sets of eyes looking at me pleading for love and human touch. I remember walking into the classroom and just being so appalled. I walked in and it was like throwing yourself against the wall when you see something so frightening. I wanted to do that so badly, because there were all these babies and children on the floor like animals. I thought, Oh God, what have I done? What am I doing? I was not prepared for it. I really was not prepared for the state these children are in. What am I getting into? I will never forget realizing how their little bones stick out. Our big bodies looking down on these little animals on the floor. We were useless. It was scary. I was scared. I really was.

a. Reentry into America. The reentry into America was a shock. As co-researcher Mary states:

> When I got home the first thing I had was a salad and a big glass of water. I was just crying. This is crazy! I had red tomatoes, and it was incredible.

Co-researcher Jeanann describes her experience:

> I thought about that first meal that I had in New York after arriving from the three months in Romania, and how I had to get up; it was at the hotel. I had to get up and leave the restaurant. We were having breakfast, and I ordered a waffle, and before it could even come, I looked around and everything was so clean, and everyone was eating these foods with all these choices, and it was just the imbalance of them with nothing there and us with so much that I just wanted . . . it was too overwhelming for me. I had to get up and go up to the room and just cry and sob. I had a lot of tears that I guess I had not let go of and that's when they chose to come out, or it's when I'm just sitting there looking at all of this opulence, really, that we take for granted every day.

4. Making a difference, often with hope. People saw themselves as coming to the orphanages to make major changes, to make a difference, to help. The need to make a difference, to effect some change, however small, was paramount. Many people felt called to go and give the children love. As co-researcher Jeanann shared:

> I can go and love those kids and I just had to, and then everything fell into place. Couldn't I go and do something—anything—to make a little difference?

The desire to have awakened something, started a ripple, ignited a spark is still present in these individuals today. There is an ongoing struggle about the pull to do more, even go back.

In this context, the theme of needing to keep hope alive appears again and again. Some even say feeling the pain or thinking about the children keeps their hope alive and gives meaning to their experience.

Co-researcher Krystal states:

> If I couldn't cry, these children wouldn't be in my heart and maybe this whole thing was for naught, and it certainly wasn't. And I still dream about them, so that means there's hope and that's good.

Co-researcher Mary says:

> There's got to be a grain of hope in there somewhere, or else we should all commit hari kari and be done with it.

Co-researcher Jeanann reflects:

> I guess I have to convince myself, yes, some changes are being made. And we did make a difference. All these teams going over had awakened something over there. And it just has to start with a little ripple, you know, and if we've done that, praise God, because some things will be done, and are being done.

Co-researcher Carrie says:

> I'm sure we have done some good to comfort them in their internal and external suffering.

Co-researcher Mary recalls:

> There is hope. Hope for these children, and in that thought I can go on. I leave group 5 and begin climbing the stairs to the cries of the others.

5. Being transformed. There were various experiences that were transformative in nature: moments of renewal when all felt hopeless, anger and judgment melting into compassion, and self-transformations that opened the heart. Learning about self, others, and ways of the world occurred in different ways. There was also a growing and changing awareness regarding the suffering of all children, a rising above oneself, this orphanage, this moment in time. Co-researcher Rachel said:

> You change, you never come back the same.

a. Renewal. Even small changes in the children were renewing. The gifts received from the children are mentioned throughout the co-researchers' descriptions. Co-researcher Marcos describes his visits to one of the rooms:

> The occasional visits I paid to this group would be refreshing and I realize their energy gave me a "shot in the arm" numerous times when I needed it.

Relating one particular memory, co-researcher Mary continues:

> Ceresala looks up as I call her name and this sets her into motion like a wind-up doll. She claps her hands and rocks her head back and forth like she is keeping time to music that only she can hear . . . I stand at the edge of this insanity. I turn to go and outside the frosted windows I see the tops of white little flowers breaking the ground as they make their way through the garbage to the sun. I am renewed in this moment.

b. Compassion. Anger and frustration at those who were entrusted with the care of these children eventually gave way to compassion for the Romanians involved. The Americans came to understand how hard life really was in this country. An emerging understanding and empathy for the plight of the Romanian people led some to reflect. Consider co-researcher Marcos' questions:

> And then I wonder, are they that much different from me? If I was there, you know, would I have been the one doing that? How do people get like that?"

The co-researchers were also hard on themselves. They thought they were not doing enough and felt they were not able to respond with enough caring for the children. But, much like the change in their feelings for the Romanians, they came to have more compassion for themselves. At one point, co-researcher Lisa stated:

> I came to understand the situation was a difficult one and [my] feelings were normal for such a painful situation.

c. Self-learning. New and deeper self-realizations occurred. Some found, for example, a new strength and sense of power balancing or replacing the feelings of inadequacy and helplessness after they returned. Co-researcher Rachel describes this side as:

> I've been to Romania. I can do anything now.

Co-researcher Marcos says:

> I have confronted evil in the face and done something about it.

In addition to this self-recognition, these caregivers felt gratitude toward their own parents and expressed an appreciation for parenting skills. A new appreciation for democracy was expressed as well.

d. A global perspective. Seeing meaning in the suffering of innocent children is very hard. Yet, being with the pain and isolation of these children awakened an awareness of the needs of children all around the world. The strong desire to get others involved, to heighten their awareness, and to break one's own and others' denial around this suffering is repeated again and again. As co-researcher Carrie describes:

> I really wasn't aware of problems and I didn't care, because I didn't know and I never thought about it. You think you would know there would have to be problems in the places where there is oppression and there is war and strife, there would be problems with children. But you don't think about it because you're so busy doing whatever you have to do. But since I was there, I've been to other places and you really see how kids don't have a chance.

The orphanage conditions and general situation were often compared to some kind of death camp or to the Holocaust. There was even an expressed desire to build a memorial to these children. Caregivers had to reach more deeply into themselves to attend to the suffering they were seeing and to admit to themselves that such suffering does, in fact, exist. As co-researcher Krystal stated:

> I had to experience it. And it was a lot crueler than I wanted to believe that people are capable of being. Look at our history books; look at some of the stuff that isn't documented. This is the thing that's not documented.

Each individual's awareness regarding suffering in the world has changed, and the desire to do something about this suffering seems quite alive in them. They have come to realize that separation among groups of people is one of the main contributors to the continuance of unnecessary suffering and that the needs of humanity require a bridge that connects each of us, one to the other.

6. Experiencing an altered reality. There was a quality about the experience that was unreal, an altered reality, like watching a horror film or having a bad dream. Which world was the real world? Co-researcher Rachel, at one point, asked:

> Will the real world please stand up?

Yet there was also a quality of seeing in new ways, as described by co-researcher Mary:

> Suffering [has] this other side. Like getting a package of M & M's from home would be like the biggest thrill of your whole life. You would spread out all the colors and you would just look at them. I've never been on drugs, but I imagine that's what it would be like. Look at the yellow. It's just so incredible! Every little thing mattered, like the flower in the garbage. But I wouldn't see that otherwise. I never saw that kind of thing in my life before. My other world, my American world before I went to Romania. You just walk and you don't see things. [Now] it's almost like you're looking for anything subconsciously that could be worthwhile or redeemable. Something to say, okay, we are okay as a human being because of this flower. Okay, there's a child smiling. I won't kill myself. That smile was the most incredible smile of any child that had ever smiled or will ever smile in the world or in the Universe. It was all you needed. Everything that was good was intensified more than it was. Every joke was funnier. I don't know. There was something about being stripped down to that rawness and basicness. . . .

7. Relating to the children. The co-researchers attempted to develop a connection with the children in various ways. They described themselves as trying to empathize, to feel what the children feel, asking "What if that were me?" or noticing that a child reminded them of someone they knew. Co-researcher Lisa said:

> I tried to get into their little bodies to feel what they were or were not feeling. There were times when I [just] couldn't go that deep.

They perceived the children communicating with and through their eyes. Co-researcher Lisa noted:

> Their eyes did search for it [love]. How many times had they searched and not got it?; their eyes . . . their big, big eyes.

Every caregiver spoke of the children's eyes: "fifty sets of eyes"; "intelligence in her eyes"; "their big eyes staring at you." Co-researcher Krystal noted:

> The eyes said so much; the eyes were asking and I couldn't respond.

The children themselves were instrumental in helping the co-researchers to be with the situation they were experiencing. Recognizing the children's courage in the midst of their suffering was strengthening and inspiring. There was a sense of awe at the resiliency of the human spirit in the ability of the children to respond or simply to go on with so little.

Knowing these children was also referred to as a way of remaining sane. They were perceived as giving and, at times, even taking care of the caregivers. Co-researcher Marcos said:

> He helped me keep my sanity.

Co-researcher Lisa stated:

> I used to think I was coming to solve their problem; they helped me.

One can hear, see, and feel how the children helped in the co-researchers' descriptions. Co-researcher Marcos describes his relationship with the children:

> I didn't think of it at the time as them being my friends, but you know, I kinda disqualified them from being eligible to be friends, you know, the roles we had of each other, but they really were. They helped me to deal with the situation so that I could give them something back. It was an exchange. I'm not sure if it was equal. At times, they gave me more than I gave them.

Co-researcher Lisa describes her experience:

> I was afraid. I felt like they were the adults and I was the kid. Because their little faces just kept smiling and doing whatever. But they were helping me.

Co-researcher Krystal says:

> The children gave and taught me [that] in all the suffering and misery we can learn something from everyone if we take the time.

There were times when the connection with the children had a transpersonal quality. Co-researcher Marcos describes his experience with Radu:

> He was one of those kids you would go wow, this kid is so cute, I was almost stunned to look into his eyes. I thought, what's going on here? I don't get it. This innocent look. This and it was like he had something deep in his soul, some question. . . . I would take to laying down on the floor and I put him on my chest so we were face-to-face. Initially he cried, but after a couple of moments, he'd just get real peaceful and relax. Really tender moments. Very gentle, very gentle kid. He'd settle down and you'd feel like the connection was there, and you could hear him. I'd hear him breathing and he would just relax and I could feel him having some peace. And I would too. There was one time when I was talking with him, and he just reached over and put his hand on my leg or put his hand on my arm and I just felt this energy. This warm energy in his hand. This [boy] has . . . he has healing energy coming out of his hands. I said, "Wow." And this is between me and Radu. . . .

Touch was an important part of relating to the children for many. This was especially evident for those who expressed fear and even disgust toward the children. There would be an important moment when physical touch would break through the barrier. As co-researcher Jeanann says:

> . . . the barriers were broken, even the sickening nauseating smell started to leave and magic began to happen.

The following passage is from co-researcher Lisa, who worked in one of the most challenging rooms of the orphanage:

> Alina was a little girl who had AIDS and she had open sores all over her body, open sores all over her hands. I couldn't touch her. I couldn't touch her. She would look at me and I would feel so ashamed that I could feel that way. I didn't know how to deal with that. How could I feel this way? I'm still ashamed now of how I felt. I was sitting on the porch outside one day and she walked by and rushed over, and I felt so grossed out. I remember picking her up and putting her in the corner as far away as possible. . . . All of a sudden I felt this thing crawling across my lap and it was Alina. She had crawled all the way over from the corner. She was in my lap. And I looked down and here she is smiling at me. And I remember I started to cry. I just broke. How could I do this to her? It was horrible. It was almost like God's spirit in her. How could I do that to her? It was then I broke and it was really good. I just broke. I'll never forget that moment.

8. Communicating with God. Among the various transformative, even transpersonal, experiences described throughout the protocols, there were moments of turning to God in prayer. These prayers often asked for the grace, courage, and strength to face what was going on both internally and externally. Co-researcher Jane tells us:

> I knew of only one right thing to do in those moments of intense emotional pain and that was to pray. How many nights in the middle of the night you wake up, but then often that was when you would just pray. There was only so much you could do, try to accomplish. It was about waking up at 3 a.m. with the sky clear as a bell and not have any answers for what was going on around you.

Co-researcher Lisa stated:

> I remember praying to God. God, please give me the strength to do it. I didn't think I was going to be able to. All these little kids, I felt they were looking at me, and I had to save them. God, what have I done to myself?

While God offered comfort and solace for some, others felt forsaken. Questions were asked: "Where was God in all this?" and "Why would He allow it?" Some caregivers expressed anger at God for allowing the suffering of innocent children. Co-researcher Rachel describes the challenge she faced:

> How would a merciful and compassionate God allow such suffering? Confronted with your faith and your beliefs, you think you're just going to take care of some babies and all of a sudden you're confronting everything you love, who you are, and what you think. . . .

9. Developing ways of coping. As described above, there was a wide range of intense, at times overwhelming, emotions, anger, guilt, hopelessness, fear, helplessness, frustration, and the ever-present sadness. Some described emotions they had never felt before. Some described the feelings as so intense that they were unbearable. During these times, there was often a need to escape, to get out, a feeling of not being able to spend another minute there, even of being at the edge of insanity.

At times, there was not just a desire to leave physically, but a sense of leaving one's body, of putting up a wall and distancing. Some recognized this as an emotional defense designed to protect them from really letting the suffering in. Co-researcher Lisa said:

> I remember those times when I didn't feel anything for them. I was oblivious to their situation. . . . It was my way of keeping my sanity.

Even that which had first been intolerable became bearable over time. Remembering one particular visit, co-researcher Jeanann says:

> I was never so happy as when I was walking away from that orphanage. I could hardly walk fast enough back to our temporally home to throw myself on the bed and cry and cry and cry.

One means of coping was in knowing the stay was finite. Co-researcher Marcos states:

> One of the ways I coped was I kept the end in sight.

Co-researcher Krystal says:

> One of the ways I coped was by saying, America is not like this.

In many ways, the experience of the suffering the co-researchers encountered is still very much with them in their lives such that they continue to cope with and integrate their experiences from this time. Feelings of sadness are still pervasive, and pain regarding the children and their situation continues to arise. Co-researcher Jane describes how she currently deals with these feelings:

> I shut off and don't think of it unless there is something constructive to be done.

Knowing such cruelty first-hand has led the caregivers to suffering, although different in kind, with the children. When they allow themselves today to think of the children, they describe and reexperience the pain. Many saw more than they ever wanted to see; a loss of innocence occurred along with the disbelief that human beings could be treated this way.

These feelings are currently being processed in different ways. Co-researcher Marcos describes how he continues to cope:

> My intellect is real strong, and I think it is an indication of how I've kind of steeled myself [to] the emotions that go with what I saw there. I intellectualize it as a result of it. I'm listening to music [music is playing in the background], sometimes, and it's like . . . there's no words. I don't have any explanation for when I feel this deep, soul level kind of sadness. And as you were reading, I wasn't really listening. I was listening to the music, I started to well up from that sound. It speaks to the softness and the beauty of the human soul that suffers. That music is the soul speaking. The music is so pure. It really sets me free to let my soul feel.

10. Reality orientation: Coming to terms. At some point, the realization hits that there is only one thing the caregivers have to give, and that is themselves. Experiencing what was real often involved a more direct coming to terms with what could or could not be done. The breaking down of tasks to small doable, achievable goals gave many the courage to go on. Co-researcher Carrie states:

> All we could give them was [our] attention.

Co-researcher Rachel describes the second day:

> . . . the realization of the whole situation began to set in, . . . simply all we had were children.

Following this initial realization, a number of co-researchers reported issues involving their perspective regarding plans and actions. Co-researcher Jan states:

> We started making posters; we tried to bring some color into the room . . . and we started to do classes.

Co-researcher Jane says:

> I was feeling frustration with an urgency to change the situation.

Co-researcher Jeanann describes her process:

> . . . my enthusiasm was bashed so many times over the next few days. . . . It was like we were doing so little. . . . I sure didn't know if I was even penetrating, if any of my love or ideas or anything I came to offer was even making a difference.

Co-researcher Rachel expressed her feelings:

> You think you can make a difference and when I looked at these children I knew there wasn't a whole lot I could ever do.

Co-researcher Marcos states:

> I had to override my recognition that I was singularly inadequate to meet their needs and offer what I could. I understood that I made a valuable difference in their lives though it was far short of what I wanted for them.

While many had thought they could save all the children, they now realized that working to save one child took all of their focus, all of their energy. A child must be able to walk and feed himself or herself to even be allowed in a normal orphanage. When these more concrete and attainable goals were accomplished, this accomplishment seemed to replenish the caregivers, allowing them to carry on their work. Small changes were often experienced as giant steps, offering some hope. It was these moments that provided some answer to the repeatedly asked question "Why am I here?" Co-researcher Mary's words offer an integrated description of these different themes related to first realizing and then coming to terms with the hard nature of their situation:

> It was like trying to save a drowning man who clawed at you and held you under the water trying to climb out on top of your body. Only one of you would survive. Emotions ran wild. At first, I was emotionally available to everyone, including the dogs on the street. Then, over the months, I slowly began to pull away. By the end of my stay I was not even emotionally available for myself. And, a year and half later, in some ways I'm still not. I came as a hopeful savior and left as a frightened child. Exhausted, defeated, and disillusioned. How can I describe my feeling of pain and reaction to the suffering around and in me. In the beginning you don't believe what you are seeing, smelling, and feeling. Everything is played out in some cheap surreal movie. Then, as the sickening reality sinks in, you feel such pity and sadness for the children. Sadness for yourself. You are that frightened child again abused by your surroundings, abandoned emotionally by your mother, and left alone to turn inward to the only safety you know. The pain of your childhood is multiplied a hundred times in the lives of these children. The walls you have built to protect you from the cruel reality stand in your way now as you try to reach into the souls of these children. You want to bring them back to themselves. This all seems really worthwhile and meaningful when you see a smile on a once sullen face. The day four year old Costinel fed himself for the first time, I was elated. I was high on life and hope and promises! I was renewed in this major accomplishment.

As a synthesis and structural summation, I offer the following phenomenological psychological description of the experience of being with the suffering of orphaned children for whom one had cared:

The experience of being with the suffering of others was a developmental process for these co-researchers. It was a humbling experience. The participants initially all had altruistic ideas of how they would respond to being with the suffering of others. Each participant felt that he or she would be able to offer significant relief to the suffering Romanian orphaned children.

The initial contact with the suffering of others elicited shock and bewilderment. The participants were not prepared for the sensory assault they faced. Suffering does not take place in a vacuum; it is an interactive process between the individual and his or her immediate environment. One can physically leave the scene of suffering and still emotionally feel its presence. The experience of being with suffering seems to possess a lingering quality.

Throughout the experience of being with suffering one engages in periods of pulling back from the suffering and then moving toward it again. Being with the experience of others' suffering does not fit with one's picture of how people want the world to be. This experience provoked questions such as, "How and why is this happening?" There were no humanitarian answers to these questions. This experience of being with suffering was even more painful because it represented man's inhumanity to man. Caregivers can experience cognitive dissonance when being with the experience of suffering. Participants found themselves in various uncomfortable states arising from thoughts such as, "I came to love and care for these children, but I am also afraid and disgusted by them."

Being with the experience of suffering can challenge the previously held constructs of a person. This experience challenged the participants' sense of the world, and some co-researchers' faith in God. This research suggests that being with suffering in an effective way requires moving away from the suffering. Being with suffering can drain a person of his or her physical and psychological reserves. Moving away from the suffering allows one to reconnect with oneself and tolerate being with the suffering person.

The experience of the co-researchers suggests that an empathic connection is crucial, but not easy to achieve. This connection may be initiated by the one who is suffering. In this case, the orphaned children initiated contact with the participants. The participants found that physical touching, hugging, and playing with the children created a powerful emotional bond. This initiation by the child and the acceptance by the co-researcher melted the co-researcher's fear. This represents strong evidence that physical contact is a key to a relationship between the caregiver and his or her experience of being with suffering and the suffering person.

This study suggests that relationships with the person suffering and one's colleagues in the shared experience are important to everyone. Strong personal bonds, teamwork, and a sense of community developed among the co-researchers. Findings also indicate that the experience of being with suffering lives on with these co-researchers. Physically leaving the immediate environment did not relieve their internal conflicts about their experience of being with suffering. Two to three years after returning from working with the suffering children, they are still trying to make sense of their experiences.

For these participants the experience of being with suffering adds another dimension to life. Once one has been exposed to the suffering of others it is not easy to forget. There is a continual call to respond to the needs of those who are suffering—an ongoing internal struggle for not doing more, and a feeling of guilt for not taking action to relieve the suffering. Having been with the suffering of others elicits an expansion of one's world view that includes the suffering of others as a harsh reality.

Each held the desire to break the secrecy around suffering and involve others in the responsibility of caring for those in need. The internal conflict of living one's own life—being present for oneself—and being present for the needs of others created a moral dilemma.

It is painful to have broken through the denial of suffering, but it may be freeing, too. The freedom comes from confronting one's fear of suffering, breaking through one's isolation, and being touched by the suffering of others.

THOUGHTS, FEELINGS, REFLECTIONS

The initial shock of the first day at the orphanage was a powerful surprise, as the co-researchers all felt that they were prepared for anything until that first encounter. As they went home on that first day to vent their feelings, regroup, and face the reality of the situation, the formation of a true team or community began to develop. It was clear from the beginning that the support they would need in order to serve these children would have to come primarily from each other. I remember participating many years ago in a five-day team building exercise, a program designed to help business managers learn to work together. The "program" in Romania, however, did not end at five in the evening. Everyone could not leave to have a nice meal, call home, check office messages, or have a drink. Romania was the real thing.

Throughout this chapter, I have referred to the co-researchers as caregivers, "caregiver" being the official description of the position they were signing up for. Sherwood (1992) describes caregivers' responses to the experience of suffering and what it means to care. According to Sherwood (1992, p. 106), caring produces positive change, encourages growth of the other, and involves both a mutual connectedness and a being present to the other: "Our purpose as caregivers is to improve the human condition, to enable others to live life to its fullest, and to alleviate suffering."

Regarding the study described in this chapter, producing positive change, improving the human condition, and encouraging growth were the exceptions, not the rule. While there were moments of mutual connection, and even of being present, these moments represented only brief interludes in an otherwise overwhelming situation and experience. The ideas of enabling others to live life to its fullest and to alleviate suffering are most likely notions the caregivers had when they first left for Romania. They believed these contributions were possible, but soon realized this belief to be a naive dream. Over and over again, the caregivers speak of "not making a difference" and even express the fear of "making it worse." Their initial expectations of major change ended with the hope that they had added something, anything, to this dismal environment, that they had at least been "a drop in the bucket" or "a single spark." As co-researcher Carrie said, "I think we all thought we would go there and change things right away in the first week, right? Walk in there, and all of a sudden miracles would happen. It didn't happen like that."

Given their knowledge of the extent of the children's suffering, there was guilt for not doing or being able to do anything to reduce it. The experience had opened them wide, and now there was no way to close that door. Existential philosophers speak of existential guilt, the guilt of not living up to one's potential or, as Yalom (1980, p. 277) states, "[feeling] guilty to the extent that one has failed to fulfill authentic possibility." Since they saw themselves as caregivers, it became a transgression of self not to take action to ease suffering once it was known. They felt guilt not only for not fulfilling their own potential, but also for walking away from children who already in life have little or no chance at even beginning to fulfill their own. This self-perception of "caregiver" was challenged daily in Romania, as well as later when they reentered American life.

Is there an answer to ease the pervasive guilt these individuals are still facing in their lives? Perhaps. Becker (1992, p. 227) offers a perspective that addresses this very question:

> Faced and accepted, existential guilt can clarify values and ground us in commitments by which we can live with self and others in a shared world. Ultimately, it enables us to face ourselves and others

> with integrity, to stand with human values we can live by, to stretch beyond the confines of our personal concerns, and to join hands with like-hearted people. Facing existential guilt transforms it from a limitation into a strong basis for inter-human living.

What if the caregivers had been presented with information about how people respond to such suffering? What if they had been able to read what had happened to other people who had experienced these kinds of circumstances? Would it have mattered? Would they have felt more prepared? Would it have seemed less of an assault on their physical and emotional beings? As one co-researcher stated, "I guess you can't prepare for such suffering."

For this group, we will never know. I do believe, however, that anyone "back home" who had received letters from any of these caregivers describing their experiences would now be better prepared to face the situation they will encounter than the first individuals who went in without such information. One person who returned from Romania after her first visit to the orphanage stated quite frankly why she did not stay: "I could not stand the smell." What better way to prepare than listening to the words of others who have had the experience?

Since there was no program offered to debrief and reintegrate this particular group of caregivers as they returned to the United States, and no ongoing support of any kind, this research provided a first opportunity for some to discuss their feelings about their experiences in Romania, thereby breaking the isolation they were feeling. For others, participating in the research allowed them to feel as though they were continuing to do something to help these otherwise forgotten children.

The experience of analyzing the descriptions of those who worked with me in the orphanages brought me into their experiences in a very deep and powerful way. In a close and committed group, there is a very real sense of connection, even power, when another speaks and you know you're not alone, when you realize that there are others who feel the same way you do, or when you see that they are struggling with issues that touch your heart. As a caregiver, I too felt these various feelings, including the often pervasive guilt about leaving and not doing enough. The question "Could I have done it differently?" kept me awake on many nights. I now see that when I had first decided to do this research, prior to leaving for Romania, I did not realize the full implications of what I was about to do. As it turned out, I had unknowingly committed myself to personally integrating a powerful, self-transformative experience.

The beauty of this investigation was in experiencing the dance of my own emerging responses and reactions as I carefully read and worked with the co-researchers' descriptions. I am not separate from that which I am investigating. I am a true participant in this phenomenon as its meaning and form became clear. The process of bracketing, in which I acknowledged my participation by observing and noting my initial expectations and feelings, facilitated my going more deeply into the meaning mosaic. This research, in the end, provided a forum for me to work through an experience that was far more complex than I had originally believed or imagined. In fact, at this very moment as I write the ending of this chapter, its effects are evident. I am aware in the present of desires, fears, and hopes (i.e., "biases") arising as I move to write the next phrase or sentence. Let me share some of these thoughts with you as evidence, if nothing else, of the continuing effects this whole process is still having on my life:

> I find myself wanting to write a "happy ending." Trying to sort through some way of ending this study on an upbeat note. Is there a way to let the reader down gently, and with hope? Was there

> some great spiritual transformation in the caregivers, or some great relief which resulted from what they had set out to do? Surely, in all this data, I can find some evidence that will not leave the reader with a heavy heart. Even as I close this chapter, another bias becomes apparent: the need to make it better, to protect others from the truth of how difficult and exceedingly painful this experience truly was. The experience speaks of a loss of innocence and the removal of a veil that, in the past, kept us from acknowledging and truly feeling the suffering in the world. The truth is there is no happy ending. There is only the endless suffering of these and other children. The only difference is that this suffering has now been shared with you.

As I cradle my own son in my arms, I see the unknown faces of all those children who do not have the love of even one person. I feel the pain of mothers whose children are hungry in their arms. I cry for the mothers and fathers who leave their children at the doors of an institution believing those walls have more to give their children than they do. The needs of the world's babies overwhelm me, and I observe myself stopping these thoughts. They are too much as I hold my own child. These children will always be with me. They are in my heart, but, at the same time, I see that this does not nourish, hold, or protect them. My own father, a World War II veteran who suffered his entire life from the ravages of that war, tried to protect me from feeling the effects of this suffering. He had asked me not to go, saying "Patricia, you will never forget; you will be forever changed." It appears that he was right.

CONCLUSION

If we are to offer support to those who suffer, then we need to deepen our understanding of the experience of all those involved, Perhaps because of its intense and emotionally painful nature, the experience of being with suffering has not been investigated, prior to this study, in any formal, systematic way. There is a true need for more qualitative research from psychologists, sociologists, anthropologists, and the medical professions, from those who have the ability to listen with their hearts. This work is ultimately the work of the soul. It is therefore through being present and listening, not through rational or experimental analysis, that these dimensions will come to be understood. In this way, phenomenological research offers a way of understanding the experience of being with suffering on its own terms.

The pain of this experience will not go away. It remains a continual reminder for those who opened to it. In this way, opening to suffering serves us by keeping us aware of and not separate from the suffering in the world. I will never forget arriving back in America and reading that in my small but wealthy California community, there were an estimated 350 homeless school-age children. Romania could attribute the suffering of its children to a dictator, poor economic structure, and the fact that almost everyone was poor. Did I respond to those homeless children in my own community? No, I did not. Although the same basic question that we asked in Romania, "How could this be happening?" also arises in our lives here in the United States, for me, observing my own lack of response, the central question has become: "Why do I not respond to those who suffer?" There are many factors, intrapersonal, interpersonal, and external, in need of further study regarding the nature of suffering. It is obvious that suffering is all around us and that those who suffer "call" to us. This phenomenon is complex, our response or lack of response has far-reaching consequence for those who suffer, for those who care for them, and for all sentient beings.

REFERENCES

Archer-Copp, A. (1990). The spectrum of suffering. *American Journal of Nursing*, August, 1990, 35–39.
Becker, C. (1992). *Living and relating*. London: Sage Publications.
Bockle, F. (1987). The anthropological challenge of pain. *Acta Neurochirurgical Supplementum, 38*, 194–195.
Colaizzi, P. F. (1978). Psychological research as the phenomenologist views it. In R. S. Valle & M. King (Eds.), *Existential–phenomenological alternatives for psychology* (pp. 48–71). New York: Oxford University Press.
Frankl, V. (1959). *Man's search for meaning*. New York: Washington Square Press.
Giorgi, A. (1971). A phenomenological approach to the problem of meaning and serial learning. In A. Giorgi, W. F. Fischer, & R. von Eckartsberg (Eds.), *Duquesne studies in phenomenological psychology: Volume I* (pp. 88–100). Pittsburgh, PA: Duquesne University Press.
Giorgi, A. (1975). An application of phenomenological method in psychology. In A. Giorgi, C Fischer, & E. Murray (Eds.), *Duquesne studies in phenomenological psychology: Volume II* (pp. 82–103). Pittsburgh, PA: Duquesne University Press.
Metz, J. B. (1995). *Faith and the future*. Maryknoll, NY: Orbis Books.
Polkinghorne, D. E. (1989). Phenomenological research methods. In R. S. Valle & S. Halling (Eds.), *Existential–phenomenological perspectives in psychology: Exploring the breadth of human experience* (pp. 41–60). New York: Plenum Press.
Sherwood, G. (1992). The responses of caregivers to the experience of suffering. In P. L. Starck & J. P. McGovern (Eds.), *The hidden dimension of illness: Human suffering* (pp. 105–113). New York: National League for Nursing Press.
Starck, P. L., & McGovern J. P. (1992). *The hidden dimension of illness: Human suffering*. New York: National League for Nursing Press.
Valle, R. S., King, M., & Halling, S. (1989). An introduction to existential–phenomenological thought in psychology. In R. S. Valle & S. Halling (Eds.), *Existential–phenomenological perspectives in psychology: Exploring the breadth of human experience* (pp. 3–16). New York: Plenum Press.
von Eckartsberg, R. (1986). *Life-world experience: Existential–phenomenological research approaches in psychology*. Washington, DC: University Press of America.
Watson, J. (1985). *Nursing: The philosophy and science of caring*. Niwot, CO: University Press of Colorado.
Webster's new collegiate dictionary (1976). Springfield, MA: G & C Merriam.
Yalom, I. (1980). *Existential psychotherapy*. New York: Basic Books.

17

The Experience of Being with a Dying Person

Thomas B. West

INTRODUCTION

Fifteen years ago, I sat at my aunt's bedside and held her hand as she died. This was my first experience of being with someone at the moment of death. It was a moment of great intimacy. At the time, I was aware of how common an experience this was for the hospital personnel around me. Yet, I felt a sense of awe and privilege at having been able to accompany my aunt and care for her during her last month of life, right up to the moment of her death. Since that time, I have been with many people throughout the last days of their lives. And I have had the privilege to be at the bedside of a number of them as they died. Caring for the dying and being with them as they die is a common human experience. This chapter describes an existential–phenomenological study of this common human experience and the meaningful characteristics that make it up.

Existential–phenomenological research is based on the philosophical presuppositions concerning human nature found in the philosophical schools of both existentialism and phenomenology. Kierkegaard (Bretall, 1946), in examining his own personal struggles, found in them examples of the universal struggles of being human. Existentialists followed his lead in looking at what is fundamental to lived human experience. Husserl (1962) also was concerned with opening up the phenomenon of concrete lived experience in and of itself. Heidegger (1962) investigated phenomenologically the experience of authentic human beingness, which he termed *Dasein*. For him, an awareness of death was a fundamental aspect of authentic beingness. Heidegger (1962, p. 354) says:

Thomas B. West • Franciscan School of Theology, Berkeley, California 94709.

Phenomenological Inquiry in Psychology: Existential and Transpersonal Dimensions, edited by Ron Valle. Plenum Press, New York, 1998.

> Death is not "added onto" Dasein at its 'end'; .... The nullity by which Dasein's Being is dominated primordially through and through, is revealed to Dasein itself in authentic Being-towards-death.

In discussing Heidegger's thoughts on death, Macquarrie (1965) points out that the person who lives an authentic life always anticipates death and lives resolutely in the light of this possibility. This understanding that all living is also dying allows for a form of transcendent awareness that gives a sense of unity and meaning to life.

Existential–phenomenological research does not focus on explanations of causality, but rather provides a descriptive structural analysis of any lived experience and thereby describes qualities of being. Johnston (1988) points out that a phenomenological investigation of the immediacy of lived experience is one that is deeply concerned with questions related to the meaning of life. He states that an examination of death, which represents nonbeing, is central to such an analysis.

Heidegger (1962), seeing that an essential aspect of being human was "Being-towards-death" (p. 354), noted that this awareness of one's own death is revealed to one through the experience of "Being-with-Others" and experiencing their death, that is, the death of the Other (p. 379). Being with and caring for others is essential to the nature of Dasein. Indeed, Heidegger says that "it is as care that Dasein's totality of Being has been defined" (p 370). Caregiving and being with the dying are very much a part of the immediacy of lived human experience.

Herzog (1966), a Jungian psychologist, also felt that the ability to acknowledge the presence of death in life is decisively important for authentic human living. He found that those who have gone through the actual experience of the death of another are then able to open themselves to a deeper aspect of "becoming" that is "the very stuff of life" (p. 9). It is also worth noting that individuals who have themselves come close to dying, those who have had a near-death experience or who have survived a suicide attempt, often report dramatic changes in how they live following this experience (Heckler, 1994; Ring, 1984, 1991; Chapter 19 in this volume).

This chapter will discuss an existential–phenomenological study that focused on the experience of those who have been with a dying person through the process of that person's dying. What is this experience, inherent in the human condition, for those who have experienced it?

A REVIEW OF THE LITERATURE ON BEING WITH THE DYING

Rituals surrounding the process of dying are part of all cultures (Campbell, 1959). In the West, institutions to care for the dying began with the spread of Christianity and can be traced back to the second century. They were often built in connection with a Christian church and were called Houses of God. As early as the ninth century and continuing into the present, Christian religious orders of men and women have been formed specifically to care for the sick and the dying (L. Butler, 1967). A recent example is the Dominican Sisters of St. Rose of Lima, which was founded in 1896 to care for those dying of cancer. The order currently has six hospices in New York City. Another is Mother Teresa's Missionaries of Charity, which was founded in India in 1948 and works primarily with the dying in hospices the order has established there and elsewhere throughout the world.

During the Middle Ages, there developed a literary genre called the *ars moriendi,* the art of dying. These books were religious tracts of instruction both for those facing death and for those who were supporting and attending them through this process. They focused on death as a transition to eternal life. Books of this type were best-sellers in their day, even as late as the 17th century in England (Kastenbaum & Costa, 1972). Today, we can see a revival of interest in this area of books that address the spiritual aspects of dying and caring for those who are dying (Cassidy, 1991; Kubler-Ross, 1975, 1978, 1981; Levine, 1979, 1982, 1984).

Although Mechnikov introduced the term *thanatology* in 1901, there was not much scientific research in the area of death and dying until Freud postulated his theory of a death instinct as part of his metapsychology (Alexander & Adlerstein, 1960). Psychoanalysts, however, either ignored or rejected this part of psychoanalytical theory because they could not apply it in their clinical practice (Lifton, 1979). It was not until Eissler (1955) wrote his seminal work, *The Psychiatrist and the Dying Patient,* that a psychoanalytical treatment approach to working with a dying client was developed.

Eissler understood a dying person to be in a biological process determined by the death instinct. He suggested that a different kind of therapeutic relationship was necessary in such a case. Eissler called this "the gift situation" (p. 126). He advised the therapist with such a client to be totally present to the needs and wishes of the dying person and to give himself or herself as a gift to the dying other without charging a fee. He said that the peculiar difference in psychotherapy with a dying person was the necessity of providing such a person with an experience of love, and that the giving of oneself as a gift to the dying other was itself an act of love. He comments that "love seems to be an antidote against the agony of death" (p. 139). Charon (1963, 1964) found that both the existential psychology of Binswanger and the existential philosophy of Marcel also saw love as the appropriate human response to being with the dying.

Yet what Eissler advocated was a very unorthodox psychotherapeutic relationship that even today not many professionals would willingly undertake, given its scope of involvement. On the other hand, he was describing a familiar and common aspect of human relationship, that of caregiving, of attending to the needs of, comforting, and being present with a dying person. In our society, entering into such a relationship has often been the role of women. The final caregivers who attend to the needs of the dying, both in the family and in hospitals and hospices, are usually female. It should not be surprising, therefore, that the early work of psychotherapy with dying clients, as reported in the literature, was done primarily by women (Joseph, 1962; Norton, 1963; Rosenthal, 1983).

Joseph (1962) had not read Eissler's book when her client, Alice, developed cancer. However, she adjusted their therapeutic relationship according to Alice's needs. In doing so, she felt that she was not being a good analyst because she was acting in a manner that was too emotional and subjective, and yet she states that she could not have acted differently under the circumstances (p. 27). She began to visit Alice at home, bringing her flowers, holding her hand, stroking her hair, and comforting her. Joseph made herself available at all times and did not charge a fee for being with Alice in this way. Later, she attended Alice's funeral and commiserated with her family on their loss. In retrospect, Eissler's work on being with the dying gave Joseph a sense of relief in regard to her own professional behavior. It enabled her to gain a consistent psychotherapeutic interpretation for all aspects of her relationship with Alice.

Norton (1963) noticed how reluctant therapists and even medical personnel are to build a relationship with a dying person. She comments that in such a relationship, the professional involved must be willing to be constantly available, empathetic, and appropriately responsive to the dying person's needs (p. 559).

Eissler (1955) noted how emotionally difficult it was to do this kind of work with the dying. Renneker (1957) examined the countertransference reactions of therapists working with cancer patients. He noted a cycle of reactions that began with a wish to terminate treatment and to avoid the client, followed by feelings of guilt and shame, which in turn were covered over by a reaction-formation of seeing oneself as the Good Samaritan. He described the narcissistic investment these therapists had in their clients' health and the subsequent threat to their feelings of omnipotence when their client's disease progressed and became terminal.

Burton (1962) also commented on the reluctance of therapists to work with the dying because of their own fear of death. He found that the major countertransference defenses such therapists used were denial, displacement, and compensation. Clarke's (1981) 20-year follow up study on Burton's research had very similar results. Feifel (1969) found a tendency in all health professionals to withdraw from dying patients because of their own issues about death and its threat to their professional narcissism. Weisman and Hackett (1961) said that professionals would rationalize in a defensive manner their own behavior in pulling away from a dying person. In such situations, they would note a retraction of libidinal interest and a turning away from others by the person in the dying process. It was actually the professional, however, who first began to withdraw from the patient (p. 251).

Although it can be, and often is, very difficult to be with a dying person, it can also be highly rewarding. Like Joseph (1962), Sadowy (1991) felt her way in her therapeutic work with her dying client, Dee. She too adjusted her boundaries so she could be totally present and available to Dee's needs. She comments that in empathetically being present to Dee, she was able to "enter a living vibrant internal reality" (p. 197), and that ultimately Dee helped her to differentiate her own spiritual values. Robbins (1989) also comments on the aspect of spirituality that becomes part of the psychotherapeutic work of being with a dying client. He recognizes to do so is to cross a boundary line as a therapist, but asks (p. 283):

> If we do believe that death is but one more transition to another level of consciousness, then joining our patients in this "no boundary" level of existence may be a choice for every therapist to make.

Feifel (1990) described the need of a new *ars moriendi* that would help the dying come to a more "human death" (p. 540). Such a treatment model must speak to the soul and have a spiritual dimension as well as a biological and behavioral one. For him, this meant that the dying need the support and presence of those who represent "psychological security" for them. It is their support, their sustaining presence and ability and willingness to communicate with the dying, that eases and facilitates the dying process. Feifel implied that it was the dying who would choose the support people whom they wanted to be with during this time.

In a similar way, Weisman (1977) saw the need of someone "skilled in human interactions" to work with a dying person to facilitate an "appropriate death" (p. 119). He described this as a collaborative relationship and not one that was primarily professional in nature. Such a caregiver would be there to support and guide the patient through the dying process, providing an experience of a "safe conduct" or passage "through peril and the known" (p. 116).

In an effort to distinguish this type of relationship from that of psychotherapy, Feigenberg and Shneidman (1979) called it *clinical thanatology*. In their model, it is the dying person who chooses who will fill this supportive role. Such a caregiver does not need professional credentials, but must be familiar with the dying process and have empathy and self-awareness. The relationship this caregiver undertakes is a committed one that lasts until the client dies. The aim of such a relationship is to support and help the dying person in any way possible.

As early as 1962, Burton (1962) noted a shift in psychotherapists' approach to working with dying clients from that of a privileged professional position to one that was more a "true encounter" between equals in an "I–Thou" relationship (p. 18). Perhaps no one exemplifies this attitude better than Elizabeth Kubler-Ross (1969, 1970, 1975, 1978, 1981 1983) in her extensive work and research with the dying. She says (Kubler-Ross, 1970, p. 165):

> Counseling the dying is one of the most intense, most personal, the most intimate kind of encounter between two human beings. . . . It is a listening to—and at times sharing—the innermost feelings and concerns, without layers and layers of trimmings around it. . . . There is no other counseling with any other type of patient where it is so important to take off one's professional white coat and just be a human being. Those who have experienced such moments will agree with me that working with dying patients is not depressing at all and can be an instructive, gratifying, and at times even beautiful experience.

R. N. Butler (1991), a physician, noted the continuing need for a "comprehensive therapeutic approach to dying" (p. 11). He wondered (p. 12):

> Considering the critical place that dying and death play in all our lives, especially in medical practice, it is surprising that there are few studies concerning the various dimensions of dying. There are few direct in-depth studies of the inner experience and immediate conditions in which people die and perhaps fewer still of the thoughts, feelings and practices of physicians.

Of course, Kubler-Ross (1969, 1970, 1975, 1978, 1981, 1983), in her books and with her Life, Death, and Transition workshops, has been directly addressing these needs in a nonacademic manner for 30 years. She emphasizes the spiritual needs of both the dying and those who care for them. Mother Teresa (1975) also understands the dying process to be a spiritually significant time both for persons who are dying and for those who are with them. Therefore, she reminds her sisters to always be grateful to God for the spiritual privilege of caring for the dying. The books of Stephen Levine (1979, 1982, 1984, 1987) and the workshops he gives with his wife, Ondrea, also address, in a spiritual way, the inner experience and immediate conditions of both the dying and those who are supporting them through this process. Levine has been influenced by the spiritual teachings of Ram Dass (Ram Dass & Gorman, 1985) and his work with the dying. Valle (1996) describes the relationship between caregivers and the dying as a "spiritually auspicious time" for both, and says (p. 3):

> Serving others selflessly in life and surrendering one's self-identity as a unique body and personality in death are, therefore, two manifestations of the same process—spiritual awakening.

Two recent phenomenological psychology research projects addressed the need Butler raised to examine the inner experience of dying. Stephen (1987) investigated Weisman's concept of an appropriate death by interviewing 12 terminally ill hospice patients. She felt that what constitutes an appropriate death is a notion best understood from the dying individual's point of view. And Ross (1987) investigated the phenomenon of death directly by engaging 26 people in an open-ended discussion of death. His study describes the phenomenon of death from the individual's experience. He analyzed his data hermeneutically.

It was to fulfill the further need for investigating the inner experience of a caregiver who is with a dying person as that person goes through the dying process that the study reported in this chapter was undertaken.

INVESTIGATING THE COMMON EXPERIENCE OF BEING WITH THE DYING

I began my existential–phenomenological research project by reflecting on my own experience of being invited to be with a dying person throughout that person's dying process. Colaizzi (1973) calls this initial procedure individual phenomenological reflection. Through it, I tried to identify my own preconceived notions about this experience before engaging co-researchers in my investigation. This process of bracketing my own presuppositions helped me to listen more carefully and understand more clearly my co-researchers' experience as they described it to me.

The results of my study were co-constituted by myself and my co-researchers. who shared with me their experience of being invited to be with a dying person. Because this process involved both their lived experience and my reflection and analysis of that experience, we did this work together. They are therefore truly co-researchers. After meeting with each of them, ten in all, and explaining my research interest and intent, I asked each one to respond in writing to the following request:

> Please describe for me as completely as you can your experience of being with a dying person, who was aware of being in the dying process, and who had asked you to be with him or her as he or she went through this process.

Having collected their written replies (protocols) and having reviewed each co-researcher's written response until I was totally familiar with it, I would arrange for an individual face-to-face follow-up walk-through interview. During this walk-through interview, I would slowly read back to my co-researcher his or her protocol. This step was included in order to allow each co-researcher the opportunity to make any corrections or additions. After this step, the tape recording of the walk-through interview was transcribed. The subsequent transcripts of my ten co-researchers, their original written responses plus whatever else was added in the walk-through interview, constituted the data that I analyzed using an existential–phenomenological method first developed by Giorgi (1975) and later elaborated on by Colaizzi (1978) and Elite (see Chapter 14 in this volume). The following is a summary of the steps used in this analysis process:

1. All the protocols were read and reread in order to get the sense and feel of the whole.
2. In each protocol, all sentences, phrases, or statements that referred to the experience being investigated were extracted to form a list of meaning units from that protocol.
3. The meaning units of each protocol were sorted into theme clusters, within each protocol.

4. The theme clusters from each protocol were then combined across all protocols to form a list of 38 overall constituent themes. During this process, the everyday language of the theme clusters was translated into psychologically coherent expressions used to describe the constituent themes.
5. Through integration, the overall constituent themes were reduced to 24 comprehensive constituent themes.
6. Further integration of these comprehensive constituent themes resulted in the list of the 14 final comprehensive constituent themes.
7. An exhaustive description of these 14 final comprehensive constituent themes was then written.
8. Disciplined reflection upon this description resulted in the statement of the fundamental structure of the experience.

WHAT THE INVESTIGATION OF THIS EXPERIENCE SHOWED

The co-researchers' protocols described, in depth and with intensity, their profoundly human experience of being with a dying person. For all of them, it had been an experience that was both painful and rewarding. Their experience was essentially one of loss through which many of them also felt enriched. The analysis of their data produced the following 14 final comprehensive constituent themes:

1. An explanation and personal description of the relationship and the invitation.
2. Anticipatory grieving: the process of denial and acceptance.
3. Painful, conflicted, and confused emotional states.
4. The course of the illness.
5. Remembering others who have died.
6. The role of family and friends.
7. Feeling a strong connection with the other.
8. Putting oneself aside to care for the other.
9. Humor: the energy of lightness.
10. The experience of gift and gratitude.
11. Spirituality, prayer, and religious ritual.
12. The dying hour.
13. The experience of grief and mourning.
14. Learning from the experience and feeling changed by it.

The results of this study, as indicated in these final comprehensive constituent themes, show that the experience of being with a dying person is, first of all, one of relationship. It is from this place of being in a relationship that the invitation to be with the other arises. This was true for all ten co-researchers: Alina, Catherine, Curt, Dick, Gregg, Jeannie, Mary, Michael, Michael Mary, and Regan. As Dick remembered:

> It was early summer of 1989, . . . when Steven asked me to walk this journey with him and beyond. He asked if I could see it all the way through and preside at his Celebration of Life ceremony. . . . The fabric of this journey is woven of many threads. There is friendship, love, suffering, courage, perseverance, hope and trust. These are not unusual threads for these kinds of journeys, but they are woven very uniquely in this one for me.

It is also an experience of anticipatory grieving. This process of grieving in advance begins in a place of emotional denial. The reality of the death to come is avoided or its impact is minimized. Gradually, this denial gives way to a place of acceptance. Throughout this process, the caregivers experience a wide variety of painful, confused, and conflicted emotional states. Among these are feelings of anger, guilt, fear, and anxiety. Jeannie reflected about this time in this way:

> I was very confused. I wanted to be with him, but I didn't want to upset him by being there and showing a disregard of his feelings. . . . It was probably my most trying time with him. I didn't understand what he was going through. I was besieged by emotions that I'd never dealt with before. The emotions I felt, I still can't even put into words. It was almost like not knowing where do I turn? What do I do? How do I help? How do I stay out of the way? What do I do?

The experience of being with someone who is dying is one of being with them through the course of their illness. This means being with them during times of pain, suffering, and distress. It is necessarily an experience of being with another who is progressively getting sicker and more debilitated. It is a very trying experience, as Mary attests:

> At one very critical time, I stayed in the hospital for seven nights. I took care of my husband during the night. He was in agony. This was the hardest time for me. . . . He was in absolute misery. Absolute misery. And I was, too. We were both suffering a lot right at that point.

Such caregiving also evokes memories of having been with others who have died. It is, as well, an experience of being with others and forming a sense of community around the dying person. This community is made up of family, friends, and other support-givers.

The caregivers develop a strong emotional connection to the dying other that is often expressed by physical touching. Accompanying this connection is a willingness and effort on the part of the caregivers to put their own needs aside in order to care for the dying other. This selflessness is often spoken of as an expression of love for the other. Alina describes this:

> I experienced myself as setting aside my emotions with regard to Richard and his illness in order to be present for him in another way. I felt myself stripping off any emotion that Richard could experience as emotional need, any thought or impulse that might draw one drop from Richard's inner resources, which he needed all for himself. . . . That was how I experienced myself then, a friend not detached but taking on detachment as a favor to a friend, gently but firmly laying my immediate emotions aside for a different kind of love. . . . I see all the complex subterfuge that went into that supposedly simple laying aside. It was loving subterfuge. . . .

Unexpected moments of humor shared with others during the dying process are likewise a constituent of the experience. Catherine found insight into the naturalness of dying when her friend's dying request was received in wonder, joy, and mirth by her family and friends. She remembers:

> The air was still and sacred. And then we laughed. Cathy broke the silence with an almost inaudible request for a new flavor of popsicle. . . . Her parents and I looked at each other in delight . . . we've talked about it since that time . . . it was like Cathy played a joke on us with our incredible seriousness about death, so that suddenly things became beautiful in a new way. . . . It was like dying is not just about serious profundity and trying to be so perfect. It's just about laughing too and wonder. . . .

Many co-researchers describe their relationship through this time in terms of a gift they have received from being able to be with the other in this process. A strong sense of gratitude and appreciation for having been part of this experience with the dying other is a common response. Michael said:

> I could let go and say good-bye, and also say thank you, with a tremendous, deep abiding sense of gratitude. That's just what gratitude is basically—it's the ability to be present to the love that's there.

Another aspect is a sharing in prayer and religious ritual, both during the course of the illness and after the death of the other. The caregivers felt their whole experience was a spiritually significant time in their lives. Remembering his friend before his death, Dick said:

> . . . he really afforded me the opportunity to see God in even broader ways than I've ever experienced God. . . .

Later, discussing his friend's funeral service, he said:

> It was just amazing to look out at that group of people congregated there. . . . I never felt so empowered by the experience of the people I was experiencing this with . . . we loved and had been loved. . . . We sought understanding in his death. . . . We were reduced to that base we call Faith. Several of us had been chosen to make this journey with him. I, for one, have a new understanding of healing. Healing is what happens in the soul, the innermost part of our existence, when we surrender to the love of God. . . .

There is a recognized terminal phase in the process of dying, a dying hour, that is an important part of the caregiver's experience. It includes final communications, the behavior of the dying other and those present at the moment of death, and the memory of either being there or of receiving the announcement of the death from others after it occurred. Here is how Gregg described the moment of his friend's death:

> Sam continued to breathe intermittently until just after midnight, when he took a single last breath. This time we could see that the color had drained from his face, and within a couple of minutes were sure that he was dead. I held him a while longer, remembering the last two years and especially the last two months, saying good-bye to him inside.

There is a time of postmortem grief and mourning that is shared with others, usually with the help of rituals. And there is the sense, after having come through this process, of having learned from it and of being changed by the experience. Jeannie said of her brother's death:

> He carved a path for me with his death. I think until we understand perhaps what life is about, we're fearful of death, and I think I understand that now. I certainly am not frightened of death at all. I tell you, I love life. I am very much like my brother. I love living—even the sad times of life. But there are times I think that death will also be welcomed, and there is no fear in it for me at all. There's none at all.

From the analysis of this data, a final summary of this unique human experience can be stated in this way:

The fundamental form or structural definition of the experience of being with a dying person, who was aware of being in the dying process, and who had asked for your company through this process, is essentially one of relationship. It is within the context of this relationship that the agreement to accompany the other through death arises. The experience of being with the dying other is also characterized both by feeling a strong connection with the other and by a sense of putting oneself aside in order to care for the other. It is an experience of anticipating loss and of grieving that loss in advance, and of feeling painful, confused, and conflicted emotional states. It is an experience of being with the dying other through the course of the other's illness and of remembering other people who have died. It is an experience of being close to or present with the dying other for the other's dying hour and, after it, of experiencing postmortem grief and sharing in mourning rituals. It is an experience shared with family, friends, and other support caregivers It is often characterized by moments of humor and feelings of gratitude. It is, likewise, an experience of receiving gifts, of learning about oneself, about life, and about death, and feeling changed by all of this. It is also experienced as a spiritually significant time, involving both individual and collective prayer and religious ritual.

CONCLUSION

This study supported Eissler's (1955) insight that to be with and helpful to people who are in the dying process, it is essential to give them an experience of love. For the co-researchers in this study, doing so was a mutually shared gift, one they gave as well as received. And through receiving it, they were changed. They came to a more profound understanding of life and of themselves. Carse (1980) sees the most fundamental aspect of being human as being embedded in a "web of interconnectedness with other persons" (p. 4). It is death that reveals to us the paradox central to our reciprocal relationships. Carse says (p. 5):

> A personal relationship does not exist until we respond to others. . . . What death reveals therefore is something of a paradox: *we have our life from others, but only to the degree that we freely participate in our relationship to those other persons* . . . our life is not our own in the sense that it belongs exclusively to us; however, it *becomes* our own to the degree that we share it, make a gift of it to others.

A major theme in the co-researchers' experience was also gratitude for being able to share this experience with a community of others drawn together as caregivers. Bartlett (1972) notes that the crisis of grieving the death of a loved one can break the pattern of isolation in our lives and force us to come close to one another. Experiencing community in this way can bring a sense of serenity and peace with the knowledge that we belong to one another.

Benjamin (1988) and Surrey (1991) see the human condition as essentially one of intersubjectivity. In this way, they join the feminist critique of Western psychology's ideal of autonomy. Since we are always in relationship, a sense of mutuality is the most common human experience. Surrey (1991) talks about a continuous sense of relationship in life and says (p. 61):

> By relationship I mean an experience of emotional and cognitive *intersubjectivity:* the on-going, intrinsic inner awareness and responsiveness to the continuous existence of the other or others and the expectation of mutuality in this regard.

In a like manner, Kunkel (1962) believes that there is a oneness that underlies and informs all human experience. He calls it the "we-experience" (p. 44). He sees it as the essential human reality that is at the root of all the sciences, including psychology. The co-researchers all felt a strong fundamental and spiritual connection with their dying companion. Their experience of connectedness and of community supports this interconnected view of human nature.

Aries (1981) ends his historical survey of Western cultural attitudes toward death with a criticism of our contemporary society and our avoidance of death and neglect of the dying. Science and medical technology have depersonalized the dying process. He points to a breakdown in the sense of community in our society that has interfered with our ability to mourn. People are less and less involved when someone dies because there is not a sufficient sense of mutual solidarity, a feeling of being a community. Instead, we have become "an enormous mass of atomized individuals" and, as such, we are a society that both denies and is ashamed of death (p. 613).

This loss of caring for each other, especially at the time of death, is a loss of shared humanity. Heidegger (1962) saw caring for the other as an essential aspect of being human. As we lose a sense of ourselves as belonging to each other, and being responsible for each other in community, we become less human because we lose a spiritual awareness of ourselves. Mother Teresa, in an address she gave at the College of Marin, commented that the United States is the poorest country she has ever visited because we are the most spiritually destitute.

Kubler-Ross (1975) also sees this loss of community involvement with the dying as a symptom of our society's spiritual death. In our avoidance of death and those who are dying, we hide from experiencing the fullness of our own life. For her, death is the key to the door of life. The emptiness and sense of purposelessness seen in our contemporary Western culture are the result of our denial of death. And for Kubler-Ross, the denial of death is a major problem for all of human evolution and the ultimate survival of our race. She says (p. 165):

> Humankind will survive only through the commitment and involvement of individuals in their own and others' growth and development as human beings. This means the development of loving and caring relationships in which all members are as committed to the growth and happiness of the others as they are to their own. Through [this] commitment . . . individual human beings will also make their contribution to the growth and development—the evolution—of the whole species to become all that humankind can and is meant to be. Death is the key to that evolution. For only when we understand the real meaning of death to human existence will we have the courage to become what we are destined to be.

There are many responses today to Feifel's (1990) call for evolving a new *ars moriendi,* one that would emphasize the whole human condition and would care for both the body and the soul. We can see this in the work of Elizabeth Kubler-Ross and those who have been inspired by her, as well as in the work of Mother Teresa and her Missionaries of Charity, and in that of Stephen and Ondrea Levine, and of Ram Dass, and of Valle, to name but a few. These and others who participate in the difficult task of being with and caring for the dying are helping, not just the dying individual alone, but themselves and all of our society. Such caregivers are witnessing the truth of our inherent mutual and spiritual interconnectedness. Their experience affirms the core human value of relationships and the need we all have for communities of mutual support, especially in helping the dying.

As a Franciscan Friar, I am encouraged by the many ways working with the dying is ministering to the building up of an awareness of our mutual spiritual interconnectedness. Eight hundred years ago, as he was dying, St. Francis joked with his friends and invited them to rejoice with him at the approach of "Sister Death," whom he praised, along with all of creation, in his song the "Canticle of Creatures" (Fortini, 1981, p. 601).

Commenting on this song, Doyle (1981, p. 40) notes that St. Francis ". . . revealed that all beings are held in unity through a vast and intricate network of love relationships."

Boff (1989) points out the responsibility for modern humanity that is inherent in St. Francis' cosmic vision of the mutual interdependence of all creation. He says that we have forgotten that at our deepest level (pp. 46, 95):

> We do not simply live in the world, we co-live. . . . We cannot achieve our identity while denying a friendly and fraternal relationship with our natural world . . . a right understanding of the basic structure of humanity, to-be-in-the-world-with-all-things . . . in a cosmic democracy . . . one must

live fraternally with the birds, fire, water, the lark, the wolf, the worm on the road, treating all with respect and devotion, gentleness and compassion. We all belong mutually to one another. . . .

Part of this interconnectedness of all life is the process of dying. Death is a reality that we are always living with. Being with and caring for another person who is dying is a fundamental human experience. This study has examined the impact such an experience has on the individual caregiver involved in that process. It is hoped that this research will add to our understanding of death and the dying process and thus contribute to our shared task of human development.

REFERENCES

Alexander, I., & Adlerstein, A. (1960). Studies in the psychology of death. In H. David & J. Brengelmann (Eds.), *Perspectives in personality research* (pp. 65–92). New York: Springer Publishing.
Aries, P. (1981). *The hour of our death.* New York: Alfred A. Knopf.
Bartlett, G. (1972). Grief and humanity. In A. Kutscher & L. Kutscher (Eds.), *Religion and bereavement* (pp. 123–127). New York: Health Services Publishing.
Benjamin, J. (1988). *The bonds of love.* New York: Pantheon Books.
Boff, L. (1989). *St. Francis: A model of human liberation.* New York: Crossroad.
Bretall, R. (Ed.). (1946). *A Kierkegaard anthology.* New York: The Modern Library.
Burton, A. (1962). Death as a countertransference. *Psychoanalytic Review* (Winter), *49,* 3–20.
Butler, L. (1967). Hospitallers and hospital sisters. *New Catholic encyclopedia.* New York: McGraw-Hill.
Butler, R. N. (1991). Toward a therapy of dying. *Geriatrics* 46(7), 10–12.
Campbell, J. (1959). *Primitive mythology.* New York: Penguin Books.
Carse, J. P. (1980). *Death and existence.* New York: Wiley.
Cassidy, S. (1991). *Sharing the darkness.* New York: Orbis Books.
Charon, J. (1963). *Death and western thought.* New York: Macmillan.
Charon, J. (1964). *Death and modern man.* New York: Collier Books.
Clarke, P. J. (1981). Exploration of countertransference toward the dying. *American Journal of Orthopsychiatry, 51*(1), 71–77.
Colaizzi, P. F. (1973). *Reflection and research in psychology: A phenomenological study of learning.* Dubuque, IA: Kendall/Hunt.
Colaizzi, P. F. (1978). Psychological research as the phenomenologist views it. In R. S. Valle & M. King (Eds.), *Existential–phenomenological alternatives for psychology* (pp. 48–71). New York: Oxford University Press.
Doyle, E. (1981). *St. Francis and the song of brotherhood.* New York: Seabury Press.
Eissler, K. R. (1955). *The psychiatrist and the dying patient.* New York: International Universities Press.
Feifel, H. (1969). Perception of death. *Annals of the New York Academy of Sciences, 164,* 669–677.
Feifel, H. (1990). Psychology and death. *American Psychologist, 45*(4), 537–543.
Feigenberg, L., & Shneidman, E. S. (1979). Clinical thanatology and psychotherapy: Some reflections on care for the dying person. *Omega, 10*(1), 1–8.
Fortini, A. (1981). *Francis of Assisi.* New York: Crossroad.
Giorgi, A. (1975). An application of phenomenological method in psychology. In A. Giorgi, C. Fischer, & E. Murray (Eds.), *Duquesne studies in phenomenological psychology: Volume* II (pp. 82–103). Pittsburgh, PA: Duquesne University Press.
Heckler, R. A. (1994). *Waking up alive: The descent, the suicide attempt, and the return to life.* New York: G. P. Putnam's.
Heidegger, M. (1962). *Being and time.* New York: Harper & Row.
Herzog, E. (1966). *Psyche and death.* Hodder & Stoughton.
Husserl, E. (1962). *Ideas.* New York: Collier.
Johnston, R. C. (1988). *Confronting death: Psychoreligious responses.* Ann Arbor: U.M.I. Research Press.
Joseph, F. (1962). Transference and countertransference in the case of a dying patient. *Psychoanalytic Review* (Winter) *49,* 21–34.
Kastenbaum, R., & Costa, P. T. (1972). Psychological perspectives on death. *Annual Review of Psychology, 28,* 225–249.
Kubler-Ross, E. (1969). *On death and dying.* New York: Macmillan.
Kubler-Ross, E. (1970). The dying patient's point of view. In O. G. Brim, H. E. Freeman, S. Levine, & N. A. Scotch (Eds.), *The dying patient* (pp. 156–170). New York: Russell Sage Foundation.

Kubler-Ross, E. (1975). *Death, the final stage of growth.* New York: Simon & Schuster.
Kubler-Ross, E. (1978). *To live until we say goodbye.* Englewood Cliffs, NJ: Prentice Hall.
Kubler-Ross, E. (1981). *Living with death and dying.* New York: Macmillan.
Kubler-Ross, E. (1983). *On children and death.* New York: Macmillan.
Kunkel, K. (1962). *Notes and lectures: Volume* 2. Berkeley, CA: Graduate Theological Library.
Levine, S. (1979). *A gradual awakening.* Garden City, NY: Anchor Books.
Levine, S. (1982). *Who dies? An investigation of conscious living and conscious dying.* Garden City, NY: Anchor Books.
Levine, S. (1984). *Meetings at the edge: Dialogues with the grieving and the dying, the healing and the healed.* Garden City, NY: Anchor Books.
Levine, S. (1987). *Healing into life and death.* Garden City, NY: Anchor Books.
Lifton, R. J. (1979). *The broken connection.* New York: Simon & Schuster.
Macquarrie, J. (1965). *Studies in Christian existentialism.* Philadelphia: Westminster Press.
Mother Teresa. (1975). *A gift for God.* San Francisco: Harper & Row.
Norton, J. (1963). Treatment of a dying patient. *The Psychoanalytic Study of the Child, 18,* 541–560.
Ram Dass & Gorman, P. (1985). *How can I help?; Stories and reflections on service.* New York: Alfred A. Knopf.
Renneker, R. E. (1957). Countertransference reactions to cancer. *Psychosomatic Medicine 19(*5), 409–418.
Ring, K. (1984). *Heading toward omega.* New York: William Morrow.
Ring, K. (1991). Amazing grace: The near death experience as compensatory gift. *Journal of Near Death Studies, 10,* 11–39.
Robbins, A. (1989). The *psychoaesthetic experience.* New York: Human Sciences Press.
Rosenthal, H. R. (1983). Psychotherapy for the dying. In H. M. Ruitenbeek (Ed.), *The interpretation of death* (pp. 87–95). New York: Jason Aronson.
Ross, L. M. (1987). *The experience of death and dying: A phenomenological study.* Unpublished doctoral dissertation, University of Tennessee, Knoxville, TN.
Sadowy, D. (1991). Is there a role for the psychoanalytic psychotherapist with a patient dying of AIDS? *Psychoanalytic Review, 78*(2), 199–207.
Stephen, D. L. (1987). *Appropriate dying: Some phenomenological aspects of the dying process.* Unpublished doctoral dissertation, University of Manitoba, Winnipeg, Manitoba, Canada.
Surrey, J. L. (1991). The self in relationship: A theory of women's development. In J. Jordan, A. Kaplan, J. B. Miller, L. Stiver, & J. Surrey (Eds.), *Women's growth in connection* (pp. 51–66). New York: Guilford Press.
Valle, R. S. (1996). *Awakening to life and death.* Unpublished manuscript. Walnut Creek, CA: Awakening: A Center for Exploring Living and Dying.
Weisman, A. D. (1977). The psychiatrist and the inexorable. In H. Feifel (Ed.), *New meanings of death* (pp. 108–122). New York: McGraw-Hill.
Weisman, A. D., & Hackett, T. P. (1961). Predilection to death. *Psychosomatic Medicine, 23*(3), 232–256.

18

The Experience of Feeling Grace in Voluntary Service to the Terminally Ill

Paul Gowack and Valerie A. Valle

But Grace is there all along.
Grace is the Self.
It is not something to be acquired.
All that is necessary is to recognize its existence.
—Sri Ramana Maharshi (1988)

There exist in many cultures some powerful and distinct human experiences we have little scientific understanding of. One such experience is the experience of feeling grace. In this chapter, the nature of the experience of feeling grace in voluntary service to the dying is explored from a phenomenological perspective. Twelve individuals who described themselves as having experienced grace while working with the dying were interviewed, their descriptions were analyzed, and seven constituent themes were identified. The relationship of these themes to previous writings on the nature of grace is then explored.

THE NATURE OF GRACE

Throughout history, from multiple cultures at various times, human beings have spoken of positive experiences that appear to have emanated from beyond themselves. In the

Paul Gowack • 1339 Milvia Street, Berkeley, California 94709. **Valerie A. Valle** • St. Alban's Episcopal Church, Brentwood, California 94513.

Phenomenological Inquiry in Psychology: Existential and Transpersonal Dimensions, edited by Ron Valle. Plenum Press, New York, 1998.

West, these have been called experiences of grace. The word *grace* is the anglicized form of the Latin *gratia,* a translation of the Greek *charis,* whence comes the word "charisma." In ordinary Greek usage, it meant "gracefulness," "charm," "favor," or "kindness," especially when it was favor and kindness shown without obligation, as by a superior to an inferior (Watson, 1959). As the word has come to be understood in religious usage, it refers to "an influence emanating from God and acting for the spiritual well-being of the recipient" (Gove, 1966).

Self-described experiences of grace often have a life-transforming effect on the individual, yet they have rarely been systematically studied by either religious systems or modern psychology. A review of the literature related to the experience of grace follows.

Religious Perspectives on Grace

The concept of grace suffuses the Hebrew scriptures and the Christian New Testament. God is seen as showering favor upon the people of God, and the people, who are filled with grace, do gracious deeds by showing kindness to the poor or generosity to all living things. Christian theologians filled volumes with definitions and classifications for grace, yet little has been said about the nature of the experience of grace.

The Christian understanding of the concept of grace was first systematically developed by St. Paul. He saw God's grace as the gift of preserving, loving, purposeful generosity that becomes visible in a climatic way in the life, teaching, and resurrection of Jesus (Fransen, 1965).

St. Augustine wrote in *Confessions* (Liderbach, 1983) that individuals may be able to free themselves from entrapments and turn to God, but humans are wholly dependent upon God for such conversions. To Augustine, the human species is born into sin, and the only release from this condition is through the help of God. Such freedom comes exclusively as a consequence of the love or grace of God. The person who has received grace receives as well the gift of God dwelling within the person, the ability to love other persons, and the will to do virtuous acts (Liderbach, 1983).

St. Thomas Aquinas wrote an analysis of grace in *Summa Theologiae* (Dreyer, 1990). His basic premise is that grace is primarily the gratuitous gift of God, given out of the abundance of God's merciful love. Grace within a person is the outcome of God's mercy, and God alone is the cause. Through grace are given the gifts of love, wisdom, and understanding for our enjoyment.

In Hinduism, specifically Vedanta, divine grace is ever abundant (Frenz, 1975). Lord Krishna declares in the *Bhagavad Gita* (9:29) (Vedakkekara, 1981): "I am the same to all beings. To Me there is none hateful, none dear." God is seen as all-merciful with infinite generosity, making no distinctions and blessing human beings with liberation. Grace showers over all equally, but according to Hinduism, one has to rise to avail oneself of it. To the extent one surrenders the self, one enjoys the experience of achieving peace of mind, tranquility, and the grace of God (*prasada*). This self-surrender is the culmination of all yoga.

Sri Aurobindo (1989) writes that in the spiritual knowledge of self, there are three parts of the one knowledge. The first is the discovery of the soul, the divine element within us. When we are consciously aware of the soul, we become aware of a guide within that knows the truth, the good, the true delight and beauty of existence. The next step is to become aware of the eternal self in us, unborn and one with the self of all beings. The third step is to know the divine being who is at once our supreme transcendent self, the cosmic being, and the foundation of our universality.

When we are in contact with our divine consciousness, writes The Mother (1972), everything is tinged with grace and presence, and things that usually seem dull and uninteresting become charming, pleasant, and attractive. When opened to the self, one feels stronger, freer, happier, and full of energy, and everything has meaning.

Psychology and Grace

William James (1977) was the first psychologist to speak of the individual as having a "personal identity" as well as a "spiritual me," the personal identity being the ego and the spiritual identity being one's entire collection of states of consciousness. According to James, we pass into mystical states from out of ordinary consciousness, as from a less into a more, as from a smallness into a vastness, and from an unrest to a rest. These experiences are felt as reconciling and unifying states that may render the soul more energetic and inspired. The effect of the religious experience is somehow to unify the discord within the self.

Meissner (1984), a Catholic priest and psychiatrist, writes that understanding the depth's of one's spiritual nature is the work of a psychology of grace and that a meaningful psychology of grace would help the modern individual realize the relationship of the spiritual life to psychological development and maturity. Meissner uses St. Augustine's basic premise of the nature of grace, which equates grace with divine favor and love, and asserts that in grace, it is God who begins, God who works, and God who finishes.

Meissner (1987) believes that grace works in and through the resources of the ego. Its influence is maintained in the vital capacity of the ego to perform its proper and autonomous functions. In this framework, therefore, grace can be regarded as a dynamizing activation of the energy resources latent within the ego. Grace does not force the ego to act, nor does it replace the ego's proper function with a divine activity; rather, its healing effect is to enable the ego to mobilize its own latent resources and direct them to purposeful action.

Meissner contends that our capacity to relate to God as an object is related to our capacity to relate to other human beings as objects. The limitations on our ability to relate to humans limit our ability to relate to God; therefore, the capacity to fully enter into a relationship with the divine must be given as a special gift coming from God's loving initiative. As our ego develops in its ability to relate to others, however, its response to God's grace becomes more spontaneous and complete. It is only with the development of ego-autonomy that results from a high level of inner development that we can fully respond to the divine. Love of God implies complete surrender, and surrendering is an act of unrestrained autonomy; it is also how we achieve the highest expression of our freedom.

Psychologist A. H. Almaas uses object relations theory to understand grace. According to Almaas (1986), the personality is false while essence represents the real person, the real and true self, the spirit. He sees our ego identity as masking our true self by taking the place of essence in our personality. This false self develops through the organization of the individual's early experience from smaller units into larger more comprehensive ones concurrently with the development of object representation.

Almaas (1988) contends that for the realization of essence, the first step is to disidentify, to see that we are not whatever self-image we have. This lessening of identification loosens the rigid structure of the personality and creates more space within us. The final process of disidentification is the experience of the dissolution of the psychic structure of the self-image and the experience of space or what is called the void (Almaas, 1986). When self-image is dissolved, the individual will experience the loss of boundaries, both physical and mental, and the nature of the mind is then revealed as an emptiness, a void, an immaculate empty space. Through this experience, essence emerges as the source, the life, and the fullness of our being. When essence is reached, it begins the process of freeing itself from the personality (Almaas, 1984). The consciousness or awareness of the individual begins to develop, and one becomes more sensitive and aware of emotions and feelings. Almaas believes the essence is reached through the practice of meditation and service. Service without regard to boundaries of self and others is seen as necessary for realization and development.

Maslow (1971) also explores what it means to be a whole and psychologically healthy person. He sees the "fully-human" or self-actualized individual as one who devotes his or her life to the search for ultimate values. For Maslow, self-actualization implies being able to see the sacred, the eternal, the symbolic in the world. Self-love, self-worth, and positive self-regard are also essential components of the self-actualizing process.

Solimar (1986) also stresses the importance of self-love. She writes that the necessary factor in establishing a mature and genuine capacity to relate to others is through self-love. She describes self-love as "an emotional embodied experience of well-being, gladness, self-confident motivation, and unconditional acceptance and validation of self, other, and life; it is an experience of greater inner potential and wholeness based on an expanded sense of self-identity" (p. 115). Those individuals who are self-loving are more loving and connected with friends, family, and humanity as a whole.

Maslow (1964) writes that peak experiences are transient moments of self-actualizing. The concept of peak experience is similar to what theologians describe as the experience of grace, yet peak experiences can be secular, religious, mystical, or transcendent and are not necessarily dependent on churches or specific religions, nor do they necessarily imply the supernatural. Maslow believed that they are well within the realm of nature. Peak experiences (Maslow, 1962) can be varied, yet contain common characteristics. During peak experiences, the whole universe is often perceived as an integrated and unified whole, while visual perception, hearing, and feelings are described as most true and complete. Often there is a disorientation in time and space, including the experience of both universality and eternity. Perception can be relatively ego-transcending, self-forgetful, and unselfish. The peak experience is felt as a self-validating, self-justifying moment that carries its own intrinsic value. The world is seen as good, desirable, worthwhile, beautiful, and sacred.

The spontaneous mystical experience is in many ways similar to the peak experience. Pahnke and Richards (1973), through a historical survey of the literature on spontaneous mystical experience, derived the following nine interrelated categories that attempt to describe the core of that experience:

1. Unity: The experience of an undifferentiated unity, either internal or external, is the hallmark of mystical consciousness.
2. Objectivity and reality: This second category has two elements: (a) insightful knowledge that is felt at an intuitive, nonrational level and gained by direct experience and (b) the feeling of certainty that such knowledge is ultimately real in contrast to the feeling that the experience is a subjective illusion.

3. Transcendence of space and time: This refers to the loss of a person's orientation as to where he or she is during the experience and the radical change in his or her perspective during which the individual may feel outside of time.
4. Sense of sacredness: An acute awareness of finitude is reported, as though one had stood before the infinite in profound humility.
5. Deeply felt positive mood: Feelings of joy, love, blessedness, and peace are inherent in this experience.
6. Paradoxicality: Significant aspects of mystical consciousness are felt to be true even though they violate the laws of Aristotelian logic.
7. Alleged ineffability: One who tries to communicate about mystical consciousness often claims that the available linguistic symbols are inadequate to contain or even begin to reflect the experience.
8. Transiency: This refers to the temporary duration of the experience and is an important difference between mystical awareness and psychosis.
9. Positive changes in attitude and/or behavior: People with such experiences tend to report changes in attitudes toward themselves, toward others, toward life, and toward mystical consciousness itself.

Even though these categories came from a review of the literature on mystical experience, the authors did not personally study individuals who had had such experiences. Phenomenological research provides the opportunity to investigate any meaningful experience systematically and directly. A study by Elite (1993) (see Chapter 14 in this volume) investigated the experience of individuals who were voluntarily silent for a period of four or more days. In some ways, the experiences they described had a similarity with the peak experiences described by Maslow and the core categories of mystical experience described by Pahnke and Richards. Elite found the following nine constituent themes:

1. Experiencing the essence of one's being.
2. Experiencing one's inner life with a heightened sense of awareness.
3. Experiencing more acutely through the senses.
4. Experiencing auditory, visual, perceptual, and/or other sensory alterations.
5. Feeling connected and/or unified with various aspects of existence.
6. Feeling intensely a wide range of feelings and emotions.
7. Feeling rejuvenated.
8. Perceiving a change in the ontological meaning and/or significance of ideas and the nature of personal reality.
9. Perceiving the experience as ineffable.

The Power of Service

In many religious traditions, charitable service to one's fellow human beings is considered the most perfect form of devotion to the divine. Christ is quoted as saying "Truly I tell you, just as you did it to one of the least of these who are members of my family, you did it to me" (Matthew 25:40) as a call to serve those who have less then we do (Boff, 1979). The Christian saints have often demonstrated a selfless love for others expressed through service to those with the greatest need.

Charitable service, or Karma Yoga, is considered one of the paths to enlightenment within the Hindu tradition. Through Karma Yoga, one realizes the oneness of one's individual soul with the universal soul by doing one's duty without attachment to the fruits thereof. This self-surrender through service leads to greater union with the divine (Reichenbach, 1990).

Buddhism also encourages selfless service. Although Buddhism sees suffering as a universal reality, suffering may be relieved through the application of three principles: *metta,* loving-kindness actively pursued (see Chapter 15 in this volume); *karuna,* compassionate mercy that does not repay evil with evil; and *mudita,* the feeling and expression of approval for other people's good deeds. These principles find their expression in works of social welfare, including public works projects and the maintenance of hospitals and shelters (King, 1952). Islam encourages people to serve Allah by means of good works, including almsgiving, kindness, and good treatment of parents, the dying, orphans, and the elderly (Frenz, 1975).

In Maslow's (1959) description of self-actualized individuals, he states that for them there is no line or wall between the self and other, between inside and outside. Without a differentiation between self and other, there is no distinction between personal pleasure and service to others. For Maslow, selfless service is a natural outgrowth of human development.

For Almaas (1984) as well, service is useful and necessary work as part of self-realization and development. Essence manifests as compassion and service as well as peace, strength, consciousness, and contentment.

Very little has been done to study the actual experience of those who choose to be of service to others. The studies of volunteers who work with the dying have found that those who become the greatest resource to the family involved, and who themselves are most fulfilled, are often motivated by a desire to serve, to share, and to learn. These volunteers find value in what they do and the circumstances they find themselves in (Shoenberg, 1972).

The first author of this chapter volunteered time on a weekly basis for three years in an AIDS hospice. From the rich experiences he had during this time, he grew to feel that there is a particular "grace" that is experienced by those who extend themselves in service to those who are in the dying process, and that the experience of grace through voluntary service is an identifiable experiential phenomenon awaiting a deeper understanding. His research reported here came from this realization that through selfless service one can have an experience of grace and from the desire to know more about the nature of this experience.

THE RESEARCH: BACKGROUND AND PROCEDURE

The research involved a phenomenological investigation of the experience of "feeling grace" while being of service on a volunteer basis to the terminally ill. Existential–phenomenological psychology represents a marriage between existentialism and phenomenology. Søren Kierkegaard, a founder of existential philosophy, addressed the fundamental themes that human beings struggle with in order to bring meaning to their existence (Haecker, 1937). Edmund Husserl (1962), the primary proponent of phenomenology, placed great importance on the unbiased study of things as they are actually lived and experienced so that one might be able to come to an essential understanding of human consciousness and experience. In brief, existentialism is the philosophy of existence and phenomenology the philosophy of experience.

Martin Heidegger (1962) brought together existential concerns and the study of human experience, laying the groundwork for an existential–phenomenological psychology. Valle, King, and Halling (1989, p. 6) summarize this approach: "*Existential–phenomenological psychology* [is] that psychological discipline that seeks to explicate the *essence, structure,* or *form* of both human experience and human behavior as revealed through essentially *descriptive* techniques including disciplined reflection."

This discipline rests on the concept that the individual and his or her world co-constitute one another. In this sense, the research subject is seen as a co-researcher because the researcher and the subject together manifest or constitute the meaning of the experienced situation.

Again, the research phenomenon under investigation was the experience of feeling grace while being of service on a volunteer basis to the terminally ill. Twelve individuals or co-researchers (eight women and four men) who had had this experience were selected. Six co-researchers had served as volunteers in a hospice or hospital setting, and six had volunteered to be in the home or other settings. They were instructed to describe their experience as follows:

> Select a time and space where you can be alone, uninterrupted and relaxed. Recall a time/s when you were "feeling grace" while being of service on a volunteer basis to the terminally ill. Please describe how you felt during that time/s. Try to describe your feelings just as they were, so that someone reading or hearing your report would know exactly what that experience was like for you. Keep your focus on the experience, not just the situation itself. Please do not stop until you feel that you have described your feelings as completely as possible. Take as long as you would like to complete your description.

After the written protocols were received and read, a face-to-face "walk-through" interview was conducted to further elucidate and deepen the description given in each protocol (Mishler, 1986). The author read what had been written back to the co-researcher sentence by sentence, giving him or her the opportunity to confirm and elaborate on what had been written. Each interview was tape recorded, transcribed, and added to the original written protocol.

Upon receipt of the protocols, the meanings contained in the different descriptive accounts of the phenomenon under study were extracted while the researcher exercised a reflective attitude toward each life-text or descriptive report of the experience (von Eckartsberg, 1986). Central to this reflective process is bracketing, an inner discipline by which the researcher's assumptions, biases, and preconceptions about the phenomenon under study are suspended in awareness in order to minimize the effects these preconceived notions would have on the process of revealing the meaning structure of the experience being investigated (Giorgi, 1975).

Focusing on the subjects' experienced meaning as revealed in these life-texts, the objective was to find the fundamental structure of the phenomenon as it appeared throughout the many diverse accounts of the phenomenon itself. This was accomplished through the method of explication (von Eckartsberg, 1986), whereby the life-text is transformed into a universally understood language using psychological and phenomenological terminology. What has been implicit becomes explicit through the form of interpretive reading that reveals the underlying prereflective structure of the experience that is being investigated. This results in an understanding of the lived human meaning of the phenomenon.

The specific analysis used was based on an adaptation of Colaizzi's (1973, 1978) six-step procedure. The steps were:

1. All of the final protocols were read carefully in order to acquire a feeling for them.
2. The researcher then returned to each protocol and extracted phrases or sentences that directly pertained to the investigated phenomenon, creating a list. This process is known as extracting significant statements. Repetitions were eliminated, and the remaining relevant statements were classified into naturally occurring categories.
3. Each categorized statement was then translated for the purpose of communicating them more clearly in psychological terms. Colaizzi (1978, p. 59) remarks: "This is a precarious leap because, while moving beyond the protocol statements, the meaning he arrives at and formulates should never sever all connection with the original protocols; his formulations must discover and illuminate those meanings hidden in the various contexts and horizons of the investigated phenomenon which are announced in the original protocols." This step was repeated for each protocol.
4. All of the formulated meanings were then organized into theme clusters across all protocols. These theme clusters were then referred back to the original protocols in order to validate them by searching for anything that was contained in the original protocols that was not accounted for in the clusters of themes, or to see whether the theme clusters proposed anything that was not implied in the original protocols. At this point, even if discrepancies were noted between the various clusters, or if themes flatly contradicted each other or appeared totally unrelated, they were not eliminated. The researcher proceeded with the conviction that what is logically inexplicable may be existentially real and valid. The researcher was careful to not prematurely generate a theory that would eliminate the discordant findings or to ignore data that did not appear to fit.
5. All of the results were integrated into an exhaustive description of the investigated topic by tying together and integrating the themes as a whole:
   a. The theme clusters from all 12 protocols were combined across protocols to form 52 constituent themes.
   b. From the 52 constituent themes, the ones that were found to contain the same meaning were then combined to form 16 comprehensive constituent themes.
   c. Upon further reduction, 7 final comprehensive constituent themes emerged.
6. An effort was then made to formulate the exhaustive description of the investigated phenomenon into a statement identifying its fundamental structure. This was the essential structural definition of the phenomenon under study.

THE CONSTITUENTS

For the experience of feeling grace while in volunteer service to the terminally ill, 7 final comprehensive constituent themes were identified. These 7 themes represent the results of the research:

1. Feeling present in the moment, often with heightened awareness.
2. Feeling blessed and/or loved.
3. Feeling oneness or connection, often without fear.

4. Feeling guided.
5. Feeling energized.
6. Feeling peace.
7. Feeling joy.

Let us look at each of these in turn.

1. Feeling present in the moment, often with heightened awareness. This theme was evident in the descriptions of all 12 co-researchers, and is illustrated in their words:

> Grace grounded me in the present and surrounded me as though always being part of my dominion, my environ.
>
> Mentally, it [grace] is an experience of alertness at an elevated level. I feel merged with the moment, feeling the complete rightness of now. There is no sense of serving another, but only of being served my this moment.
>
> Grace has a feeling of its own. I am very aware of the other person, their needs and what they need to do, and what I need to be with that person, and what I need to do to help. I am also aware of God in that place.
>
> My focus seems intensely sharpened on my duties of making him more comfortable.

2. Feeling blessed and/or loved. Ten co-researchers reported this feeling, usually as a blessing or love bestowed upon the one serving. Some examples are:

> I think the countenance of the Lord shone upon us . . . my father and every member of my family who focused his or her attention on caring for my father during the last weeks of his life.
>
> I am aware of a boundless sense of receiving what I giving. A boundless and deep compassionate power through a quiet, loving Divine Connection
>
> It was an honor to take care of both of these people, and I feel blessed by the experience.

The blessing received was sometimes felt as a transfer of beneficent power, a power that emanates from the supernatural world and confers a new quality on the object of the blessing. One example is:

> My ministry at the church is musical. For me the hour is a state of grace. I believe that music heals, that God's love flows through me and out my fingers, and everyone there is bathed in the vibrations coming from the piano strings.

For others, the feeling of love did not so much come from an outside power as emerge from the service being offered, and it included a sense of self-love. One co-researcher said:

> It seems that in serving my father we served ourselves; in loving him, one another.

3. Feeling of oneness or connection, often without fear. This theme was reflected in 10 of the 12 descriptions. The co-researchers expressed feeling connected to all human beings and to all there is. In addition, this experience was described as being spiritual or sacred by those who felt connected with God, the universe, a higher power, or the inner self. Examples include:

> I seem to be connected to an energy and power much greater than myself and my friend.
>
> Sometimes I would touch or kiss his forehead, sometimes quietly hold his hand in mine, but always with a holiness for all of life. I combed his hair, washed his face, swabbed his mouth, all with the touch I would imagine belonging to an angel. I did not feel like I was important, that I was like an angel, but that I had transcended my usual self, and was in touch with the sacredness of all things on this earth.
>
> I just suddenly found myself connected into and acting from a very "deep" place. I was aware of oneness.

> Grace for me is that connecting to a Higher Power.

For several co-researchers, the feeling of being connected was an experience of being of service without fear. Two examples are:

> I have had so many gifts from being with the dying. One is the knowledge that comes with grace, that there is no need to fear. That life is infinitely precious. That what I do, think, feel and say every minute of every day affects everything else. Everything feels connected to everything else.
>
> I continue to do my work for Alan. I have never disconnected from that activity at all. I carry on caring for my friend, caring for myself. I work in a kind of stillness with no fear.

4. Feeling guided. Nine of the co-researchers felt guidance during their service to the terminally ill. Some felt they had assistance from a higher power and that what they did while in service to the dying was the "correct" action because of this guidance. The following are examples of this theme:

> When I am filled with grace I am empowered by God, the Universe, a Supreme Being, to carry the necessary information and assistance to someone who is dying. This is usually in the form of sending clairvoyant pictures telepathically to the dying person. Pictures that show them what is left to do, say, release, so that they may leave their bodies in peace.
>
> Divine grace is intimate, immediate, and personal. While in the grace I am not aware of consciously making decisions in my head. I am aware that I am intuitively responding out of my soul connection with God; whatever I do or say, and however I respond, it is out of response to God.
>
> I felt as if my every intention was being guided by a Higher Power.

5. Feeling energized. Nine co-researchers felt energized in particular ways while serving as a volunteer. This aspect was described in various ways, including a feeling of buoyancy, of being held or carried, of being physically invigorated, and of being permeated with light. Examples include:

> My experience of it [grace] is that it surrounds quietly, seeping into your cells. This feeling is of great buoyancy. This feeling of buoyancy filled me and carried me to the threshold of his hospital room door.
>
> Physically it [grace] manifests as a rush of energy.
>
> It [grace] was so energizing—but it doesn't last.
>
> In my body, the warmth and glow of grace unfolds from above and behind my solar plexus. It eventually feels like there is a balloon inflated there inside me. There is a feeling of buoyancy and lightness.
>
> It [grace] always begins with a warm glow, and then the realization that the warmth and glow come from the light, the golden light that pulses throughout the universe. I have seen the light and words evade a complete description. The light shimmers, pulses, permeates, pervades, fills, surrounds, and sustains everything.

6. Feeling peace. Eight of the co-researchers experienced the feeling of peace, an experience that allowed them to be of service without exhausting themselves physically or emotionally. They reported an ease when working with the dying. Examples include:

> I feel calm, suffused with life force.
>
> I feel peaceful despite a stressful life and I am not worried or anxious.
>
> Nothing about the experience was sanctimonious, nothing felt religious, just a feeling of continuous steady peace.
>
> The moment of grace is—graceful, peaceful.

One co-researcher described this peace as a prevailing experience during the time spent with the dying:

> I have often thought of what it was I felt. What described the experience of being with Tom that last year. My ever-present sense was that I could and would do nothing else but take care of him. How can I describe the feeling, the enveloping aura of serenity in the midst of death?

7. *Feeling joy.* Seven co-researchers expressed a feeling of joy that stemmed from simply giving and sharing what was needed in the moment. Joy came from seeing gratitude and through the intention of being available and freely serving the one who was dying. It would arise out of loving action toward the one in need, and was connected to the felt grace. The following examples illustrate this theme:

> I go about my work in a kind of sacred joy.
>
> Joy is the closest I can describe it. Joy without exuberance.
>
> I stayed with him a few more minutes making him comfortable, said good-bye. I was very, very happy.
>
> Now, in this moment, tears come to my eyes and my heart fills me with sweet sadness, but that day, there was no sadness, there was only the joy of sharing his company.

The structural definition that emerged from these constituents is this: The essence of the experience of feeling grace in voluntary service to the terminally ill was a sense of being present in the moment with heightened awareness while feeling blessed and loved, connected and guided, energized and at peace, with a sense of joy.

AN EMERGING UNDERSTANDING OF GRACE

The literature on the experience of grace had suggested that through selfless service, one enhances the opportunity to experience grace. From the Christian call to serve those in greatest need to the ancient Vedic wisdom regarding selfless surrender, from Maslow's (1971) description of the self-actualized individual to Almaas's (1984) discussion of the importance of service, the need to give of ourselves, selflessly, has consistently been seen as part of the process of fulfilling human potential. This study revealed that, at least for these 12 individuals, service to the dying offered an opportunity to experience grace.

The experiences of these volunteers fit with previous descriptions of the nature of grace, from both the religious and the psychological perspective. Let us now compare the results of this study with the spiritual and psychological literature on the nature of grace by looking at each of the constituent themes in turn.

The feeling of being present in the moment, often experienced with heightened awareness, was included in many of the previous descriptions of grace. Pahnke and Richards (1973) described a transcendence of time and space, while The Mother (1989) stated that in divine consciousness, things that usually seem dull and uninteresting become charming, pleasant, and attractive. Maslow (1964) said that during peak experiences, visual and auditory perceptions are heightened and a disorientation of time and space may occur. Similarly, Elite (1993) found both alterations in perception and a more acute experiencing through the senses.

Feeling blessed and/or loved was not specifically described in any of the sources reviewed above, but was implicit in St. Augustine's sense that God dwells within (Liderbach,

1983) and in St. Thomas Aquinas' sense that grace is a gift of God's merciful love (Dreyer, 1990). In the protocols of this study, this love was often experienced as an exchange of love between individuals and/or a higher power.

The feeling of oneness or connection, often without fear, is an experience that shows up in many of the previous descriptions of grace. James (1977) describes the mystical state as an experience that is felt as reconciling and unifying. One of the primary aspects of peak experiences as described by Maslow (1962) is that the whole universe is perceived as an integrated and unified whole. Pahnke and Richards (1973) list unity as the first of their categories that describe the core of mystical consciousness. And Elite (1993) found feeling connected and/or unified with various aspects of existence to be one of the core constituents of the experience of being voluntarily silent.

The feeling of being guided by a higher power was consistent with some of the oldest descriptions of grace. St. Thomas Aquinas described grace as the gift of wisdom (Dreyer, 1990). It is also consistent with the sense of the world as sacred described by Maslow (1962) and Pahnke and Richards (1973).

Feeling energized is a concept often used in describing the nature of grace. From the Hindu tradition, The Mother (1972) describes divine consciousness as full of energy. William James (1977) felt that the mystical state rendered the soul more energetic and inspired. Meissner (1987) regarded grace as the activation of energy resources latent within the ego, and Elite (1993) found that one of the core constituents of the experience of being silent was feeling rejuvenated.

The feeling of peace gave the respondents an inner ground from which to love and serve the dying. This is consistent with St. Augustine and St. Thomas Aquinas, who suggest that one of the gifts of grace is the ability to love other persons (Dreyer, 1990; Liderbach, 1983).

The feeling of joy is described by Pahnke and Richards (1973) as a deeply felt positive mood. Solimar (1986) found that through the experience of loving self and others there is gladness.

CONCLUDING REFLECTIONS

By employing a phenomenological method to investigate the experience of grace, we have addressed, in an empirical scientific manner, an experience that, to date, has been discussed primarily in religious and theological circles. The results of our study suggest that through the giving of oneself it is possible to experience a feeling of grace, and that the essence of this experiences, as revealed through this constituent analysis, is consistent with both religious and psychological theories regarding the nature of grace.

Volunteer work with the dying has been a source of experiencing grace for both authors of this chapter. Theologians have asserted that grace is a free gift from the divine and cannot be sought or "caused," yet it appears that one can open oneself to experiencing the presence of grace through selfless service. How can we reconcile these two apparently opposing perspectives? It has also been asserted that we are always surrounded by grace, but most of the time we live our lives unaware of its presence. If this is in fact true, then the results of this study suggest that love expressed through selfless service, at least when being with the dying, can help dissolve the boundaries of ego-self that keep us from the awareness of the presence of grace.

The co-researchers in this study all experienced a sense of being present, often with a heightened awareness. The deep sense of being present that comes with a more focused kind of loving service can heighten our awareness not only of our sensory experience, but also of the presence of grace. It is through this awareness that one can find the energy to be of loving service without "burning-out." The incredible amount of energy and love often needed to care for the terminally ill can be supported by this grace, this source of energy, peace, and joy. It appears that through surrendering oneself in selfless service, one discovers within all that one needs. The results of this study reflect the wisdom of these words from a prayer attributed to St. Francis: "For it is in giving that we receive" (*The Book of Common Prayer,* 1979).

REFERENCES

Almaas, A. H. (1984). *The elixir of enlightenment.* York Beach, ME: Samuel Weiser.

Almaas, A. H. (1986). *Essence.* York Beach, ME: Samuel Weiser.

Almaas, A. H. (1988). *The void.* Berkeley, CA: Diamond Books.

Sri Aurobindo (1989). *The psychic being.* Wilmont, WI: Lotus Light Publications.

Boff, L. (1979). *Liberating grace.* New York: Orbis Books.

Colaizzi, P. F. (1973). *Reflections on research in psychology.* Dubuque, IA: Kendall Hunt.

Colaizzi, P. F. (1978). Psychological research as the phenomenologist views it. In R. S. Valle & M. King (Eds.), *Existential–phenomenological alternatives for psychology* (pp. 48–71). New York: Oxford University Press.

Dreyer, E. (1990). *Manifestations of grace.* Wilmington, DE: M. Glazier.

Elite, O. (1993). *On the experience of being voluntarily silent for a period of four or more days: A phenomenological inquiry.* Unpublished doctoral dissertation, California Institute of Integral Studies, Novato.

Fransen, P. (1965). *Divine grace and man.* New York: New American Library.

Frenz, A. (1975). *Grace in Saiva Siddhanta, Vedanta, Islam, and Christianity.* Madurai, India: Tamil Nadu Theological Seminary.

Galley, H. (Ed.). *The Book of Common Prayer* (1979). New York: Seabury Press.

Giorgi, A. (1975). An application of phenomenological method in psychology. In A. Giorgi, C. Fischer, & E. Murray (Eds.), *Duquesne studies in phenomenological psychology: Volume II* (pp. 82–103). Pittsburgh, PA: Duquesne University Press.

Gove, P. B. (Ed.). (1966). *Webster's third new international dictionary.* Springfield, MA: G. & C. Merriam.

Haecker, T. (1937). *Kierkegaard.* London: Oxford University Press.

Heidegger, M. (1962). *Being and time.* New York: Harper & Row.

Husserl, E. (1962). *Ideas: General introduction to pure phenomenology.* New York: Collier.

James, W. (1977). *The varieties of religious experience.* New York: Macmillan.

King, W. (1952). *Buddhism and Christianity: Some bridges of understanding.* Philadelphia: Westminister Press.

Liderbach, D. (1983). *The theology of grace and the American mind.* New York: Edwin Mellen Press.

Maslow, A. (1959). *New knowledge in human values.* New York: Harper.

Maslow, A. (1962). *Toward a psychology of being.* Princeton, NJ: D. Van Nostrand.

Maslow, A. (1964). *Religions, values, and peak-experiences.* Columbus: Ohio State University Press.

Maslow, A. (1971). *The farther reaches of human nature.* New York: Viking Press.

Meissner, W. W. (1984). *Psychoanalysis and religious experience.* New Haven, CT: Yale University Press.

Meissner, W. W. (1987). *Life and faith: Psychological perspectives on religious experiences.* Washington, DC: Georgetown University Press.

Mishler, E. G. (1986). *Research interviewing: Content and narrative.* Cambridge, MA: Harvard University Press.

The Mother (1972). *Surrender and grace.* Auroville, India: Sri Aurobindo Society.

Pahnke, W., & Richards, W. (1973). Religion and mind-expanding drugs. In J. Heaney (Ed.), *Psyche and spirit* (pp. 108–118). New York: Paulist Press.

Reichenbach, B. (1990). *The law of karma: A philosophical study.* Honolulu: University of Hawaii Press.

Shoenberg, B. (1972). *Psychological aspects of terminal care.* New York: Columbia University Press.

Solimar, V. (1986). *The nature and experience of self-love.* Unpublished doctoral dissertation, California Institute of Integral Studies, Novato.

Sri Ramana Maharshi. (1988). *The spiritual teaching of Ramana Maharshi.* Boston: Shambala Publications.

Valle, R. S., King, M., & Halling, S. (1989). An introduction to existential–phenomenological thought in psychology. In R. S. Valle & S. Hailing (Eds.), *Existential–phenomenological perspectives in psychology: Exploring the breadth of human experience* (pp. 3–16). New York: Plenum Press.

Vedakkekara, C. M. (1981). *Divine grace and human response.* Bangalore, India: Asirvanam Benedictine Monastery.
von Eckartsberg, R. (1986). *Life-world experience: Existential–phenomenological research approaches in psychology.* Washington, DC: Center for Advanced Research in Phenomenology; University Press of America.
Watson, P. (1959). *The concept of grace.* Philadelphia: Muhlenberg Press.

19

On the Encounter with a Divine Presence during a Near-Death Experience

A Phenomenological Inquiry

Timothy West

INTRODUCTION

This chapter presents a phenomenological study of the experience of "encountering a divine presence during a near-death experience." As a result of my own near-death experience (NDE), I was inspired to investigate what I intuitively felt was the essence and defining element of the phenomenon: the meeting with an omnipotent, all-knowing entity or what many people commonly conceive of as "God." Although there exists a relatively large body of both empirical and speculative literature concerning near-death experiences, rigorous existential–phenomenological investigations, such as the study presented in this chapter, are rare. Since the nature of near-death experience defies "objective" measurement, qualitative or phenomenological methods seem best suited to describe the essence of this remarkable event.

The phenomenon of near-death experience, as we shall see, is a multifaceted one with great breadth and detail. One study cannot hope to cover all its implications. It is hoped that this synopsis will whet your appetite to the point where you will make your own investigation into near-death experience in whatever way you see fit.

In my opinion, near-death experience has not been adequately examined at the most basic level of the experiencer's perception. The phenomenon and its meaning appear to have

Timothy West • 512 Malobar Drive, Novato, California 94945.

Phenomenological Inquiry in Psychology: Existential and Transpersonal Dimensions, edited by Ron Valle. Plenum Press, New York, 1998.

been amplified through the lenses of the respective researchers and their selected excerpts, rather than through the explication of, and engagement with, raw and primary descriptive data elicited in a nondirective way. The reports of near-death experiencers themselves have not directly created, in a true phenomenological fashion, the meanings and themes that near-death experience has come to represent in the literature. This investigation attempts, in part, to examine what shifts may take place in the overall meaning of the near-death phenomenon if the locus of the event's description lies solely in the protocols of those who have had a near-death experience.

Significance of Divine Presence in Near-Death Experience

The term *near-death experience* was coined by Raymond Moody (1975), and since that time, several hundred publications, films, and videos concerning the subject have been produced. Most researchers, however, have not directly concerned themselves with examining the nature of a divine encounter during an NDE. In fact, only one study asked its subjects whether or not they felt they had encountered "God" or an all-powerful, infinite presence (Atwater, 1994). Previous research has often confined itself to asking questions, for example, about the "light" or an "all-consuming love" or "encounters with deceased relatives" (Moody, 1975; Ring, 1980). Yet, until the investigation presented in this chapter, there has not been a rigorous, phenomenological study of the element of divine encounter in near-death experience.

Why have investigators shied away from examining this "divine presence" factor? If, as this study has revealed, a high percentage of individuals who identify themselves as having had a near-death experience also report encountering "God" or an infinite being, then why has this aspect not been emphasized in the near-death experience prototypes and rating scales (Moody, 1975; Ring, 1980)? There are probably a number of reasons for this situation.

First, near-death researchers, from very early on, certainly wanted their research to be respected and accepted by the scientific community at large. Empirical science is generally more interested in quantifiable data than in nebulous or infinite entities such as those suggested by reports of a divine presence. Investigators (Ring, 1980; Sabom, 1982), at least in the early days of this field's development, certainly did not want to appear scientifically "soft" or unrigorous. Near-death phenomena were in themselves difficult enough to quantify—researchers did not want to stray into theological or philosophical quagmires. Research therefore focused on the most measurable elements of the overall experience—for example, the number of times individuals reported a certain category of experience or how individuals were changed as a result of their NDEs.

Second, although those who have had a near-death experience usually take pains to describe themselves as having increased interests in spirituality and *not* religious activity, some religious organizations feel threatened by the information that is emerging from near-death studies (Rawlings, 1993). These more dogmatic religious groups would react strongly to any reports that ordinary individuals were encountering God during involuntary experiences outside a religious structure. Researchers did not want to invite any more pressure or interference than they were already receiving from this quarter.

Finally, there exists within the NDE research community itself an ongoing debate that centers around the differences between "radiant" or positive NDEs and the hellish or more unpleasant varieties. Presumably, although much more evidence needs to be gathered, unpleasant NDEs tend not to contain a divine presence in its positive form, even though, as at least one researcher has pointed out (Bush, 1994), the "divine" can appear as a dark or underworld entity as well. The point here is that the study of divine presence may challenge our more Pollyannaish concepts of who or what God is.

One prominent researcher (Ring, 1994) implies that he will not give ontological validity to experiences that do not contain the element of the all-embracing, all-powerful light that has been emphasized in so much of the near-death lore. This sort of stance appears to exclude from the NDE category those frightening experiences that report unpleasant affect or do not contain "the light," and yet still have powerfully transformative effects on their experiencers, quite similar to their "radiant" counterparts. If, indeed, research into divine presence in near-death experiences shows the divine encounter to be as much of an element in frightening experiences as in their more pleasant forms (as the study reported herein implies), the ontological status of "the light" as reflected in the conclusions of early research (Morse, 1992; Ring, 1984) will need to be reevaluated.

Even though there can be resistance to examining this question, it is an issue that is clearly important to near-death experiencers themselves, as evidenced by the study reported in this chapter, in which 10 of 12 individuals, responding to an advertisement for participants in the study, claimed that they had unequivocally encountered a divine presence. The goal of this research is to present an essential description of this near-death encounter or, as Ken Vincent (1994) has phrased it, our present culture's "cleanest generic vision of God." In a society dominated and often blinded by materialism and technology, perhaps it is time that we examine the essence of a common form of modern mystical experience.

Mystical Experience, God, and Self-Transformation in Near-Death Experience

Past research indicates that near-death experience bears the earmarks of what is often termed mystical experience or an enlightenment event (Vincent, 1994). This experience is often described as containing an encounter with God or some sort of divine or "higher" entity (Ring, 1984). Furthermore, the effects of having such an encounter are nearly always found to be profoundly transformative (Atwater, 1994; Morse, 1992). These themes, in my estimation, represent the most important and powerful elements of near-death experience and of what this phenomenon can teach us.

Let us first look at the term *divine presence* as it is being used for the purposes of this study. The term "presence" is meant to be understood as the condition of being present or perceived as existing or being in attendance in the immediate surroundings or vicinity. The term "divine" is intended to imply the quality of being like God or a god, given or inspired by God, or supremely and all-powerfully great, good, and loving. These definitions closely parallel those given in *Webster's New World Dictionary* (1988).

Although none of the participants in this study was confused by the meaning of the term "divine presence," it is important to be exact for the purposes of this chapter, especially concerning the concept of "God," which is inherent in the term "divine." Again from *Webster's* (1988), God is defined as "the entity perceived as being eternal, infinite, all-powerful, and all knowing, often seen as being the creator and ruler of the universe."

The first researcher to carefully examine the "divine" nature of near-death experience was Kenneth Ring (1984) in his book *Heading Toward Omega.* Ring, when writing about his investigations, is painstaking in his efforts to let the data speak for themselves, whether these data be from a statistical analysis or the actual transcripts of his interviews with NDErs. In this spirit, I quote a number of Ring's research subjects to give a flavor of what the divine element in NDE might be like. These passages from Ring are taken from the hook *Visions of God from the Near-Death Experience* by Ken Vincent (1994):

> It's something which becomes you and you become it. I could say, "I was peace, I was love." It was the brightness, it was part of me (p. 29).
>
> It was just pure consciousness. And this enormously bright light seemed almost to cradle me. I just seemed to exist in it and be part of it and be nurtured by it and the feeling just became more and more ecstatic and glorious and perfect (p. 29).
>
> It was neither man or woman, but it was both. I have never, before or since, seen anything as beautiful, loving, and perfectly pleasant as this being. An immense, radiant love poured from it. An incredible light shone through every single pore of its face. The colors of the light were magnificent, vibrant, and alive. . . . I had the overpowering feeling I was in the presence of the source of my life and perhaps even my creator (p. 31).
>
> Upon entering that light . . . the atmosphere, the energy, its total pure energy, its total knowledge, its total love, pure love—everything about it is definitely the afterlife, if you will (p. 31).
>
> I cannot begin to describe in human terms the feeling I had at what I saw. It was a giant infinite world of calm, and love, and energy and beauty (p. 31).
>
> An immense radiant love poured from it. An incredible light shone through every single pore of its face. . . . I was filled with an intense feeling of joy and awe. I was consumed with an absolutely inexpressible amount of love (p. 35).
>
> Now I know there's a God and that God is everything that exists, [that's] the essence of God. . . . Everything that exists has the essence of God within it. I know there's a God now. I have no question (p. 21).

As implied in the last quotation above, it appears that the divine encounter can have significant effects on an individual's belief system. Research has shown that near-death experience, in general, nearly always transforms the nature of one's attitudes toward the existence of a divine power, often changing a static religious acceptance or even atheism into a dynamic, spiritual orientation based on experience. In fact, Ring (1984) found all 25 of his research participants to "strongly believe" in the existence of God after their NDEs. His research sample included those who were atheists, agnostics, and only "nominally religious" prior to their experiences. Ring (1984, p. 85) states:

> Plainly, whatever adherence to a notion of God we may find in such people following their NDE cannot be attributed to their prior orientation.

After presenting case after case that contains accounts of what we have defined as *divine presence,* Ring asks the ultimate questions (p. 84):

> Isn't it obvious that what core NDErs experience when they come close to death is what the rest of us would call God, or if not God, then surely some aspect of the infinitude of God manifest to the mind or spirit of the NDEr? To what other agency is it, after all, reasonable to ascribe the attributes of the core experience: the brilliant light, the all-consuming love, the feeling of total acceptance, the sense of total knowledge. . . . If this experience is *not* of God, then what else could it possibly be?

He goes on to conclude that the core NDE contains the direct and immediate experience of God and that it is this meeting, connection, or immersion that is the essence and basis of the NDE's power to transform. He writes (p. 155):

> The implication is that the qualities of the light somehow infuse themselves into the core of the experiencer's being so as to lead to a complete union with the light. In apparently the sense in which medieval theologian and mystic Meister Eckhart spoke of man becoming God, NDErs may experience this merging of their own individuality with the divine. In any event, the testimony from more than one core experiencer indicates that there is a direct transmission of the light's energy into

> themselves and that what is absorbed in that encounter with the light in that moment outside of time remains with them when they return to the world of time. In short, the seeds of transformation appear to be planted during the NDE. . . .

Morse (1992) echoes Ring's observations. Morse clearly believes that the essence of the near-death experience is to be found in "the light." Since nearly all NDErs who experience this light equate it with what we have come to call divine presence, we may assume this light to have divine properties. Here is what Morse has to say about the effects of this light (p. 171):

> The . . . study neatly documents the fact that people who have near-death experiences are changed for life. Those changes are most profound in the NDErs who have experiences of light. This was true whether they had a vivid and powerful memory of a flower-filled heaven bursting with light, or just a brief and fleeting memory of seeing the light.

Morse in effect is saying that it is during the encounter with this light that the experiencer undergoes the changes associated with NDE. Although Morse rarely mentions God or "the divine" except through the quotations of his research participants, he obviously regards the light as all-powerful when it comes to the transformative properties of near-death experience.

Atwater (1994), a near-death experiencer as well as a researcher, is more explicit about the divine quality of the oft-reported light in NDEs. She clearly links the all-powerful quality of this presence with the essential transformative effects of the phenomenon (p. 142):

> . . . as an experiencer, I can positively affirm that being bathed in the Light on the other side of death is more than life changing. That light is the very essence, the heart and soul, the all-consuming consummation of ecstatic ecstasy. . . . You know it's God. No one has to tell you. You know. You can no longer believe in God, for belief implies doubt. There is no more doubt. None. You now know God. And you know that you know. And you're never the same again.

Vincent (1994) believes the near-death experience to be our present culture's "cleanest" and most direct form of individual contact with the divine. Along with Ring and Atwater, Vincent sees the central meaning of the NDE as emanating from its ability to afford the experiencer a direct revelation of God. From the work of these various researchers, it seems that the pivotal, transformative force within the near-death experience is an element of divine presence or power, usually in the form of "the light."

It is also known that many individuals, throughout history, have had experiences that parallel modern NDEs, encounters that have come to be termed "mystical." The question arises whether mystical experience, by definition, includes a divine encounter and whether near-death experience can therefore be associated with this category of experience. Several researchers (Atwater, 1994; Moody, 1977; Morse, 1992; Ring, 1984; Vincent, 1994) feel that the near-death experience is indeed a form of mystical experience. Moody (1977, pp. 99–100) quotes William James as listing the following characteristics of mystical visions:

> 1. *Ineffability*—The subject of it immediately says that it defies expression, that no adequate report of its contents can be given in words.
> 2. *Noetic quality*—Mystical states seem to those who experience them to be also states of knowledge. They are states of insight into depths of truth unplumbed by the discursive intellect. . . .
> 3. *Transiency*—Mystical states cannot be sustained for long. Except in rare instances, half an hour, or at most an hour or two, seems to be the limit beyond which they fade into the light of common day. . . .

4. *Passivity*—. . . when the characteristic sort of consciousness has set in, the mystic feels as if his own will were in abeyance, and indeed sometimes as if he were grasped and held by a superior power. . . . Mystical states . . . are never merely interruptive. Some memory of their content always remains, and a profound sense of their importance. They modify the inner life of the subject.

Based on the findings reviewed above, a given NDE may contain any or all of these characteristics. Moody points out, however, that near-death experiences appear to contain elements that are not classically reported in mystical visions recorded throughout history (e.g., a life review and meeting with dead relatives).

Atwater (1994) quotes another early investigator of mystical phenomena, Richard Maurice Bucke, and notes that there are striking similarities between Bucke's "pattern of mystical enlightenment" and the aftereffects of NDE. Bucke's characteristics are (Atwater, 1994, pp. 145–146):

1. *The subjective light:* A brilliant blinding flash is seen. The individual's surroundings take on colors of unearthly hues and brilliance. . . .
2. *The moral elevation:* Afterward, the individual becomes moral and upright, shunning the temptation to judge or criticize another. . . . A greater duty and service to God and humankind becomes a life priority.
3. *The intellectual illumination:* All things are made known, all knowledge is given, all secrets of the universe are revealed . . . he or she is overwhelmed by total and complete love. Glowing beings give instructions, as the "Word of God" is seen or felt; the oneness of all things shown.
4. *The sense of immortality:* Thinking is replaced by knowing. The individual realizes his or her divine identity . . . that we are all immortal and divine from "The Beginning."
5. *The loss of the fear of death:* Death loses all meaning and relevance. The individual now knows death does not end anything, is nothing but a change of awareness.
6. *The loss of the sense of sin:* Evil is understood as good misused, that all things are good in God's eyes.
7. *The suddenness, instantaneousness of the awakening:* . . . the actual moment of illumination is always unexpected, sudden, and blinding. It can last minutes or hours or days.
8. *The previous character of the person:* . . . resident characteristics are expanded and enhanced . . . latent abilities surface, including genius . . . the desire to learn and excel is strong.
9. *The added charm of the personality:* The individual becomes so magnetic that people and animals are drawn to him or her. The individual seems divinely protected and guided.
10. *The transfiguration:* There is a marked change in appearance. The individual seems to glow and have a light around him or her. There are physical changes. The face looks different, and the individual behaves like a "new" person—as if suddenly more than before.

On the basis of these criteria, the near-death experience again shares important qualities of an enlightened or mystical event.

Although it may not be clear whether all mystical experiences include a divine encounter, a number of NDE investigators are quite certain that the near-death experience involves some kind of contact with a divine element (Atwater, 1994; Morse, 1992; Vincent, 1994). Atwater (1994, p. 143) comments:

> Although not everyone speaks of God when they return from death's door . . . the majority do. And almost to a person they begin to make references to oneness, allness, isness, the directive presence behind and within and beyond all things.

In the same regard, Vincent (1994, p. 10) points out:

> It is this Being of Light or Light, which many experiencers go on to label God or an emissary of God . . . they also and most importantly provide us with a generic vision of how humans see God.

Vincent goes on to list a number of historical examples of mystical insights that mention the "light," including *The Egyptian Book of the Dead* (Vincent, 1994), *The Tibetan Book of the Dead* (Evans-Wentz, 1957), the Christian gospels, Zarathustra, St. Paul, and others. Moody (1988) gives a similar list. Bucke (Atwater, 1994) mentions "light" as the first element of mystical experience. Morse (1992, p. 175), moreover, views the "light" encounter as the sine qua non of the NDE's power to transform:

> The transformative part of the experience is seeing the light. If a person has a paranormal experience such as leaving their body but it is *not* accompanied by the light, then the experience is not usually transformative. If the light is experienced then there is a transformation. The transformative powers are in the light. That is what our research tells us.

Perennial wisdom and its scriptures contain passages that echo elements of near-death experience. Christianity's *New Testament, The Tibetan Book of the Dead,* and the spiritual science of Sant Mat from Northern India all contain written material alluding to out-of-body traveling, encounters with "light" phenomena, and the return to a divine oneness (West, 1996). The element of divine presence seems to appear, in some form, throughout the great mystical teachings of history. A modern researcher, Vincent (1994), concludes that present-day near-death experiences have become our culture's most direct form of religious revelation because of: (1) the consistency of the revelations across cultures and religious backgrounds and (2) the closeness and verifiability of the revelations' sources. If divine presence is indeed an essential element of near-death experience, as the previous literature seems to imply, what are its essential qualities and what form does it take in NDE? This is the primary question that this phenomenological inquiry was designed to answer.

THE RESEARCH

This study examined the near-death experiences of six men and four women, all of whom claimed to have encountered a divine presence during this time. A written description of the experience was obtained from each of the participants, who were subsequently interviewed in a nondirective, "walk-through" fashion. The written protocol plus the transcribed interview were then analyzed carefully for common thematic structure. The method of analysis was based on an adaptation of Colaizzi's (1978) existential–phenomenological approach. This research approach proceeded in five phases:

1. Data from the researcher's self-reflection.
2. Selection of co-researchers.
3. Data collection from the written protocols.
4. "Walk-through" interview.
5. Data analysis.

Phase 1. Data from Self-Reflection. During this phase, I closely scrutinized my own presuppositions concerning the near-death phenomenon by examining my own experience of the event and by listing any presuppositions or biases that this experience or any related information may have produced. This was done in order to minimize any inclinations to impose predeterminations on or unwarranted interpretations of the co-researchers' descriptions. This self-reflective process produced a list of 26 preconceptions or biases.

Phase 2. Selection of Co-Researchers. Co-researchers were solicited through word of mouth, a posted flyer, and several announcements. Twelve individuals responded and, over the phone, ten were determined to have undergone a near-death experience in which they had encountered a divine presence.

The ten co-researchers included six men and four women ranging in age from 29 to 67. Each co-researcher received a consent form to be signed and returned.

Phase 3. Data Collection from the Written Protocols. A questionnaire based on that of van Kaam (1966) was used to elicit the descriptions. Each co-researcher was asked to:

1. Recall the time when you experienced encountering a divine presence during a near-death experience.
2. On a separate piece of paper, write your description of how you felt during that time.
3. Try to describe your feelings just as they were, so that someone reading the report would know exactly what the experience was like for you. Keep your focus on the experience itself, not just the situation.
4. Please do not stop until you feel that you have described your feelings as completely as possible. Take as long as you would like to complete your description.

The co-researchers thus produced written narratives of the experience being investigated.

Phase 4. "Walk-Through" Interview. The face-to-face interviews proceeded with the researcher reading back the respective written description to each co-researcher in a relaxed statement-by-statement fashion, pausing at appropriate times and allowing the experiencers to deepen and detail their descriptions. In this way, the co-researchers "walked through" their written protocols to give them an opportunity to elaborate as completely as they cared to. The interviews were tape-recorded and then transcribed.

Phase 5. Data Analysis. The 10 steps used for the data analysis are based on methods used by Colaizzi (1978) and von Eckartsberg (1986):

1. The final protocols were read carefully several times to gain a feeling for their characteristic meaning and intent.
2. Significant statements were extracted from each protocol, and a list was made for each co-researcher.
3. These significant statements were sorted into theme clusters within each protocol.
4. These theme clusters were then combined across protocols to form 84 constituent themes.
5. The constituent themes were then examined to make sure that all the extracted significant statements pertained to that theme.
6. The constituent themes were then clustered in groups that appeared to share a new level of meaning. This is where I began to use intuitive judgment to generalize statements without losing the underlying essence of the particular co-researcher's experience. Elite (see Chapter 14 in this volume) notes that "Colaizzi (1978) describes this procedure as a 'precarious leap' that the phenomenologist makes in order to bring an interpretive–psychological meaning to the extracted statements."

7. The constituent themes found to have the same thematic meanings were combined to form 25 comprehensive constituent themes.
8. Further reduction revealed seven final comprehensive constituent themes. All seven of the final comprehensive constituent themes were contained in all ten protocols.
9. The final comprehensive constituent themes were then checked to make sure they included all the meaning statements that belonged in the respective category. Where appropriate, meaning statements were moved to another theme where the category was more reflective of the statement. When necessary, the themes themselves were altered to more accurately reflect their constituent meaning statements.
10. The essential structural definition of the phenomenon was then derived from the final comprehensive constituent themes.

Final Comprehensive Constituent Themes

Seven final comprehensive constituent themes emerged from the analysis. Among them, these themes contain every significant meaning statement in the protocols; in addition, each theme was found to be present in all ten of the protocols analyzed. The seven final comprehensive constituent themes are as follows[1]:

1. Leaving the body, traveling, and returning to the body.
2. Experiencing infinity and no boundaries with a surrender to and/or merging with a higher power.
3. Experiencing divine refuge and/or homecoming with extraordinary feelings of love, acceptance, peace, and joy.
4. Experiencing absolute truth and divine knowledge.
5. Perceiving the experience as ineffable.
6. Perceiving the experience as a personal message.
7. Perceiving the experience as transformational.

A discussion of each final comprehensive constituent theme follows, with several excerpted examples from the co-researchers' protocols chosen to represent the flavor and range of the statements that comprise each theme. The excerpts are presented in such a way as to give an idea of the whole of each co-researcher's unique experience.

1. Leaving the body, traveling, and returning to the body. Each co-researcher in this study had some sense of having left his or her body or of beginning to do so. This sensation was expressed in a variety of ways, ranging from simply "going through darkness somewhere," to a sense of having died and encountering a feeling of "incredible warmth" in a deep blackness, to a relatively complex journey, to a celestial landscape and city of light. From these data, it appears that the encounter with a divine presence during a near-death experience takes place, not within the perceptual field of the physical body, but rather at some other level of consciousness. The seven general ways in which co-researchers expressed the fact that they felt they were leaving or had left the physical body are as follows:

a. The sense of having died yet being alive.
b. The sense of having a detached, disembodied awareness without pain.

[1]As with all category descriptors in this study, the language of the co-researchers is preserved and included as much as possible to reflect the essence of the experience as they described it.

c. The sense of traveling without a body.
d. The experience of having abilities that are not indigenous to the physical body.
e. The experience of going to another world with different dimensions and conditions.
f. The feeling that one has left the concerns of the ordinary world completely behind one.
g. A sense of pain, loss, gain, or problematic reentry into the physical body.

Examples of the different ways that this final comprehensive constituent theme manifested include:

> Everyone thought I had died, but I knew I was still alive. . . . I was conscious of everyone's presence. I could hear them talking. . . . All that was left of me was a blue line of light spanning from my head to my toes about the circumference of a pencil, but that light was fully conscious, fully aware, fully present. . . . Then I left the room . . . there was no pain . . . I moved without any effort further and further away and progressively higher . . . whatever problems I had before or whatever troubles I had—none of it mattered anymore. . . . I moved down the dark tunnel much faster than I had gone up and felt myself again in my body . . . just like that!

> I had been traveling through an incredibly vast darkness . . . there was no reference . . . no arms. . . . I can distinctly remember that memory of no arms, no legs. . . . And then becoming aware of this light, just a dot, just a speck; and then just traveling to it, like going through this eternal space and being conscious.

> Immediately I found myself in the midst of a paralyzing electric current which locked my hands and arms all the way from my shoulders to my fingers. . . . Suddenly I was just locked into it. . . . It was the closest that I have possibly come to death. . . . I remember things got dark and I began to realize that I was going to die. . . . I experienced that I was going somewhere, I was actually, it was almost like I was going through this darkness somewhere. That was very, very scary and I knew right away that something had to happen so I could come back. . . .

2. Experiencing infinity and no boundaries with a surrender to and/or merging with a higher power. The co-researchers also experienced encountering some force or power that they deemed infinitely greater than themselves and responded to in various ways, including fear, struggle, awe, and wonder. Some co-researchers felt completely relaxed during this experience and felt that they had merged with some divine source. Others struggled and resisted out of fear before feeling a sense of communion. Another felt that he only approached this union without completing it. Several co-researchers reported a sensation of not having boundaries, of not knowing where they began or the infinite began, and of being completely permeable to a higher power. All the co-researchers invariably felt a respect and humility in facing what they all perceived to be a colossal and infinite reality far beyond the limits of this world. In the face of this infinite power, many co-researchers felt a sense of personal insignificance and even a complete loss of identity and personal control, which in some cases was seen to be a joy and a relief.

This final comprehensive constituent theme was expressed in three general ways:

a. Fear, wonder, awe, or amazement at the divine vastness or eternity.
b. A profound sense of personal insignificance or loss of identity and control; an absence of boundaries.
c. Surrender to and often immersion with a supreme oneness, sometimes in the form of an all embracing "light" or "being of light."

Examples from the co-researchers' words are:

> I became light. . . . I not just became with, but became of and became part of. . . . The darkness, it was like infinite, and I was conscious of this eternity, yet there was no reference . . . [just] pure conscious awareness of vast eternity . . . eternal space . . . going beyond your conception of speed, be-

> yond your conception of space . . . being aware of so much yet no time passed. And yet all time passed . . . the light became everything. . . . All this wholeness was there, and it was like it was radiating from the light and myself and the light were merging, evolving, coming together, or becoming. . . . I was conscious of being in awe continually and thinking there could be no grander sensation of being. . . .
>
> I simply gave up and relaxed all effort. . . . It was a wonderful feeling to give up actually . . . it's really an incredible experience. . . . I gave up, it was like, it's bigger than I am. . . . I was acutely aware of the blackest black I had ever experienced. . . .
>
> I was in awe of it . . . it is awesome. . . . I heard a multitude murmuring, but could only see and sense the light. . . . [I] felt as much as saw the shimmering light—bright yet soft and not a blinding light . . . that just seemed to permeate and pass through . . . that nothing could be hidden or not revealed. I felt exposed as I never have felt before. . . . Episodes and experiences came round one after another to my mind and I felt there was nothing about me that was unknown . . . the light was both bathing and passing through me like a slide projector. . . .
>
> I was standing in this incredibly beautiful meadow or pasture . . . almost overwhelming. . . . I was absolutely part of everything. . . . I felt such utter shock, amazingly surprised . . . there was no choice . . . like I said it hit me like a train. It was the most powerful force I have ever encountered . . . so powerful beyond anything I could have imagined . . . each time I would have a thought of amazement . . . it was just this total connection with everything. There was no hiding and there was no reason to hide . . .
>
> It was very frightening, yeah . . . and out of my control. . . . I was screaming. It was very scary . . . there was no one to help me . . . the only power that could help me was God. . . . So I just begged and prayed to that higher force to save my life. . . . I began to beg for my life and beg God to help me and untie me from the rope of death that was pulling me in, and continued to kick, beg, and gasp for life. . . . It was so powerful, I had incredible fear and yet respect and love for it.

3. Experiencing divine refuge and/or homecoming with extraordinary feelings of love, acceptance, peace, and joy. The co-researchers also reported experiencing some aspect of divine refuge; that is, they felt loved, peaceful, accepted, safe, and protected to a degree that they had never felt before. Sometimes this theme was expressed in terms of feeling like one belonged to a divine family, by being overwhelmed and permeated by an all-powerful wave of unconditional love, by simply experiencing a feeling of incredible warmth, or by having a sense of being taken care of or helped. One co-researcher whose experience was primarily terrifying and unpleasant nevertheless retained a feeling of being rescued by an all-powerful presence, a sense of having been given a second chance through a divine act of love.

The co-researchers found their encounter with a divine presence to be ultimately positive. Although other themes may express feelings of fear, doubt, and insignificance concerning one's own stance and personal abilities vis-à-vis the attributes of divine presence, this final comprehensive constituent theme shows that contact with this presence is emotionally salutary and even ecstatic. Indeed, it is this intense emotional experience that is perhaps the most memorable element of the encounter for these co-researchers and that creates the "longing" and desire to experience the encounter again. The four general ways in which this theme manifested were:

a. Feelings of unconditional acceptance and love.
b. Experiencing comfort, warmth, and safety, including a sense of homecoming or rescue.
c. Feelings of peace, contentment, and wholeness with a sense of divine plan.
d. Feelings of joy, bliss, and freedom, including a sense of extraordinary beauty.

Examples from the protocols are:

> It started with fear and ended up with huge feelings of love, of acceptance . . . there was no right or wrong. . . . It's been very important . . . to have the experience of not being judged, of not feeling that I've fucked up, or could have done it better, and the feelings I got were extraordinary feelings of acceptance the way I was and that I had done a good job.
>
> I never knew that such peace, such bliss could exist. I was totally loved, totally accepted by this light. . . . I felt so safe, and I felt so seen and known and accepted. . . . Everything was okay . . . it was like someone was there taking care of me and loving me. . . . I felt the joy, the peace, the deeply, loving caring glorious energy of that presence. . . . I was home. I was safe . . . it really felt like I had come home . . . and all my little problems were solved. . . . Everything was okay.
>
> I wasn't overwhelmed or frightened or anything else . . . I was just muy contento. I had a peaceful easy feeling; I mean I was content. . . . I'm a free man. I mean I understood, I felt totally forgiven. No more guilt . . . no guilt at all. Nothing to fear.
>
> I felt such sheer joy and love . . . as though I'd been hit by a train of unconditional love and forgiveness. As though God were saying, "Debbie, you are loved and accepted no matter what you do. I am with you always". . . . [God was] telling me everything was okay and I was completely loved and forgiven . . . there was absolutely nothing I could do to make Him love me more or any less. The love was constant, unchanging, pure and absolute. I knew I belonged and was loved no matter what . . . it was very much like going home or something.
>
> I became peaceful. . . . I was at peace and so thankful. . . . I really think something from a higher force just gave me another chance. . . .

4. Experiencing absolute truth and divine knowledge. The experience of having received some form of absolute truth or divine knowledge was reported. Usually, this theme manifested as a discrete or specific revelation such as experiencing the truth that "we are never separate from God and can contact him at any time" or experiencing the knowledge that "one's life is a choice." Several co-researchers reported that specific passages from the Bible became alive for them because of their new knowledge. The Scriptures now seemed real and based on experience. Sometimes co-researchers reported coming to a complete understanding of how all of creation works or experiencing a complete clarity all at once of everything there was to know. An example of this sense is:

> It was like all questions and answers became one. . . . I became part of yet a grander understanding . . . the light of illumination. If you could turn a light bulb on and have all the knowledge, all answers, that would definitely radiate a light of understanding [that's what it was like]. . . .

In some instances, this "knowledge" theme seemed closely allied with the "transformation" category, since several co-researchers described the way that they had been changed by their encounter in terms of the knowledge they had gained from the experience. Consider this example:

> From that day on I experience my life as a choice. I KNOW I am here by choice, that this is an experience for me to be involved in. There are occasionally times where I will not want to be here because there's a lot of work. But I'm aware that I made a choice to be here and I'm not actively looking for any way out. I just stay with it and I know it's just my choice.

This person clearly considers himself psychologically changed by the knowledge that he now has a choice.

The three main elements of the absolute truth and divine knowledge theme are:

a. The experience of coming to a complete and total understanding of existence all at once.
b. The experience of acquiring specific knowledge concerning a certain subject.

c. The experience of a new and more profound understanding of spiritual Scriptures and teachings.

Examples of these elements are:

> I know I will either be reborn here on earth upon awakening or experience once again what I felt [during my NDE]. . . . I understand that I am the internal external self, I'm the pilot and this is my earth capsule spaceship. . . . I am now supremely confident and comfortable with the knowledge that sometime in the future I will return to that light where I am once again free of my earthly flesh.

> I felt so relieved to know God was so real and so powerful beyond anything I could have ever imagined. . . . I always had that terror. You know, that kind of terror that you wake up with in the middle of the night, that—oh, I'm gonna die—you know . . . since this occurred I've never experienced that terror . . . now it's more the opposite where I'm never alone. Obviously now there's somebody there. . . . [I know] I'm this spiritual being stuck in this physical body. . . . I know I have some mission to perform and each person leads me closer to the light. . . . If I had to say which messages were the most important they were about loving unconditionally and unconditional forgiveness . . . and that all the answers I was looking for I already had within.

> I describe [it] as a wave of knowing or awareness of the truth. I received a "course" in how the universes were constructed and worked. . . . When you're part of God then you know everything God knows because you are God . . . it seemed like questions but all coming so fast that it was a matter of forming my attention towards something and realizing that I actually knew all of that . . . there isn't a separateness though we describe it in a separate fashion as we were still outside and separate from this, separate from God, separate from the knowledge. . . .

> In those infinite moments I acquired the knowledge that allowed me to go back to earth to complete my life. . . . I learned how to live with my murderer for another fifteen years by learning what I could from him and leaving the rest. I learned that the most important phenomena in the universe are love, truth, and the quest for knowledge.

5. Perceiving the experience as ineffable. The co-researchers reported difficulty in describing their experience. Several of them found their experience to be so extraordinary and otherworldly that they struggled to find a situation from this world to compare it to. Two of the co-researchers sum up this frustration:

> Being asked about my feelings when I experienced a divine presence, I am immediately at a loss for words. For my trouble with explaining what I felt is that I am truly speechless. There are no words, combination of words, or for that matter any things short of experiencing what I know I have experienced . . . as soon as I bring it out into words they're so limiting, it brings it down and it tries to package something that is boundless, endless, and eternal.

> How do you describe an unlimited universe in the limited terms of our universe? That's always been the hard part. . . . [I] just never have been able to find words to describe all that because it's not invented in our universe.

This difficulty in verbally capturing the nature of their divine encounter at times causes experiencers themselves to doubt their own memories and seek validation from others. The fact that it is hard to find other people who understand what they have gone through was mentioned in several protocols to the point that several co-researchers found themselves comparing their experiences to other NDE reports in order to have some context in which to place their own reports. Several co-researchers mentioned that thoughts rather than words were used during the encounter to communicate with other beings or the divine presence.

In sum, mere words and the normal avenues of human communication are found to be quite limiting and cannot fully describe the grandeur of this phenomenon according to most of the co-researchers in this study. If, indeed, we use language to help ourselves make sense of our experiences, the fact that this experience is largely ineffable for those who undergo it may account for the difficulty and extended time period required in integrating the encounter into one's personality and belief system. The primary ways in which this theme manifested are:

a. Experiencing difficulty in finding words to accurately describe the experience.
b. The inability to get others to understand the nature of the experience, including the sensation of doubting one's own memory or sense of the experience.
c. Experiencing the fact that thoughts rather than words are used to communicate during the experience.
d. Experiencing the need to compare one's own NDE to other NDE reports for support or validation.
e. Experiencing difficulty integrating the experience into one's belief system.

Examples from the co-researchers' descriptions include:

> I don't remember anything about any tunnels or the trip there. . . . I didn't see the tunnel or anything that anybody talks about. . . . When you ask me to reduce the experience to feelings, it is difficult because even to place the experience into words seems insulting because of the magnitude of the event. . . . I can't put it into words. . . . I'm simply at a loss as to describe how I felt at that moment. . . . There simply are no words to describe Him . . . this sounds bizarre but . . . I communicated without speaking. . . . [Things were] pointed out to me in a way no words could have ever communicated. . . .

> It is very difficult to describe the experience in words, the closest word for it is a miracle. . . . I guess I was in shock and needed some kind of assurance that this realy happened . . . telling the story to someone, telling what I went through, yeah, I was really seeking that. . . . I remember it wasn't really enough talking to the neighbor, but it was a validation that I hadn't lost my mind, this was really something that had happened. . . .

> You know, we're trying to relate it to our eyes and vision . . . there's no other words for that. And it was so much greater than anything I've ever experienced in a physical body that, you know, it's again limited terms here. . . . I realized right away that if I said anything out of the ordinary they might think I had brain damage or something. . . . I went through periods of thinking I was crazy 'cause there was no support at that time.

> I can't even describe it, I can't even describe it. . . . You know, people always try to describe these things . . . it is just nothing really works.

6. Perceiving the experience as a personal message. The co-researchers also expressed the feeling that their divine encounter was personally meant for them or that the experience had a purpose that spoke uniquely to them as an individual. For example, one co-researcher felt she had been allowed to come back to this life in order "to be a mother to my baby," that the divine presence had granted her wish so that she might fulfill a personal destiny. As she says:

> I was really grateful. I was really happy to be back. Yeah, I knew it was right.

Others also felt that their experience was personally tailored for them, that the encounter was uniquely individual and meant to speak directly to one's special needs:

> I doubt very much if you're exposed to more than you can assimilate. I imagine that you are given what you can handle and not more than you can handle.

Along these same lines, these two individuals saw their experience as a kind of "wake-up call" in which it was pointed out to them that they needed to change certain ways of doing things or to incorporate a specific piece of knowledge into their lives in order to better fulfill a divine destiny. Another co-researcher reported feeling challenged by his experience to bring more harmony and love to the world:

> I think one of the great frustrations I have is trying to work with that understanding of the unconditional love and acceptance . . . which is the challenge and the journey. . . . That's what the world needs more of and that's what I live to bring: that peace, that love, and that universal harmony . . . there's got to be action.

Several experiencers expressed feelings of profound gratitude and reported a sense that they had been personally graced by being allowed to have had the experience. Others found the encounter to be so personally profound and moving that they knew they would never forget the experience and in fact periodically found themselves accessing the memory for personal comfort, validation, or support. One co-researcher, in fact, felt his experience to be so unique and private that he asked that his protocol not be published so that its sacred nature would not be violated or disrespected. In sum, the co-researchers reported their encounters with a divine presence to have a personally directed quality that manifested in several different ways, including these:

a. Experiencing the encounter as a unique directive concerning one's personal destiny.
b. Experiencing the encounter as unforgettable.
c. Experiencing the encounter as a wake-up call.
d. Experiencing the ability to access the memory of the encounter for personal comfort and support.
e. Experiencing one's life being assessed.
f. Experiencing an intense feeling of good fortune and grace to have had the encounter.
g. Experiencing the realization that the NDE makes perfect sense in terms of one's historical background.
h. Experiencing the fact that one's prayers or wishes were granted by the divine presence.

Examples of this theme are:

> And something has driven me from that day that, even where we are sitting right now is part of that. . . . I mean for me to understand that things have happened that I don't even know about. . . . [I] distinctly remember feeling grateful. . . . I made the conscious decision to live, and for whatever purpose, it's ongoing right now and I know it. . . . Each and every day since I had this experience I awaken with the same sense of rebirth . . . it's a good strong healthy place to visit, and that's where I go when I go to bed,

> I do firmly believe that I brought back some of that love with me. Maybe it was always there but I needed a big jolt to find it . . . it was as though I was being told, you know, this is a big area that you need to change, that you need to adjust your perception of everything. . . . I know I have some mission to perform and . . . I am real grateful. . . . I know I'm doing what I'm supposed to be doing. It's like that pain had a purpose.

> I knew God bad released me from the electric power and gave me a second chance. . . . Thank you! I am living and kind of celebrating my life. I can drink orange juice again. . . . It's an experience that I will never forget. Every time I become depressed or sad about something I do remember that experience and I have a whole different perspective.

> It was communicated to me that I had to go back, that I had a great deal more to do on earth. . . . I was brought to that point for some reason, to see that perspective. . . . I've learned a lot about myself from the experience since then. . . . And I chose to accept it because it wouldn't go away, and it still hasn't.

7. Perceiving the experience as transformational. The co-researchers reported that, in some way, they felt changed by their encounter with a divine presence. As with all the final comprehensive constituent themes in this study, the details that comprise this part of their experience vary widely. Five co-researchers, however, agreed in their reports that, in general, the divine encounter had a "tremendous impact" on their lives. This impact varied from experiencing difficulty in adjusting to this world after their encounter to experiencing a new appreciation of children to being able to play the church organ with greater proficiency. Three co-researchers revealed that they had no fear of death because of their experience, a phenomenon that is often reported in the near-death literature. One co-researcher even reported that the experience prompted her to enter a recovery program for alcoholism.

In general, the aftereffects of the co-researchers' experiences appear to be positive, although two of them reported experiencing frustration and a lack of reinforcement for their new perspectives, and one experienced intense rejection and judgement during the period immediately following his encounter. Moreover, one should not forget that these co-researchers often experience physical aftereffects, frequently in terms of a painful and long-term recovery from injuries or surgery. This theme manifested itself in three general ways:

a. Difficulties readjusting to the realities of this world, including a sense of feeling "incomplete."
b. Newly acquired abilities, sensitivities, talents, or awareness.
c. Changes in attitudes, interests, and/or beliefs.

Examples of this theme are:

> It's been very important to me since I came back to have the experience of not being judged . . . the significance of right or wrong or doing a good job or not doing a good job or fucking up or evil and good is shifting, has shifted ever since then in me, and I'm not—I know, personally, I'm not as judgmental and I am less interested in hanging out with or listening to the people who are judgmental. . . .

> [There is] a peacefulness that will travel with me the rest of my life . . . there is a tear that resides next to this for not being able to express or accept the peace that I have experienced in this presence. . . . I think one of the greatest frustrations I have is trying to work with that understanding of the unconditional love and acceptance and know it's there, and at the same time, be human, which is the challenge and the journey. There's this pure love that I want to radiate outward and I know there's not too much reinforcement that will nurture it.

> I was so awestruck by my experience, it really helped me get through the first phase of recovery. . . . I was a Christian prior to this experience, but since this happened, I have an even more meaningful and personal relationship with Jesus as my Savior . . . because He was more theory before I had this experience . . . it was more theory than being real. And then I had just absolute knowledge after this experience which has really helped me get through my lifetime now. . . . I have total confidence in a literal heaven and have no fear of death. . . .

> I do firmly believe that I brought some of that love with me. . . . I don't feel the same about other people now. Many times I'll see a complete stranger and feel such love for them or I look at someone I don't know and I'll start praying for them. It's like people just don't look the same to me. . . . I just thought the Bible was a book and Jesus was some guy in history. . . . And now it's like I can't hear the Bible, the Scripture read, without just getting totally overcome.

From these protocols, we can begin to appreciate the intensity, depth, and range that characterize the encounter with a divine presence during a near-death experience. Perhaps what is most striking about this experience, among its many extraordinary qualities, is the profoundly personal and deeply transformative effects it has had on the co-researchers. It is as though all experiencers, appreciating the fact that they have been allowed contact with a reality so extraordinarily transcendent, feel as though they have been given a special "gift"—or "peek in the box," as one co-researcher put it. It is not surprising that all the co-researchers spoke either in their protocols or privately of the problems in sharing with others whom they did not know well an experience that is so uncommon and personal.

In my synthesis of the seven final comprehensive constituent themes, the following essential structural definition or fundamental essence of the experience of encountering a divine presence during a near-death experience emerged:

> One perceives an ineffable out-of-body sensation during which there occurs an awe-inspiring meeting with an all-loving and infinitely powerful atmosphere or presence. This contact with the divine is characterized by such infinite power, loving acceptance, and complete immersion in feelings of well-being or safety that one emerges with a knowledge of ultimate reality that is at odds with what one has experienced in day-to-day life. The experience engenders intense feelings of gratitude, a sense of grace, and a sense of a private and personal communication with and acquisition of knowledge from a divine source. The indescribable and extraordinary nature of the experience that accompanies the wide range of its positive effects can also make it difficult for one to readjust to normal life. The experience is characterized by producing such dramatically changed attitudes and awarenesses that one who undergoes it often develops life-changing commitments to pursuits of spiritual value, and a new, experientially grounded belief concerning divine salvation and afterlife.

SUMMARY AND CONCLUSIONS

In addition to offering support for past characterizations of NDEs in general, the results of this study suggest a number of new ways to understand the nature of a divine encounter during a near-death experience. I offer the following conclusions based on the results of this phenomenological inquiry:

1. It appears that "divine presence" comes in many forms other than "the light" and, in fact, appears to be available throughout the entire near-death experience.
2. No one event in the experience was singled out by the co-researchers as being any more powerful or transformative than any other.
3. The divine encounter in near-death experience is an exclusively private and personal communication and, as such, cannot be evaluated objectively from an outside perspective. The locus of evaluation for any measure of depth, completeness, or meaning of the experience must ultimately reside within the experiencer.
4. The themes regarding divine knowledge and experiencing grace in perceiving the experience as a personal message are underrepresented in the previous literature. In the whole of Moody's (1975) and Ring's (1980) NDE prototypes, these elements are mentioned only once.
5. The divine encounter in near-death experience, rather than showing a predictable structure or sequence of events, is best appreciated or defined in terms of themes that can embrace a wide variety of occurrences, including terrifying experiences.
6. Any speculation concerning ultimate outcomes or teleologies regarding where the near-death "continuum" leads, especially in a hierarchical sense, is best left to the experiencer and his or her learning from the event.
7. The "myths of near-death experience" as elaborated by Atwater (1994) are becoming so well known in the culture at large that experiencers often compare their encounters to other reports or hearsay of what the experiences *should* be like. This can be used either to validate or to question their own experience.

8. Traditional spiritual teachings appear to have numerous concordances with the material elicited in these co-researchers' protocols. A foundation for a unified spiritual vision or truth, much like that of the traditions rooted in the perennial wisdom, appears to exist in the reports of these experiencers.
9. The phenomenon of the divine encounter during a near-death experience belongs squarely in the category of mystical experience as defined by several researchers. This being the case, perhaps we need to consider NDE as an event having more to do with living than with dying. That is, as Bache (1994) suggests, perhaps we must separate our notions about NDE from the context of death to understand it more fully. Clearly, the message and knowledge inherent in the events reported by the co-researchers in the study have ostensibly more to do with improving how one may live one's life than with the process of death itself.

As a modern form of mystical experience that is involuntarily experienced by men and women, young and old, from all walks of life, and apparently without predilection, the divine encounter during a near-death experience can be an overwhelming, ecstatic, life-transforming, and disturbing event. It indelibly impresses upon the experiencer that one is not ultimately in control of one's destiny, and that identification with one's body, with the conditions of the physical world, and with the relationships with one's most intimate loved ones is temporary at best. The experience can be a jolting change to one's system of reference, as well as a teasing glimpse of the divine realm with its extraordinary love and acceptance. Whether this is a boon, a bane, or a mixed blessing is an assessment only the experiencer can make, perhaps in his or her most private moments.

NDErs who encounter a divine presence describe their experience in a passionate and moving way. They often claim that what they have seen has nothing to do with sectarian religions, theologies, or dogmas. Their encounter is, they would say, a direct experience of the divine without mediation or interpretation—a purely spiritual or mystical phenomenon.

This inquiry has shown, from the testimony of the co-researchers themselves, that the ultimate meaning of an encounter with a divine presence during a near-death experience lies in its intensely subjective nature and its ongoing evaluation by the individual who has experienced it. The task of eliciting this meaning, and any benefits that it may have for others, requires an approach that mirrors the reported qualities of the divine presence itself, that is, a bearing of open acceptance and nonjudgment. This is what the phenomenological approach provides.

REFERENCES

Atwater, P. M. H. (1994). *Beyond the light.* New York: Birch Lane Press.

Bache, C. (1994). The perinatal interpretation of frightening near-death experiences. *Journal of Near-Death Studies, 13,* 25–45.

Bush, N. E. (1994). The paradox of Jonah: Response to "solving the riddle of frightening near-death experiences." *Journal of Near-Death Studies, 13,* 47–54.

Colaizzi, P. F. (1978). Psychological research as the phenomenologist views it. In R. S. Valle and M. King (Eds.), *Existential–phenomenological alternatives for psychology* (pp. 48–71). New York: Oxford University Press.

Evans-Wentz, W. (Ed.). (1957). *The Tibetan book of the dead.* New York: Oxford University Press.

Moody, R. (1975). *Life after life.* New York: Bantam Books.

Moody, R. (1977). *Reflections on life after life.* New York: Bantam Books.

Moody, R. (1988). *The light beyond.* New York: Bantam Books.

Morse, M. (1992). *Transformed by the light.* New York: Ballantine Books.
Rawlings, M. (1993). *To hell and back.* Nashville, TN: Thomas Nelson.
Ring, K. (1980). *Life at death: A scientific investigation of the near-death experience.* New York: Quill.
Ring, K. (1984). *Heading toward omega.* New York: William Morrow.
Ring, K (1994). Solving the riddle of near-death experiences: Some testable hypotheses and a perspective based on *A course in miracles. Journal of Near-Death Studies, 13,* 5–23.
Sabom, M. (1982). *Recollections of death: A medical investigation.* New York: Harper & Row.
van Kaam, A. (1966). *Existential foundations of psychology.* Pittsburgh, PA: Duquesne University Press.
Vincent, K. R. (1994). *Visions of God from the near-death experience.* Burdett, NY: Lanson Publications.
von Eckartsberg, R. (1986). *Life-world experience: Existential–phenomenological research approaches in psychology.* Washington, DC: University Press of America.
Webster's new world dictionary. (1988). New York: Simon & Schuster.
West, T. (1996). *On the experience of encountering a divine presence during a near-death experience: A phenomenological inquiry.* Unpublished doctoral dissertation, California Institute of Integral Studies, San Francisco.

Name Index

Subject Index

Breinigsville, PA USA
21 July 2010
242201BV00003B/5/A

9 780306 455438